THE TOTAL SYNTHESIS
OF NATURAL PRODUCTS

The Total Synthesis
of Natural Products

VOLUME 5

Edited by

John ApSimon

Ottawa–Carleton Institute for Research
and Graduate Studies in Chemistry

and

Department of Chemistry
Carleton University, Ottawa

A WILEY-INTERSCIENCE PUBLICATION

JOHN WILEY & SONS

New York · Chichester · Brisbane · Toronto · Singapore

Library of Congress Cataloging in Publication Data:

ApSimon, John.
  The total synthesis of natural products.

  Includes bibliographical references.
  1. Chemistry, Organic—Synthesis.   I. Title.

QD262.A68   547'.2   72-4075
ISBN 0-471-09808-6 (v. 5)

Printed in the United States of America

10 9 8 7 6 5 4 3 2 1

Contributors
to Volume 5

Samuel L. Graham, Department of Chemistry, University of California, Berkeley
Clayton H. Heathcock, Department of Chemistry, University of California, Berkeley
Michael C. Pirrung, Department of Chemistry, University of California, Berkeley
Frank Plavac, Department of Chemistry, University of California, Berkeley
Charles T. White, Department of Chemistry, University of California, Berkeley

Preface

The art and science of organic synthesis has come of age. This is nowhere more apparent than in the synthetic efforts reported in the natural products area and summarized in the first four volumes of this series.

This present volume describes the synthetic activities reported for a 10-year period only in the sesquiterpene field—evidence enough for the successful efforts of the synthetic organic chemist in recent years. Professor Clayton Heathcock and his colleagues have produced a masterly, timely and important contribution, the breadth of which necessitates a complete volume in the series.

The sixth volume in this series is in an advanced stage of preparation and will contain updating chapters on the subject matter included in the first two volumes together with a description of synthetic efforts in the macrolide field. A seventh volume, covering diterpene synthesis, is in preparation.

JOHN APSIMON

Ottawa, Canada
October 1982

vii

Contents

Total Synthesis of Sesquiterpenes, 1970-79

CLAYTON H. HEATHCOCK, SAMUEL L. GRAHAM,
MICHAEL C. PIRRUNG, FRANK PLAVAC, AND
CHARLES T. WHITE

Department of Chemistry,
University of California,
Berkeley, California

1

1. INTRODUCTION

The first total synthesis of a sesquiterpene was Ruzicka's farnesol synthesis, communicated in 1923.[1] In Volume 2 of this series, we reviewed the sesquiterpene total syntheses which had been published since that time, up to the middle of 1970.[2] That review, covering a 47-year period and including about 300 papers, required 361 pages. In the intervening decade since our initial survey of the field there has been a veritable explosion of activity. In this chapter, we review a further 533 papers dealing with the total syntheses of over 260 different sesquiterpenes. We have made an effort to include all papers dealing with sesquiterpene total synthesis which appeared in the literature through the end of 1979. In addition, we have added a few papers which were inadvertently omitted from the first installment of this review, and have included a few which were either published while the review was under preparation during 1980 or were communicated to us in the form of preprints during that time. Although some of the 1970-1979 papers are improved routes to molecules previously prepared by total synthesis, most of them are new.

The general organization of the earlier review[2] has been followed, with some modification. In general, we have grouped the syntheses according to the number of carbon rings: acyclic, monocyclic, bicyclic, and tri- and tetracyclic. Compounds containing a cyclopropane ring are generally included with the class which would contain the molecule with the cyclopropane ring absent. This arbitrary decision has been made since many of these syntheses are simple extensions of syntheses of a parent with addition of the cyclopropane ring being an additional terminal step. In addition, the review now includes a separate section for sesquiterpene alkaloids.

As before, not all relay total syntheses are included. The general rule of thumb is that a relay synthesis is included only if the final product differs in carbon skeleton from the starting material. Thus, conversion of santonin into a germacrane or elemane would be included, but conversion into another eudesmane would not. The core of the review is the flow charts, which outline the syntheses. We have described the syntheses in words, sometimes rather succinctly and sometimes in more detail. We have attempted to point out novel chemistry or unusual synthetic strategy and have sometimes offered a brief critique of the synthesis.

One of the most interesting aspects of a field such as sesquiterpene synthesis is comparison of the various strategies which different workers have employed for a given target. Consequently, we have been more verbose in discussing such comparative syntheses in several cases, such as occidentalol, the vetivanes, the acoranes, the pseudoguaianolides, vernolepin, gymnomitrol, and dendrobine. For the purpose of comparing the efficiency of different syntheses, we generally use the criteria of number of steps, overall yield, and the number of isomer separations required in the synthesis.

2. ACYCLIC SESQUITERPENES

A. Farnesol and Farnesene

Corey and Yamamoto have reported the elegant synthesis of *trans*, *trans*-farnesol which is outlined in Scheme 1.[3] The synthesis features a method for stereospecific synthesis of olefins from β-oxido phosphonium ylides and aldehydes.[4] Thus, the phosphorane derived from salt **2** is treated first with aldehyde **3** at low temperature to give the β-oxido phosphonium salt **4**, which is deprotonated and treated with formaldehyde to obtain allylic alcohol **5**, uncontaminated by the *trans*, *cis*-diastereomer. The allylic hydroxyl is removed by the reduction of the bisulfate ester and the terminal hydroxyl is deprotected to obtain farnesol (**7**).

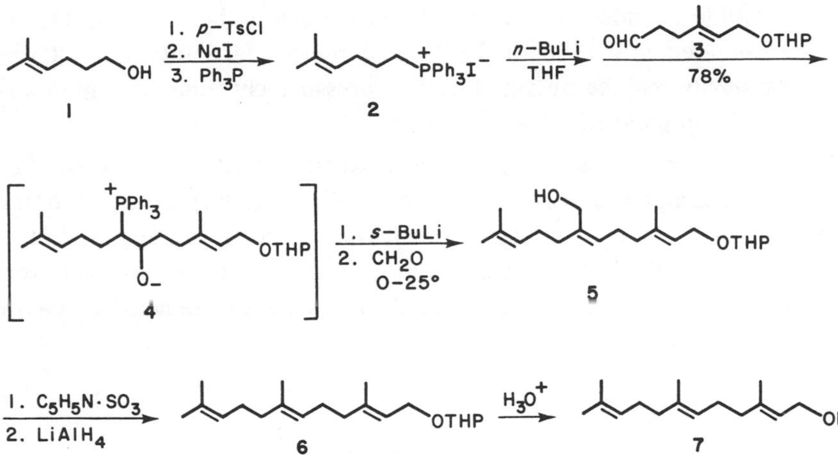

Scheme 1.   Corey-Yamamoto Synthesis of Farnesol

Pitzele, Baran, and Steinman, of Searle Laboratories in Chicago, have studied the alkylation of the dianion of 3-methylcrotonic acid (8), with geranyl bromide (Scheme 2).[5] After addition of the geranyl bromide,

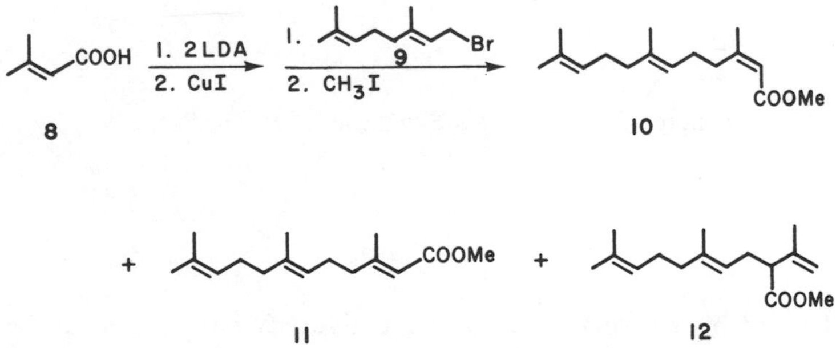

Scheme 2.   Searle Synthesis of Methyl Farnesate

methyl iodide is added to obtain the methyl esters. Isomers **10, 11,** and **12** are obtained in a ratio of 2.3:2.1:1.0; methyl farnesate (**11**) of 89% isomeric purity may be obtained by low pressure chromatography in 26% yield, based on geraniol.

O. P. Vig and co-workers report a synthesis of β-farnesene (**17**) wherein the dianion of acetoacetic ester is alkylated with geranyl bromide and the resulting β-keto ester transformed into a butadiene unit as shown in Scheme 3.[6] It is not quite clear from their paper just what they synthesized, since both geraniol and β-farnesene are depicted as having *Z* double bonds.

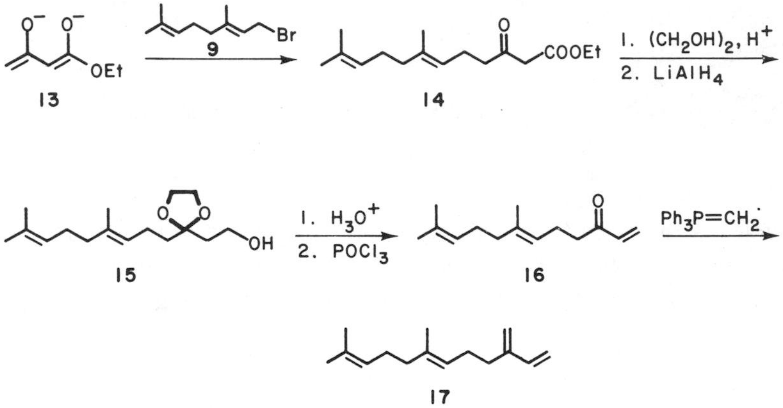

Scheme 3.   Vig's Synthesis of β-Farnesene

Otsuka and his co-workers at Osaka University have reported the most direct sesquiterpene synthesis yet—direct trimerization of isoprene (Scheme 4).[7] Several catalysts were found which give a preponderance of the linear trimers **17-19**. The best system for production of β-farnesene

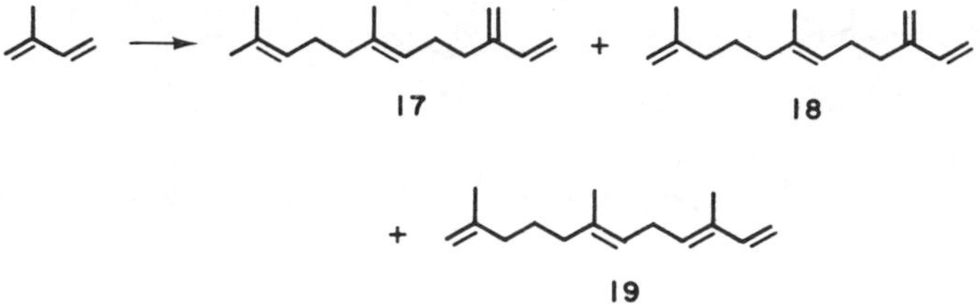

Scheme 4.   Otsuka's β-Farnesene Synthesis

(17) utilizes [NiCl(η_3-C$_3$H$_5$)]$_2$-As(n-C$_6$H$_{13}$)$_3$ and t-BuOK. If the reaction is stopped at 30% conversion of the isoprene, β-farnesene comprises 57% of the product. Unfortunately, preparative glpc is required to separate 17 from its isomers.

B. Terrestrol, Caparrapidiol, and Caparrapitriol

Terrestrol, (3S)-2,3-dihydrofarnesol (20), is the marking perfume of the small bumble bee. Caparrapidiol (21) and caparrapitriol (22) are plant sesquiterpenes which contain centers of chirality.

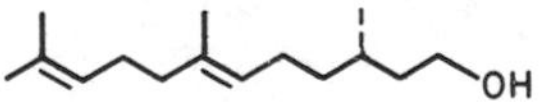

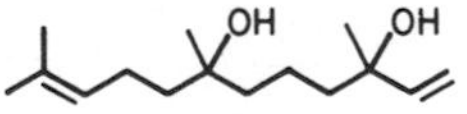

Ahlquist and Ställberg-Stenhagen of the University of Göteborg in Sweden have synthesized both enantiomers of terrestrol by way of the Kolbe electrolysis of homogeranic acid (23) with the enantiomers of monomethyl 3-methylglutarate (24, Scheme 5).[8] Ester 25 is obtained in 8% yield, based on homogeranic acid.

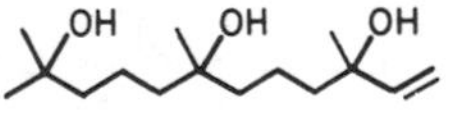

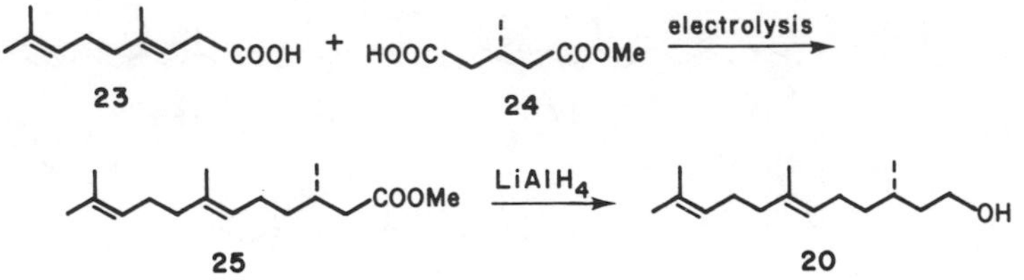

Scheme 5.   Ahlquist-Ställberg-Stenhagen Synthesis of Terrestrol

A synthesis of caparrapidiol by O. P. Vig is summarized in Scheme 6.[9] The question of diastereoisomerism in the formation of **21** is not addressed by the authors, who simply state that "...The identity of the synthesized compound was established by comparing its IR and NMR (spectra) with those reported in literature."

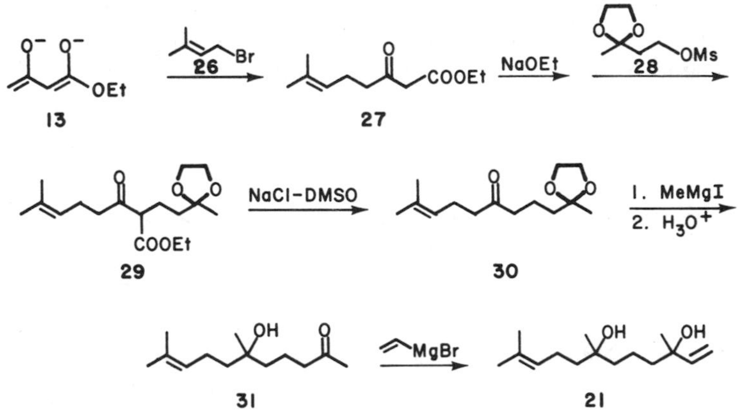

Scheme 6.   Vig's Carrapidiol Synthesis

Weyerstahl and Gottschalk, at the Technical University of Berlin, have synthesized caparrapitriol as shown in Scheme 7.[10] As in the Vig synthesis of caparripidiol, the German group makes no mention of a diastereomeric mixture in the addition of vinyllithium to methyl ketone **35**. However, in this case the final triol is obtained as a sharp-melting solid (mp 78-79°C) in 90% yield! Chromatography on starch provides one pure enantiomer of caparripitriol.

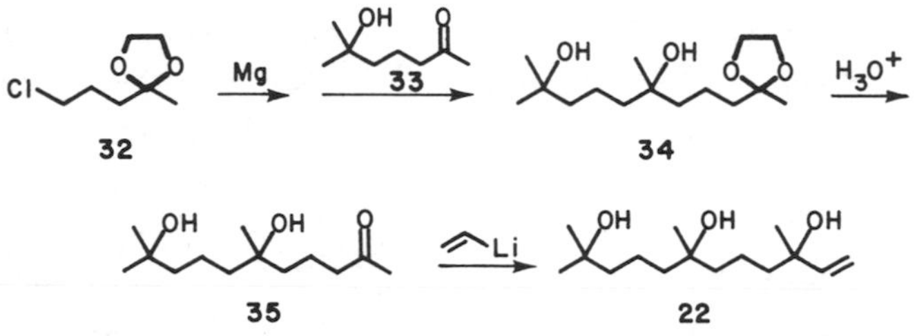

Scheme 7.   Weyerstahl-Gottschalk Synthesis of Caparrapitriol

C. Juvenile Hormones

The C_{17}- and C_{18}-*Cecropia* juvenile hormones (**36** and **37**) (JH), although not sesquiterpenes, are included because their structures are so similar to those of the acyclic sesquiterpenes. Although **37** was not characterized until 1967 and **36** until 1968, a total of 15 syntheses had been reported by 1972.

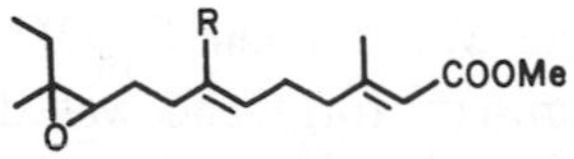

36: R = Me
37: R = Et

Corey and Yamamoto have utilized the β-oxidophosphonium ylide method for the synthesis of both C_{17}- and C_{18}-JH, as shown in Scheme 8.[3] Intermediate **40** is converted via aldehyde **41** into tetraene **42**, which

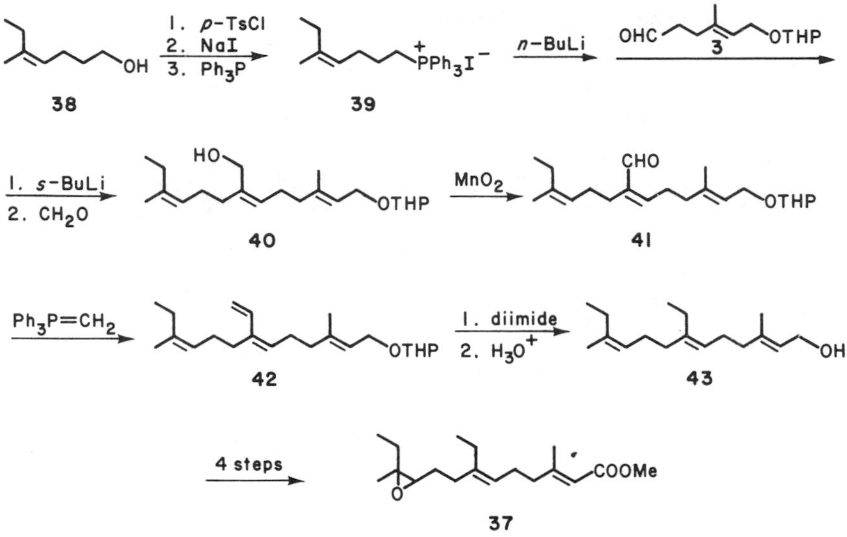

Scheme 8.   Corey-Yamamoto Synthesis of Juvenile Hormones

is selectively reduced to obtain alcohol **43**. This material has previously been converted into C_{18}-JH.[11] The C_{17}-JH **36** is prepared from **40** along the same lines as are used to convert alcohol **5** into farnesol (see Scheme 1).

Findlay and MacCay at New Brunswick, and Bowers at the Agriculture Research Service in Beltsville have reported full details of stereorandom syntheses of both **36** and **37**.[12a] Their C_{18}-JH synthesis had previously been published in preliminary form and was discussed in Volume 2 of this series.[12b] The New Brunswick-Beltsville C_{17}-JH synthesis is essentially the same as the Schering synthesis of C_{17}-JH.[13]

Cochrane and Hanson of the University of Sussex have reported two C_{18}-JH syntheses.[14] Their first, summarized in Scheme 9, is modeled closely on the Julia nerolidol synthesis.[15] Bromide **46** is obtained as a 3:1

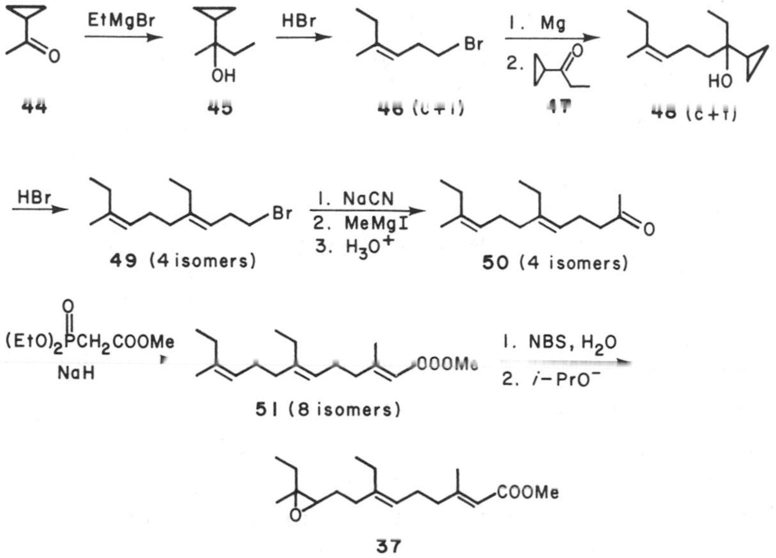

Scheme 9.   Juvenile Hormone: Sussex Synthesis A

mixture favoring the unnatural E stereoisomer. The second cyclopropyl carbinol solvolysis (**48→49**) also produces a bad stereoisomer mixture, giving 59% of *3E* and 41% of *3Z* compounds. Analysis at the stage of dienone **50** showed the *ZZ, ZE, EZ,* and *EE* stereoisomers to be present in a ratio of 16:43:11:30. A final Horner-Wadsworth-Emmons olefination (**50→51**) affords a mixture of all eight stereoisomers, of which the natural *EEZ* isomer is less than 10%. The Sussex group also reports a somewhat more stereoselective synthesis (Scheme 10). The starting unsaturated bromide **53** is prepared as a 3:1 mixture favoring the

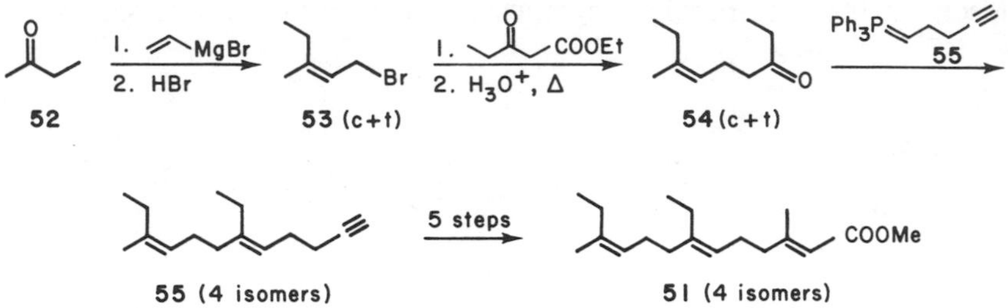

Scheme 10.   Juvenile Hormone: Sussex Synthesis B

undesired E stereoisomer. The second double bond is introduced by a Wittig reaction, which proceeds in an essentially stereorandom fashion, as expected. The final double bond is introduced by the Corey procedure.[16] Analysis of ester 51 showed it to be an approximately equimolar mixture of the four stereoisomers having $2E$ stereochemistry. The desired isomer comprised 22% of the mixture.

A Zoecon group headed by C. A. Henrick has prepared the C_{17}-JH from trans-geranylacetone (56) as shown in Scheme 11.[17] This substance is converted into methyl farnesate (11), which is then degraded to aldehyde 58. The epoxide moiety is introduced via chloroketone 60 by a method adapted from Johnson's earlier C_{18}-JH synthesis.[18] Since this synthesis starts with trans-geranylacetone (56), the C_6 double bond is homogenous. The C_2 linkage is established in the Wadsworth-Emmons reaction. The reaction gives a 2:1 mixture favoring the desired $2E$ stereoisomer which is obtained in pure form by distillation. Although the Stanford group originally reported that the epoxide construction occurs with 92% stereoselectivity,[18] Henrick and co-workers were only able to obtain 36 as an 82:18 mixture with its C_{10}-C_{11} trans isomer.

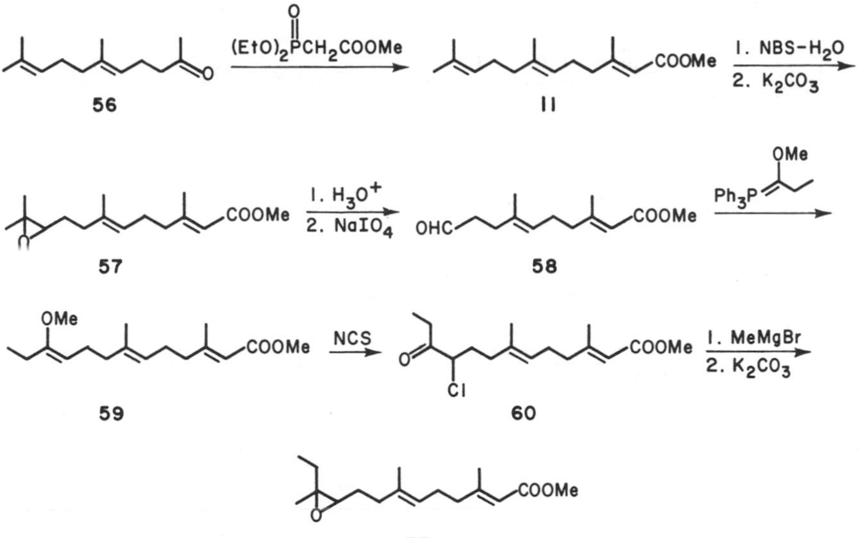

Scheme 11.   Zoecon Synthesis of C_{17}-JH

The Zoecon group has reported two methods for synthesis of C_{18}-JH.[19] The first (Scheme 12) begins with methylheptenone (**61**), which is converted into methyl geranate (**62**). Although this reaction shows only modest stereoselectivity, the *2E* stereoisomer is conveniently isolated in pure form by distillation of the crude product. The terminal double bond is cleaved and the resulting aldehyde is treated with the Grignard reagent derived from 2-bromo-1-butene to obtain allylic alcohol **64**. The C_6 double bond stereochemistry is established by Claisen rearrangement (96% stereoselectivity). After selective reduction of the saturated ester function, the synthesis is completed as in Scheme 11. Again, the final hormone is obtained as an 82:18 mixture of cis and trans isomers.

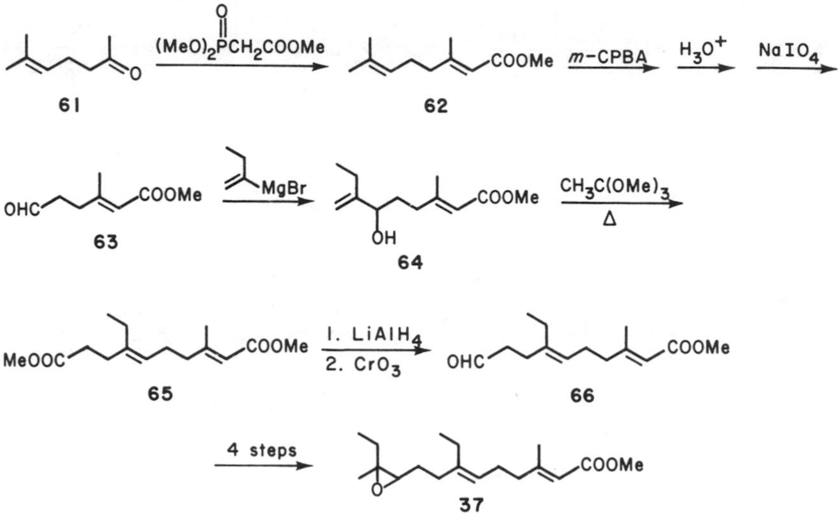

Scheme 12.   First Zoecon Synthesis of C_{18}-JH

The second Zoecon synthesis (Scheme 13) starts with cyclopropyl carbinol **45**, which is solvolyzed to unsaturated chloride **67** as a 3:1 mixture of *E* and *Z* isomers. The mixture of isomers is oxidized by singlet oxygen to obtain allylic alcohol **68** as the major product of a 55:39:6 mixture of isomers. After separation of the mixture, **68** is subjected to Claisen rearrangement using the orthoacetate method to obtain chloroester **69**. As usual, the stereoselectivity in this reaction is excellent, only 4% of the *Z* stereoisomer is produced. The C_2-C_3 double bond geometry is established by adding the cuprate derived from **71** to methyl 2-butynoate to obtain **66**.

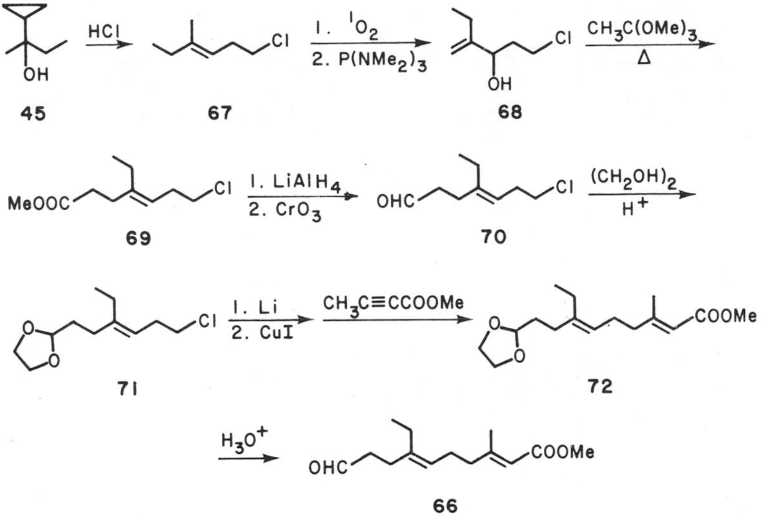

Scheme 13.   Second Zoecon Synthesis of C_{18}-JH

D. Sinensals

The sesquiterpene aldehydes α- and β-sinensal (**73** and **74**) are important contributors to the aroma and taste of Chinese orange oil. Büchi and

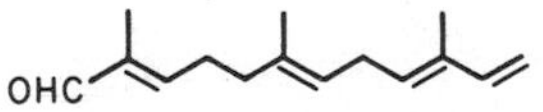

Wuest have reported the stereorational synthesis of the α isomer (**73**) which is outlined in Scheme 14.[20] The stereochemistry of the C_9 double bond is assured by the use of the diene alcohol **75** as the starting

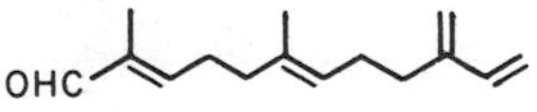

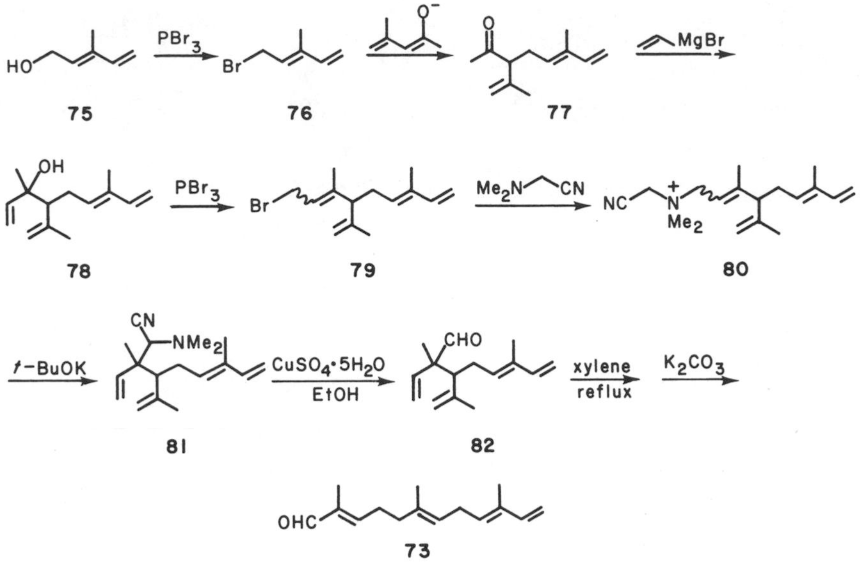

Scheme 14.   Büchi's α-Sinensal Synthesis

material. The synthesis features a novel [2,3]-sigmatropic rearrangement of the ammonium ylide derived from **80** to form amino nitrile **81** (3:2 mixture of diastereomers). Stereochemistry at the C_6 double bond is established in the final Cope rearrangement; **73** and its *2Z* diastereomer are produced in a 2:3 ratio. The latter isomer is quantitatively isomerized to the more stable *2E* isomer **73** by heating with potassium carbonate.

A BASF group headed by Werner Hoffmann has reported a synthesis which affords a mixture of the two sinensals, as well as modifications which allow the production of either pure isomer.[21] The first synthesis (Scheme 15) begins with chloroaldehyde **83**, which contains the eventual C_2 double bond. The chain is elaborated to **88** by two cycles of the basic Nazarov-Ruzicka-Isler synthesis (vinyl Grignard, Carroll reaction).[22]

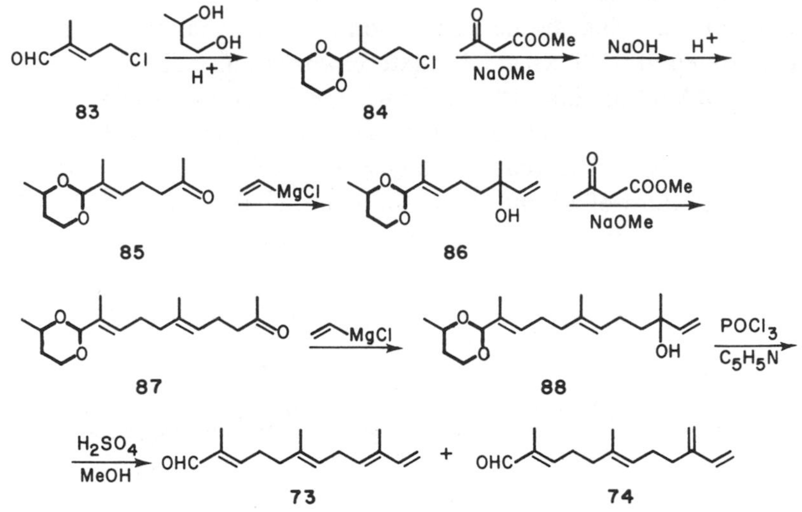

Scheme 15.   First BASF Synthesis of Sinensals

Dehydration of **88**, followed by deprotection of the aldehyde affords α- and β-sinensals in a ratio of 2:1. The E/Z ratio at the C_9 double bond in **73** is not stated. A modified synthesis which yields no β-sinensal is shown in Scheme 16. The C_7-C_{12} segment is assembled as shown, using

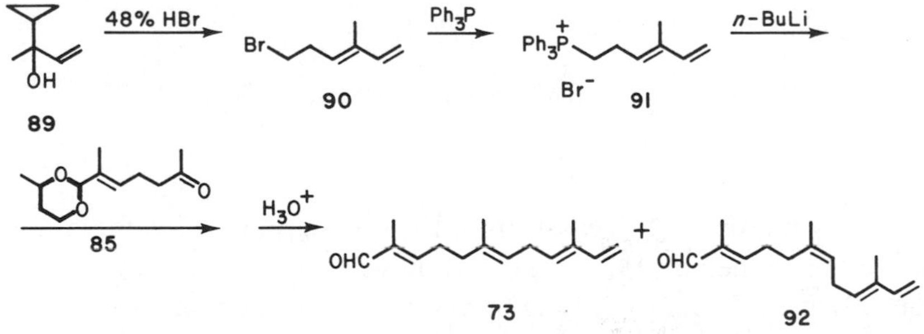

Scheme 16.   BASF Synthesis of Pure α-Sinensal

the Julia method. Bromo diene **90** is obtained as an 85:15 mixture of stereoisomers favoring the desired *E* isomer. The final Wittig coupling affords α-sinensal (**73**) as a 1:1 mixture with its C_6 diastereomer **92**. The other BASF modification (Scheme 17) leads to β-sinensal (**74**), uncontaminated by α-sinensal, again as a 1:1 mixture with the C_6-diastereomer (**99**). The required bis-unsaturated halide **96** is isolated from a 7:3 mixture of **95** and **96** by formation of the sulfolene **97**. The remainder of the synthesis follows the same lines as are used to prepare the α isomer.

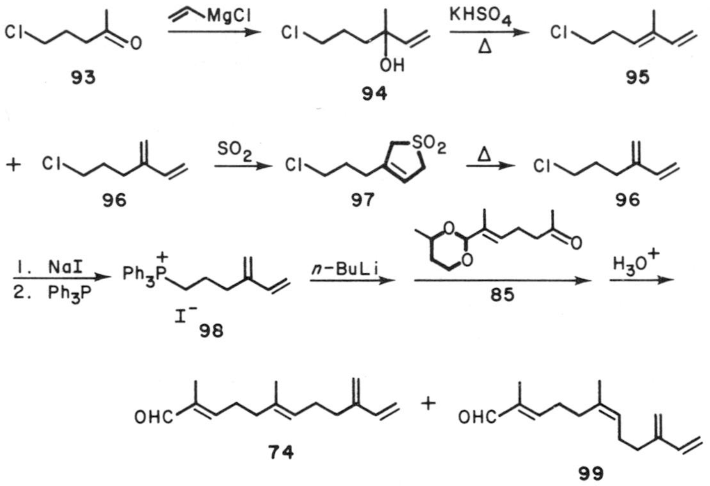

Scheme 17.   BASF Synthesis of Pure β-Sinensal

A final synthesis of β-sinensal, from Hiyama's group in Kyoto, is summarized in Scheme 18.[23] The synthesis starts with myrcene, a frequently-used precursor for the preparation of β-sinensal. After oxidation of the more reactive trisubstituted double bond, epoxide **101** is subjected to Crandall-Rickborn isomerization to an allylic alcohol, which is

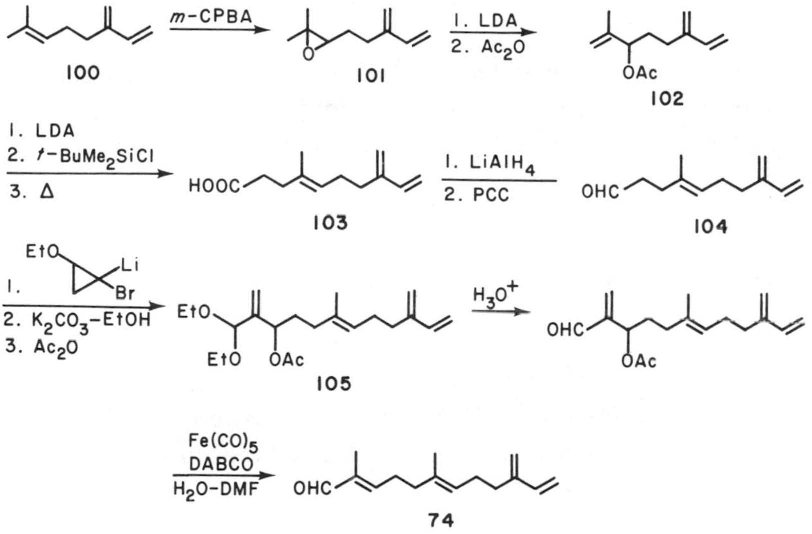

Scheme 18.   Hiyama Synthesis of β-Sinensal

acetylated and subjected to Claisen rearrangement to obtain **103**, apparently with good stereoselectivity. The terminal unsaturated aldehyde function is introduced by a method developed in Hiyama's group, whereby the carbanion derived from 1,1-dibromo-2-ethoxycyclopropane is added to aldehyde **104** at low temperature (-95°C). The resulting adduct (**106**) is solvolyzed in basic ethanol to obtain the unsaturated acetal:

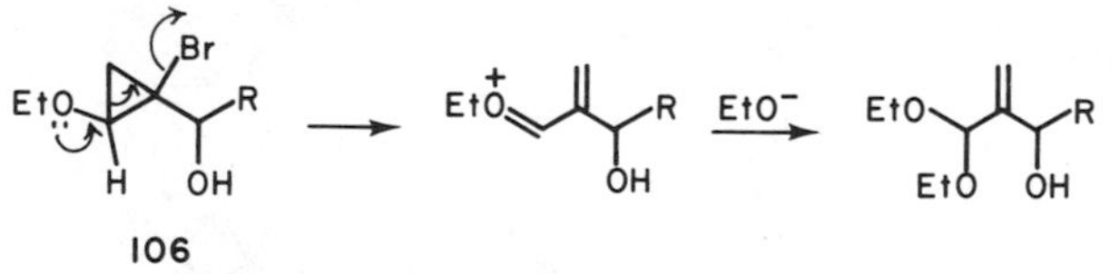

The synthesis of β-sinensal is completed by reductive removal of the acetoxy function, which occurs with double bond isomerization to the more stable position.

E. Fokienol, Oxonerolidol, and Oxodehydronerolidol

Fokienol (**107**), 9-oxonerolidol (**108**), and 9-oxo-5,8-dehydronerolidol (**109**) are relatives of the simpler nerolidol.[24] Fokienol is a stereochemically more complex problem than is nerolidol, since it may exist as four racemates.

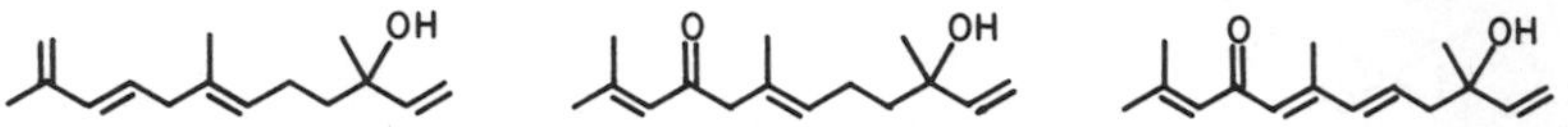

O. P. Vig's synthesis of racemic fokienol is outlined in Scheme 19.[25] The synthesis begins with a Wadsworth-Emmons reaction of keto acetal **110**, which affords unsaturated ester **111** as a 60:40 mixture of *E* and *Z*

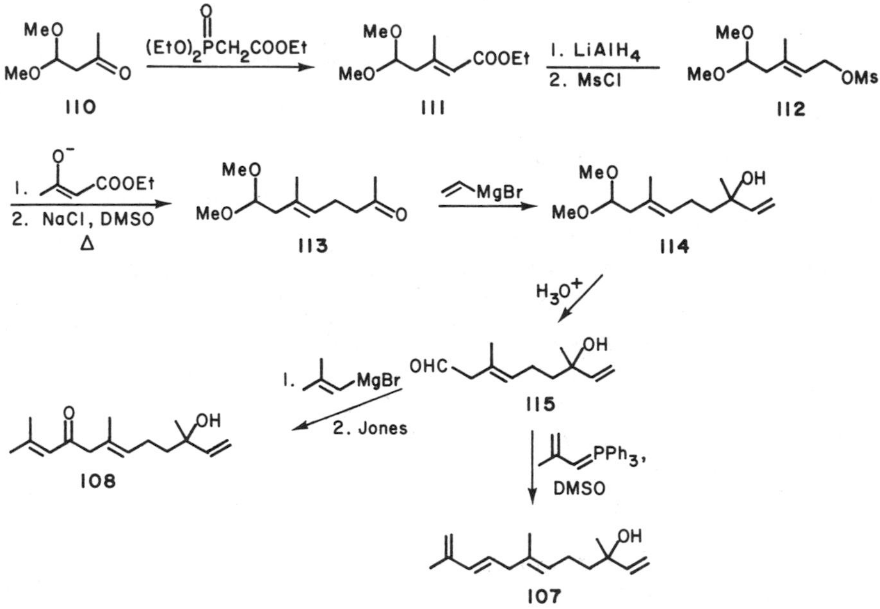

Scheme 19.   Vig's Synthesis of Fokienol and Oxonerolidol

stereoisomers. Vig separates the mixture and carries on only with the correct $2E$ diastereomer, which is elaborated by straightforward steps into aldehyde **115**. The most impressive step in this synthesis is the Wittig reaction on the β,γ-unsaturated aldehyde **115**, which is reported to occur stereospecifically, in good yield, and without enolization or prior conjugation of the unsaturated aldehyde, to give **107**. Aldehyde **115** is converted into oxonerolidol (**108**) by Grignard addition and oxidation.

Bohlmann and Krammer have synthesized 9-oxo-5,8-dehydronerolidol (**109**) as shown in Scheme 20.[26] The synthesis starts with the protected unsaturated alcohol **116**, which was used as a mixture of cis and trans isomers. The eventual C_5 double bond is established by reduction of acetylene **117** by chromous hydroxide. The mixture of C_7 stereoisomers is separated by chromatography after preparation of acid **122**. The synthesis is completed by reaction of the correct stereoisomer with 2-methyl-1-propenyllithium.

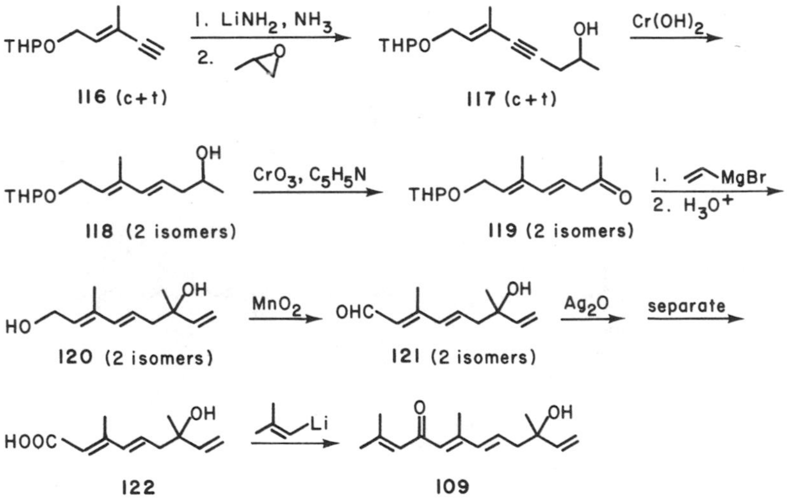

Scheme 20.   Bohlmann-Krammer Synthesis
of 9-Oxo-5,8- dehydronerolidol

F. Gyrindal

The norsesquiterpene gyrindal (**123**) is a defense secretion of the whirligig water beetle. Its synthesis has been reported by Meinwald, Opheim, and Eisner, of Cornell[27] and by Miller, Katzenellenbogen, and Bowles,

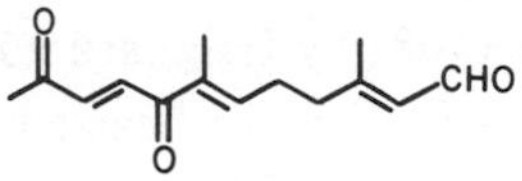

123

of Illinois.[28] The two syntheses, which are essentially identical, are outlined in Scheme 21. They differ only in the protecting group used for geraniol—the Cornell group employed the acetate whereas the Illinois group used the mesitoate—and in the method used for reducing the triple bond—the Cornell group used Li/NH_3 whereas the Illinois group used $LiAlH_4$—NaOMe. The Illinois team reports a much higher overall yield (9.6% vs. 1.7%).

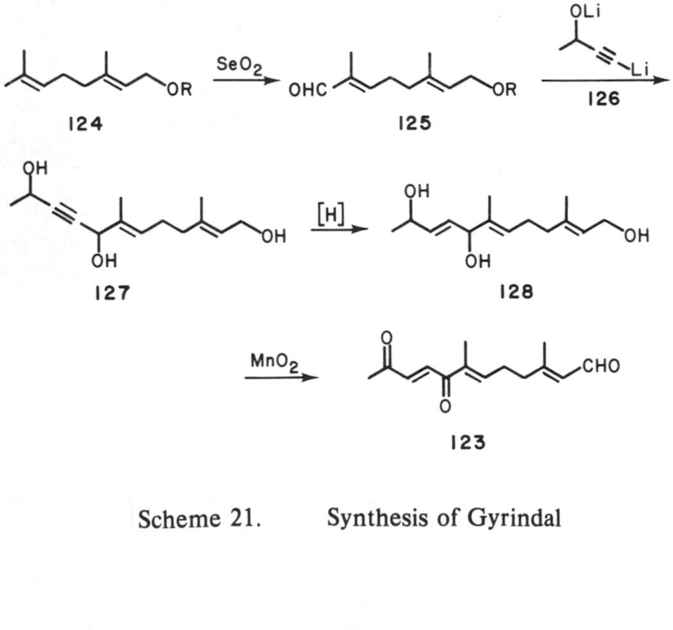

Scheme 21.   Synthesis of Gyrindal

Kato et al., have described the synthesis of oxocrinol (**129**),[29] a nor-sesquiterpene from marine algae (Scheme 22). The synthesis is conceptually identical to the synthesis of geranylgeraniol by Altman, Ash, and Marson.[30] Altman's (*E,E*)-allylic chloride **131** was coupled with sulfone anion **132**. After reductive cleavage of the sulfone moiety and deprotection, oxocrinol (**129**) was obtained in 18% overall yield.

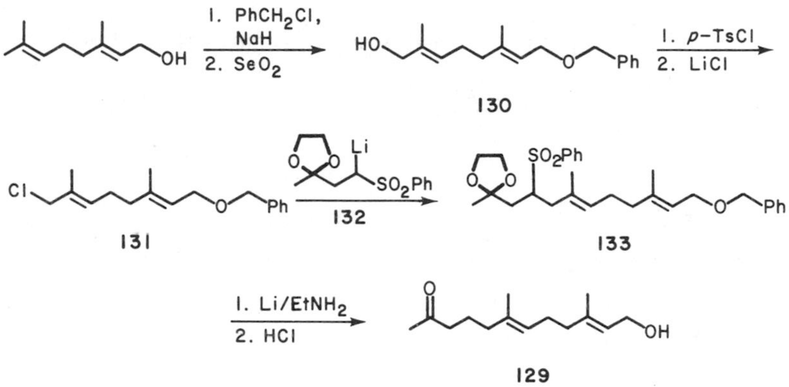

Scheme 22.   Kato's Synthesis of Oxocrinol

G. Sesquirosefuran and Longifolin

Sesquirosefuran (**134**) and longifolin (**135**) are the first 2,3-substituted furans in the sesquiterpene series. Sesquiterpene **134** has been prepared by three groups.[31-33] All three syntheses rely on coupling geranyl

134   135

bromide with a 3-methyl-2-furylmetallic reagent (Scheme 23). The exact reagents employed have been the furyllithium,[31] the *bis*-furylmercury,[32] and the furylmagnesium bromide.[33] The highest coupling yield (40%) is

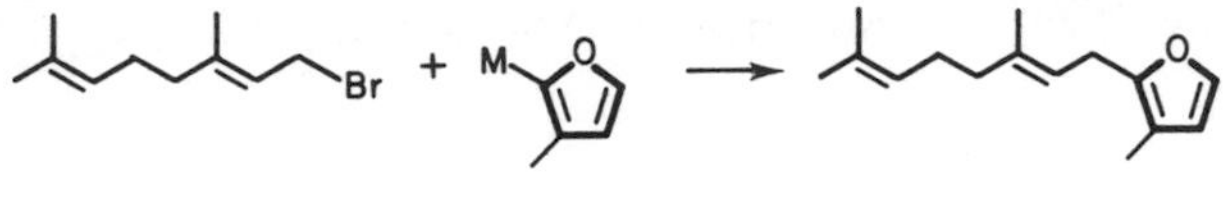

Scheme 23.   Synthesis of Sesquirosefuran

reported for the furyllithium reagent. A related synthesis of longifolin (135) is outlined in Scheme 24.[34] The final coupling proceeds in only 4% yield.

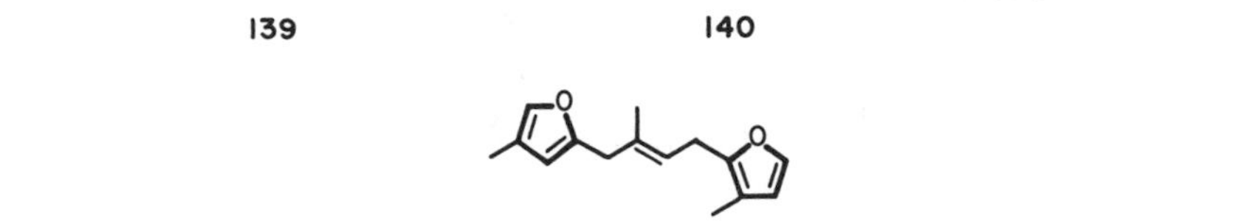

Scheme 24.   Synthesis of Longifolin

H. Davanafurans

The davanafurans (**141-144**) are a set of stereoisomeric farnesene derivatives which contribute to the characteristic odor of Davana oil. The principal component is isomer **141**. Thomas and Dubini have synthesized

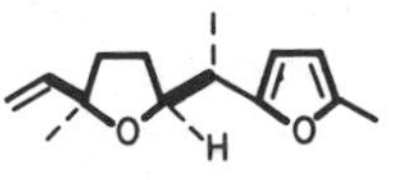

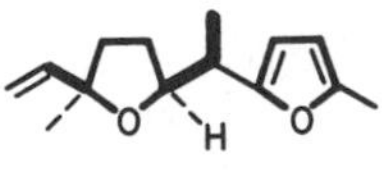

the four isomers starting with linalool oxides of known stereochemistry.[35] The synthesis of isomers **141** and **142** is summarized in Scheme
25. The (−)-*cis*-linalyl oxide **145** is converted by standard methods into
the cis ketone **146**, which is treated with 5-methyl-2-furyllithium to
obtain a mixture of unstable tertiary alcohols. The alcohol mixture is
hydrogenolyzed using $LiAlH_4$-$AlCl_3$ to obtain a mixture of **141** and **142**
in a ratio of 4:1. Similar transformation of (+)-*trans*-linalyl oxide affords

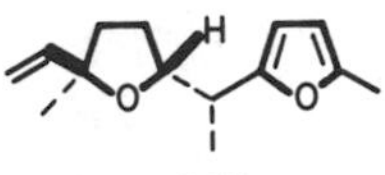

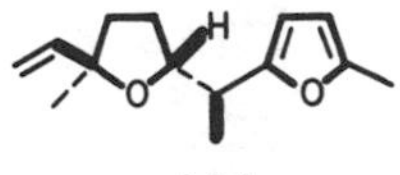

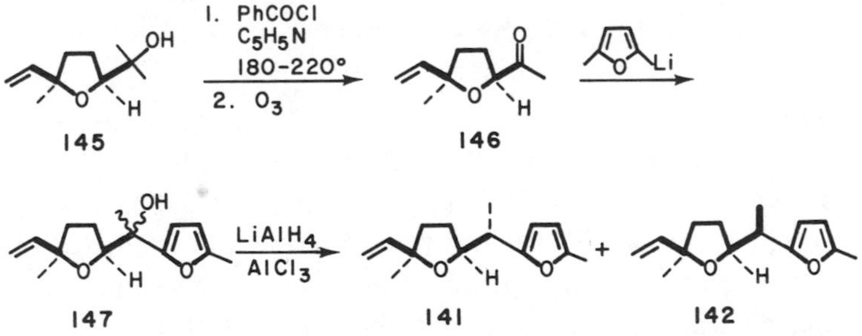

Scheme 25.   Thomas-Dubini Synthesis of Davanafurans

143 and 144 in a ratio of 3:1. Although the absolute stereostructures of the davanafuran group are established by their synthesis from linalool of known absolute configuration, the relative stereochemistry within the cis and trans pairs is still open to question. That is, does the major natural diastereomer correspond to 141 or 142? The major product of the reduction shown in Scheme 25 is identical with the major natural davanafuran. Thomas and Dubini argue that the major product in this reduction should have structure 141 on the basis of transition state 148. However, the alternate formulation 149, which leads to 142, would seem to be more consistent with the Cram-Felkin model for asymmetric induction.

I. Dendrolasin, Neotorreyol, Torreyal, Ipomeamarone, Freelingyne, and Dihydrofreelingyne

The common structural unit in this group of sesquiterpenes is the β-substituted furan ring. Several syntheses of dendrolasin (150), neo-torreyol (151), torreyal (152), and ipomeamarone (153) were recorded in our original review.[36] The first three present the challenge of double bond stereochemistry. In ipomeamarone there is the similar problem of achieving cis,trans selectivity about the tetrahydrofuran ring. Freelingyne (154) and dihydrofreelingyne (155) are much more challenging synthetic targets. In addition to their highly unsaturated nature, both have two trisubstituted double bonds and **155** has a disubstituted double bond.

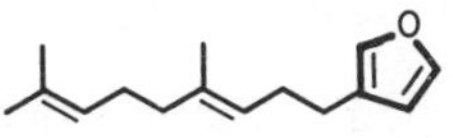

150

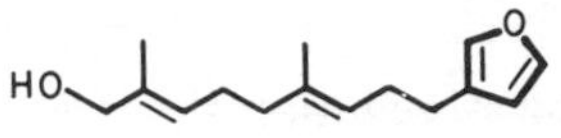

151

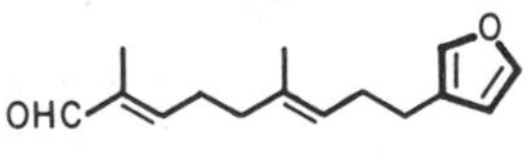

152

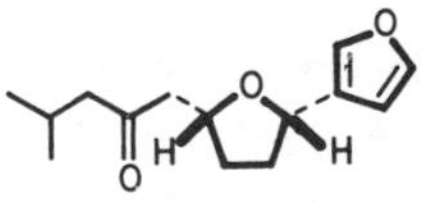

153

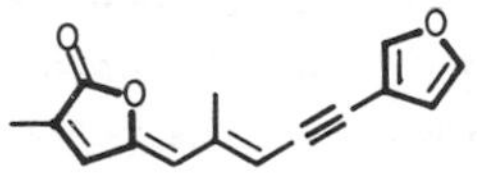

154

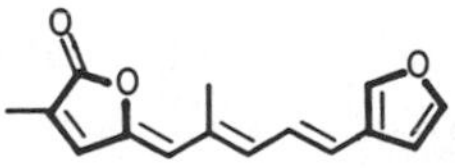

155

Takahashi has reported a synthesis of dendrolasin which is summarized in Scheme 26.[37] The synthesis features a novel method for construction of the furan ring, in which intermediate **159**, a mixture of cis and trans

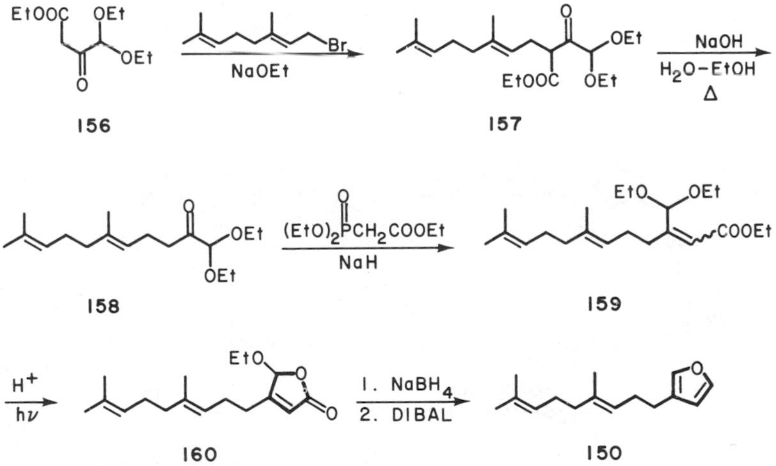

Scheme 26. Takahashi Synthesis of Dendrolasin

isomers, is irradiated under acidic conditions to produce the ethoxybu-tenolide **160**. A two-stage reduction process converts **160** into dendro-lasin.

Kondo and Matsumoto have accomplished the interesting syntheses of **150-152** which are outlined in Scheme 27. The starting point is myrcene (**100**), which is converted by two consecutive singlet oxygen oxidations into the endoperoxide **163**. The initial oxidation provides **162** and its isomer **161** in a ratio of 2:3; the desired isomer is obtained in 36% yield by chromatography. The furan ring is formed from endoperoxide **163** by base-catalyzed scission of the peroxide, followed by acid-catalyzed cycli-zation. Reaction of allylic alcohol **164** with thionyl chloride gives *trans*-ω-chloroperillene (**165**), which is used to prepare **150-152** as shown.

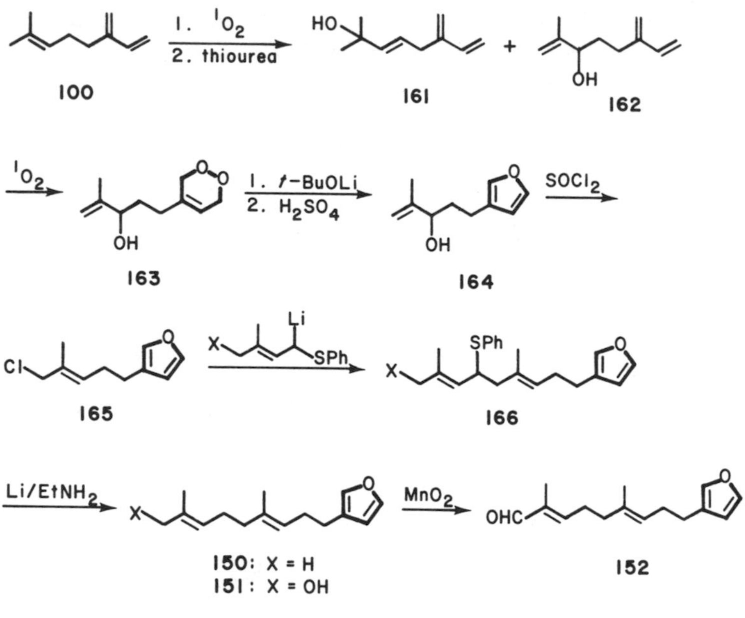

Scheme 27.   Kondo-Matsumoto Synthesis of Dendrolasin, Neotorreyol, and Torreyal

Ipomeamarone (153) and its trans isomer epi-ipomeamarone (167) are toxic metabolites produced by molds which infect sweet potatoes. The

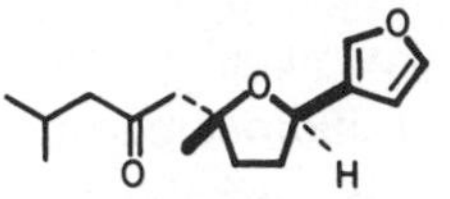

167

chief synthetic problem in this series, control of stereochemistry about the tetrahydrofuran ring, has still not been solved, as all syntheses to date are stereononspecific. A synthesis by Burka, Wilson, and Harris of Vanderbilt is summarized in Scheme 28.[39] Beginning with ethyl 3-furoylacetate (168), keto acetate 172 is constructed in a straightforward manner. Wadsworth-Emmons reaction of 172 affords enone 173. Hydrolysis of the acetate occurs with concomitant cyclization to an equimolar mixture of epi-ipomeamarone (167) and ipomeamarone (153). Although the equilibrium constant for isomerization of ipomeamarone to epi-ipomeamarone is 1, it was found that the formation of the isomer in

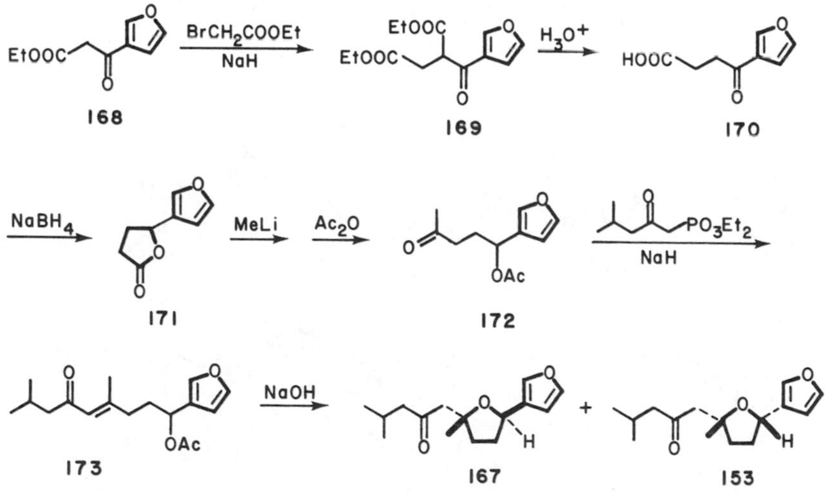

Scheme 28.   Burka-Wilson-Harris Synthesis of Ipomeamarone

this synthesis is not under thermodynamic control, but rather that the two isomers are formed from the hydroxy enone precursor with equal rates.

Kondo and Matsumoto have applied their method for synthesis of furans to the preparation of ipomeamarone and epi-ipomeamarone as shown in Scheme 29.[40] In this case the required diene is not commercially available, so it is synthesized by the addition of 2-butadienyl-magnesium chloride to aldehyde **174**. The singlet oxygen oxidation and two-stage conversion of the resulting endoperoxide to the furan proceed well (70% for the three steps). However, it should be noted that simple addition of 3-furylmagnesium bromide to **174** would probably provide **176** directly. The tetrahydrofuran ring is established by treatment of **176** with N-iodosuccinimide. The cyclization shows essentially no stereoselectivity; the two cis and trans isomers are produced in a ratio of 2:3. The synthesis is completed in a conventional manner, although

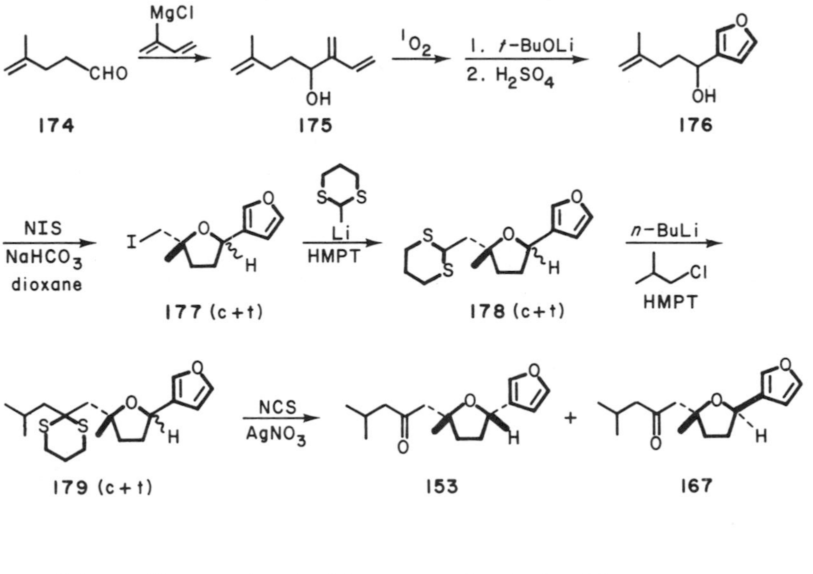

Scheme 29.   Kondo-Matsumoto Synthesis of Ipomeamarone

alkylation of 2-lithio-1,3-dithiane by the neopentyl iodide **177** is noteworthy. The final product, a 2:3 mixture of **153** and **167** may be equilibrated by base to a 1:1 mixture of the two sesquiterpenes.

The first synthesis of freelingyne (**154**) was reported by Ingham, Massy-Westropp, and Reynolds, of the University of Adelaide.[41] This synthesis is summarized in Scheme 30. Reaction of citraconic anhydride with acetylmethylene triphenylphosphorane affords a mixture of isomeric enol lactones **180-183**. Although isomer **182** can be obtained in pure form, the desired isomer **183** cannot easily be separated from **181**. However, isomerization of pure **182** affords an easily separable 2:1 mixture of **182** and **183**. The 3-furyl acetylenic ester **187** is prepared from 3-furoic acid as shown in the scheme, and is then converted into propargylic

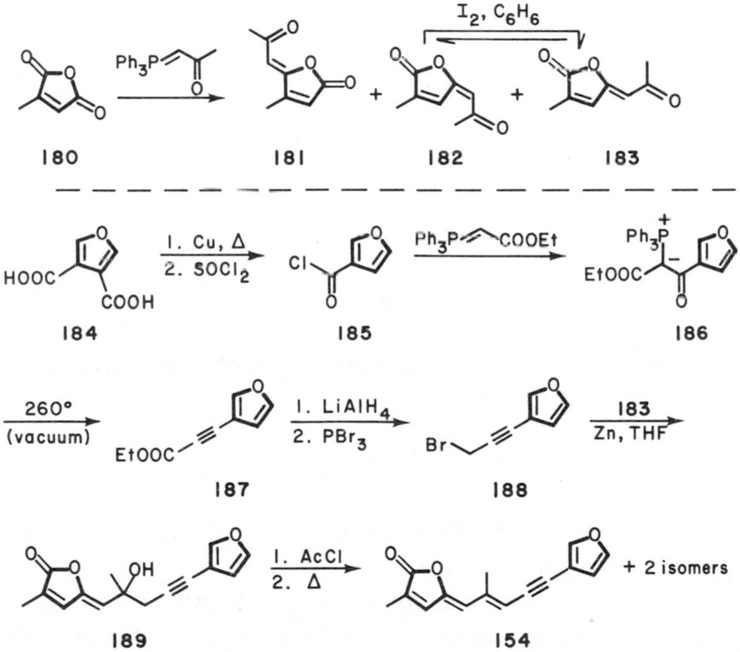

Scheme 30.   Adelaide Synthesis of Freelingyne

bromide **188**. Reformatsky coupling of **188** with **183** affords an alcohol (**189**), which is dehydrated by vacuum pyrolysis of the derived acetate. Freelingyne (**154**) is produced, along with two stereoisomers. Compound **154** may be obtained in 24% yield by chromatography.

The Adelaide group has also prepared dihydrofreelingyne (**155**) as outlined in Scheme 31.[42] The synthesis starts with propargylic ester **187**, which is elaborated in a conventional series of steps to **155**. The synthesis is highly stereoselective, as both Wittig reactions give only a single isomer.

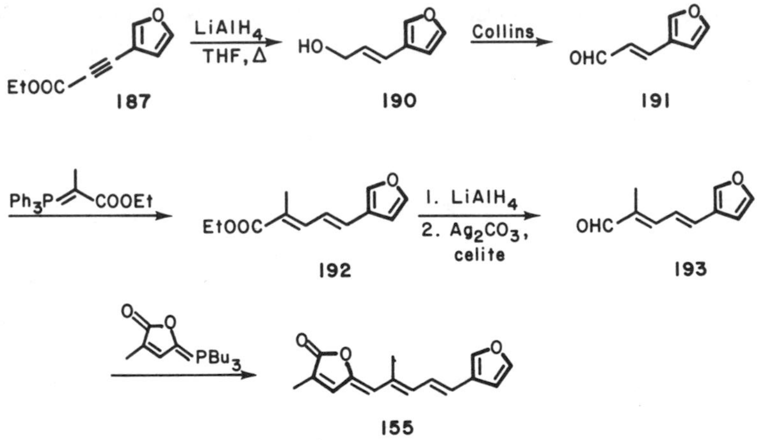

Scheme 31.   Adelaide Synthesis of Dihydrofreelingyne

Knight and Pattenden of Nottingham have reported a synthesis of freelingyne which employs a final step similar to that used to form the final double bond in the Adelaide dihydrofreelingyne synthesis.[43] The Nottingham synthesis is summarized in Scheme 32. Chloroepoxide **194** reacts with sodium acetylide to produce alcohol **195** as a single isomer. The stereochemistry of **195** was proven by X-ray analysis of a derivative.

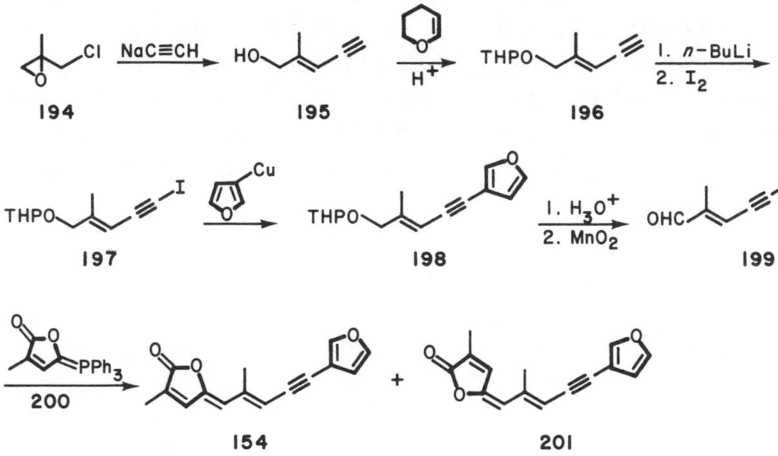

Scheme 32.   Nottingham Synthesis of Freelingyne

The hydroxyl is protected and the acetylene is iodinated. Iodo acetylene **197** is coupled with 3-furylcopper by a Stephens-Castro reaction. After deprotection and oxidation, aldehyde **199** is condensed with phosphorane **200** to obtain freelingyne (**154**) and its stereoisomer **201** in a 3:2 ratio. The lack of stereoselectivity in this reaction is in sharp contrast to the high selectivity observed by Massy-Westropp in his dihydrofreelingyne synthesis.

3. MONOCARBOCYCLIC SESQUITERPENES

A. α-Curcumene, Dehydro-α-curcumenes, Curcuphenol, Xanthorrhizol, Elvirol, Nuciferal, ar-Turmerone, Curcumene Ether, and Sydowic Acid

Perhaps the simplest monocarbocyclic sesquiterpenes are the aromatic members of the bisabolane family. Examples of those that have been synthesized in this period are α-curcumene (**1**), the dehydrocurcumenes **2** and **3**, *ar*-tumerone (**4**), nuciferal (**5**), curcuphenol (**6**), xanthorrhizol (**7**), elvirol (**8**), curcumene ether (**9**), and sydowic acid (**10**). For most of these compounds the synthetic task is merely one of assembling the carbon skeleton. Compounds **3** and **5** present a stereochemical problem in their double bonds. Although several are chiral, none have been synthesized in optically active form.

Krapcho and Jahngen have synthesized α-curcumene (**1**) as shown in Scheme 1.[44] The synthesis serves only as a vehicle to demonstrate the utility of olefination via a β-lactone. However, it seems more unwieldy than simple olefination of aldehyde **11** by the Wittig reaction.[45]

Hall and co-workers of Rutgers (Newark) have worked out a one-pot synthesis of α-curcumene wherein *p*-tolylmagnesium bromide is added to methylheptenone (**14**) in THF, ammonia is added, excess lithium is

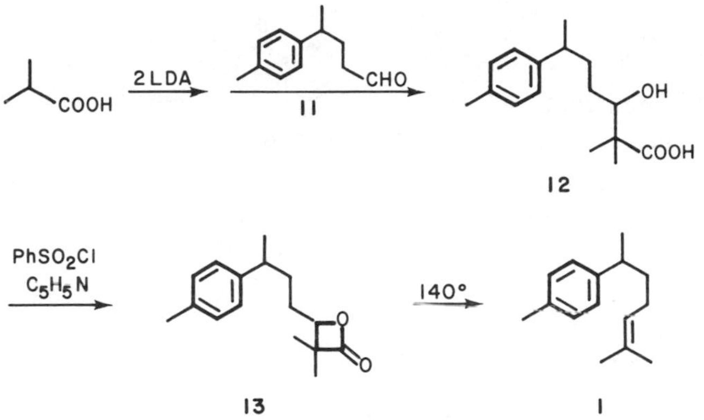

Scheme 1.   Krapcho-Jahngen α-Curcumene Synthesis

added, and the resultant solution is rapidly quenched with NH_4Cl (Scheme 2).[46] α-Curcumene is obtained in 90% yield. The synthesis is an adaptation of one introduced earlier by Birch,[47] which requires two operations and affords α-curcumene in only 35% yield.

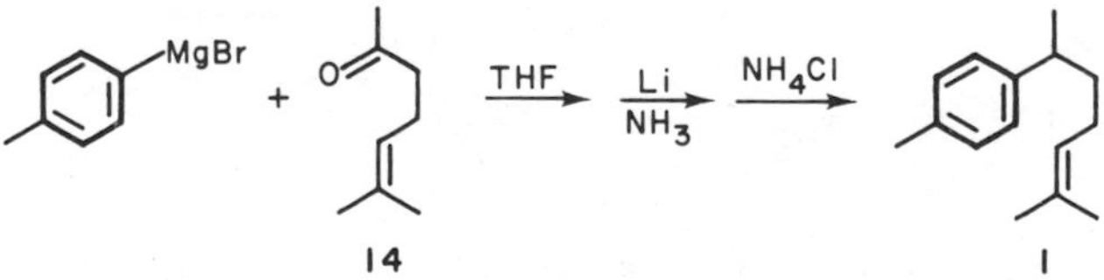

Scheme 2.   Hall's Synthesis of α-Curcumene

Dehydro-α-curcumene (**2**) has been prepared by Vig et al. in the unexceptional route shown in Scheme 3.[48] Vig and his co-workers have also

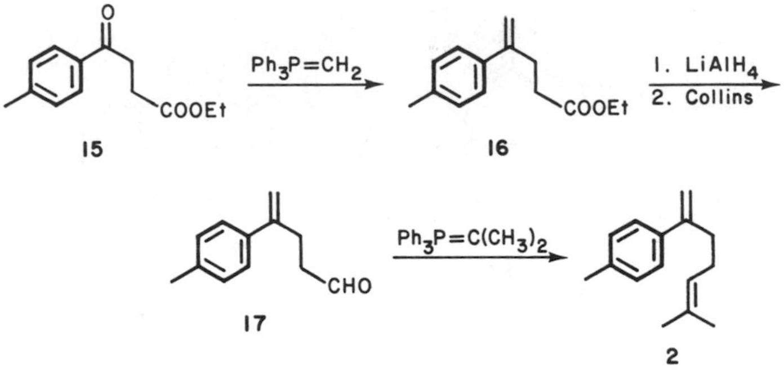

Scheme 3.   Vig's Synthesis of Dehydro-α-curcumene

synthesized the dehydrocurcumene isomer **3**, as shown in Scheme 4. The key step is Wittig olefination of aldehyde **19**; no mention is made of cis,trans isomerism in the product.

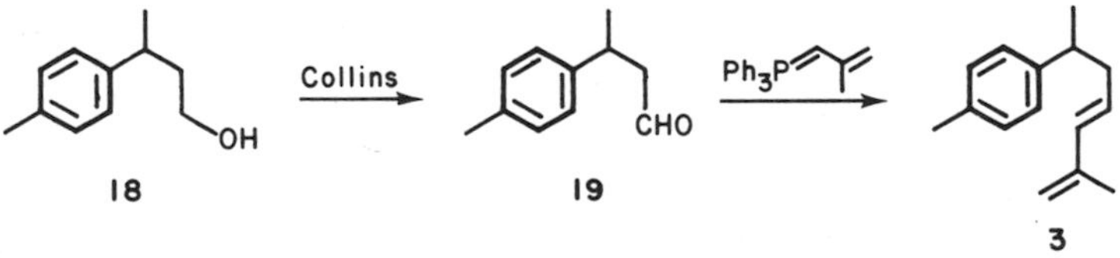

Scheme 4.   Vig's Synthesis of Dehydrocurcumene 3

Grieco and Finkelhor have reported a nice two-step synthesis of *ar*-turmerone (Scheme 5).[50] The dianion of keto phosphonate **20** is alkylated with benzylic bromide **21** and the resulting keto phosphonate is condensed with acetone to give the sesquiterpene.

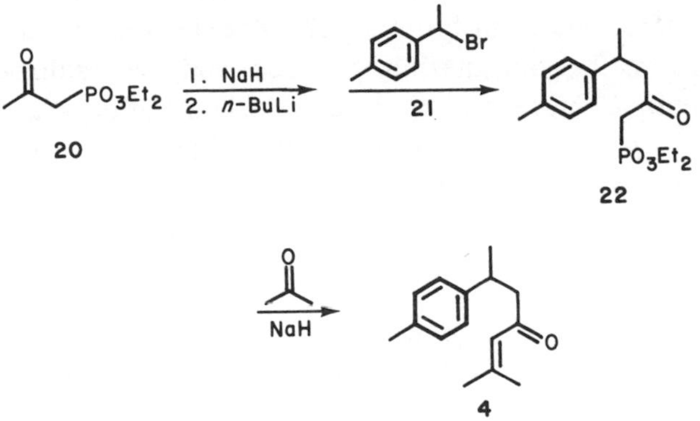

Scheme 5.   Grieco's Synthesis of ar-Turmerone

Park, Grillacea, Garcia, and Maldonado, of the University of Mexico City, reported the synthesis of ar-turmerone shown in Scheme 6.[51] The interesting feature of this synthesis is reaction of dienone **26** with lithium dimethylcuprate; addition occurs only at the less hindered double bond.

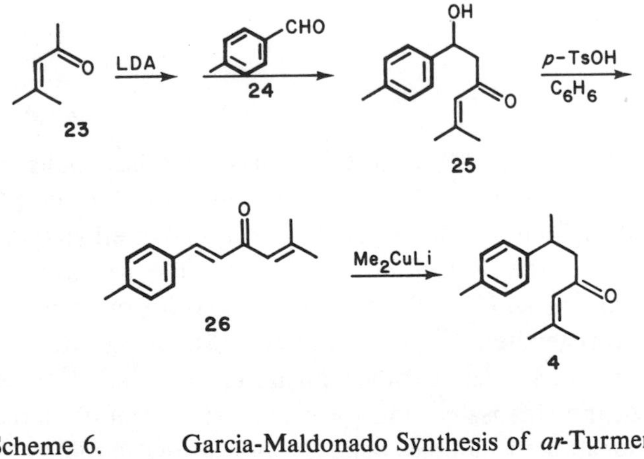

Scheme 6.   Garcia-Maldonado Synthesis of ar-Turmerone

Gosselin, Massoni, and Thuillier, of the University of Caen, have synthesized *ar*-turmerone from bromide **27**. The derived dithio ester **28** is alkylated with methallylmagnesium bromide, and the resulting dithioketal hydrolyzed. Conjugation of the double bond affords *ar*-turmerone (Scheme 7).[52] It should be noted that Mukherji has previously converted bromide **27** into **4** in a single step.[53]

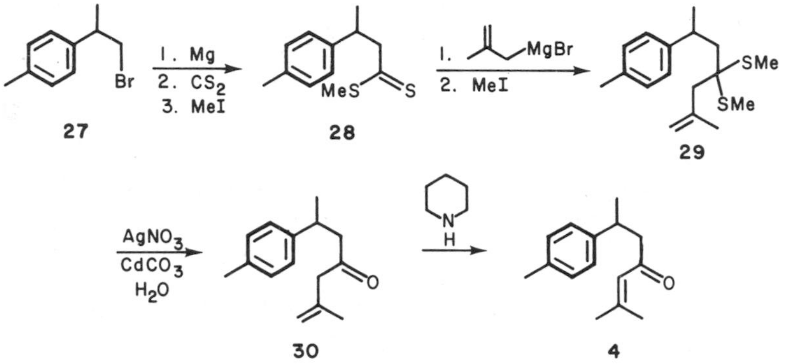

Scheme 7.   Thuillier Synthesis of *ar*-Turmerone

An interesting synthesis of nuciferal has been accomplished by Gast and Naves (Scheme 8).[54] The synthesis begins with an imaginative synthesis of phosphonium salt **32** from 1-ethoxy-2-methylbutadiene (**31**). Formation of the stabilized ylide in the presence of α-*p*-tolylpropionaldehyde (**33**) affords a 45:55 mixture of dienes **34** and **35**, which is hydrogenated to give nuciferal (**5**), along with 15% of the tetrahydro product. As in other nuciferal syntheses,[55] as well as in syntheses of the sinensals[56] and torreyal[57,58], control of stereochemistry at the trisubstituted double bond is not a real problem, as the *E*

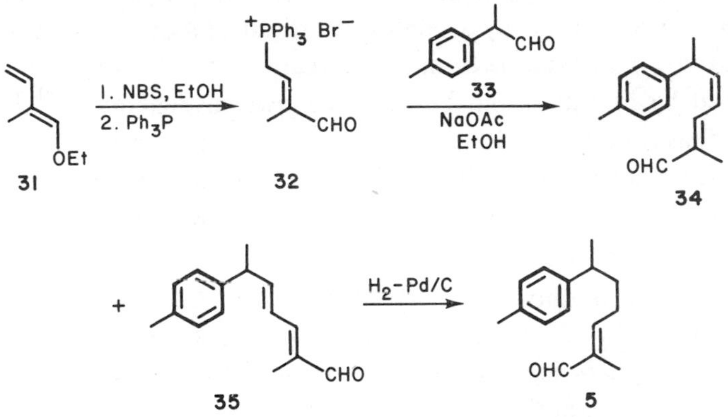

Scheme 8.   Gast-Naves Synthesis of Nuciferal

stereoisomer is apparently much more stable than the Z stereoisomer. This apparent steric bulk of the aldehyde functional group is surprising when one considers the probable preferred conformations of the two iso-mers:[59]

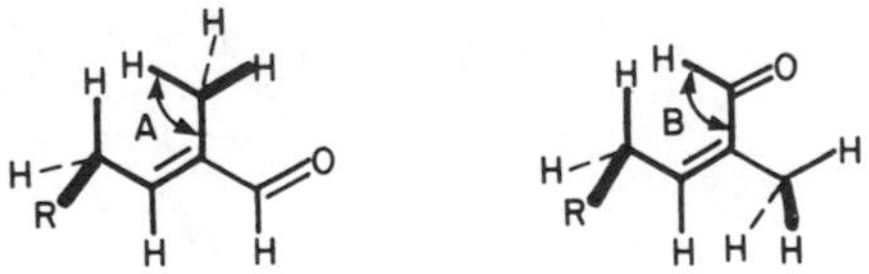

Since angle A is much more acute than angle B (109°C vs. 120°C), one might expect the E stereoisomer to experience more crowding. How-ever, whatever the explanation, the difference in energy between the two stereoisomers seems to be substantial, since there has never been a report of any Z stereoisomer under equilibrium conditions.

Phenols **6-8** do not pose any stereochemical problems. They have mostly been synthesized by straightforward adaptations of methods previously used for the synthesis of the curcumenes. Curcuphenol (**6**) was synthesized by McEnroe and Fenical as shown in Scheme 9.[60] The synthesis is characterized by its excellent overall yield (86%). An interesting chemical transformation is the quantitative dehydration of the tertiary alcohol **37** upon treatment with acetic anhydride in pyridine at room temperature.

A synthesis of xanthorrhizol (**7**), an isomer of curcuphenol, had been recorded earlier by Mane and Krishna Rao.[61] This synthesis, outlined in Scheme 10, requires eleven steps beginning with ketone **39**. It is clear that the Fenical approach (Scheme 9) is greatly superior.

Elvirol (**8**) is interesting because of its divergence from the head-to-tail isoprene rule. It has been synthesized twice, by Bohlmann and Köring,[62] who first isolated the compound, and also by Dennison, Mirrington, and Stuart.[63] The Bohlmann synthesis is outlined in Scheme 11. Although it is a good deal longer than other syntheses of bisabolene-type sesquiterpenes, it does employ rather different methodology. Unfortunately, many of the steps proceed in low yield. For example, the

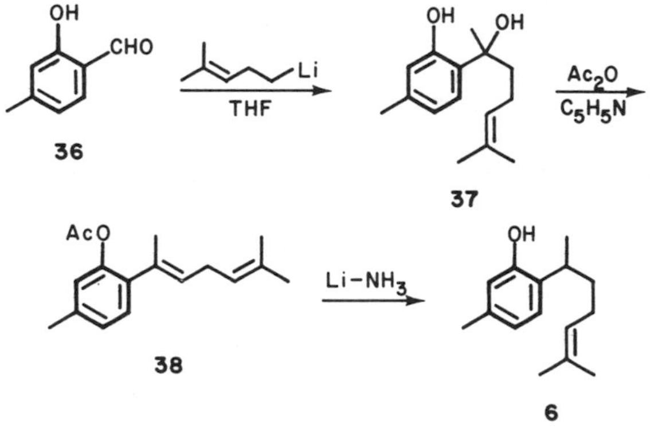

Scheme 9.   McEnroe-Fenical Synthesis of Curcuphenol

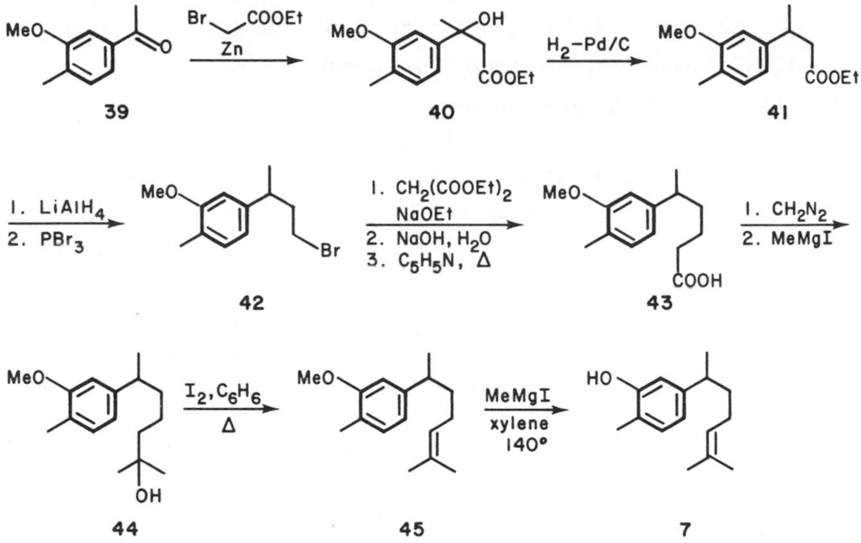

Scheme 10.   Mane-Kirshna Rao Synthesis of Xanthorrhizol

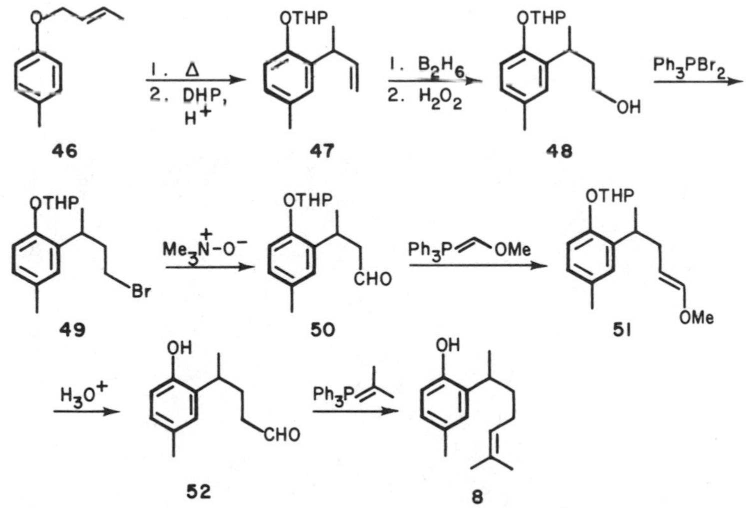

Scheme 11.   Bohlmann Synthesis of Elvirol

overall yield for the last four steps is only 0.3%. It is not clear why the two-stage Khronke process for converting alcohol **48** into aldehyde **50** must be employed rather than direct oxidation.

The Dennison-Mirrington-Stuart synthesis, outlined in Scheme 12, employs the method first introduced by Birch[47] and subsequently employed by Hall and co-workers[46] for the synthesis of curcumene. The synthesis is efficient and was easily adapted to a synthesis of isomer **56**, thus allowing an unambiguous assignment of structure for the natural product.

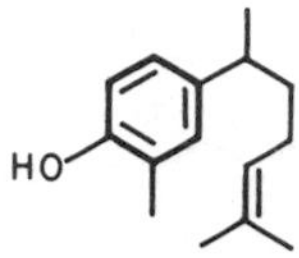

56

The final members of this class which have been synthesized are cur-cumene ether (**9**) and sydowic acid (**10**). Curcumene ether was synthesized by Vig et al. as outlined in Scheme 13.[64] Keto ester **57**, obtained in the straightforward manner shown, is treated with excess Grignard reagent to obtain diol **58**, which is cyclized by treatment with sulfuric acid. The final step proceeds in 50% yield. Vijayasarathy, Mane,

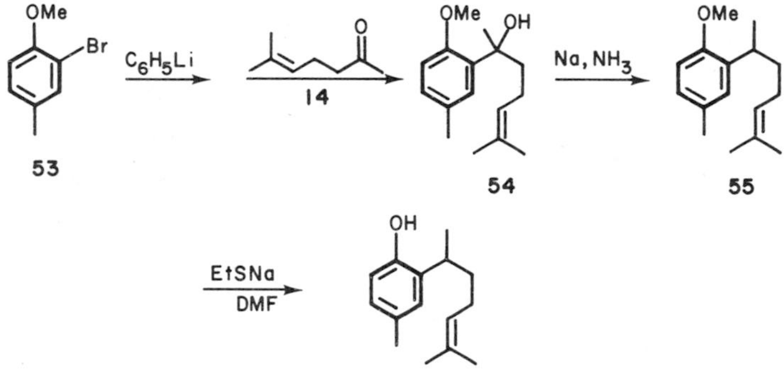

Scheme 12.   Dennison-Mirrington-Stuart Synthesis of Elvirol

and Krishna Rao employed essentially the same method to synthesize sydowic acid (**10**, Scheme 14).[65] The crucial oxidation to give **62** proceeds in only 21% yield. A number of other oxidation methods were examined, but none provide the acid in better yield.

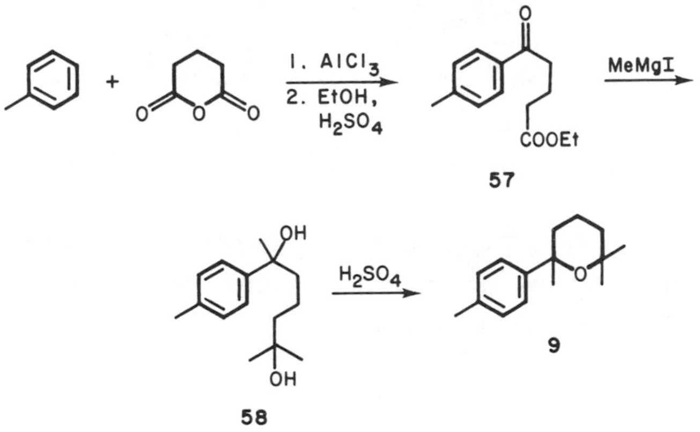

Scheme 13.   Vig's Synthesis of Curcumene Ether

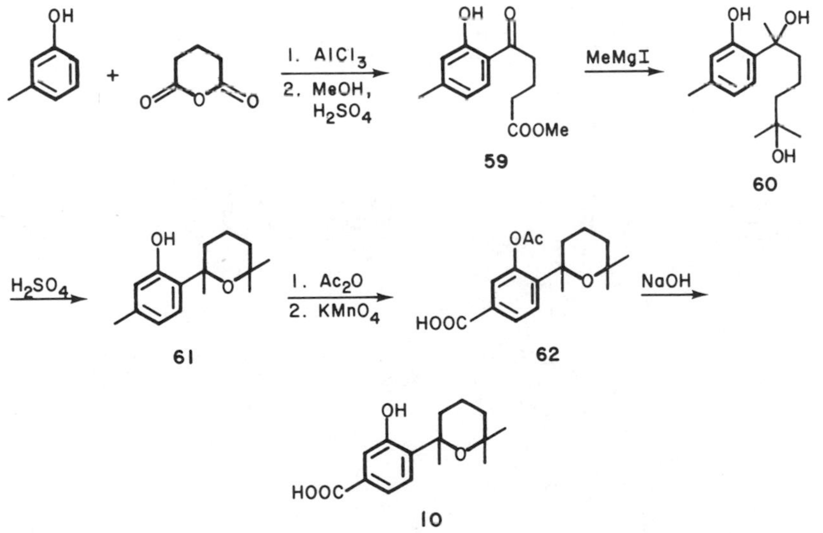

Scheme 14.   Krishna Rao Synthesis of Sydowic Acid

B. Sesquichamaenol

This interesting seco-cadinane (**63**) is the only member of its skeletal

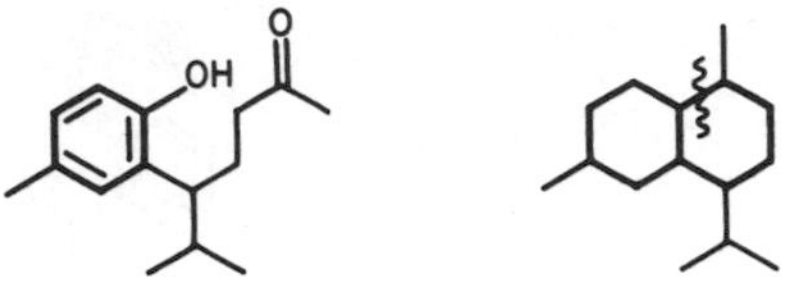

63

class. A synthesis by Takase and co-workers[66] is outlined in Scheme 15. The synthesis follows well-precedented lines and provides the racemic natural product in reasonable yield.

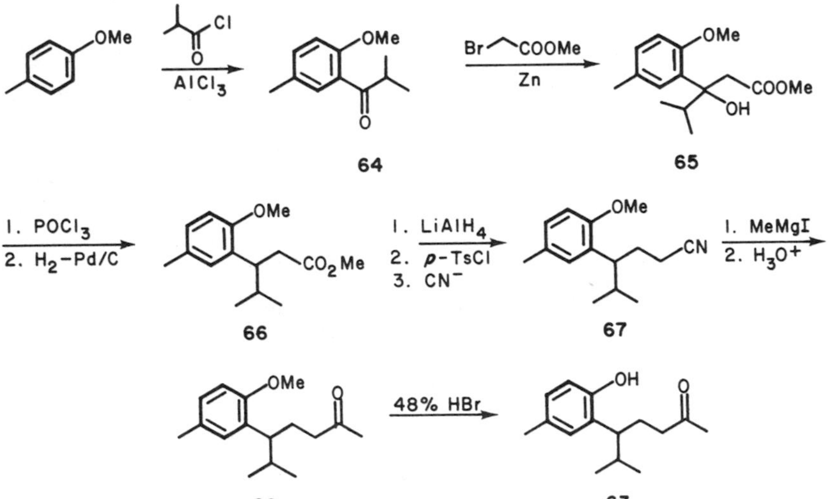

Scheme 15.   Takase Synthesis of Sesquichamaenol

C. Bisabolenes, Lanceol, and Alantone

A number of isomeric bisabolenes occur in nature. Those which have been synthesized are α-bisabolene (**69**), β-bisabolene (**70**), Z-γ-bisab-olene (**71**), E-γ-bisabolene (**72**), and isobisabolene (**73**). The related oxygenated bisabolenes lanceol (**74**) and α-alantone (**75**) will also be dis-cussed in this section. Within this group, the simpler synthetic targets

are **70** and **73**, as there is no diastereomer problem. With α-bisabolene (**69**), lanceol (**74**), and α-alantone (**75**), there is a trisubstituted double bond which complicates any synthesis. With γ-bisabolene, it has recently been found that both stereoisomers, with regard to the tetrasub-stituted double bond, occur in nature. A problem in interpreting much of the synthetic literature is that questions of stereochemistry have often been ignored. Several syntheses have been published which should lead to a mixture of diastereomers, but the authors seem to beg the issue of isomeric purity or even of isomer identity.

Two syntheses of α-bisabolene (**69**) have been published. These syntheses, by Vig and co-workers[67] and Teisseire and co-workers[68] typify the stereochemical problem. The synthesis by Vig et al. is outlined in Scheme 16. It was reported that ketone **76**, the Diels-Alder product from butadiene and methyl vinyl ketone, reacts with triethyl phospho-noacetate anion to give *only* the Z diastereomer **77**. This compound is transformed into a substance purported to be α-bisabolene as shown in the scheme.

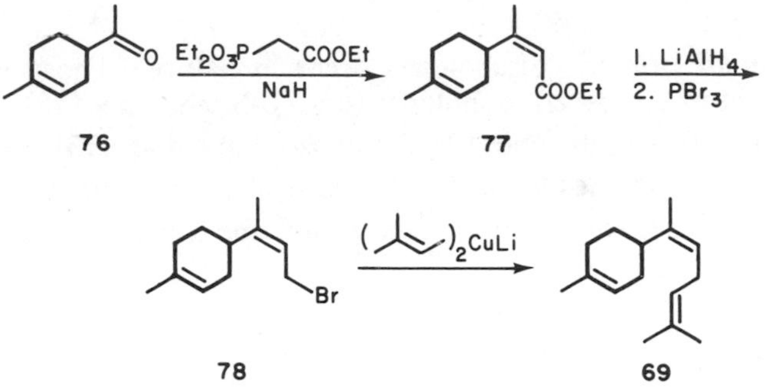

Scheme 16.   Vig's Synthesis of α-Bisabolene

The synthesis of α-bisabolene by Teisseire et al. is a simple Wittig coupling of ketone **76** with phosphorane **79** (Scheme 17).[65] Their paper notes that the Wittig reaction gives only a single isomer, which they presumed to be **80**, since that was the then-accepted structure of α-bisabolene. Delay and Ohloff reinvestigated both the Vig and Teisseire

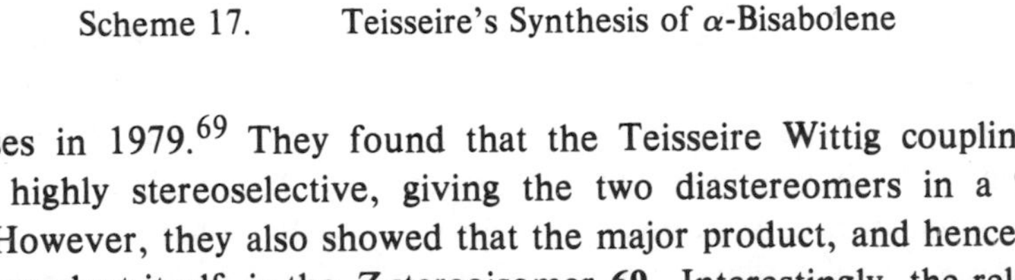

Scheme 17.   Teisseire's Synthesis of α-Bisabolene

syntheses in 1979.[69] They found that the Teisseire Wittig coupling is indeed highly stereoselective, giving the two diastereomers in a 95:5 ratio. However, they also showed that the major product, and hence the natural product itself, is the *Z* stereoisomer **69**. Interestingly, the related Wittig coupling of ketone **76** with saturated phosphorane **81** gives mainly the *E* stereoisomer **82**. In their reinvestigation of the Vig synthesis,

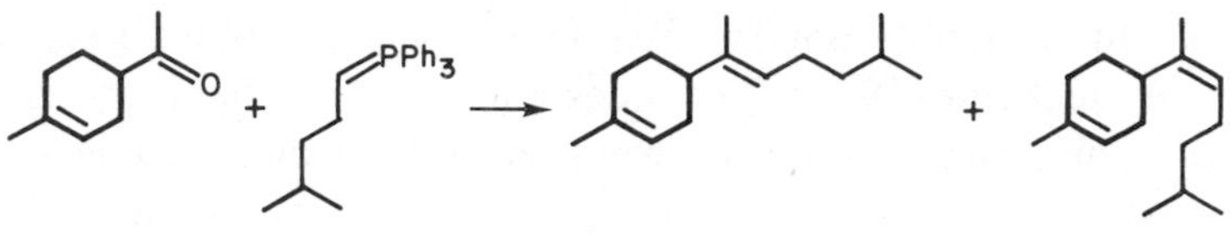

76   81   82   +   83

3:1

Delay and Ohloff found that the major isomer produced in the triethyl phosphonoacetate reaction on ketone **76** is the *E* isomer (80%), rather than the *Z* isomer as alleged by Vig and co-workers. Thus, it seems that the synthesis outlined in Scheme 16 provides the natural product as the minor component of a 4:1 mixture.

Vig has also published a synthesis of β-bisabolene (**70**) during the period of this review;[70a] the same investigator has previously published two other syntheses of the same terpene.[70b] The synthesis (Scheme 18) follows a path which is essentially identical to one of Vig's earlier syntheses of **70**; the only difference is the use of a β-keto sulfoxide (**86**) rather than a β-keto ester. Isobisabolene (**73**) was also prepared by Vig through a simple modification of the synthesis shown in Scheme 18, where **84** is methylenated by the Wittig reaction, rather than subjected to the Grignard-dehydration sequence.

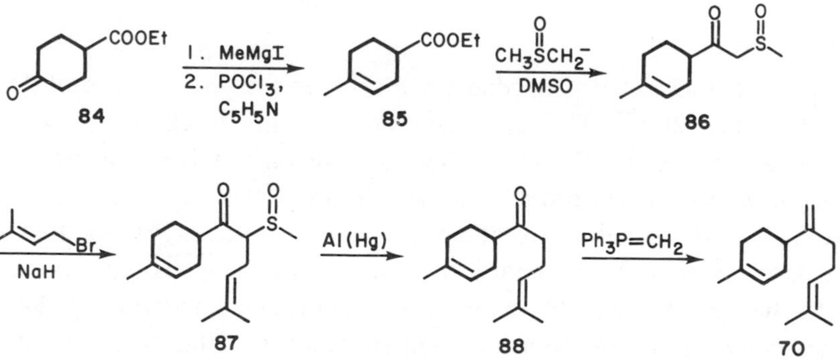

Scheme 18.   Vig's Third Synthesis of β-Bisabolene

The γ-bisabolenes (**71** and **72**) are interesting because of their tetrasubstituted double bonds. The first reported synthesis came from Vig et al. (Scheme 19).[71] The question of stereochemistry is totally ignored. The initial reaction probably gives both diastereomers. Aside from the problem of stereochemistry, the most interesting feature of this synthesis is the first step: Wadsworth-Emmons reaction on the sensitive β,γ-unsaturated ketone **89**.

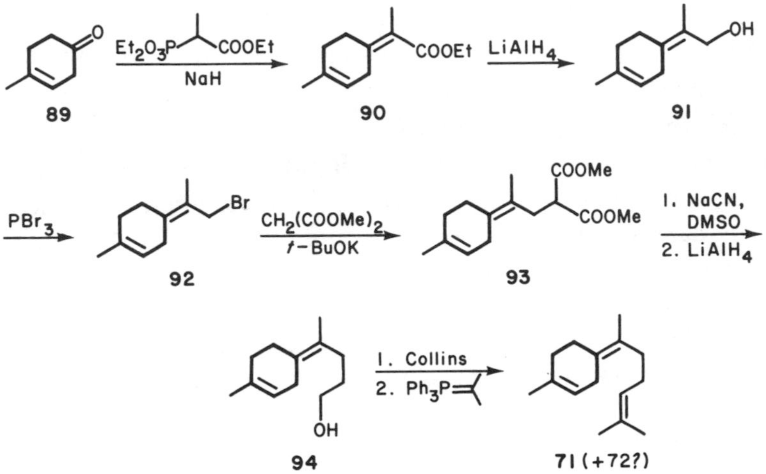

Scheme 19.   Vig's Synthesis of the γ-Bisabalols

Teisseire and co-workers reported a synthesis of the mixture of **71** and **72** (Scheme 20).[68] The synthesis begins with terpinolene (**95**) and proceeds via alcohol **97**. The double bond geometry is set in the Claisen rearrangement, which provides **98** and **99** in a 60:40 ratio. Faulkner and Wolinsky independently executed the same synthesis, except that they prepared alcohol **97** by addition of isopropenylmagnesium bromide.[72] Both the Teisseire and Faulkner teams report the separation of the final γ-bisabolene stereoisomers, but neither team was able to assign stereostructures to either the synthetic products or the natural compound. Subsequently, Faulkner and Wolinsky developed the alternate synthesis

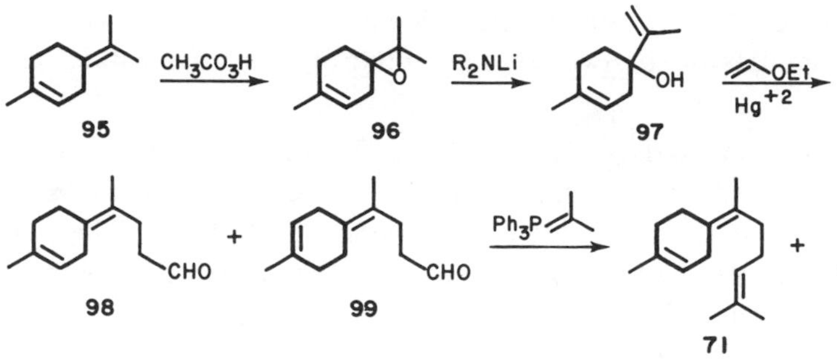

Scheme 20.   Teisseire's Synthesis of the γ-Bisabolenes

of Scheme 21.[73] The Krapcho method[44] was applied to acid **100** and methylheptenone to obtain a mixture of diastereomeric β-hydroxy acids **101** and **102** which was separated by fractional crystallization. Each isomer was converted into the corresponding γ-bisabolene. The structure

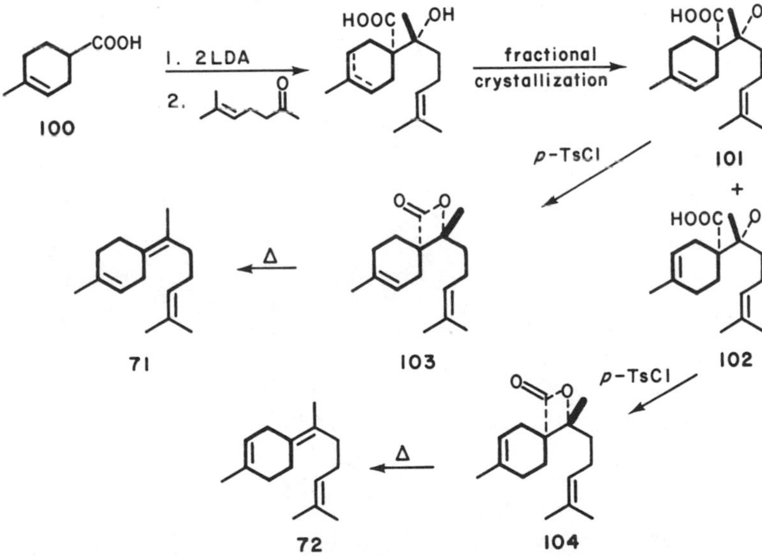

Scheme 21.   Faulkner-Wolinsky Synthesis
of the γ-Bisabolene Stereoisomers

of **101** was determined by single-crystal X-ray analysis, thus establishing the stereostructures of the two terpenes, both of which are natural products.

Crawford, Erman, and Broaddus, of the Procter and Gamble Company, have studied the metallation of limonene with a complex of *n*-butyllithium and tetramethylethylenediamine (TMEDA).[74] The reaction provides simple syntheses of several bisabolene class sesquiterpenes, as shown in Scheme 22. The Wadsworth-Emmons reaction used in the synthesis of lanceol (**74**) actually gives the *E* and *Z* stereoisomers in a ratio of 8:1. This ratio can be changed to 5:3 by ultraviolet irradiation. The two isomers were separated and each was converted into the corresponding lanceol isomer. The authors cast a shadow on the assigned stereostructure of the natural alcohol, since it was determined by oxidation to the aldehyde. As discussed earlier, it is known that such

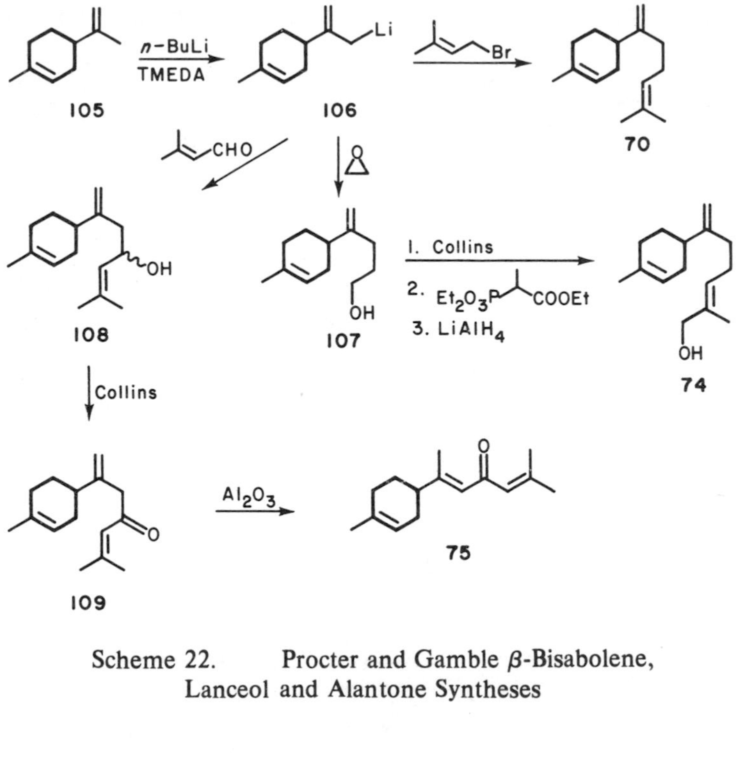

Scheme 22.   Procter and Gamble β-Bisabolene, Lanceol and Alantone Syntheses

aldehydes show a strong preference for the *E* configuration and readily isomerize under the conditions used to accomplish the lanceol oxidation. Unfortunately, spectra for natural lanceol were not published.[75]

A synthesis of (*E*)-lanceol has also been reported by Cazes and Julia of the C. N. R. S. in Paris. [76] The synthesis (Scheme 23) employs the Crawford method for activation of limonene and utilizes the novel ketene dithioacetal **110** as an isoprene equivalent. The synthesis provides lanceol in an overall yield of 30%.

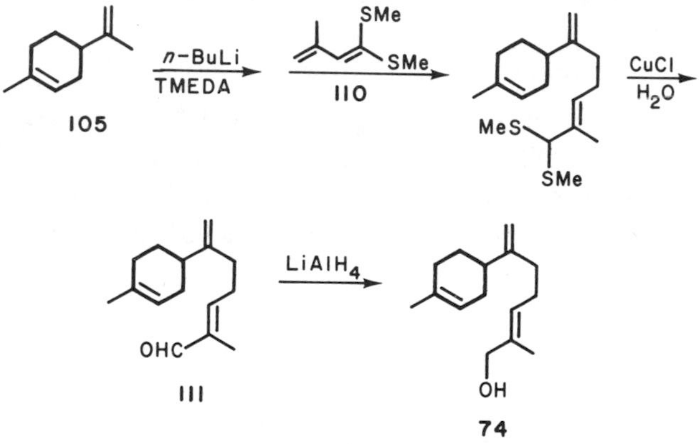

Scheme 23.   Cazes-Julia Synthesis of Lanceol

Krishna Rao and Alexander have reported the efficient synthesis of alantone which is summarized in Scheme 24.[77] The synthesis begins with free radical addition of carbon tetrachloride to limonene; the product, a result of addition, followed by elimination of HCl, is **112**. Vigorous hydrolysis of this material affords the crystalline acid **113**, which is itself a natural product. Acid **113** reacts with a large excess of isobutenyl-lithium to give α-alantone (**75**). Although the conversion of **113** to product is low (only 35%) the yield of **75** based on unrecovered starting material is good (79%).

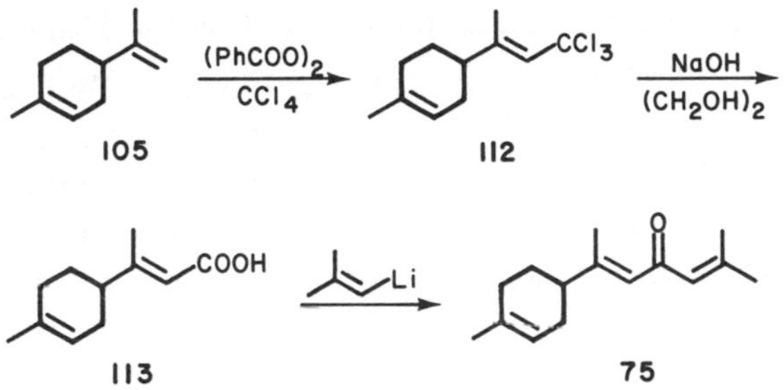

Scheme 24.   Kirshna Rao Synthesis of Alantone

Another synthesis of α-alantone, by Babler, Olsen, and Arnold, begins with addition of vinyllithium to ketone **76** (Scheme 25).[78] The resulting allylic alcohol is rearranged to the allylic acetate **115**, which is said to be at least 90% E stereoisomer. The synthesis is completed in a straightforward manner.

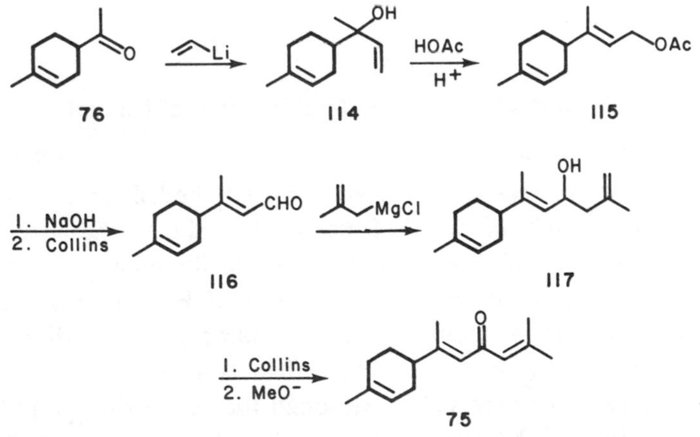

Scheme 25.   Babler-Olsen-Arnold α-Alantone Synthesis

D. α-Bisabolol, α-Bisabololone, Deodarone, Juvabione, and Epijuvabione

Of the bisabolene family of sesquiterpenes, the most challenging are those which contain two asymmetric carbons, one in the cyclohexene ring and one in the side chain. This remains a major problem, although considerable progress has been made toward stereoselective syntheses of this type of compound. Five natural products of this kind have been synthesized—α-bisabolol (118), α-bisabololone (119), deodarone (120), juvabione (121), and epijuvabione (122). Four syntheses are done

under conditions of good stereocontrol—Ficini's 1974 epijuvabione synthesis, Evans' 1980 epijuvabione synthesis, and the 1979 syntheses of α-bisabolol by Schwartz and Kakisawa. Several other syntheses follow routes which require separation of mixtures of diastereomeric intermediates. In addition, there have been the normal quota of syntheses which blithely ignore the stereochemical problem altogether. There is yet to be a synthesis which produces diastereomerically homogeneous deodarone (120). Indeed, the relative configuration of this sesquiterpene is still unknown.

α-Bisabolol (118) was first prepared by Ruzicka in 1925 by a method which undoubtedly gives a mixture of the two diastereomers.[79] The next report of the synthesis of alcohol 118 was by Manjarrez and Guzman.[80]

However, it has been noted that the material prepared by these workers is probably a mixture of double bond isomers, methyl group positional isomers, and diastereomers.[81] Rittersdorf and Cramer prepared α-bisabolol as shown in Scheme 26, which is essentially the same as the earlier Ruzicka synthesis.[82] The initial Diels-Alder reaction of isoprene and methyl vinyl ketone gives ketones 76 and 123 in a ratio of 2:3. However, as in the Ruzicka synthesis, the mixture was not separated but was subjected to Grignard addition to obtain 124 and 125, each as a mixture of diastereomers. Rittersdorf and Cramer were able to separate the diastereomers by glpc of their trimethylsilyl derivatives. As expected, they are formed in equal amounts. The cyclization of (2Z,6E)-farnesyl phosphate (126a) and (6E)-nerolidyl phosphate (126b) was also studied. Among the products formed are the two α-bisabalol isomers 118 and 127. The 118:127 ratio is 39:61 from phosphate 126a and 45:55 from phosphate 126b.

Gutsche, Maycock, and Chang[81] studied the acid-catalyzed cyclization of farnesal and nerolidol. To prepare an authentic specimen of α-bisabolol, they modified the earlier Ruzicka and Manjarrez-Guzman synthesis. Because pure ketone 76 could not easily be obtained from the Diels-Alder reaction of isoprene and methyl vinyl ketone, they prepared it as shown in Scheme 27. Preparation of 76 by this method is tedious and gives the product in poor yield, but it is isomerically pure. The Grignard alkylation of 76 gives a mixture of diastereomers 118 and 127.

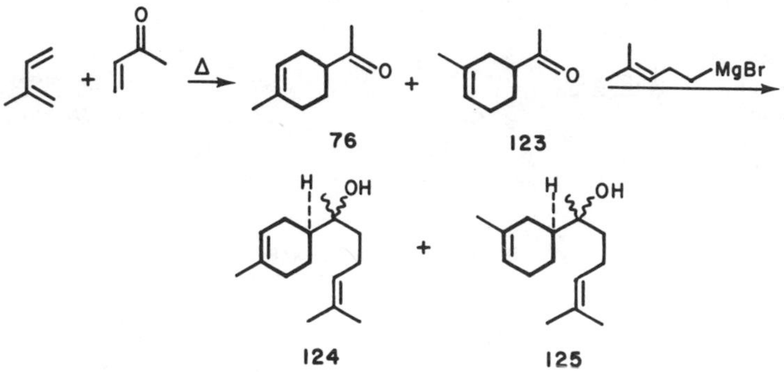

Scheme 26.   Ruzicka-Ritterdorf-Cramer Synthesis of α-Bisabalol

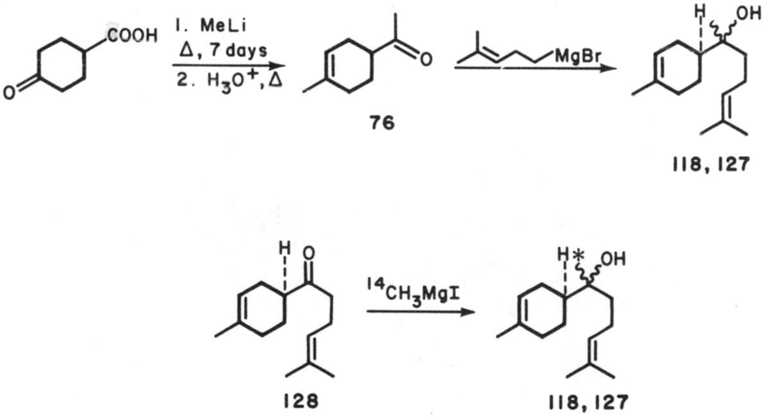

Scheme 27.   Gutsche-Maycock-Chang Synthesis of α-Bisabolol

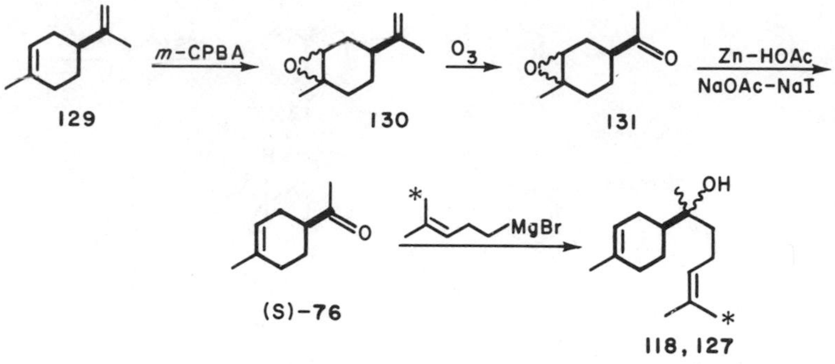

Scheme 28.   Knöll-Tamm Synthesis of α-Bisabolols

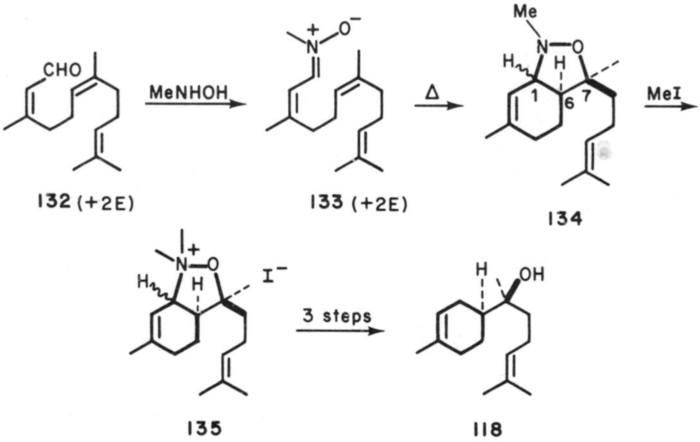

Scheme 29.   Kakisawa-Schwartz Synthesis
of (6SR,7SR)-α-Bisabolol

Two groups have prepared radio-labeled α-bisabolols for biosynthesis experiments. Forrester and Money prepared a mixture of **118** and **127** having the ^{14}C in the C-6 position.[83] Knöll and Tamm prepared a mixture of the diastereomers starting with the two pure enantiomers of 4-acetyl-1-methylcyclohexene (**76**), which were prepared from optically active limonenes (**129**).[84] The synthesis of one of the optically active diastereomer mixtures is shown in Scheme 28.

The first syntheses of the pure diastereomers of α-bisabolol were reported in 1979 by Schwartz and Swanson[85] of Florida State University, and Iwashita, Kusumi, and Kakisawa,[86] of the University of Tsukuba. The syntheses are identical and employ intramolecular cycloaddition of a nitrone derived from farnesal. Synthesis of the 6*SR*,7*SR* diastereomer from (6*Z*)-farnesal is outlined in Scheme 29. The nitrone **133** is prepared by treating (6*Z*)-farnesal (**132**) with N-methylhydroxylamine. Both **132** and **133** are mixtures of 2*E* and 2*Z* stereoisomers. Cycloaddition of **133** gives a mixture of diastereomeric oxazolidines **134**. Although **134** is a mixture of diastereomers, the relative stereochemistry at C-6 and C-7 is the same, since this is determined by the stereochemistry of the C-6 double bond. The two groups complete the synthesis by different routes, although each involves reductive cleavage of the C_1-nitrogen bond. The 6*SR*,7*RS* diastereomer **127** was prepared by applying the same sequence to (6*E*)-farnesal. The natural α-bisabolol is identical with **118**. Since Knöll and Tamm had prepared α-bisabolol from (*S*)-limonene (**129**), the natural alcohol is shown to be 6*S*,7*S*.

In 1974, Gopichand and Chakravarti reported the synthesis of α-bisabololone (**119**) and deodarone (**120**) which is summarized in Scheme 30.[87] The side chain is attached to ketone **76** by an interesting method discovered by Suga, Watanabe and Fujita.[88] Although it was not determined, the product is certainly a mixture of diastereomers. Synthetic deodarone, also a mixture of diastereomers, is obtained by acid-catalyzed cyclization of **119**.

Vig and co-workers reported another stereorandom synthesis of **119** and **120** in 1976 (Scheme 31).[89] The synthesis utilizes β-keto ester **138**, which has previously been used as a precursor to β-bisabolene and lanceol.[90]

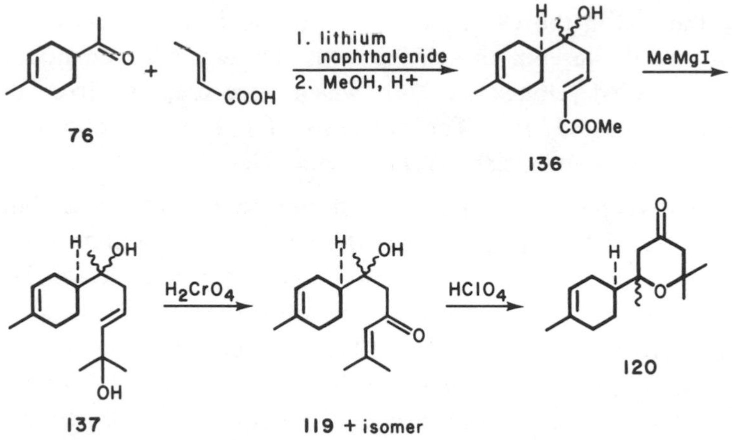

Scheme 30.   Gopichand-Chakravarti Synthesis
of α-Bisabolol-3-one and Deodarone

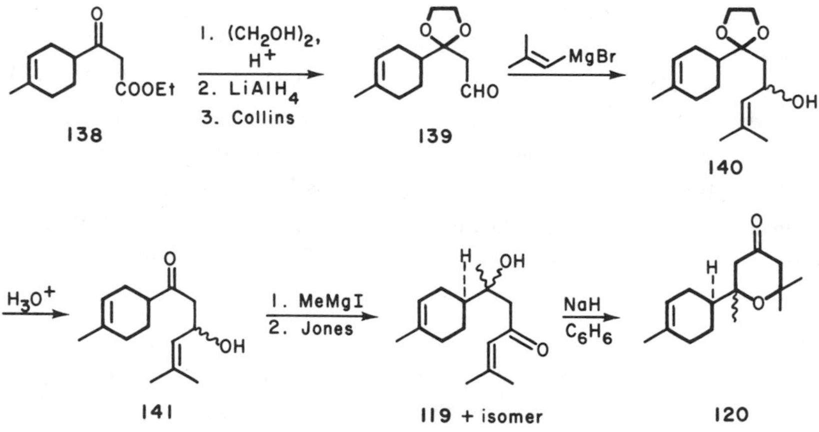

Scheme 31.   Vig's Synthesis of α-Bisabololone and Deodarone

A third stereorandom synthesis of **119** was reported by Malanco and Maldonado in 1976 (Scheme 32).[91] The synthesis begins with enone **76**,

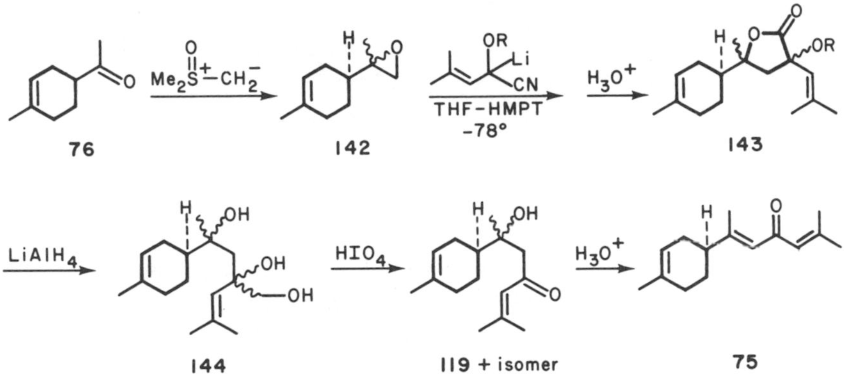

Scheme 32.   Malanco-Maldonado Synthesis
of α-Bisabolene and Alantone

which is converted into a diastereomeric mixture of limonene epoxides (**142**) of unknown composition. The epoxide ring is opened by Stork's protected cyanohydrin method to obtain lactone **143**, which is converted into bisabololone (**119**) by the method shown in the scheme. Dehydration of the mixture of diastereomeric α-bisabololones provides racemic alantone (**75**).

Kergomard and Veschambre also prepared the α-bisabololones from the limonene epoxides, but they separated the isomers first.[92] Thus, they obtained each α-bisabololone diastereomer in a pure state and determined the configuration of the natural product. The synthesis of the natural isomer is shown in Scheme 33. An identical sequence starting with the (*SR*)-epoxide afforded the other diastereomer of **119**, which was clearly distinguishable from the natural product. Unfortunately, the two diastereomers have not been cyclized to obtain the two deodarone diastereomers, which would determine the stereostructure of **120**.

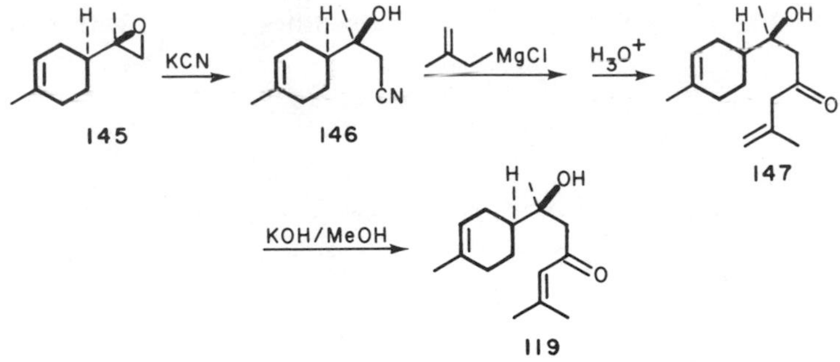

Scheme 33.   Kergomard-Veschambre Synthesis
of (-)-α-Bisabololone

The last reported α-bisabololone synthesis, due to Park, Grillasca, Garcia, and Maldonado,[51] is similar to their synthesis of *ar*-turmerone (see Scheme 6). In this case the crossed aldol condensation is applied to ketone **76** to afford directly a mixture of diastereomeric α-bisabololones.

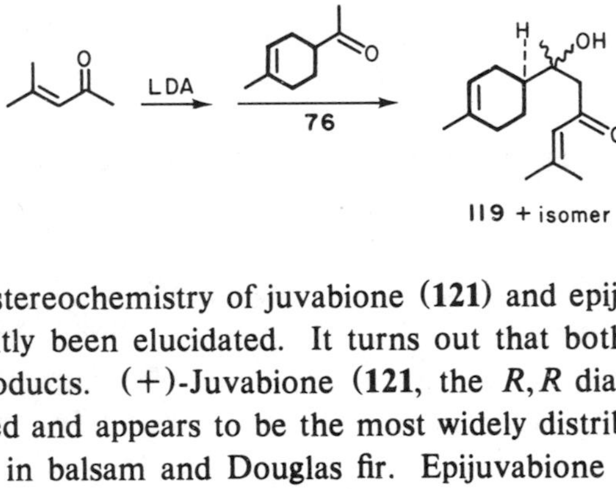

The correct stereochemistry of juvabione (**121**) and epijuvabione (**122**) has only recently been elucidated. It turns out that both diastereomers are natural products. (+)-Juvabione (**121**, the *R,R* diastereomer) was the first isolated and appears to be the most widely distributed stereoisomer occurring in balsam and Douglas fir. Epijuvabione (**122**) has been isolated only once—from a tree of uncertain species by Cerny and coworkers.[94] Most of the earlier syntheses of the juvabiones were not stereocontrolled, although some allowed separation of diastereomers at an intermediate stage.[95] The two isomers are exceedingly difficult to dis-

tinguish spectroscopically, since they have virtually identical infrared and
^{1}H-NMR spectra. As reported in Volume 2 of this series,[95] Pawson and
co-workers synthesized all four stereoisomers of the juvabiones from the
enantiomeric limonenes. An intermediate in one of the syntheses was
shown by X-ray analysis to be alcohol **148.** Since this alcohol was converted into a compound identical with a sample obtained from Cerny,[94]
the stereochemistry of the Cerny natural product was established as **122.**

148   122

However, it was not appreciated at that time that the Cerny sample was
epijuvabione, and not juvabione itself. Thus, the original structural
assignment to juvabione (R,R) was correct. This situation provides a
good deal of confusion in the literature, as all the published work
between 1968 and 1976 have the structures reversed.

The first stereocontrolled synthesis of one of the diastereomers was
achieved by Ficini and co-workers in 1973 (Scheme 34).[96] The synthesis
begins with cycloaddition of ynamine **149** to cyclohexenone to give the
aminocyclobutene **150.** Treatment of this compound, first with water
and then with aqueous acetic acid, results in clean conversion to
diastereomerically pure keto acid **151.** The stereochemistry is presumably established by kinetic protonation of **150** from the convex face to
yield the immonium ion **155.** The side chain is attached by the Stork-
Maldonado method, alkylating lithiated α-alkoxy nitrile **153** with

150   155

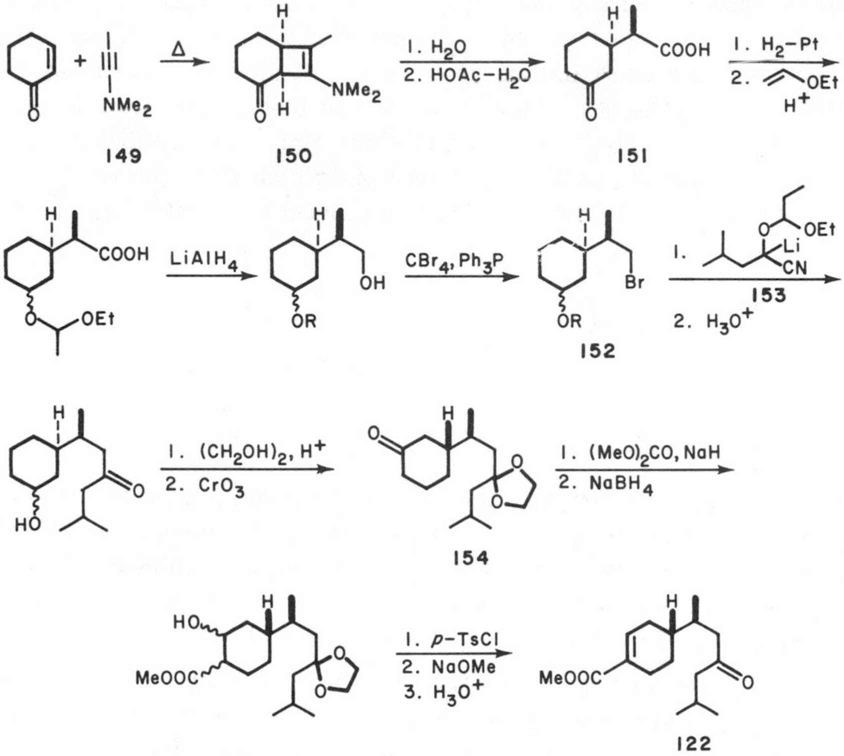

Scheme 34.   Ficini's Stereocontrolled Epijuvabione Synthesis

bromide **152**. After regeneration of the cyclohexanone carbonyl function, the synthesis of (±)-epijuvabione (**122**) is completed by standard transformations.

In 1973 Farges and Veschambre reported a synthesis of a mixture of juvabione and epijuvabione (Scheme 35)[97] which begins with (−)-perillic alcohol (**156**). The synthesis is essentially identical to the earlier Hoffmann-La Roche synthesis[95] and provides a 2:1 mixture of (−)-epijuvabione (**122**) and (−)-juvabione (**121**), the enantiomers of the natural products.

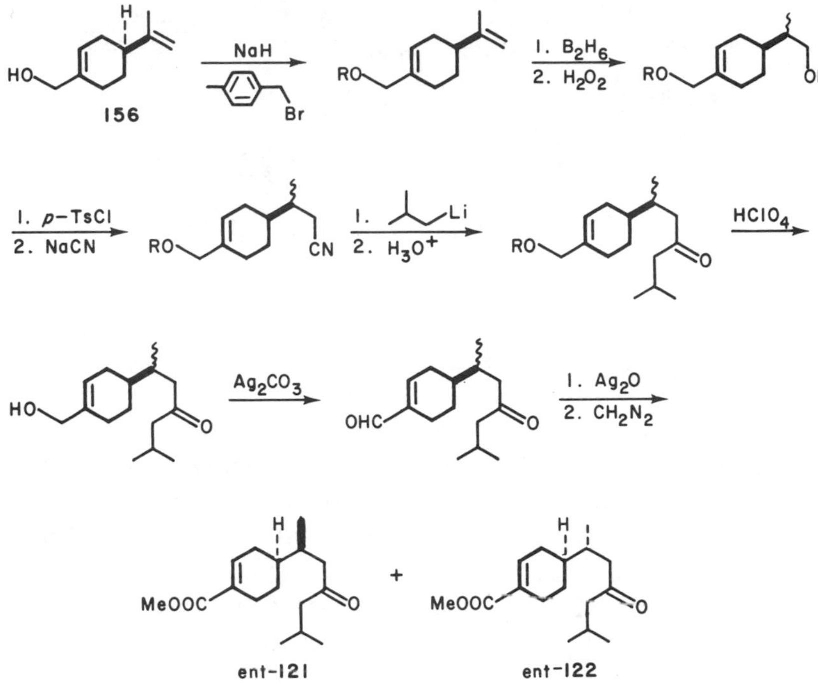

Scheme 35.   Farges-Veschambre Synthesis of the Juvabiones

An interesting juvabione synthesis was reported in 1976 by Negishi, Sabanski, Katz, and Brown (Scheme 36).[98] The synthesis starts with the commercially-available perillartine (157), which is converted via the corresponding nitrile into methylperillate (158). This compound is hydroborated using thexylisobutylborane and the intermediate adduct is carbonylated to obtain a 1:1 mixture of juvabione (121) and epijuvabione (122). Although stereorandom, the synthesis is remarkably short (four operations) and produces the mixture of 121 and 122 in 53% overall yield.

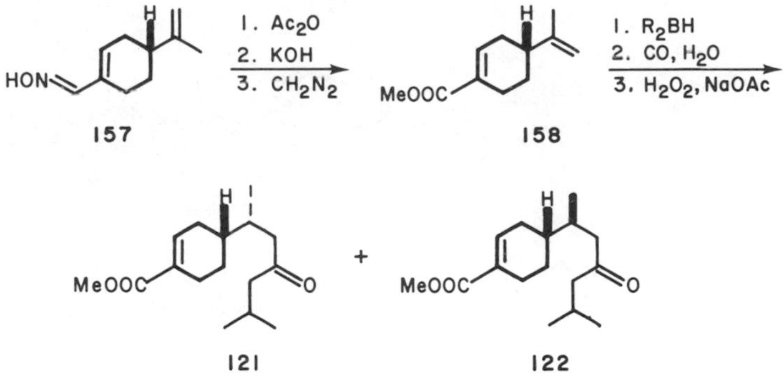

Scheme 36.   Negishi-Brown Synthesis of the Juvabiones

An elegant stereocontrolled synthesis of epijuvabione was reported by Evans and Nelson early in 1980.[99] The synthesis (Scheme 37) begins with addition of an organozinc reagent derived from 1-methoxy-2-butyne to cyclohexenone; diastereomeric methoxy alcohols **159** and **160** are produced in a ratio of 2:1. After *anti*-reduction of the triple bond, the 2:1 mixture of **161** and **162** is rearranged as the potassium alkoxide. Diastereomers **163** and **164** (each a mixture of double bond isomers) are obtained in a ratio of 10:1. The mixture is carbomethoxylated and the resulting mixture of β-keto esters deoxygenated by a variant of the Bamford-Stevens reaction to obtain a β-γ-unsaturated ester. Base-catalyzed conjugation of the latter affords **165**, which is converted into (±)-epijuvabione (**122**) by straightforward steps. The final product is contaminated by about 4% of (±)-juvabione (**121**).

An approach to a stereospecific epijuvabione synthesis has recently been disclosed by Larsen and Monti.[100] The method has been applied to the synthesis of alcohol **148**, an intermediate in the Pawson synthesis.[95] The Monti approach (Scheme 38) establishes the relative stereochemistry by kinetic protonation of the enolate derived from bicyclic ketone **168**. After hydrochlorination of the double bond, the ring is opened by Grob fragmentation. The intermediate aldehyde is reduced *in situ* to give racemic alcohol **148**.

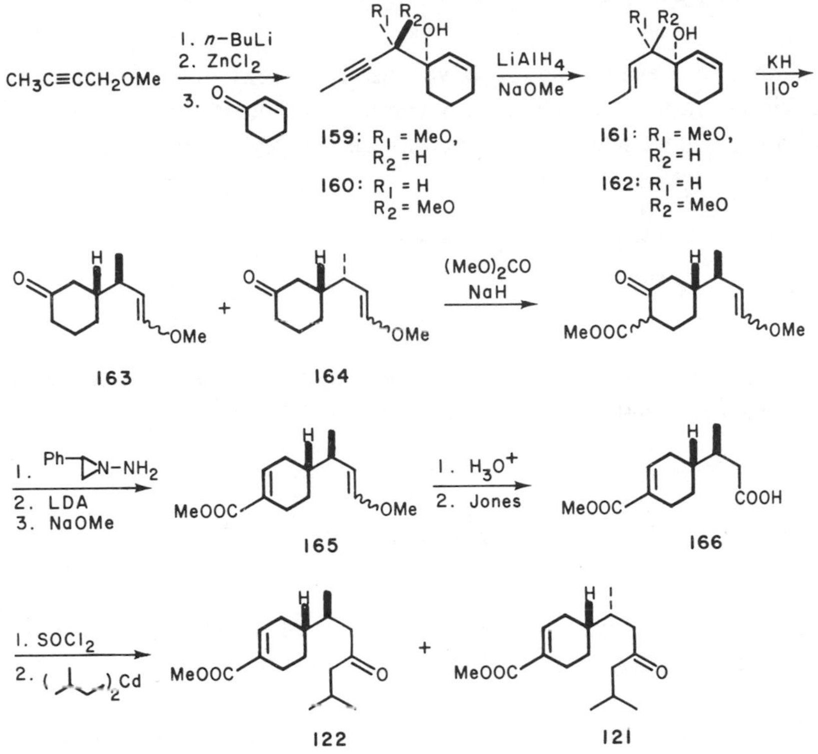

Scheme 37.   Evans-Nelson Synthesis of (±)-Epijuvabione

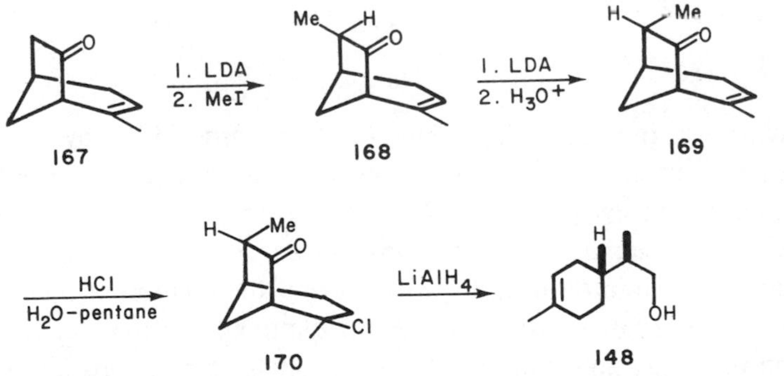

Scheme 38.   Monti's Synthesis of an Epijuvabione Precursor

E. Deoxytrisporone, Latia Luciferin, and Abscisic Acid

A number of sesquiterpenes have the 1,3,3-trimethyl-2-cyclohexenyl (cyclocitral) system, which is also common to the retinoids and caro-tenoids. Biogenetically, this skeleton presumably arises by a cationic mechanism such as:

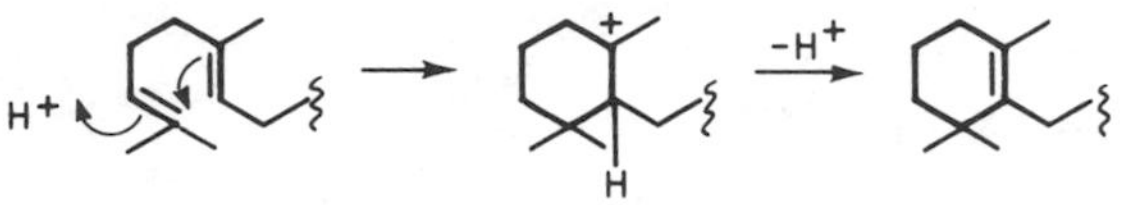

A number of marine sesquiterpenes have recently been isolated which have the same skeleton, but with bromine at C-4:

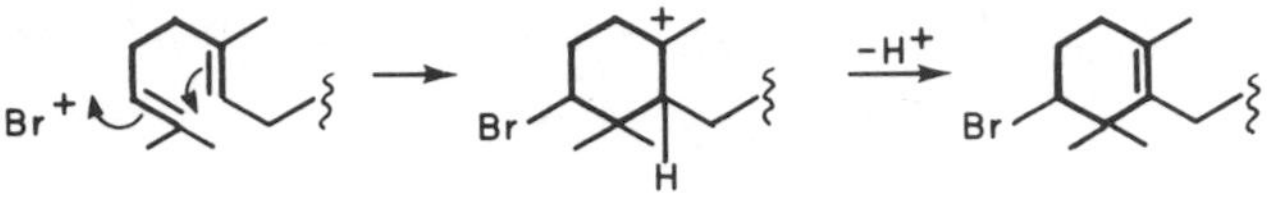

In this section, we will discuss the synthesis of deoxytrisporone (172), latia luciferin (173), and abscisic acid (174). A great deal of closely related synthetic work has been done in the retinoid and carotenoid areas. For a treatment of this research, please refer to other sources.

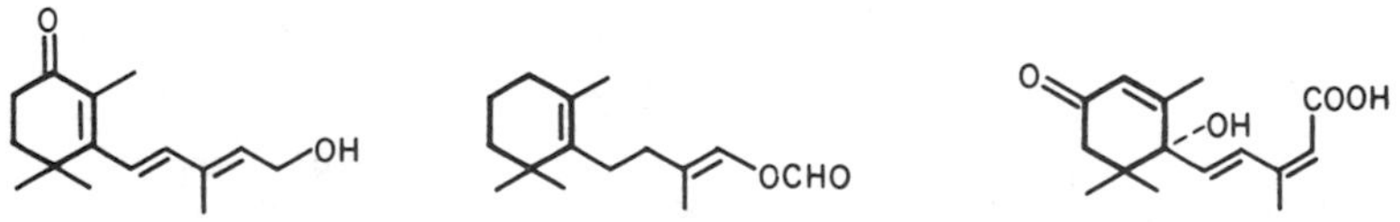

A synthesis of deoxytrisporone (172, Scheme 39), by Torii and Uneyama [101] utilizes the novel cyclocitral synthon 176, which is formed by acid-catalyzed cyclization of 175. The ring oxygen is introduced in good yield by selenium dioxide oxidation and the remaining carbons of the side chain are introduced by alkylation of the dianion of hydroxy sul-fone 177 with bromo ester 178. After elimination of benzenesulfinate ion, the oxidation states of the two functionalized positions are adjusted to obtain deoxytrisporone (172).

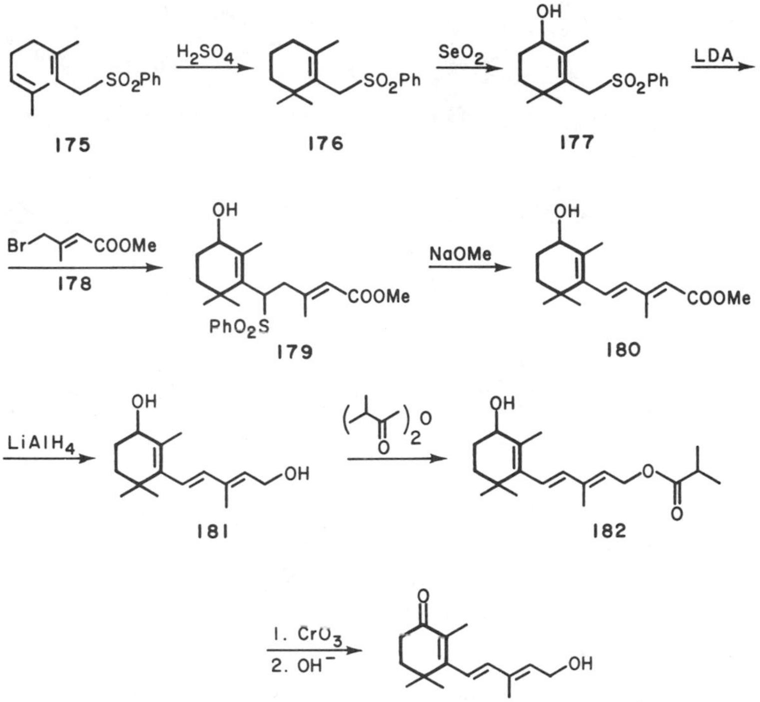

Scheme 39.   Torii-Uneyama Synthesis of Deoxytrisporone

Latia luciferin (**173**) has been synthesized by Sum and Weiler[102] as shown in Scheme 40. The synthesis employs an interesting method for repetitive introduction of isoprene units, alkylation of acetoacetic ester dianion, followed by cyclization and reaction of the derived enol phosphate with lithium dimethylcopper. The sequence is utilized twice, first in the synthesis of methyl cyclogeranate (**187**), and again in attaching the side chain (**188**→**190**). The synthesis is completed by Baeyer-Villiger oxidation of aldehyde **191** to obtain enol formate **173**.

Abscisic acid (abscisin II, **174**) is an important plant hormone. It was first synthesized in 1965 by Cornforth and co-workers[103] in order to

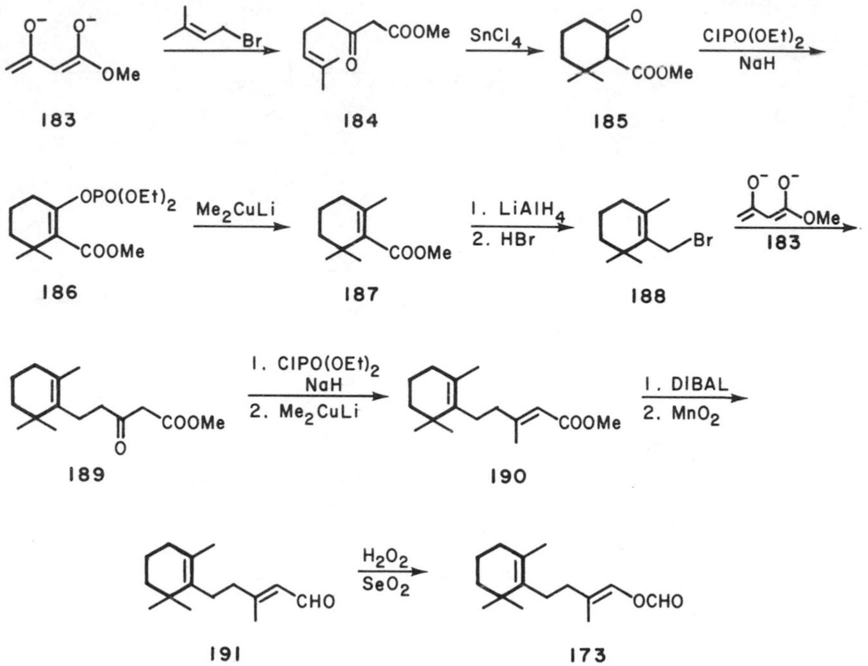

Scheme 40.   Sum-Weiler Synthesis of Latia Luciferin

confirm its assigned structure. The synthesis consisted of photooxygenation of acid **193** (obtainable from β-ionone, **192**), followed by isomerization of the resulting endoperoxide. The hormone is obtained in 7% yield from acid **193**. Mousseron-Canet and co-workers[104] examined the same procedure and made some modifications which improved the yield of **174** to 50%.

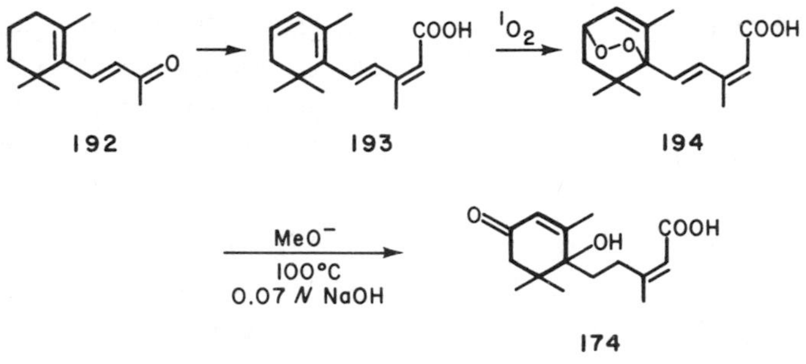

Roberts and co-workers discovered that α-ionone (**195**) undergoes oxidation by *t*-butyl chromate in *t*-butyl alcohol to afford hydroxy dione **196** in 23-27% yield.[105] This substance is subjected to Wadsworth-Emmons reaction and hydrolysis to obtain abscisic acid and its isomer **197**.

Findlay and MacKay worked out an alternative preparation of intermediate **196** beginning with β-ionone (**192**).[106] This modification produces **196** in 43% yield.

The trans,trans isomer **197** has also been prepared by Okuma as outlined in Scheme 41.[107] Emmons reaction on ketone **202** affords esters **203** (minor) and **204** (major). Hydrolysis and reduction of isomer **203** gives the abscisic acid isomer **197**, but hydrolysis of the cis isomer **204** affords the enol lactone **205**.

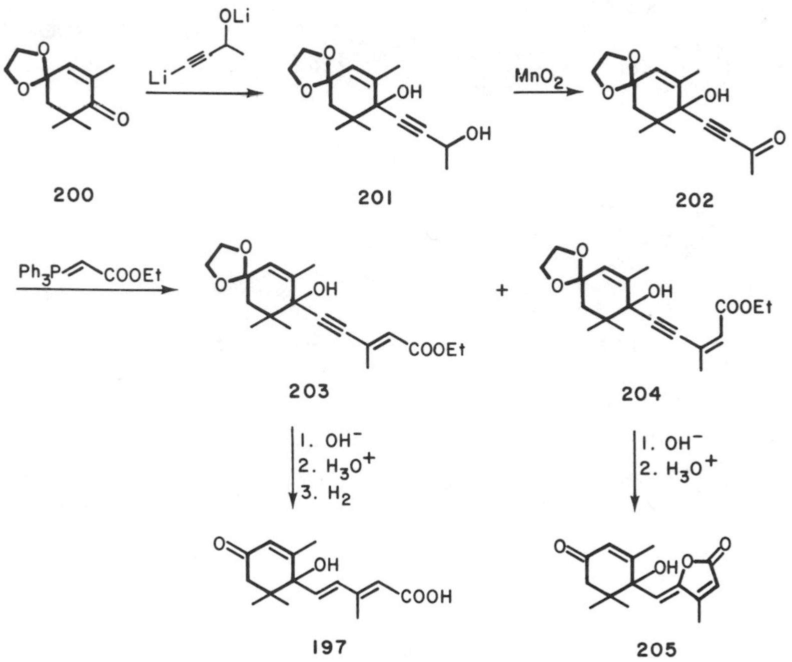

Scheme 41.   Okuma's Synthesis of an Abscisic Acid Isomer

As part of a project to determine the absolute configuration of abscisic acid, Mori carried out the synthesis with optically active hydroxy dione **196** (Scheme 42).[108] The synthesis begins with a surprising selective reduction: dione **206** is hydrogenated selectively at its more hindered carbonyl. What is even more surprising is that the product is the trans isomer **207**. This ketol is resolved by formation of an ester with 3β-acetoxyetienoyl chloride. After separation of the diastereomeric esters, isomer **208** is converted by the multi-step route shown into unsaturated alcohol **214**. Epoxidation of the derived acetate affords cis epoxide **215** (20% yield) and trans epoxide **216** (6.4% yield). The latter product is converted into the dextrorotatory enantiomer of hydroxy dione **196**, which had previously been converted into (+)-abscisic acid.

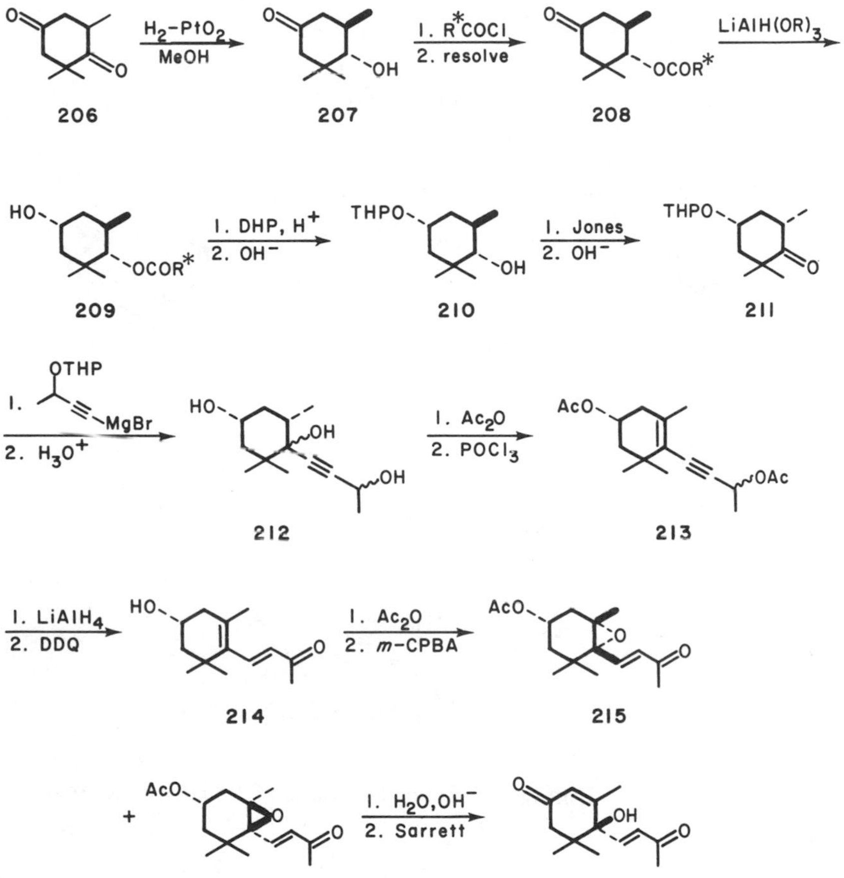

Scheme 42.   Mori Synthesis of (+)-196

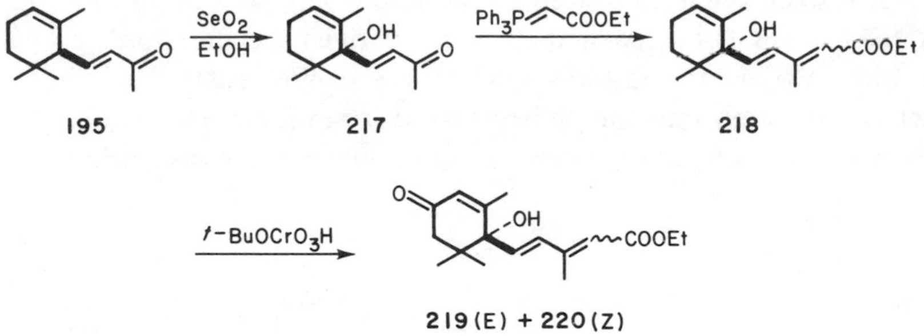

Scheme 43.   Oritani-Yamashita Synthesis of Abscisic Acid

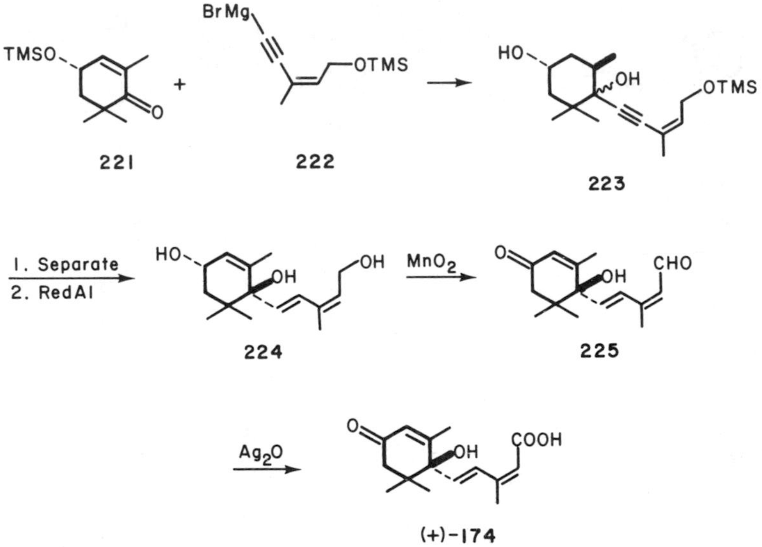

Scheme 44.   Hoffman-LaRoche Synthesis of (+)-Abscisic Acid

A synthesis of optically active ethyl abscisate (**220**) has also been reported by Oritani and Yamashita (Scheme 43).[109] The synthesis begins with optically active α-ionone (**195**), which undergoes oxidation by selenium dioxide with retention of optical activity to give keto alcohol **217**. The optical purity of the product has not been published. This material is converted to ester **218** (2:1 mixture of *E* and *Z* isomers), which is oxidized by *t*-butyl chromate to obtain esters **219** and **220** in yields of 57% and 6%, respectively. Unfortunately, the minor product corresponds to abscisic acid.

Kienzle, Mayer, Minder, and Thomas, of Hoffman-LaRoche in Basel, have reported the synthesis of the pure enantiomers of abscisic acid, as outlined in Scheme 44.[110] The synthesis begins with optically active keto ether **221**, which is condensed with the acetylide **222**. The diastereomeric diols **223** are separated and independently converted into the two enantiomers of abscisic acid. The major isomer of **223** (80%) corresponds to natural abscisic acid.

F. Caparrapi Oxide, 3β-Bromo-8-epicaparrapi Oxide, Ancistrofuran, Aplysistatin, and α- and β-Snyderols

The syntheses of these compounds are a bit more complex than those discussed in the previous section because of stereochemical complications. Caparrapi oxide (**226**), 3β-bromo-8-epicaparrapi oxide (**227**), and ancistrofuran (**228**) each contain three stereocenters within the fused ring system. These are, therefore, the easier targets for synthesis. Aplysistatin (**229**) contains these three centers as well as a fourth. Although α-snyderol (**230**) and β-snyderol (**231**) have only three stereo-

226   227   228

229   230   231

centers, one is in the side chain, well-removed from the ring. Consequently, these are the most challenging targets of this group. Unfortunately, neither the relative nor absolute stereochemistry of the snyderols is known. Thus, the two syntheses recorded to date do not address the issue of stereochemistry.

A synthesis of caparrapi oxide (226) reported in 1975 by Cookson and Lombardi (Scheme 45)[111] begins with dihydro ionone (232), which is condensed with sodium acetylide and cyclized to obtain a 1:4 mixture of 234 and 235. Semihydrogenation of 234, the minor isomer, affords racemic caparrapi oxide.

A subsequent complete study of the crucial cyclization was carried out by Lombardi and Cookson in collaboration with workers at Sandoz and Firmenich in Switzerland.[112] Their results are shown below. Unfortunately, the maximum amount of the caparrapi oxide stereoisomer

	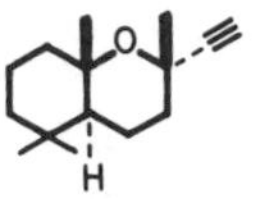 234	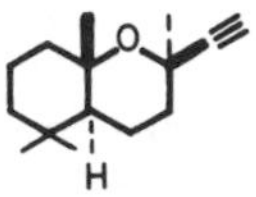 235	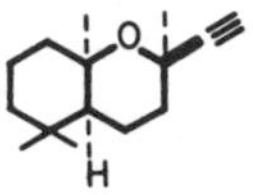 236	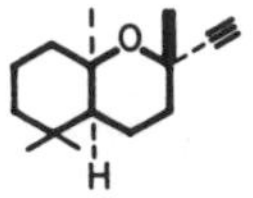 237
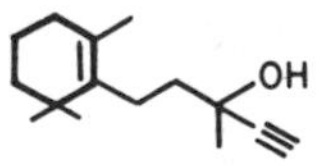 233	16%	80%	1%	3%
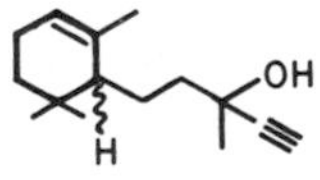 238	6%	25%	40%	29%
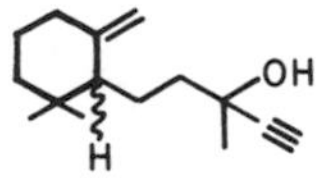 239	20%	24%	29%	27%

which may be realized is 20%. That there is any difference in the product ratio at all is of interest. In isomers 238 and 239 the relative stereochemistry of two of the asymmetric carbons is already established. However, in cyclization of 233, all the relative stereochemistry is established

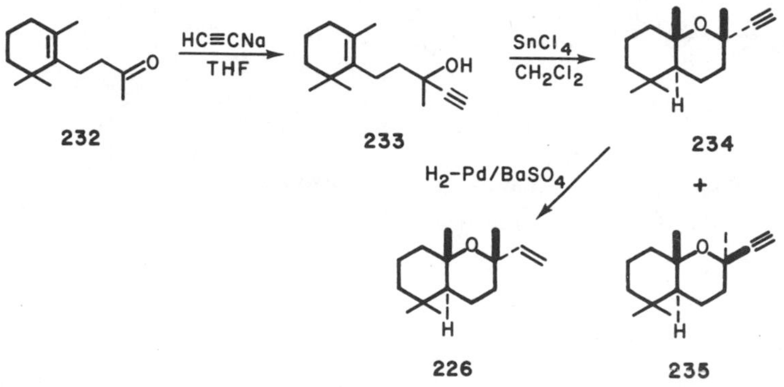

Scheme 45.   Cookson-Lombardi Caparraapi Oxide Synthesis

in one step. The ratio (235 + 236):(234 + 237) = 4:1, suggests substantial asymmetric induction in the initial protonation of 233:

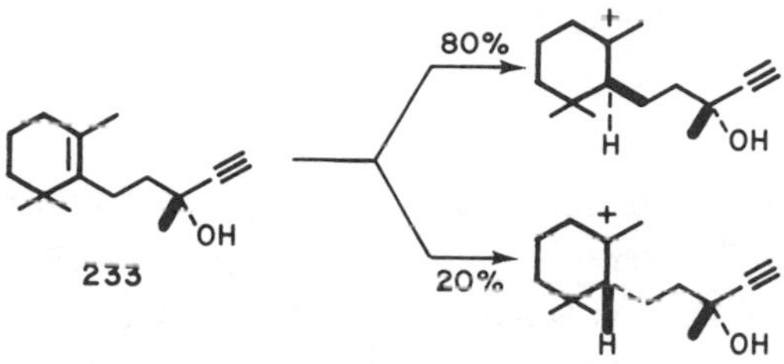

Hoye and Kurth have reported the synthesis of 3β-bromo-8-epicaparrapi oxide, which is summarized in Scheme 46.[113] The known β-hydroxy ester 240 is cyclized by treatment with mercuric trifluoroacetate. The cyclization is specific with regard to the stereochemistry of addition. That is, the E alkene 240 yields only trans products 241. However, there is no diastereoface selectivity, as 241 is an equimolar mixture of two diastereomers. The isomers are separated and the correct one is converted into 3β-bromo-8-epicaparrapi oxide by conventional methods. The side chain stereocenter does not give any asymmetric induction in attack by the mercury cation in the cyclization reac-

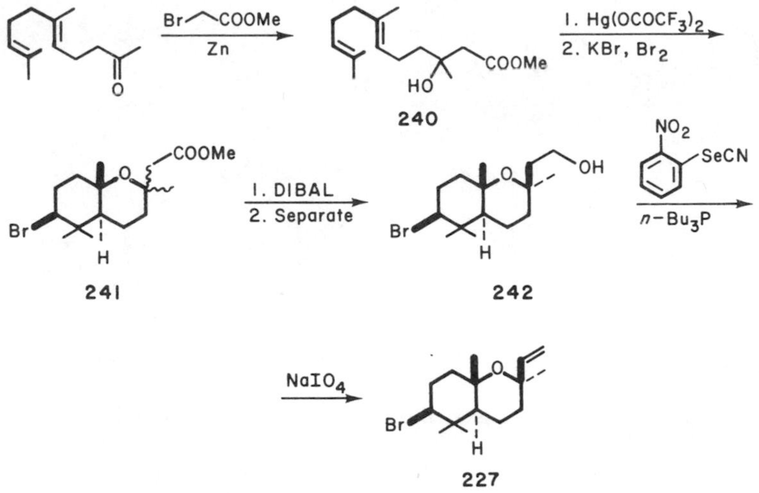

Scheme 46.   Hoye-Kurth Synthesis
of 3β-Bromo-8-epicaparrapi Oxide

tion. This is not surprising, but it is unfortunate that advantage cannot
be taken of the inherent asymmetric induction involved in formation of
the tetrahydropyran ring which was observed by Lombardi and Cookson
in their caparrapi oxide synthesis, as this would yield the correct stereochemistry for 3β-bromo-8-epicaparrapi oxide.

Ancistrofuran (228) is the principal component of the defensive secretion by soldiers of the West African termite *Ancistrotermes cavithorax.* It
has been synthesized by Baker, Briner, and Evans as shown in Scheme
47.[114] Homocitral reacts with α-furyllithium to give a 3:2 mixture of
alcohols 243 and 244. The diastereomers are separated and the double
bond is epoxidized, the resulting epoxide is reduced, and the tetrahydrofuran ring is closed by treatment of the mixture of diols 246 with tosyl
chloride. Ancistrofuran is produced along with a single cis-fused isomer,
247.

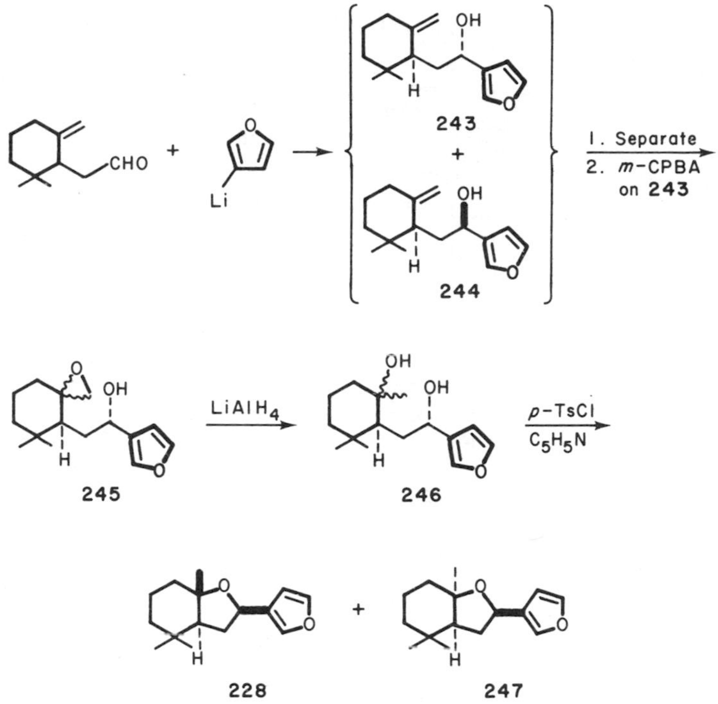

243

+

244

1. Separate
2. m-CPBA on **243**

245

LiAlH₄

246

p-TsCl
C₅H₅N

228   +   **247**

Scheme 47.   Baker-Briner-Evans Synthesis of Ancistrofuran

Aplysistatin (**229**), a brominated sesquiterpene from the sea hare, has been synthesized by Hoye and Kurth (Scheme 48).[115] Homogeranyl tosylate (**248**) is used to produce the α-phenylthioester **249**. Aldol condensation of this ester with O-benzyllactaldehyde or benzyloxyacetaldehyde affords diastereomeric esters **250** (2:1 ratio). The isomers are separated and the major one cyclized by the same method employed for the synthesis of 3β-bromo-8-epicaparrapi oxide. As in that synthesis, the cyclization shows no diastereoface selectivity; each pure diastereomer of **250** affords a 1:1 mixture of diastereomeric cyclization products (**251**). After oxidative elimination of the α-phenylthio group, the primary carbinyl ether is eliminated by treatment with trityl fluoborate. The reaction occurs with concomitant lactonization to afford aplysistatin directly.

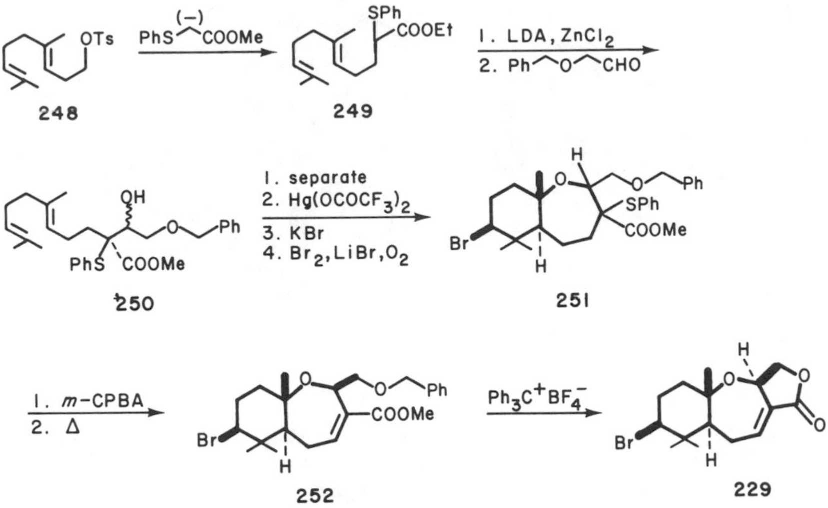

Figure 48.   Hoye-Kurth Synthesis of Aplysistatin

The snyderols (**230** and **231**) present a fascinating synthetic problem. Although two syntheses have been reported, the problem of stereochemistry has yet to be addressed. Kato and co-workers have prepared the snyderols as shown in Scheme 49.[116] Treatment of nerolidol (**253**) with the interesting brominating agent **254** affords α- and β-snyderols (**230** and **231**) each in 2% yield.

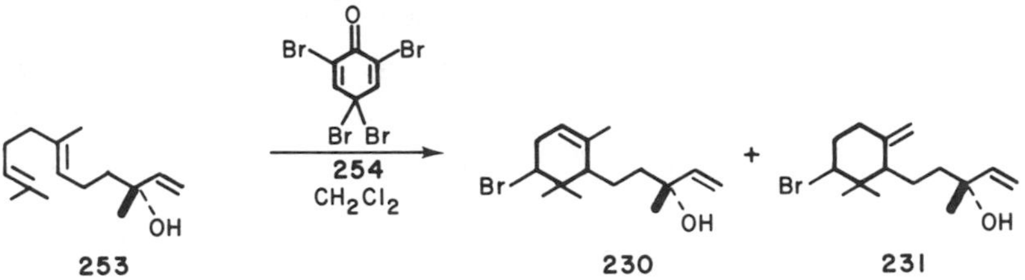

Scheme 49.   Kato Synthesis of the Snyderols

A synthesis by González and co-workers of the University of La Laguna, in Tenerife, Spain, is summarized in Scheme 50.[117] Methyl farnesate is cyclized by NBS and cobaltous acetate to give bromo ester **255** in 12% yield. The side chain is elaborated as shown. A noteworthy reaction is the clean double bond isomerization accompanying the conversion of allylic bromide **256** to alcohol **231**, which is presumably formed as a diastereomeric mixture.

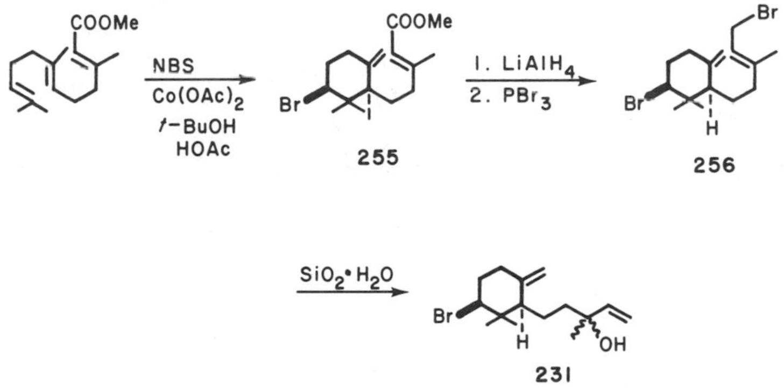

Scheme 50.   González Synthesis of β-Snyderol

G. Isocaespitol

Isocaespitol (**257**) and caespitol (**258**) are halogenated sesquiterpenes which are produced by the marine seaweed *Laurencia caespitosa*. Each of the compounds possess six stereocenters, three in each ring, thus presenting an interesting problem for synthesis which has yet to be solved. A nonstereoselective synthesis of **257** has been reported by

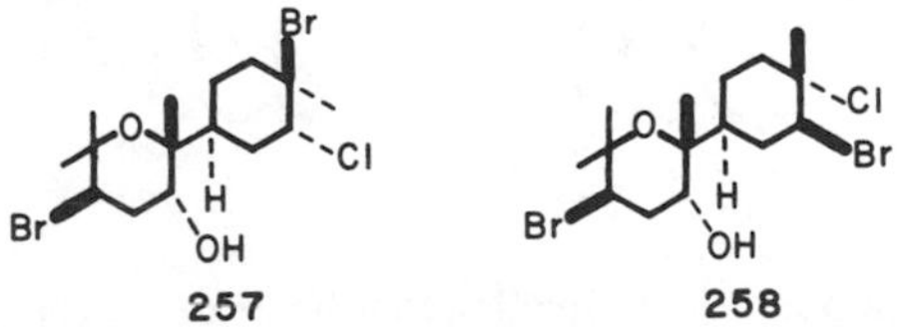

González and his co-workers (Scheme 51).[118] The synthesis begins with *trans*-β-terpineol (259), which is ozonized. The resulting ketone is condensed with trimethyl phosphonoacetate to obtain a 1:1 mixture of 261

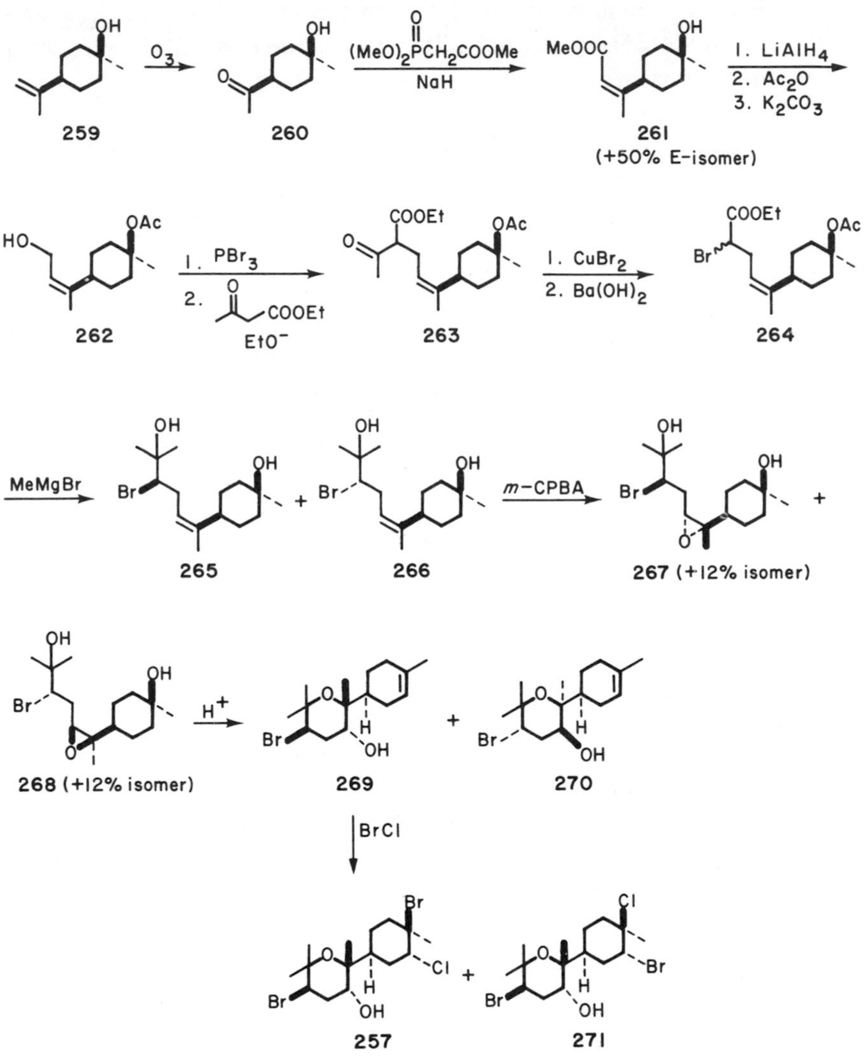

Scheme 51.   González Synthesis of Isocaespitol

and its E stereoisomer in 78% yield. The separated Z diastereomer is converted into α-bromo ester **264**, which is a 50:50 mixture of stereo-isomers at the brominated position. Epoxidation of alkenes **265** and **266** is remarkably stereoselective; each isomer gives a 7:1 mixture of products. The authors proposed that the alkenes each adopt a preferred conformation (**265a,266a**) in which the bromine shields one face of the double bond. The epoxides are opened by the tertiary hydroxy when the

265a   **266a**

mixture of **267** and **268** is treated with acid-washed alumina. The terti-ary ring hydroxy undergoes concomitant dehydration, leading to a mix-ture of alkenes **269** and **270**. Isomer **269**, didehalocaespitol, is obtained by fractional crystallization of the mixture. Compound **269** reacts with BrCl to give isocaespitol (**257**) and its isomer **271** in a ratio of 1:3. The observed regiochemistry of the halogenation step presumably reflects a preference of the electrophilic bromine to coordinate from the less hin-dered face of the double bond in **269**:

The isocaespitol synthesis is interesting, but it leaves much room for improvement in terms of reaction selectivity. First, the preparation of **261** proceeds with only 50% stereoselectivity, as does the preparation of **264**. The oxidation to **267** and **268** shows good stereoselectivity (88%). Thus, to this point the stereochemical yield is 0.5 × 0.5 × 0.88 = 22%. Relative stereochemistry is maintained in the last two steps, but the final halogenation gives a regiochemical yield of only 25%.

H. Lactaral

Lactaral (**277**) is interesting because its novel carbon skeleton departs from the biogenetic isoprene rule. It has been synthesized by Froborg, Magnusson, and Thorén[119] as outlined in Scheme 52. The synthesis begins with the commercially available diester **272**, which is converted into the monoprotected diol **273** (42% yield). The mesitoate ester **274** is coupled with bromide **275** by treating the mixture with lithium metal. The mixed coupling gives lactarol (**276**) in poor yield (44 mg of **276** from 1.8 g of ester **274**).

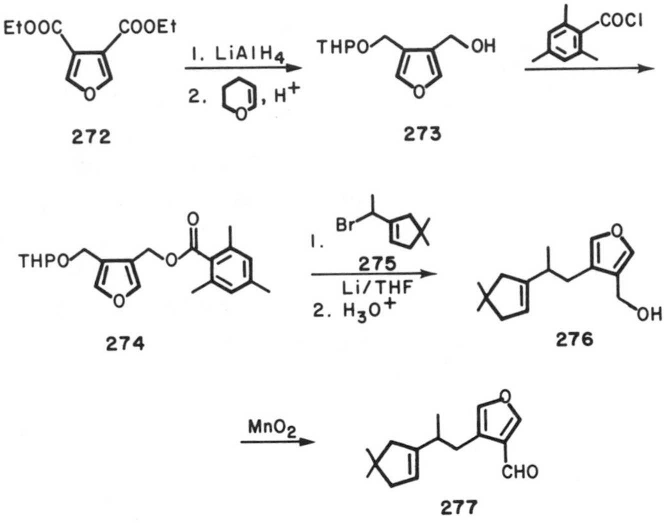

Scheme 52.   Froberg-Magnusson-Thorén Synthesis of Lactaral

I. γ-Elemene, β-Elemenone, Shyobunone, and Isoshyobunone

A number of elemanoid sesquiterpenes were synthesized in the 1970s. The simplest examples are γ-elemene (**278**), β-elemenone (**279**), shyobunone (**280**), and isoshyobunone (**281**).

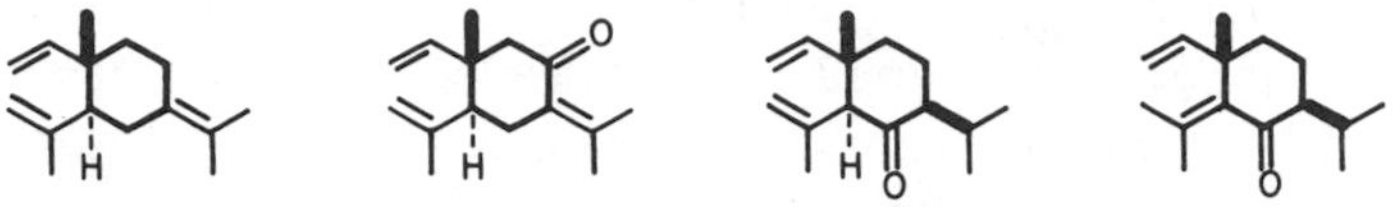

278   279   280   281

Kato, Kurihara, and Yoshikoshi have synthesized γ-elemene as summarized in Scheme 53.[120] The known unsaturated ketal ester **282**, obtainable in three steps from the Wieland-Miescher ketone, is allylically oxidized and the resulting β-acetoxy enone transformed by reductive alkylation and mesylation into the β-mesyloxy ketone **285**. This material undergoes smooth reduction and Grob fragmentation upon being

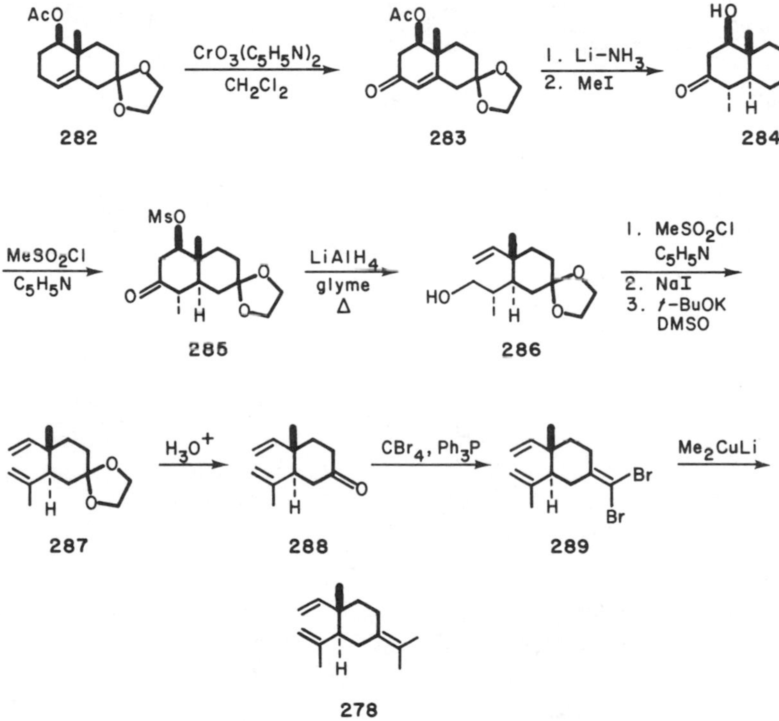

Scheme 53.   Yoshikoshi's Synthesis of γ-Elemene

refluxed with LiAlH$_4$ in 1,2-dimethoxyethane. The resulting unsaturated alcohol **286** is converted by precedented steps into geigerenone (**288**), which is transformed into γ-elemene by way of the dibromomethylene derivative **289**.

Majetich, Grieco, and Nishizawa reported the synthesis of β-elemenone which is outlined in Scheme 54.[121] The synthesis begins with the monoketal **290**, which is converted into alkene **293** by established procedures. The double bond is cleaved by ozone and the bis primary alcohol **294** is converted into diene **296** by the selenoxide elimination method. The synthesis employs a novel method for introducing the

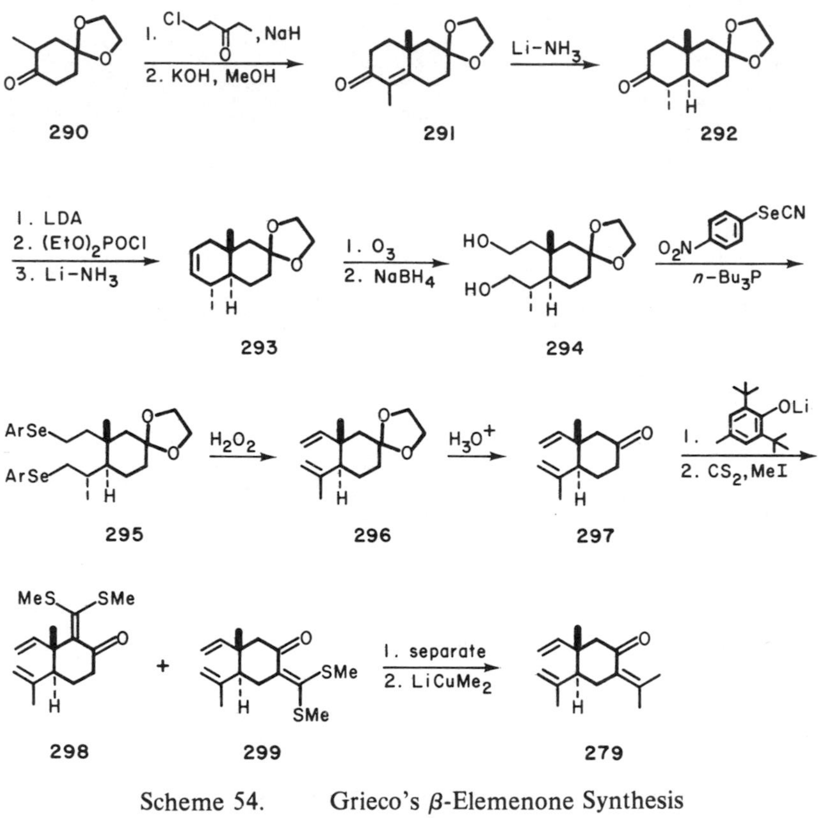

Scheme 54.   Grieco's β-Elemenone Synthesis

isopropylidene group. The lithium enolate of **297** is treated successively with CS_2 and methyl iodide to give a 1:3 ratio of **298** and **299**. After separation of the isomers, **299** is treated with lithium dimethylcuprate to obtain β-elemenone.

The first synthesis of shyobunone was recorded in 1970 by Kato, Yamamura, and Hirata.[122a,b] The synthesis (Scheme 55) begins with α-santonin (**300**), which is converted by standard methods into ketal diol **301**. This substance is converted by straightforward steps into ketone **303**. A noteworthy feature of this synthesis is the reported exclusive formation of enol acetate **304** upon treatment of **303** with isopropenyl acetate. The remainder of the synthesis is straightforward. Another

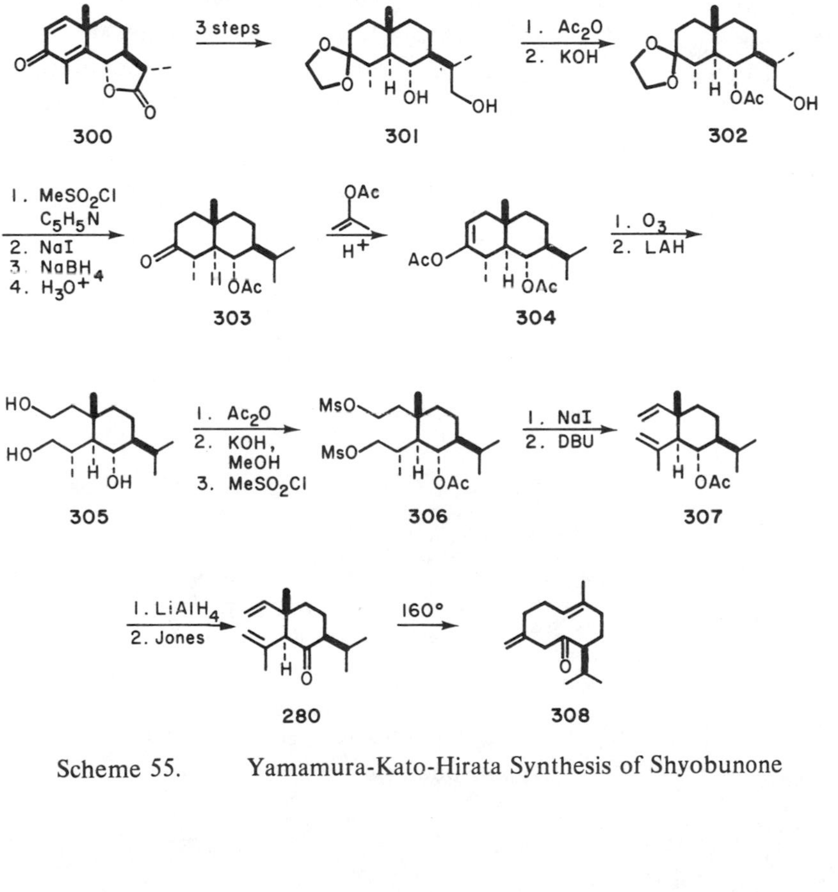

Scheme 55.   Yamamura-Kato-Hirata Synthesis of Shyobunone

noteworthy feature is the thermal conversion of shyobunone (**280**) into preisocalamendiol (**308**). Although the equilibrium between a divinylcyclohexane and a 1,5-cyclodecadiene normally favors the former isomer, in this case, equilibrium is apparently shifted by thermal 1,5-hydrogen transfer in the initial Cope rearrangement product **309**:

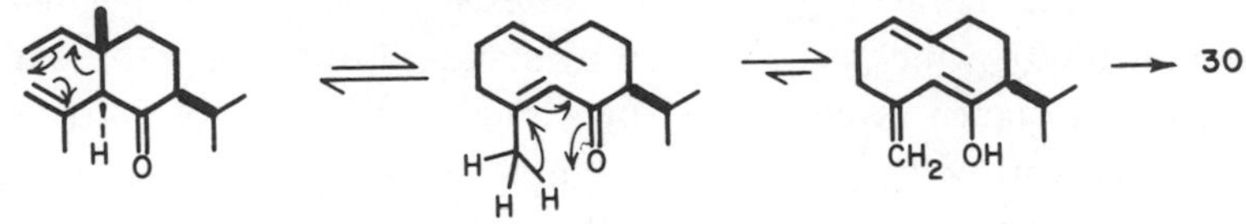

Fráter has reported the synthesis of shyobunone which is summarized in Scheme 56.[123] The synthesis begins with a clever use of the ester enolate Claisen rearrangement on neryl senecioate (**310**), which affords unsaturated acids **311** and **312** in a 4:1 ratio. The cyclohexane ring is formed by cyclization of the acid chloride with stannic chloride and the resulting β-chloro ketones are dehydrochlorinated to obtain a mixture of

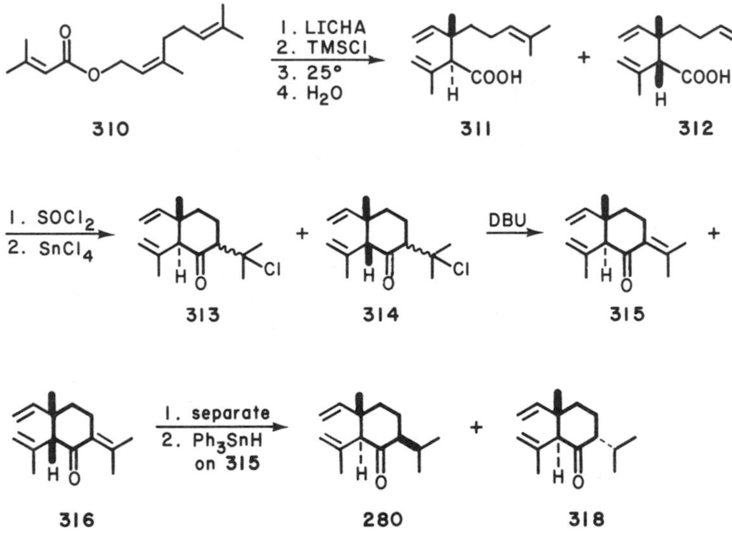

Scheme 56.   Fráter's Synthesis of Shyobunone and 6-Epishyobunone

enones **315** and **316**. The major isomer is reduced by triphenyltin hydride to obtain shyobunone (**280**) and 6-epishyobunone (**318**) in a ratio of 7:3. The sequence may by carried out with geranyl senecioate to obtain a 1:4 mixture of **311** and **312**.

Williams and Callahan have published a synthesis of (−)-shyobunone, which begins with [2+2] cycloaddition of 1-methylcyclobutene and (S)-piperitone (Scheme 57).[124] The photoadduct **319** is reduced to alcohol **320**, which is thermolyzed as a xylene solution in a sealed tube at

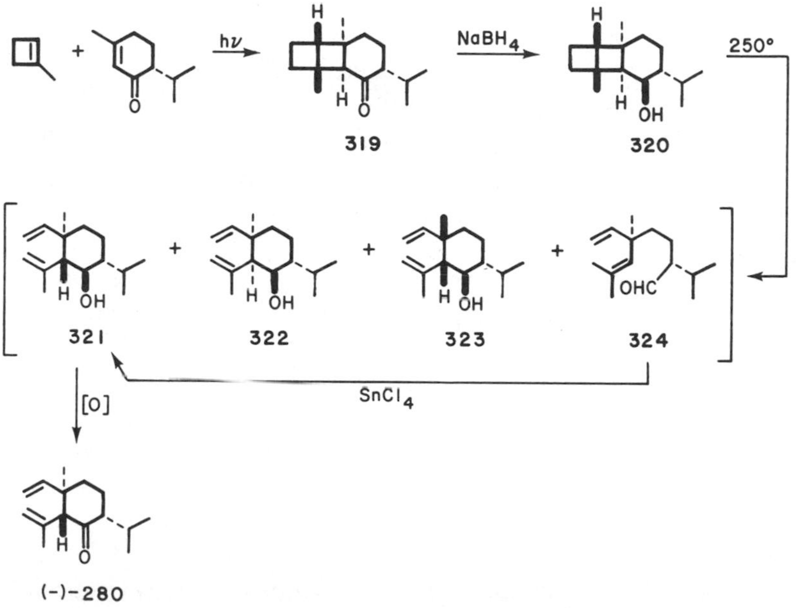

Scheme 57.   Williams-Callahan Synthesis of (-)-Shyobunone

250°C. The thermolysis product is a complex mixture consisting of the elemanoids **321** (10%), **322** (13%), **323** (27%), and the aldehyde **324** (43%). Treatment of the latter isomer with stannic chloride affords alcohol **321**, which is oxidized to enantiomeric shyobunone. Aldehyde **324** presumably arises from **321-323** by a retro-ene reaction:

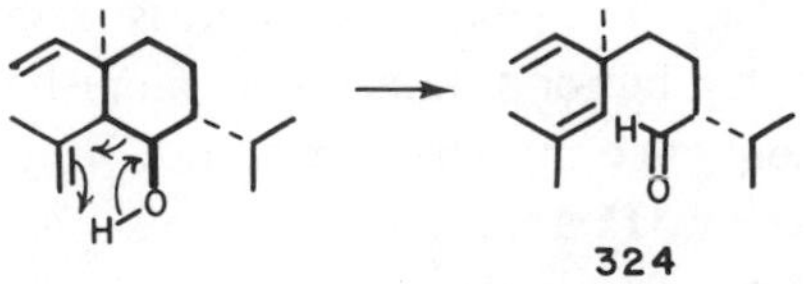

324

Isoshyobunone has been prepared by Alexandre and Rouessac as out-lined in Scheme 58.[125a,b] Enedione **326**, prepared as shown, is cyclized to a 2:3 mixture of isomers **327** and **328**. The major isomer is separated and converted into **330**, which adds vinylmagnesium bromide under copper catalysis to afford a 4:1 mixture of isoshyobunone (**281**) and its 6-epimer **331**.

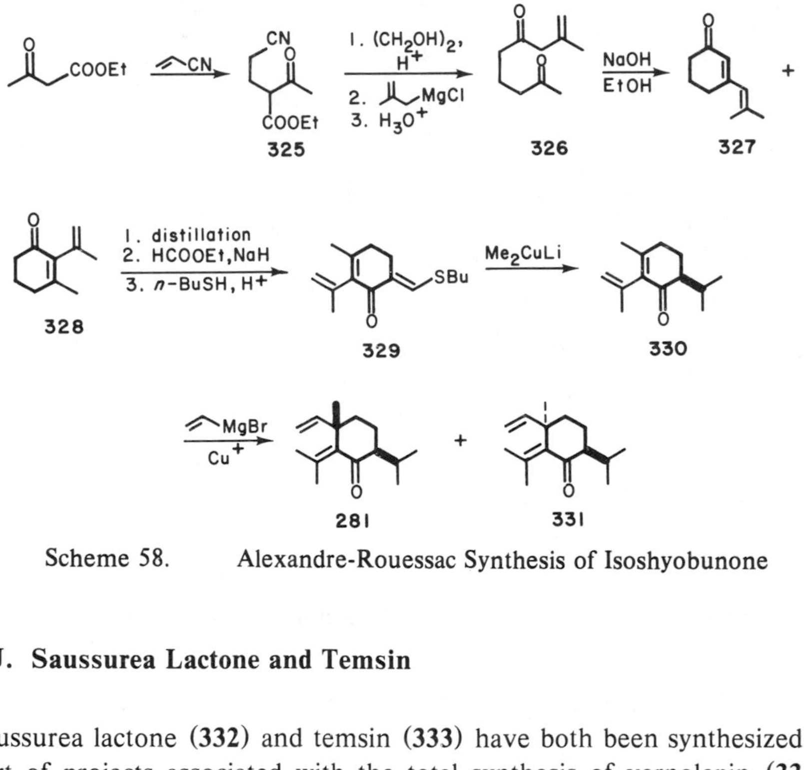

Scheme 58.   Alexandre-Rouessac Synthesis of Isoshyobunone

J. Saussurea Lactone and Temsin

Saussurea lactone (**332**) and temsin (**333**) have both been synthesized as part of projects associated with the total synthesis of vernolepin (**334**). The synthesis of saussurea lactone by Ando and co-workers is summar-ized in Scheme 59.[126a,b] It was carried out to test the epoxy mesylate

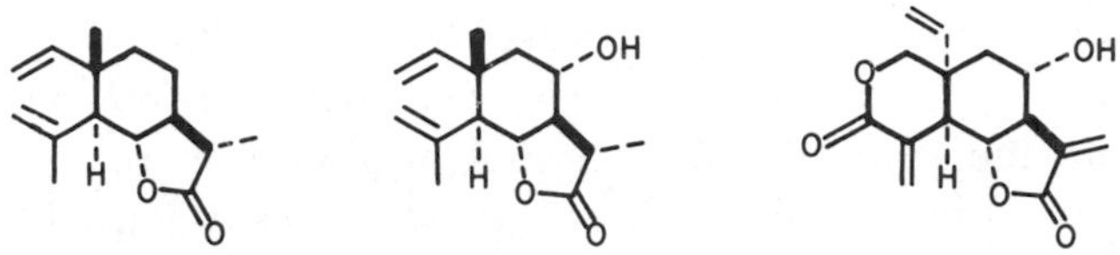

332   333   334

fragmentation (338→339) as a possible method for construction of the vernolepin A-ring. Unsaturated ketal **336**, obtained in six steps from α-santonin (**300**), is hydrolyzed and the resulting enone is reduced with lithium tri-*t*-butoxyaluminohydride to give a 4:1 mixture of isomeric alcohols. Epoxidation of the mixture gives comparable amounts of two epoxides (**337**). These are separated and the major isomer is converted into mesylate **338**. The fragmentation is carried out by boiling a toluene solution of **338** and aluminum isopropoxide. The lactone ring undergoes concomitant trans-esterification. The allylic hydroxy group is removed by lithium-ammonia reduction, and after reoxidation of the resulting hemiacetal, saussurea lactone is obtained.

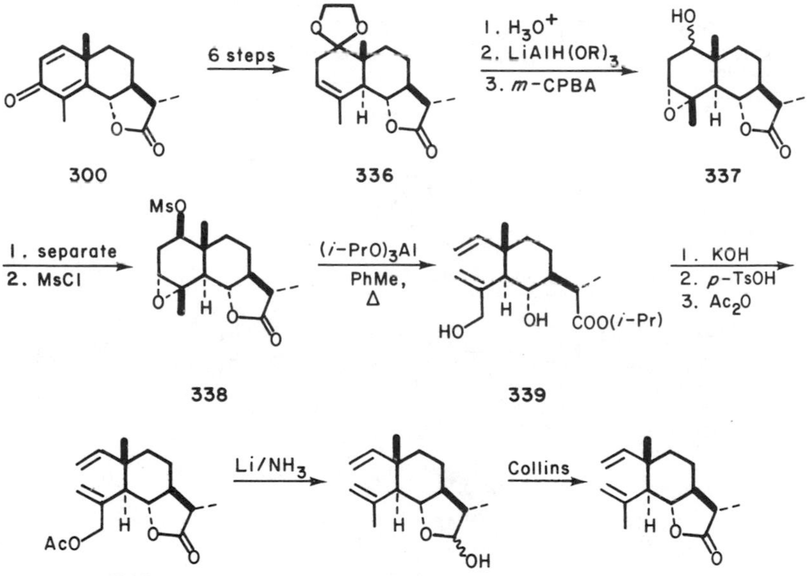

300   336   337

338   339

340   341   332

Scheme 59.   Ando's Synthesis of Saussurea Lactone

Grieco and co-workers synthesized temsin by the route shown in Scheme 60.[127b] The synthesis begins with octalone **342**, which is converted into the keto ketal **343** in straightforward steps.[127a] The ring-B functionality is introduced in the same manner as was used for Grieco's earlier vernolepin synthesis (see Scheme 61). Thus, epoxidation of dienone **344** gives the α-epoxide **345**. Lithium-ammonia reduction of **345** gives an enolate, which is quenched with ammonium chloride while

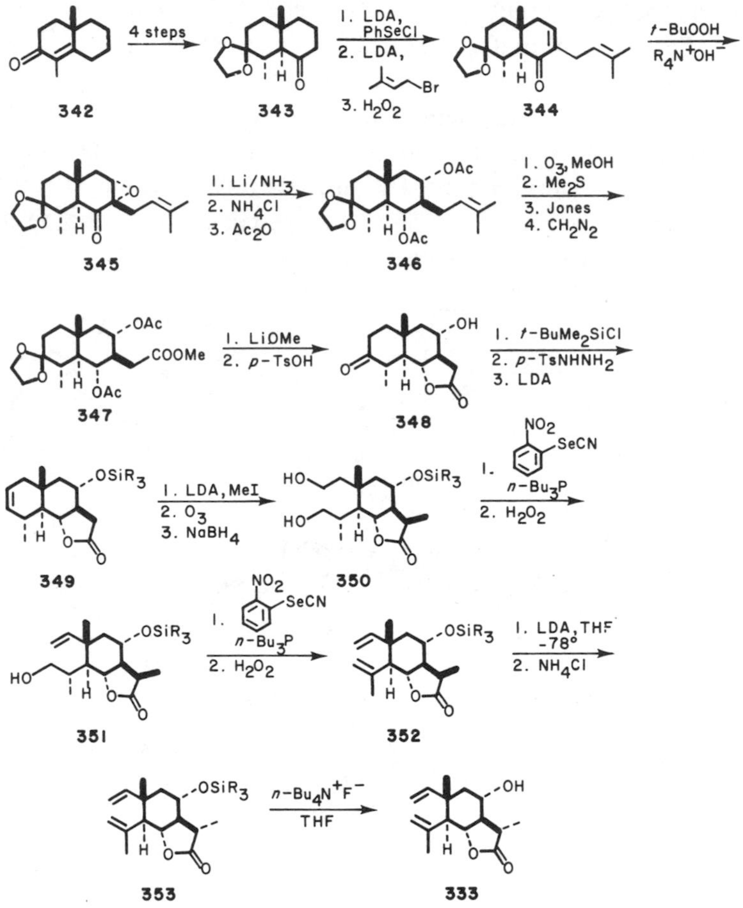

Scheme 60.   Grieco's Synthesis of Temsin

excess reducing agent is still present. The β-hydroxyketone so produced undergoes subsequent reduction to give the diequatorial diol, which is acetylated to give **346**. The trans-fused lactone ring is formed and alkylated to give the β-methyl isomer. After ozonolysis of the A-ring double bond, the resulting diol **350** is dehydrated to obtain **352**. The eliminations are accomplished by successive formation of the alkyl o-nitroselenoxide, which undergoes *in situ* elimination. This is a weakness of the synthesis because simultaneous bis elimination fails, and the four-step successive process proceeds in only 33% yield. Finally, the C-10 stereochemistry is corrected by formation of the lactone enolate, which is protonated from the β-face to afford the correct stereoisomer. The complete synthesis requires 26 steps from octalone **342**.

K. Vernolepin and Vernomenin

In 1968, the late Professor S. M. Kupchan reported the isolation and characterization of the interesting elemanoids vernolepin (**334**) and vernomenin (**354**). In addition to its interesting structure, compound **334** originally attracted attention because of its cytotoxic and antitumor

334   **354**

activity. It soon became apparent that many natural products having the α-methylenebutyrolactone unit show comparable biological activity, and that most of the members of the class are not significant medicinal leads. Nevertheless, vernolepin captured the imaginations of synthetic chemists, and for the next decade, the intense synthetic effort which was directed at this one compound was out of all proportion to its importance.

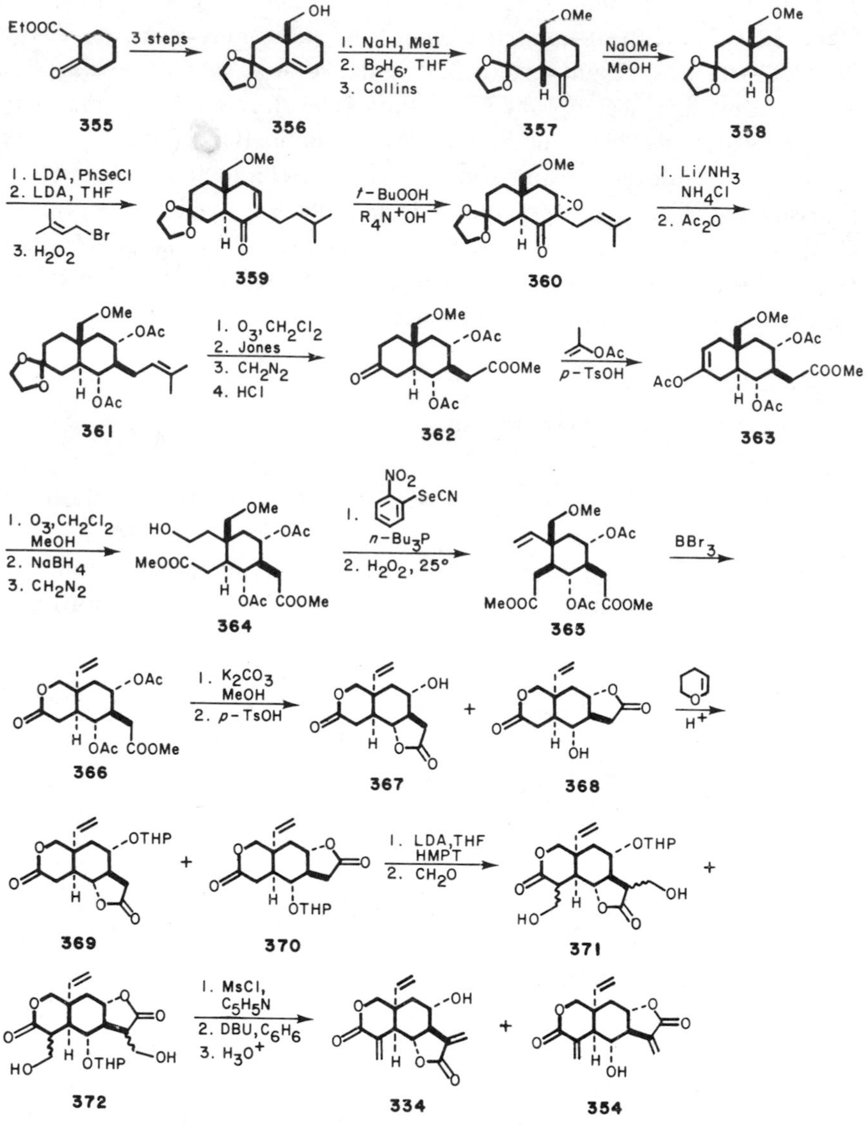

Scheme 61.   Grieco's Vernolepin Synthesis

The first synthesis to be completed by the end of the decade was due to Grieco and co-workers (Scheme 61).[128a,b] The unsaturated ketal **356**, obtainable in three steps from 2-ethoxycarbonylcyclohexanone, is methylated and hydroborated to obtain the cis-fused decalone **357**, which is isomerized to the more stable isomer **358**. The ring-B functionality is introduced using the method previously discussed in connection with the Grieco temsin synthesis (Scheme 60). After cleavage of ring A, and dehydration of the resulting hydroxyethyl side chain, the interesting intermediate **365**, which has five contiguous chiral centers, is obtained. Cleavage of the methoxy group affords the δ-valerolactone **366**. Hydrolysis of the esters, followed by relactonization, gives a 3:1 mixture of prevernolepin (**367**) and prevernomenin (**368**). After protection of the free hydroxy group, the two methylene groups are introduced by hydroxymethylation, mesylation, and elimination. The Grieco synthesis requires a total of 31 steps. A key feature was the demonstration that the two methylene groups can be introduced simultaneously at the end of the synthesis.

The synthesis of Danishefsky and co-workers is summarized in Scheme 62.[129a,b] It begins with a Diels-Alder cycloaddition of methyl 1,4-cyclohexadienecarboxylate and the versatile diene **373**. The ring-B functionality is introduced by making use of the angular carboxy group. The first maneuver in this regard is iodolactonization, which introduces the eventual free hydroxy group. After dehydroiodination, the lactone ring is saponified and the allylic hydroxy group used to direct epoxidation to the α-face of the double bond. Direct epoxidation of lactone **376** affords the diastereomeric epoxide in poor yield. The ring-A cyclohexenone is converted into the δ-valerolactone of vernolepin by oxidation to aldehyde ester **378**, which is reduced and lactonized to afford dilactone **379**. A similar process has been used by Chavdarian and Heathcock to prepare vernolepin analogs.[130] The first key to the Danishefsky synthesis is the ability to protect the δ-valerolactone as an ortholactone (**380**) while the γ-butyrolactone is reduced and the resulting aldehyde is methylenated to obtain epoxy alcohol **381**. Compound **381** reacts with a large excess of dilithioacetate to give a single dihydroxy acid which is esterified to obtain **382**. This regiospecific ring opening is the second important feature of

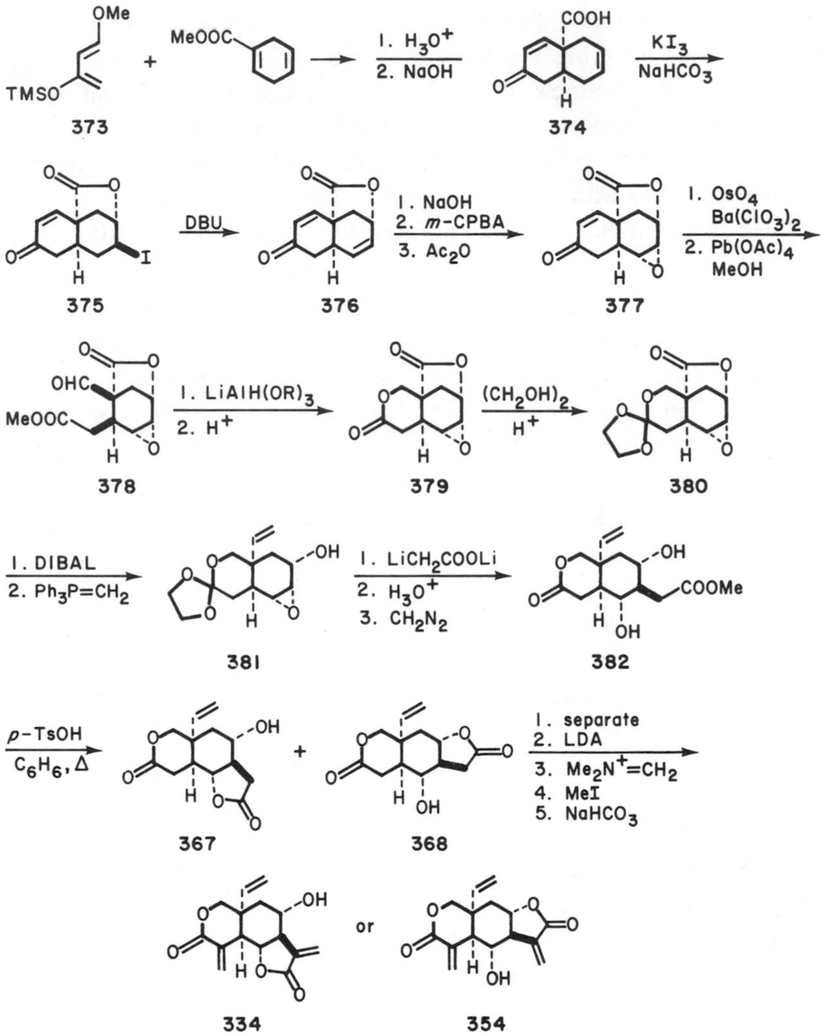

Scheme 62.   Danishefsky's Vernolepin Synthesis

the Danishefsky synthesis. If one assumes diaxial opening of the epoxide ring, then the reactive conformer of **381** is **381a**. It has been suggested that conformation **381a** is favored over the alternative **381b** mainly because the incoming nucleophile must encounter the axial oxygen of the ortholactone function in its attack upon the latter conformation. However, chelation of a lithium cation between the epoxide and

381a   381b

alkoxide oxygens may play an important role in stabilizing conformation **381a**, for it has been found that the tetrahydropyranyl ether of **381** fails to react under conditions which suffice to convert **381** into **382**. This failure is unfortunate, as a successful opening would have allowed regiocontrol of the subsequent lactonization. As it is, lactonization of **368** yields a 2:1 mixture of prevernolepin (**367**) and prevernomenin (**368**), which is separated chromatographically and then methylenated using Eschenmoser's Mannich salt to obtain either vernolepin (**334**) or vernomenin (**354**). This process for introducing the methylene groups represents an improvement over the Grieco method, as it does not require prior protection of the free hydroxy group. The overall yield for the bis-methylenation by Danishefsky's method is 31%, compared to 12% in the Grieco synthesis. In all, Danishefsky's synthesis requires 22 steps from diene **373**.

Schlessinger's vernolepin synthesis (Scheme 63)[131] differs from the Grieco and Danishefsky syntheses because it does not proceed through decalin intermediates. The synthesis begins with construction of the symmetric β-diketone **386**, which is protected as the methyl enol ether and reduced to alcohol **387**. A feature of this sequence is protection of the ketone in **386** by conversion into its enolate ion while the ester is reduced. Acetal **388** is obtained as a 1:1 diastereomeric mixture by the

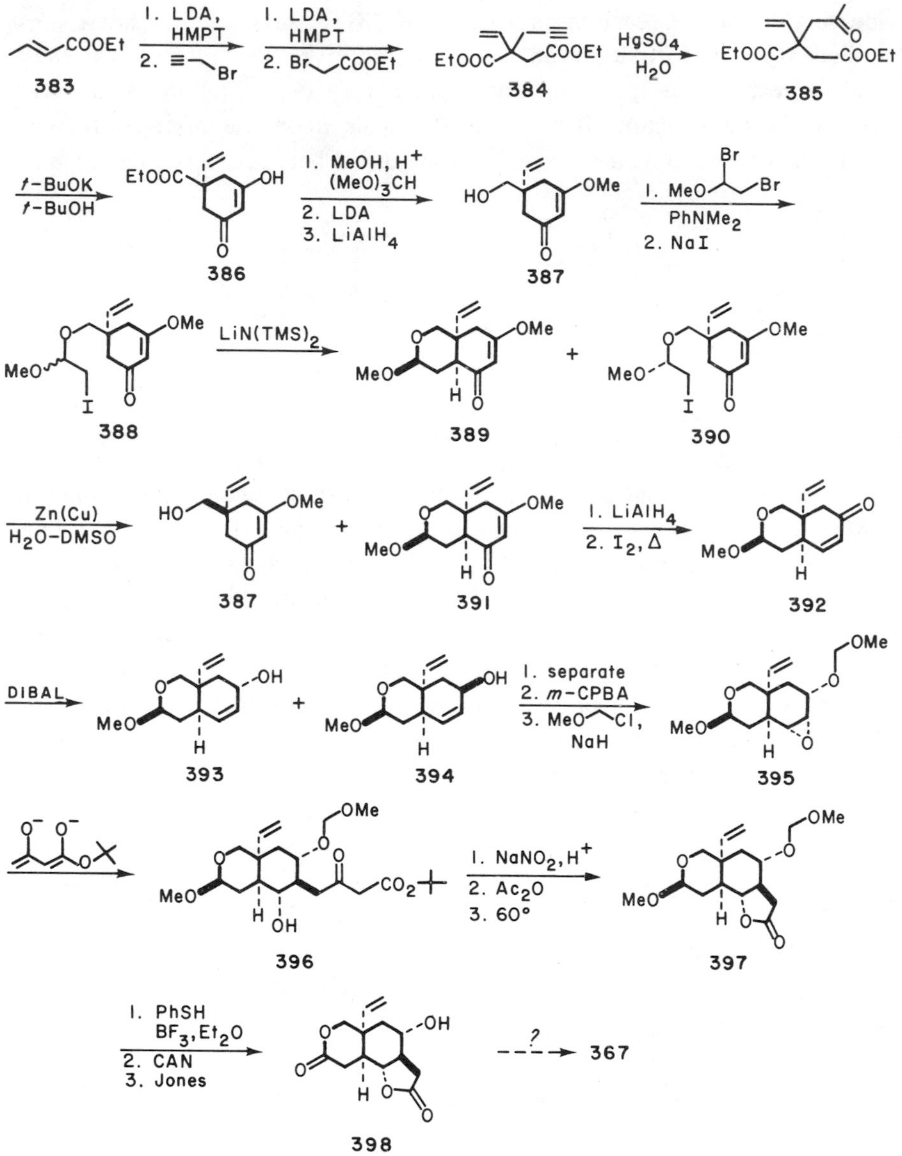

Scheme 63.   Schlessinger's Vernolepin Synthesis

method indicated in the scheme. The intramolecular alkylation of **388** is remarkably stereoselective; one diastereomer of **388** reacts and one does not. The result has been explained on steric grounds. Treatment of the mixture of **389** and **390** with zinc-copper couple in aqueous DMSO gives alcohol **387** and unreacted **389**. This specific vicinal elimination is surprising, as it appears that methanol could just as easily have been eliminated from iodo acetal **390**, yielding enol ether **399**. The ring-B functionality is introduced by reduction of enone **392**, which affords

399

alcohols **393** and **394** in a ratio of 3:2. The desired isomer is epoxidized and the secondary hydroxy group is protected to obtain epoxy ether **395**. The epoxide ring is opened by dilithioacetoacetate to obtain **396**. The successful opening of the hydroxy-protected **395** is an important feature of Schlessinger's synthesis, as it allows the first regiospecific synthesis of vernolepin. Recall that in Danishefsky's synthesis, the tetrahydropyranyl ether of **381** does not undergo epoxide opening. The synthesis is completed by a clever degradation of the β-keto ester side chain (Beckman fragmentation) and a novel oxidative conversion of the methoxy tetrahydropyran ring A into the δ-valerolactone ring. The endpoint of the Schlessinger synthesis is the protected form of prevernolepin (**398**), which can presumably be converted into the natural product by deprotection and bis methylenation. Although the Schlessinger synthesis is the first to achieve regio-control with regard to the ring-B functionality, it is longer than Danishefsky's (25 steps, allowing 3 steps to convert prevernolepin into vernolepin) and suffers from lack of stereocontrol in two steps **387→388 392→393**.

Isobe's synthesis[132] is summarized in Scheme 64. Aldehyde **403**, obtained from *p*-methoxybenzyl alcohol by Birch reduction, ketalization, and oxidation, is alkylated with allyl iodide and the resulting aldehyde is reduced to alcohol **404**. Esterification with ethyl malonyl chloride, hydrolysis, and intramolecular Michael addition affords the bicyclic keto

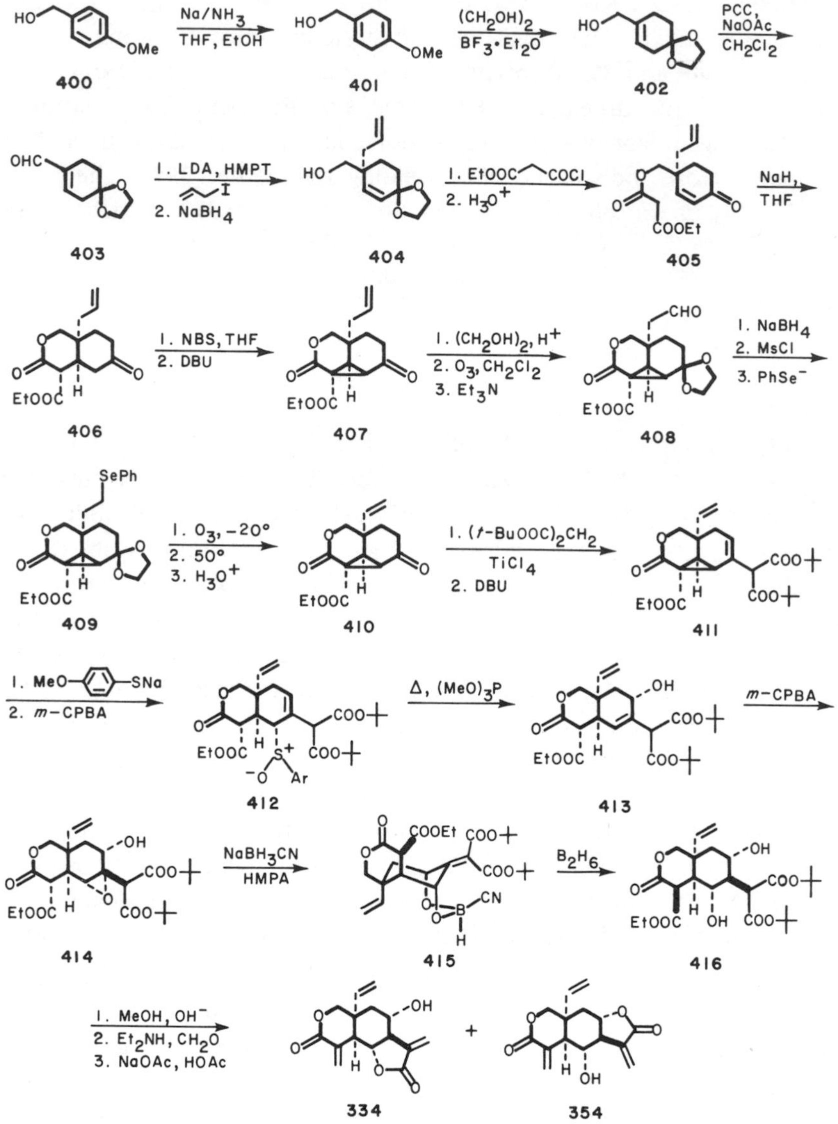

Scheme 64.   Isobe's Vernolepin Synthesis

lactone **406**. Careful monobromination, followed by base-catalyzed cyclization affords the cyclopropyl ketone **407**, which is ketalized for the sequence of steps necessary to excise a carbon from the angular group. The resulting angular vinyl compound **410** is subjected to Knoevenagel condensation with di-*t*-butyl malonate and the double bond isomerized to the β,γ isomer **411**. The ring-B functionality is now established by a wonderfully intricate and convoluted sequence involving opening of the cyclopropyl ring with inversion of configuration, Mislow-Evans rearrangement of the resulting allylic sulfoxide and epoxidation, affording **414**. Treatment of **414** with sodium cyanoborohydride effects epoxide opening providing the cyclic cyanoborate ester **415**. Addition of diborane to **415** leads to a metathesis reaction in which the cyano ligand is replaced by hydride. Subsequent intramolecular conjugate reduction leads to **416**. If **415** is treated with excess sodium cyanoborohydride in wet HMPA, the C-7 epimer of **416** is produced, apparently by an intermolecular reaction. The synthesis is completed by hydrolysis of the ester functions, Mannich condensation, and elimination to give a mixture of vernolepin (22%), and vernomenin (<5%). The Isobe synthesis is similar to Schlessinger's in that it begins with a 4,4-disubstituted cyclohexanone. It requires a total of 28 steps.

As of this writing, other vernolepin syntheses have not been completed. However, a number of model studies and partial syntheses have been published. Since these demonstrate alternative approaches to the problem, they shall be discussed briefly.

Clark and Heathcock[133] added lithium divinylcuprate to hydrindenone **417** and silylated the resulting enolate to obtain ether **418** (Scheme 65). Selective ozonolysis, reduction of the initial ozonide by sodium borohydride, and lactonization affords δ-valerolactone **419**, which is methylenated by Grieco's procedure to obtain the simple ring-A analog **420**. An alternative approach by Heathcock and co-workers is summarized in Scheme 66.[134] Alkylation of enol ether **421** by *t*-butyl bromoacetate, followed by Wittig methylenation yields the dienyl ether **423**, which is oxidized by *m*-CPBA in aqueous ethanol to obtain allylic alcohol **424**. Protection of the hydroxy group, conjugate addition of vinylmagnesium bromide, and lactonization yields keto lactone **425**.

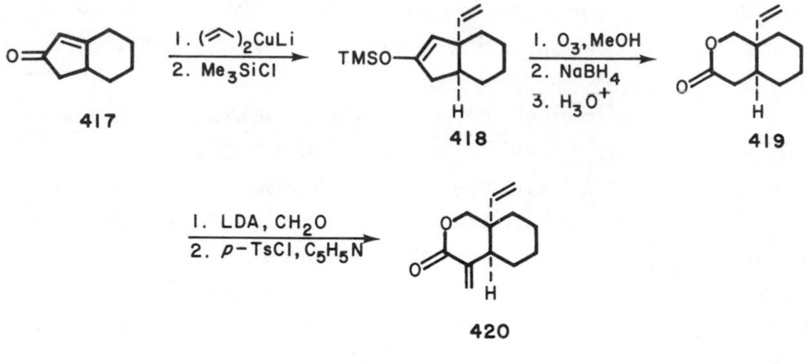

Scheme 65.   Clark-Heathcock Approach to Ring-A

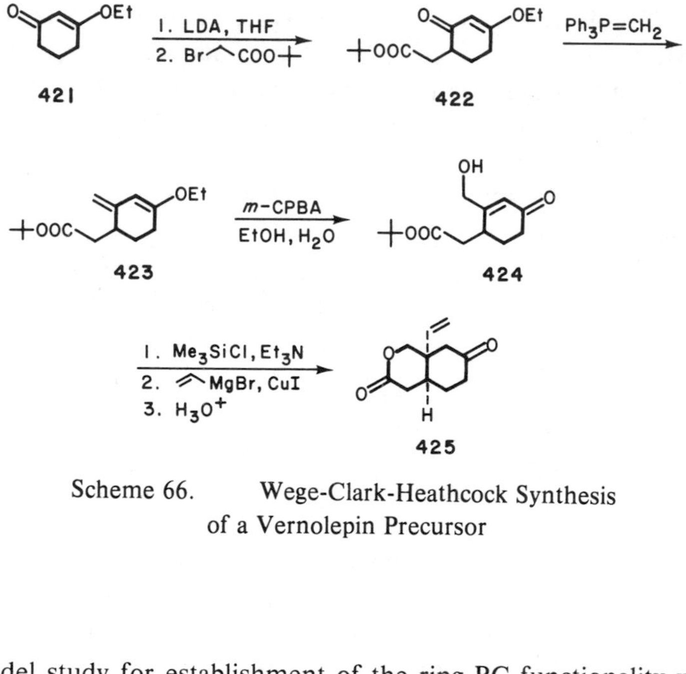

Scheme 66.   Wege-Clark-Heathcock Synthesis
of a Vernolepin Precursor

A model study for establishment of the ring-BC functionality was carried out by the Heathcock group and is outlined in Scheme 67.[135] Epoxidation of enone **427** is stereospecific, as is reduction of epoxy ketone **428**. The hydroxy group is protected and the epoxide ring opened by an

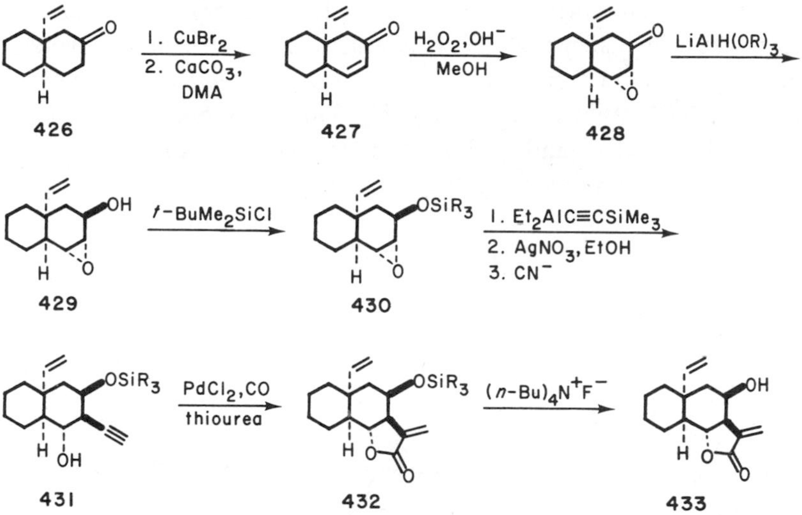

Scheme 67.   Heathcock's Ring-BC Model

acetylenic alane, to give the *trans*-ethynyl alcohol **431**, which is transformed into the trans-fused α-methylenebutyrolactone **432** using Norton's cyclocarbonylation process. Although the reaction yield is low (only 21%), this approach avoids the necessity of introducing the α-methylene unit at a final stage of the synthesis. Note that this synthesis yields the wrong stereochemistry at the secondary hydroxy position, which must be inverted in order to obtain products of the correct series. However, the stereochemistry at this position is important for the epoxide ring opening. The isomeric epoxy ether **434** does not react, presumably for conformational reasons.

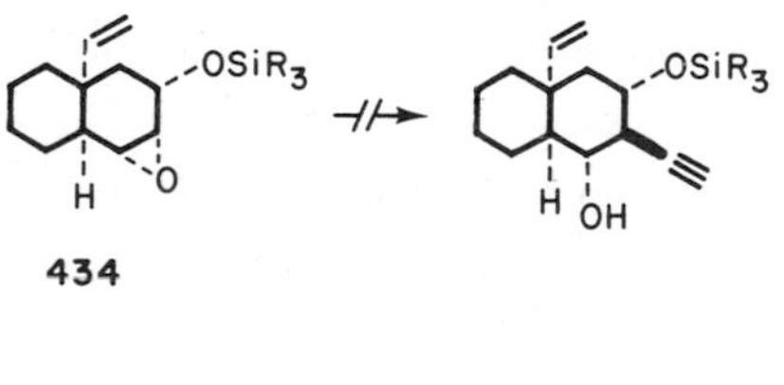

An approach by Torii, Okamoto, and Kadono of Okayama University is summarized in Scheme 68.[136] It is essentially a streamlined version of Isobe's synthesis, except that the angular vinyl group is introduced from the beginning, rather than by employing an allyl group as in Isobe's original work. Intermediate **440** could presumably be converted into vernolepin and vernomenin as is **406** in Isobe's synthesis.

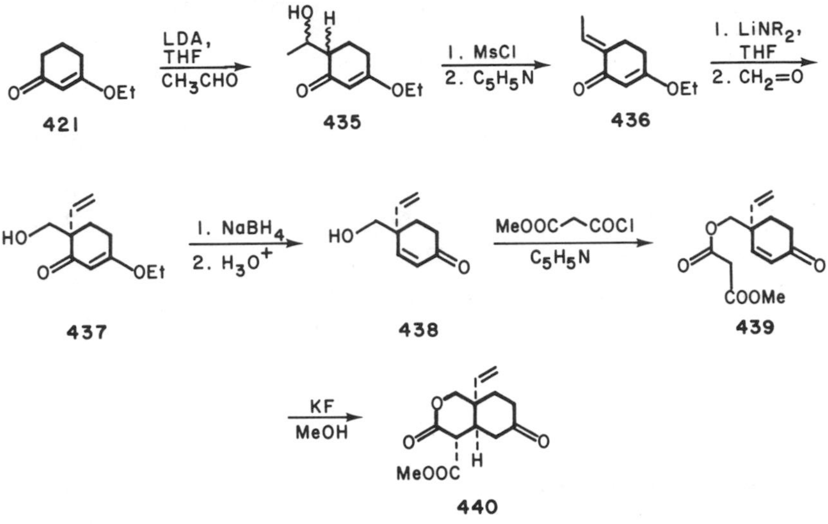

Scheme 68.   Torii-Okamoto-Kondo Approach to Vernolepin

Scheffold's approach is outlined in Scheme 69.[137] The Scheffold synthesis starts with 3,5-dimethoxybenzoic acid (**441**), which is transformed by Birch reduction and alkylation into bromide **443**. This substance is transformed into triene **445** by Cope elimination of amine oxide **444**. Reduction of the carboxy group and hydrolysis of the enol ether functions yields the symmetrical diketone **446**, which is similar to an intermediate in the Schlessinger synthesis (**387**, Scheme 63).

Zutterman, de Clerq, and Vandewalle have explored a similar approach outlined in Scheme 70.[138] The Birch reduction product of benzoic acid

serves as the starting material, leading eventually to the diazomalonate
451, which is converted into the cyclopropane **452**. Opening of the
cyclopropane ring, as in the Isobe approach, affords **453**.

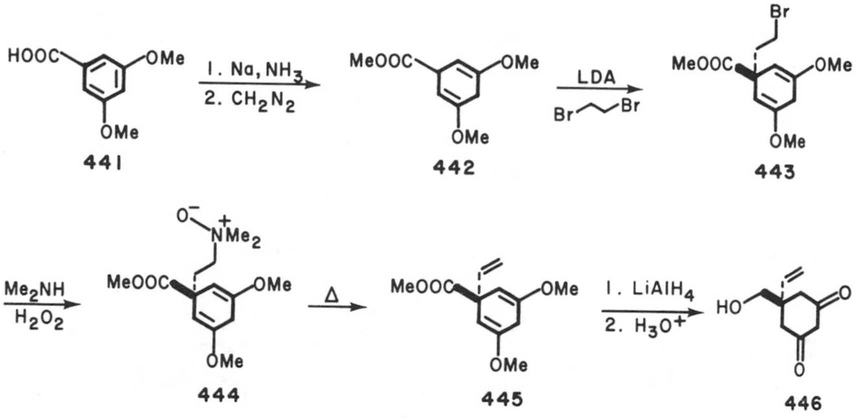

Scheme 69.   Scheffold's Approach to Vernolepin

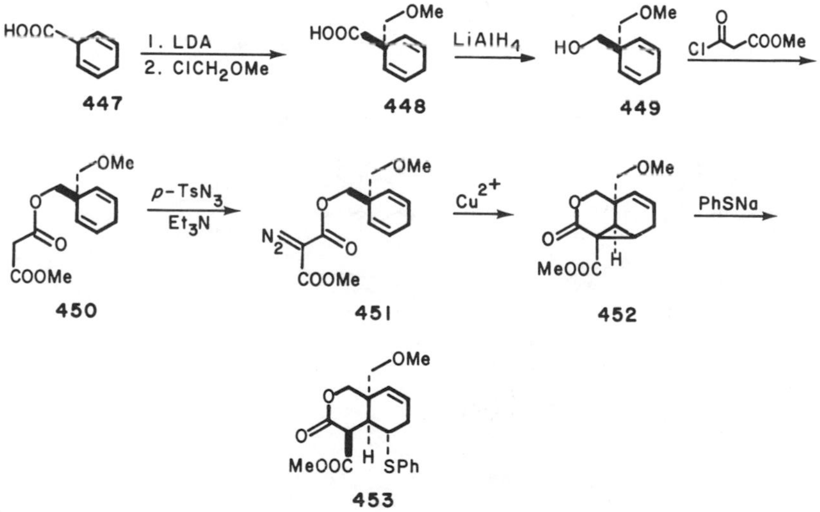

Scheme 70.   Vandewalle's Approach to Vernolepin

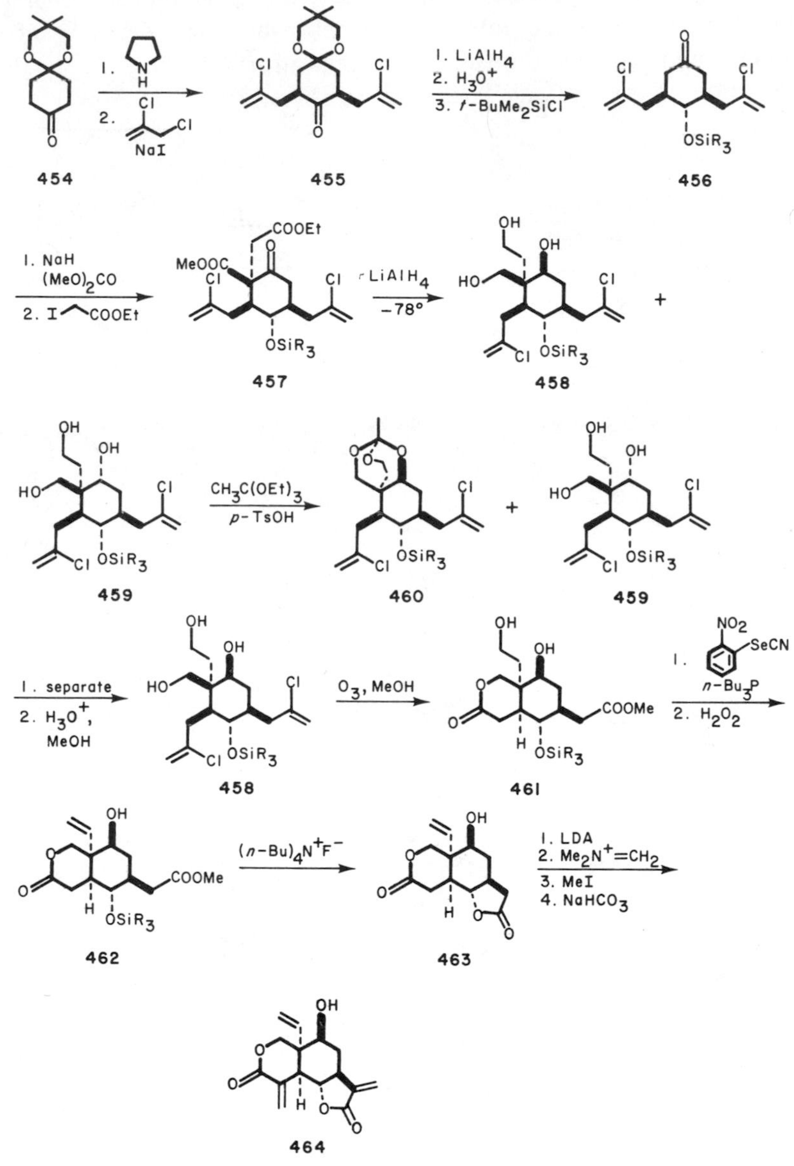

Scheme 71.   Marshall-Flynn Synthesis of a Vernolepin Isomer

Marshall and Flynn have reported a synthesis (Scheme 71) which leads to compound **464**, a positional isomer of vernolepin.[139] The monoketal of 1,4-cyclohexanedione is bis alkylated by way of its piperidine enamine to obtain the symmetrical dialkylated ketone **455**. Reduction of the ketone of **455** is stereospecific; after deketalization and protection, the symmetrical intermediate **456** is produced. This intermediate is converted into a β-keto ester, which is alkylated stereospecifically from the face opposite the vicinal chloroallyl group to obtain diester **457**. Reduction with lithium aluminum hydride at low temperature gives a 7:3 mixture of isomers **458** and **459**. The mixture is separated by treatment with triethyl orthoacetate; the major isomer forms a cyclic orthoacetate (**460**), while the minor isomer does not react. The resulting mixture of **460** and **459** is easily separated and the orthoacetate **460** is hydrolyzed to obtain pure **458**. The primary alcohol is dehydrated by Grieco's method, oxidative elimination of the o-nitrophenylselenide. After deprotection of the hydroxy group, the γ-butyrolactone ring spontaneously forms and the resulting dilactone **463** is bis methylenated by Danishefsky's method. Several attempts to transform **464** into vernolepin by transposition of the hydroxy group were not successful.

L. Fraxinellone and Pyroangolensolide

Fraxinellone (**465**) and pyroangolensolide (**466**) are probably not actual sesquiterpenes, but metabolic products related to the diterpene limonin (**467**). Nevertheless, the syntheses of **465** and **466** which have been

465   466   467

reported are typical of work in the sesquiterpene area and will be discussed in this section. Tokoroyama and associates have reported syntheses of both compounds. Their fraxinellone synthesis is summarized in Scheme 72.[140] Treatment of methyl vinyl ketone with ethoxyethynylmagnesium bromide affords a 5:4 mixture of diene esters **468** and **469**. The mixture is heated with methacrolein to obtain adduct **470** and unchanged cis ester **469** (which may be isomerized to an equilibrium mixture of **468** and **469** and recycled). Aldehyde **470** is treated with 3-furyllithium (prepared from 3-bromofuran) to obtain a 7:5 mixture of lactones **471** and **472**. Chromatographic separation and double-bond isomerization yields (±)-fraxinellone (**465**).

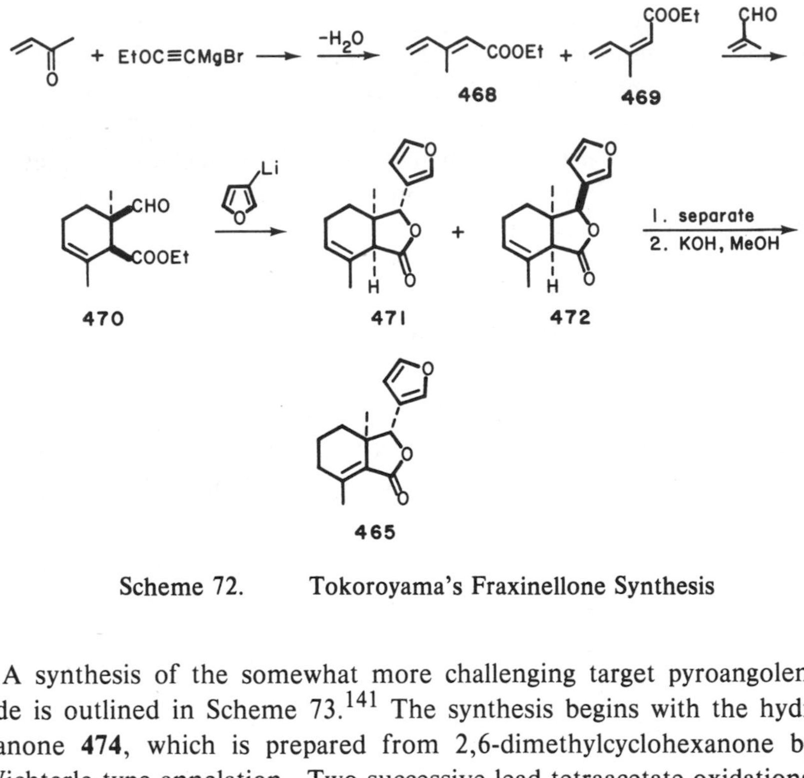

Scheme 72.   Tokoroyama's Fraxinellone Synthesis

A synthesis of the somewhat more challenging target pyroangolensolide is outlined in Scheme 73.[141] The synthesis begins with the hydrindanone **474**, which is prepared from 2,6-dimethylcyclohexanone by a Wichterle-type annelation. Two successive lead tetraacetate oxidations of **474** serve to cleave the cyclopentenone ring and establish the hemilactol

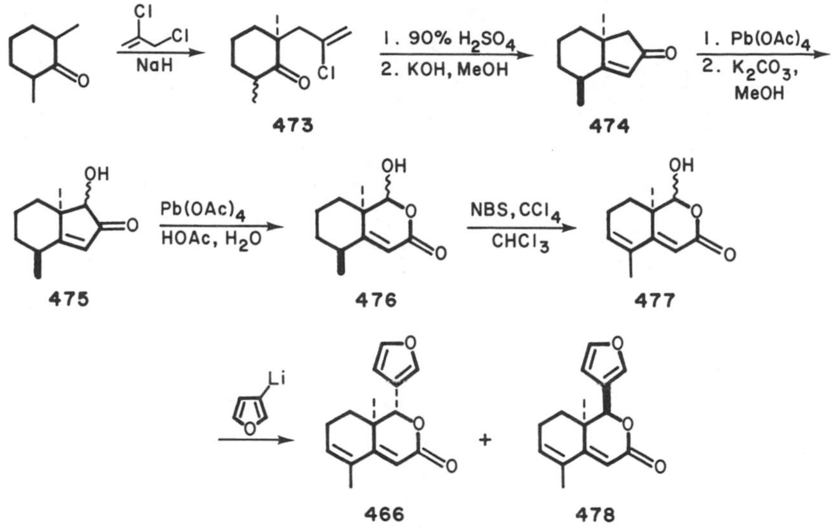

Scheme 73.   Tokoroyama's Pyroangolensolide Synthesis

functionality. Bromination of **476** with N-bromosuccinimide in a mixture of CCl_4 and $CHCl_3$ proceeds with concomitant elimination of HBr, giving the diene **477** in 64% yield. Treatment of this substance with excess 3-furyllithium gives a 7:3 mixture of pyroangolensolide (**466**) and its stereoisomer **478**.

M. Ivangulin, Eriolanin, and Phytuberin

Ivangulin (**479**), eriolanin (**480**), and phytuberin (**481**) are seco-eudesmanes which have been synthesized in the period of this review. The principal challenge in the synthesis of ivangulin and eriolanin is relating the stereochemistry of the methyl group located in the acyclic portion to the stereocenters in the six-membered ring. A similar problem exists in the synthesis of phytuberin, as can be seen in the open-chain form **481a**. In each case, the problem has been solved by first establishing the relative stereochemistry in an intermediate decalin, followed by cleavage of one ring.

479

480

481

481a

Grieco's ivangulin synthesis is summarized in Scheme 74.[142] Simmons-Smith methylenation of the homoallylic alcohol **483** occurs from the α face of the molecule. Electrophilic cleavage of the cyclopropane establishes the methyl group required for the eventual side chain. The relative stereochemistry of this center is governed by the thermodynamic preference for the β (equatorial) configuration. The elements of the eventual γ-butyrolactone ring are introduced in an interesting way, by [2+2] cycloaddition of dichloroketene to diene **486**. The addition is regio- and stereospecific, giving, after dechlorination, only cyclobutanone **487**. After a series of straightforward functional group manipulations, the A ring is opened by Baeyer-Villiger oxidation, giving hydroxy ester **489**. The tertiary alcohol is eliminated and the resulting unsaturated epoxide **490** is reduced with lithium in ammonia. Unfortunately, the desired **491** and **492** are produced in a ratio of 5:4. However, the major product may be converted into alkene **492** by acetylation and reduction. The only remaining noteworthy step is the Baeyer-Villiger conversion of the cyclobutanone ring into the γ-butyrolactone ring, which is accomplished using alkaline *t*-butyl hydroperoxide. The synthesis, although long (27-29 steps), is marked by two inventive

maneuvers, introduction of the side-chain methyl group as a cyclopropane ring, and the method used to establish the γ-butyrolactone functionality.

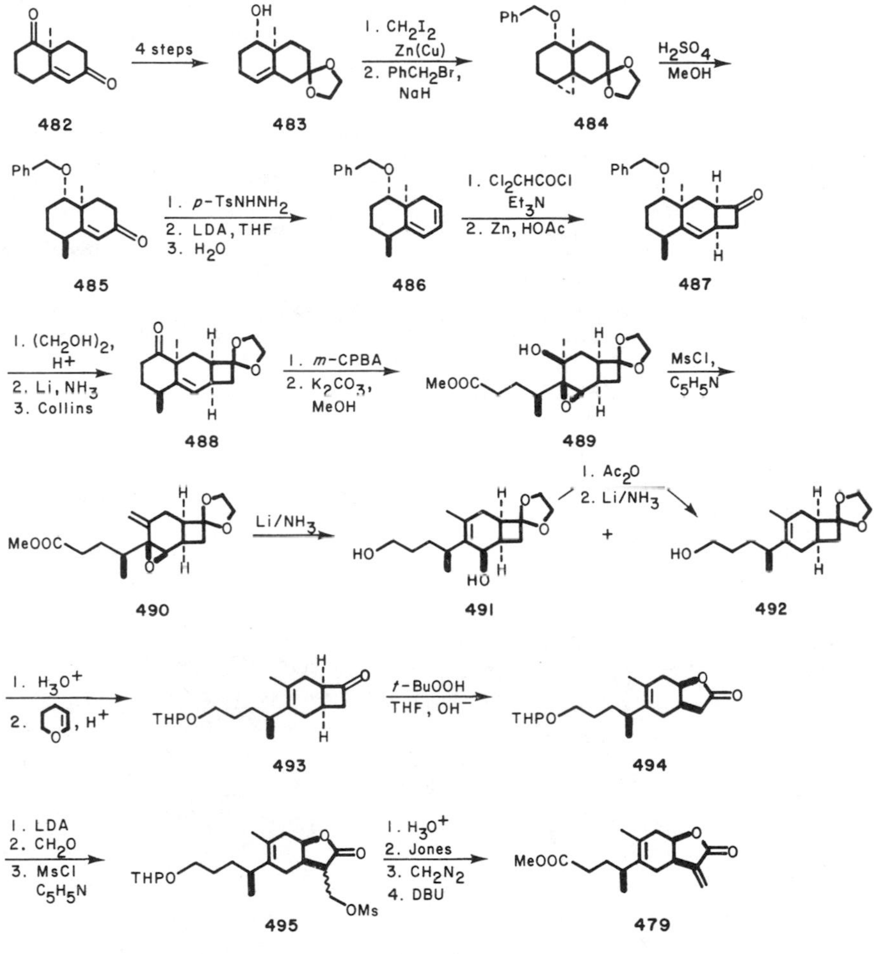

Scheme 74.   Grieco's Ivanqulin Synthesis

Grieco's eriolanin synthesis (Scheme 75) is a modification of the ivangulin synthesis.[143] It begins with alcohol **483**, which is converted into diketone **499** by a sequence of operations similar to that employed in the ivangulin synthesis. It was hoped that both Baeyer-Villiger oxidations and the epoxidation could be combined into one step. However, this did not prove possible, as peroxyacid oxidation affords the wrong epoxide (see **488→489** in Scheme 72). Nevertheless, the four-step sequence **499→501** was found to give the epoxy dilactone with good selectivity. The remainder of the synthesis is straightforward, although

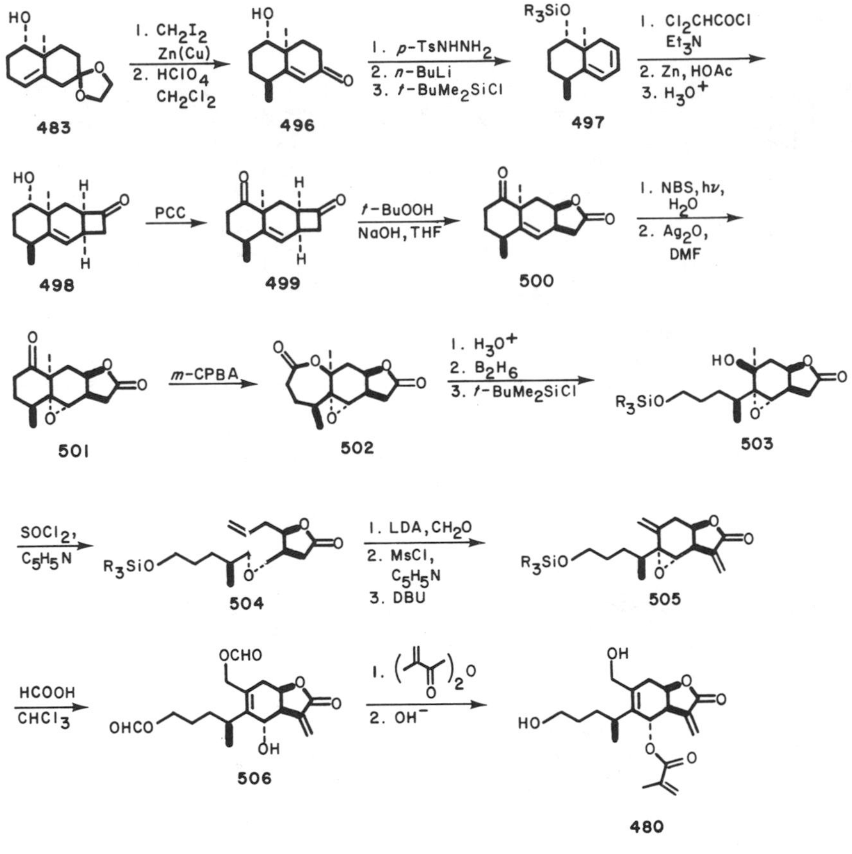

Scheme 75.   Grieco's Eriolanin Synthesis

the events accompanying the epoxide opening are of some interest. The transformation is carried out with formic acid in chloroform, which also removes the *t*-butyldimethylsilyl group from the side-chain hydroxy, which is subsequently formylated. The secondary alcohol, which undergoes formylation more slowly, is thus left for conversion to the methacrylate ester. The synthesis requires a total of 27 steps from the Wieland-Miescher enedione (**482**).

T. Masamune and co-workers, of Hokkaido University in Sapporo, have synthesized phytuberin (**481**), beginning with α-cyperone (**507**, Scheme 76).[144] Oxidation of the kinetic enolate of **507** under Vedejs' conditions, followed by protection of the resulting acyloins, gives an inseparable mixture of keto ethers **508** and **509**. Lithium

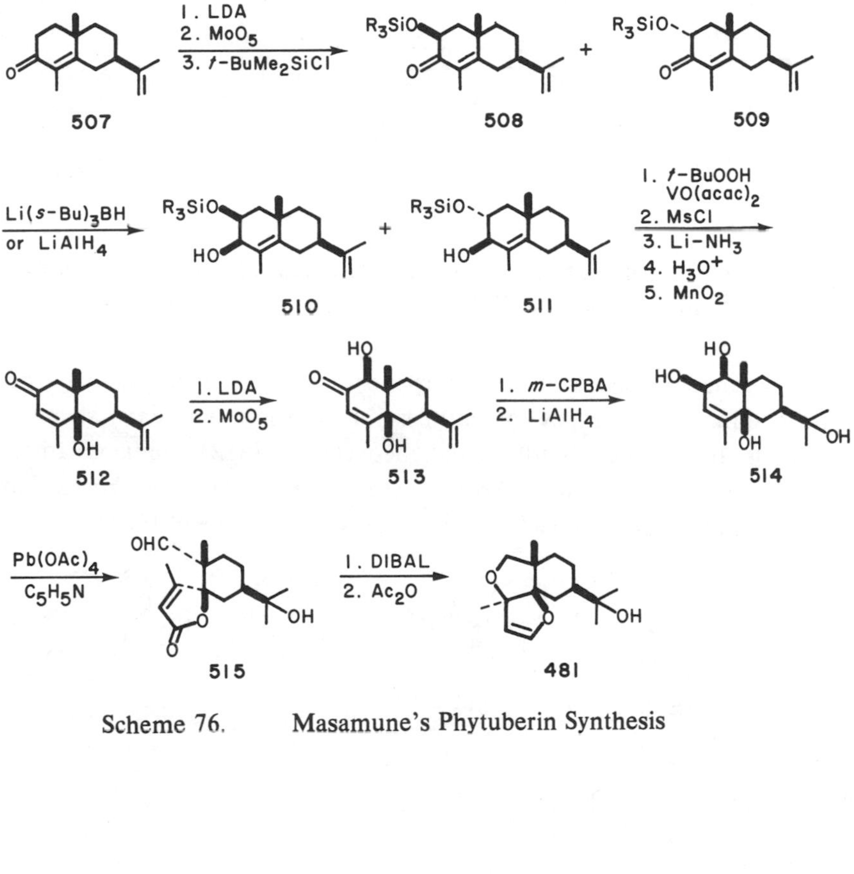

Scheme 76.   Masamune's Phytuberin Synthesis

tri-*sec*-butylborohydride selectively reduces **508** to **510**, leaving **509** unchanged. The mixture of **509** and **510** is separated and **509** is reduced by lithium aluminum hydride to obtain **511**. Both **510** and **511** are epoxidized with Sharpless' reagent, the epoxy alcohols mesylated, and the epoxy mesylates reduced under Birch conditions. After deprotection of the C_2 alcohols, and oxidation of the resulting allylic alcohols, (+)-β-rotunol (**512**) is obtained. The six-step sequence from **508** or **509** to **512** proceeds in about 50% overall yield. The A-ring is cleaved after a second application of the Vedejs oxidation, affording the unsaturated lactone **515**. Reduction of the two carbonyl groups and dehydration yields phytuberin. The multistep sequence for converting **507** into **512** is necessary because diene **516**, obtained from **507** in 72% yield via the Bamford-Stevens reaction, reacts with singlet oxygen from the α-face, yielding α-rotunol (**517**, 88%) after rearrangement of the resulting peroxide. The synthesis requires 14 steps overall, beginning with α-cyperone.

N. Hedycaryol, Preisocalamendiol, Acoragermacrone, Costunolide, Dihydrocostunolide, Dihydroisoaristolactone, and Periplanone B

The germacrane class of sesquiterpenes continues to present one of the most difficult challenges to synthesis. In the last ten years, seven compounds have been synthesized: hedycaryol (**518**), preisocalamendiol (**519**), acoragermacrone (**520**), costunolide (**521**), dihydrocostunolide (**522**), dihydroisoaristolactone (**523**), and periplanone B (**524**).

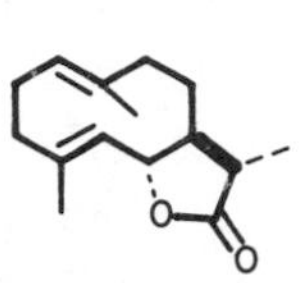

522 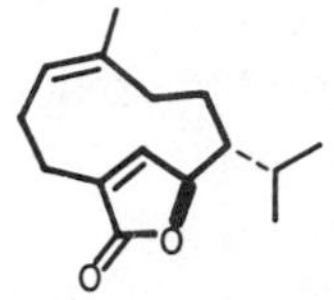   523 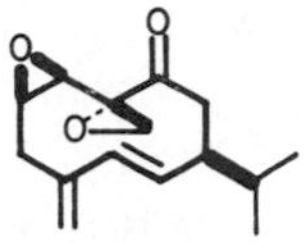   524

Hedycaryol (**518**) was first synthesized in 1968 by Wharton and co-workers.[145] The synthesis (Scheme 77) starts with dimethyl 4-hydroxy-isophthalate (**525**), a readily available by-product of the manufacture of salicylic acid. Hydrogenation of the ring followed by acetylation gives a mixture of isomeric triesters which is heated at 260°C to obtain unsaturated diester **527** in 32% overall yield for the three steps. Diester **527**

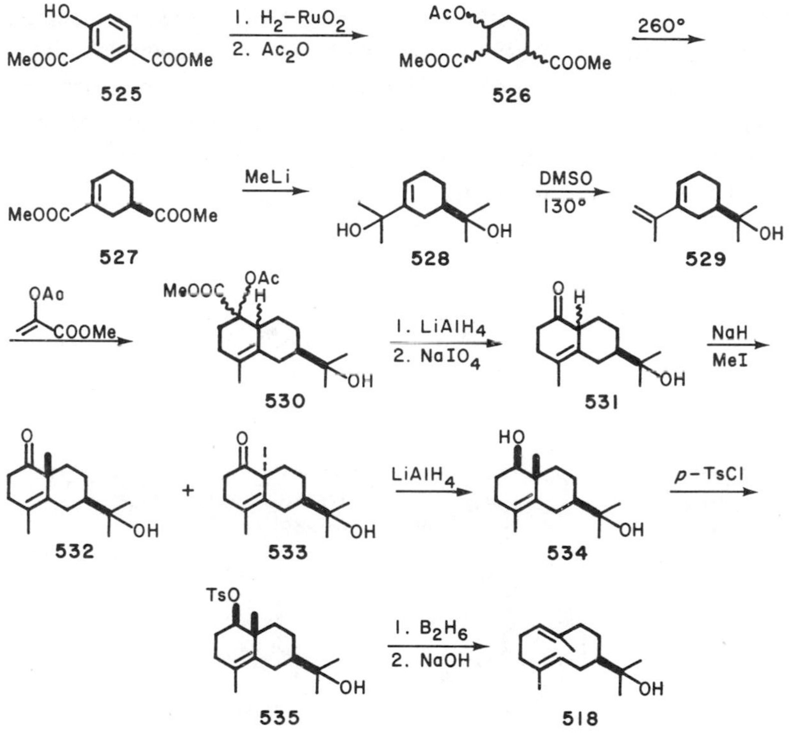

Scheme 77.   Wharton's Synthesis of Hedycaryol

reacts with methyllithium to give a diol (528), which is partially dehy-drated by heating in dimethyl sulfoxide. Diels-Alder addition of diene 529 to methyl α-acetoxyacrylate affords a mixture of diastereomeric adducts (530), which is reduced and cleaved to obtain a mixture of octalones 531. Methylation of the thermodynamically-favored enolate derived from 531 affords the trans- and cis octalones 532 and 533 in a ratio of 4:1. The mixture is reduced and the major reduction product (534) converted into tosylate 535. The decalin ring is fragmented by Marshall's method, giving hedycaryol (518) in 60% yield.

Ito has prepared hedycaryol starting with (2E,6E)-farnesol as shown in Scheme 78.[146a] Oxidation of the terminal double bond of sulfide 537 by van Tamelen's method gives epoxide 538 which undergoes a remarkably efficient cyclization to a 7:5 mixture of cyclodecadienes 539 and 540. After chromatographic separation, the major isomer (539) is subjected to reduction by lithium in ethylamine to obtain an inseparable mixture of 541 and 518 (2:3 ratio). Thermolysis of the mixture affords *trans*-elemenol (542) stereospecifically. Although the sequence is plagued by lack of specificity in two steps, the direct cyclization of 538 to 540 and 539 is nevertheless an impressive reaction that could point the

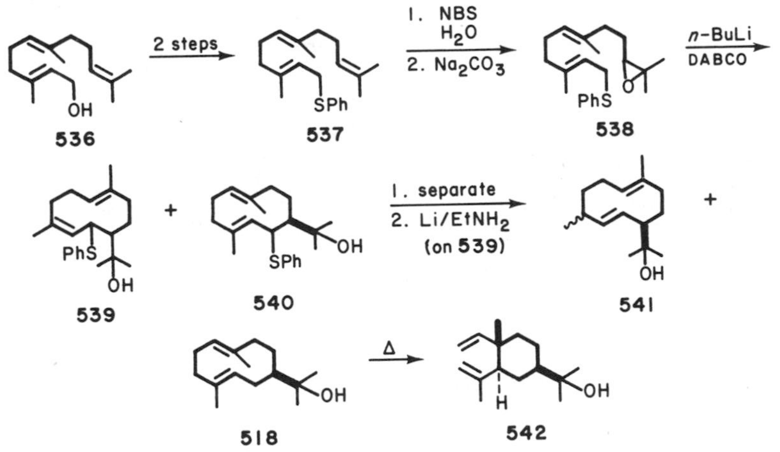

Scheme 78.   Itô's Conversion of Farnesol into Hedycarol

way to more effective germacrane syntheses in the future. In a subsequent investigation, Ito and his collaborators have studied the conversion of (2*E*,6*Z*)-farnesol (**543**) into the (6*Z*)-hedycaryols **544** and **545**.[146b]

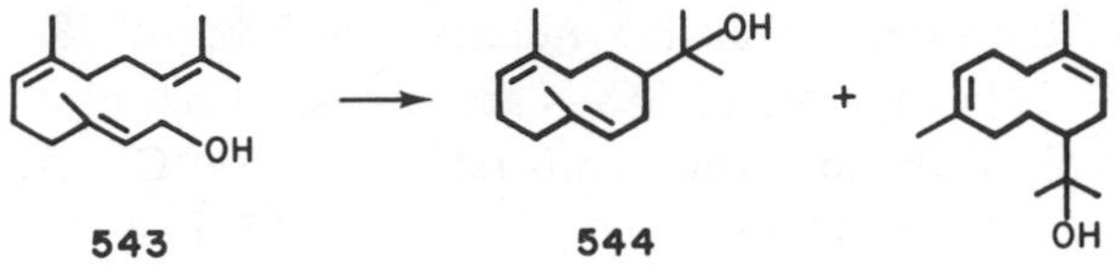
OH
OH
+
OH

543 544 545

W. C. Still has reported the concise synthesis of preisocalamendiol (**519**) and acoragermacrone (**520**) outlined in Scheme 79.[147] The monoterpene isopiperitenone (**546**) is converted into the *trans*-1,2-divinylcyclohexenol **547**, which is subjected to oxy-Cope rearrangement under Evans' conditions to obtain isoacoragermacrone (**548**). Formation of the kinetic enolate of **548**, followed by acetic acid quench, yields acoragermacrone (**519**) in only three steps from isopiperitone! The cis double bond in **548** is isomerized to the less stable trans geometry of **520** by an ingenious method involving conjugate addition of trimethylstannyllithium, followed by trapping of the resulting enolate with trimethylsilyl chloride to obtain enol ether **549**. Mild oxidation of **549** yields preisocalamendiol (**520**). The reader is referred to the original article for a discussion of the rationale for the isomerization of **548** to **520**.

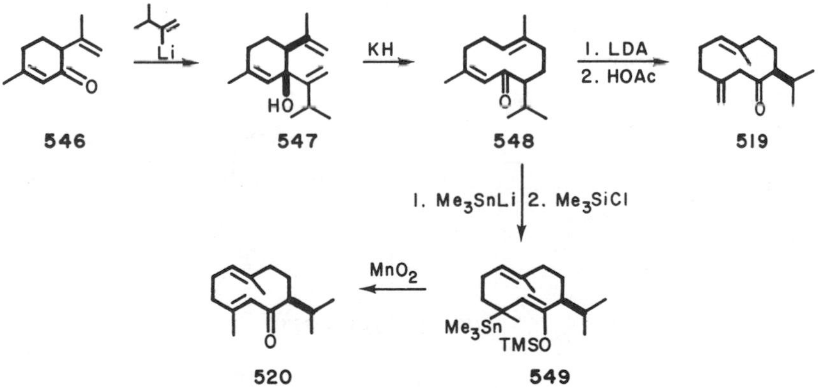
O
Li
HO
KH
O
1. LDA
2. HOAc
O
546
547
548
519
1. Me3SnLi
2. Me3SiCl
MnO2
O
Me3Sn
TMSO
520
549

Scheme 79. Still's Preisocalamendiol and Acoragermacrone Syntheses

Grieco's synthesis of costunolide is recorded in Scheme 80.[148] The well-known keto lactone **550**, obtained in two steps from santonin, is subjected to a sequence of reactions more or less identical to that employed in the Grieco temsin synthesis (see Scheme 58, **348→352**) to obtain (**554**). Thermolysis of **555** is accomplished by injection into a gas chromatograph with an oven temperature of 200°C. An equilibrium mixture is produced from which **521** and **555** may be obtained by preparative tlc in 20% and 42% yield, respectively. This appears to be a rather unusual example of a cyclodecadiene-divinylcyclohexane equilibrium in which the equilibrium does not overwhelmingly favor the six-membered ring isomer.

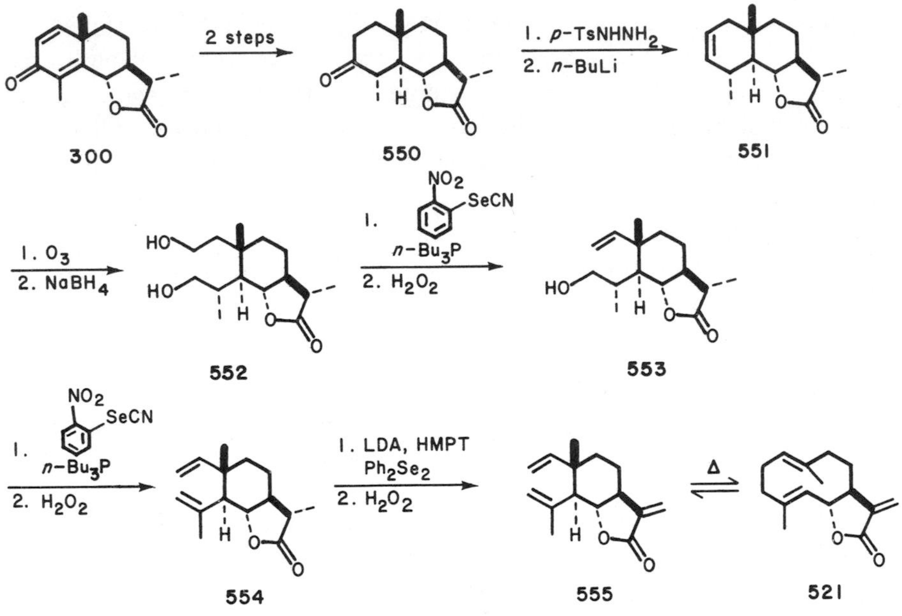

Scheme 80.   Grieco's Synthesis of Costunolide

Fujimoto, Shimizu, and Tatsuno converted santonin into dihydro-costunolide as shown in Scheme 81.[149] Chloro epoxide **556**, obtained in three steps from santonin (30% yield), is reduced by zinc to diene **557**,

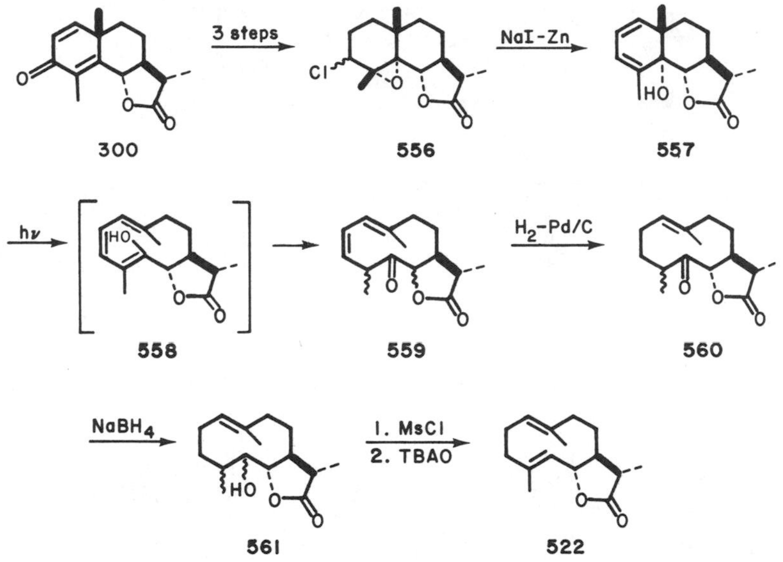

Scheme 81.   Fujimoto-Shimizu-Tatsumo Synthesis
of Dihydrocostunolide

which is photolyzed. The initially formed cyclodecatrienol **558** isomerizes to the cyclodecadienone **559**, which is obtained as a 5:1:1 mixture of
stereoisomers. Hydrogenation of the more accessible double bond of the
major isomer affords a keto lactone (**560**), which is reduced and dehydrated to obtain dihydrocostunolide (**522**). The elimination of the
secondary mesylate derived from **561** is noteworthy, in that DBU and
other amine bases are ineffective. The only base which was found to
accomplish the desired elimination was tetra-*n*-butylammonium oxalate
(TBAO).

Lange and McCarthy have investigated the use of the photo adduct
564, obtained from cyclobutenecarboxylic acid and (−)-piperitone (**563**),
for preparation of germacranes (Scheme 82).[150] Reduction of the ketone
carbonyl of **564** gives a hydroxy ester which spontaneously lactonizes to
give **565**. Thermolysis of this compound affords the dihydroisoaristolactone **523** (see also Scheme 57).

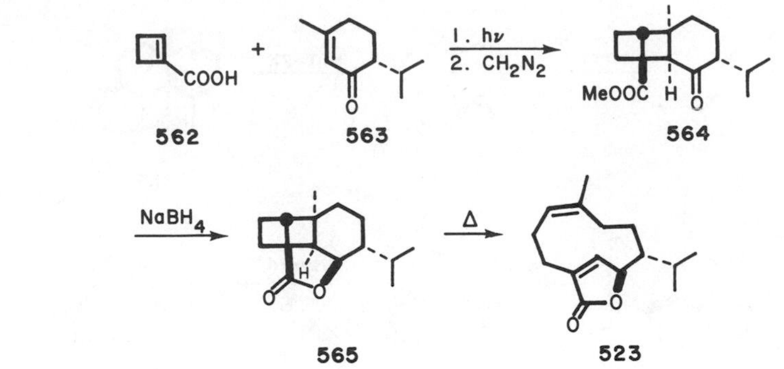

Scheme 82.   Lange-McCarthy Synthesis of Dihydroisoaristolactone

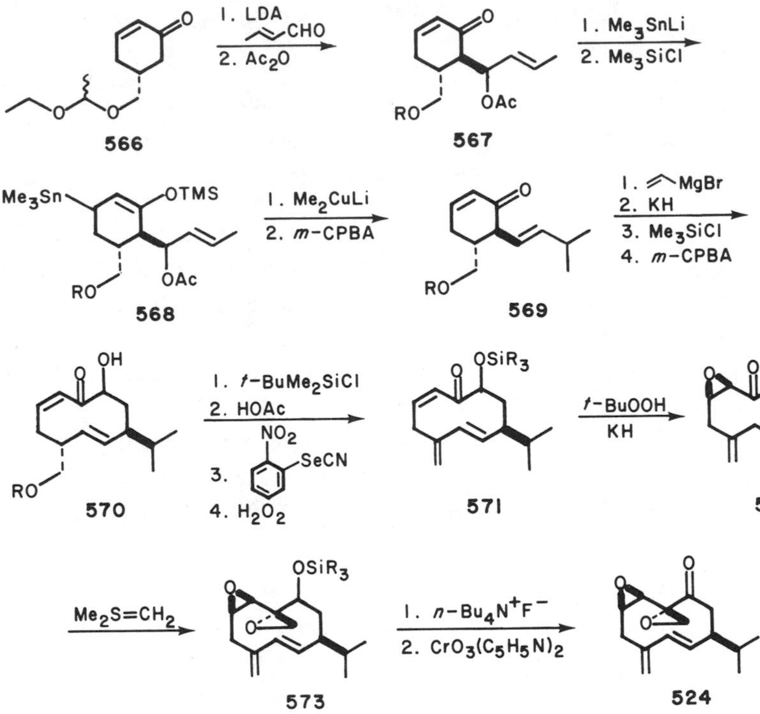

Scheme 83.   Still's Periplanone B Synthesis

Periplanone-B, the sex excitant of the American cockroach, was isolated (200 μg) and characterized in 1976 as a germacrane of structure **524**, but of unknown stereochemistry. In 1979, Still reported a brilliant series of syntheses in which he prepared three of the four possible diastereomers and determined the correct stereochemistry for the phermone.[151] The synthesis of the actual natural product is summarized in Scheme 83. Aldol condensation of cyclohexenone **566** with crotonaldehyde affords, after acetylation, adduct **567**. The enone is protected by addition of trimethylstannyllithium and the derived trimethylsilyl enol ether **568** is alkylated by lithium dimethylcuprate. After regeneration of the cyclohexenone, vinylmagnesium bromide is added and the resulting 1,2-divinylcyclohexenol is subjected to oxy-Cope rearrangement. The crude reaction product is silylated and the silyl enol ether oxidized to obtain α-hydroxy ketone **570**. The secondary hydroxy group is protected and the primary hydroxy deprotected by mild acidic hydrolysis. Dehydration of the primary alcohol is accomplished by Grieco's method, elimination of the derived o-nitroselenoxide. The resulting enone **571** is epoxidized to **572** and the ketone methylenated using Corey's reagent to obtain *bis*-epoxide **573**. Deprotection of the hydroxy group and oxidation affords periplanone-B. The stereochemical outcome of both the epoxidation and methylenation steps (**571→572, 572→573**) is understandable in terms of preferred conformations for the cyclodecadienone (**571a**) and epoxycyclodecenone (**572a**) which suffer only peripheral attack. It is assumed that the inherent preference of a 1,3-butadiene for

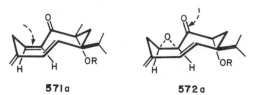

571a   **572a**

the s-trans conformation plays an important role in assuring that **571** and **572** adopt the required conformation shown. When the oxidation and methylenation steps are carried out on **574** (the protected form of intermediate **570**) isomer **575** is produced, presumably as a consequence of another preferred conformation.

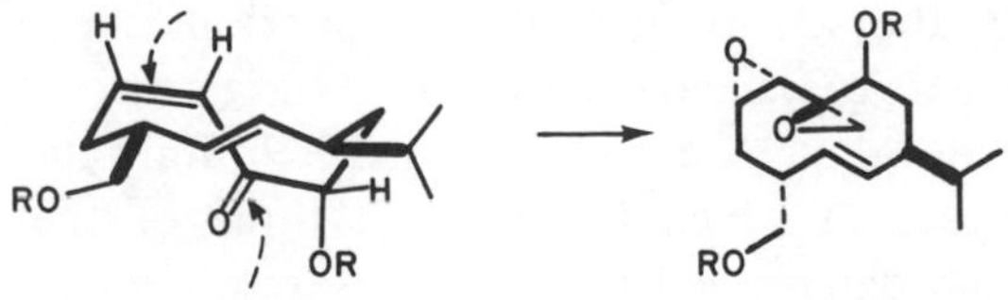

574   575

O. Humulene

Vig and co-workers have reported a synthesis of humulene (**585**) which
utilizes nickel carbonyl coupling of a bis-allylic bromide to form the
eleven-membered ring (Scheme 84).[152] Although Corey had earlier used

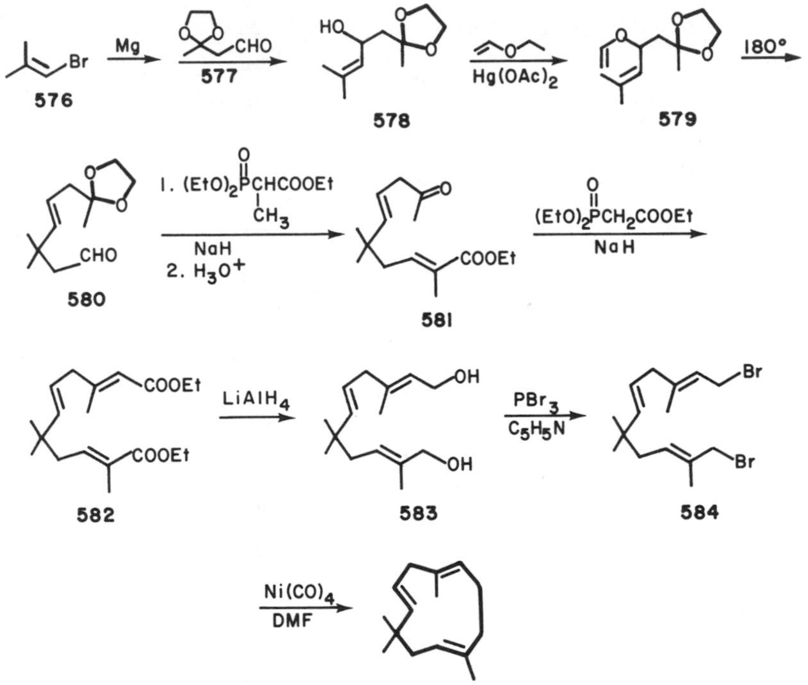

Scheme 84.   Vig's Synthesis of Humulene

this method in the first humulene synthesis,[153] the Corey synthesis proceeds through the 5Z stereoisomer of **584** and therefore yields 4,5-*cis*-humulene, which must then be isomerized to humulene. The Vig synthesis yields the correct stereoisomer without necessity of equilibration and is also much more efficient (8 steps rather than 14).

Yamamoto and co-workers have reported an extremely interesting synthesis of humulene (Scheme 85).[154] Bromo-ester **586**, obtained in two steps from geranyl acetate, is used to alkylate the dianion of ethyl 2-methyl-3-oxobutanoate to obtain keto diester **587**. The sodium salt of **587** is cyclized by Trost's method to obtain cyclic β-keto ester **588** in 45% yield. The carbonyl groups are reduced and the resulting diol converted into oxetane **589** by way of a monotosylate. The second key reaction of this imaginative synthesis is base-catalyzed ring opening of the oxetane, which yields trienol **590**. The primary alcohol is deoxygenated in a straightforward sequence of steps to obtain humulene.

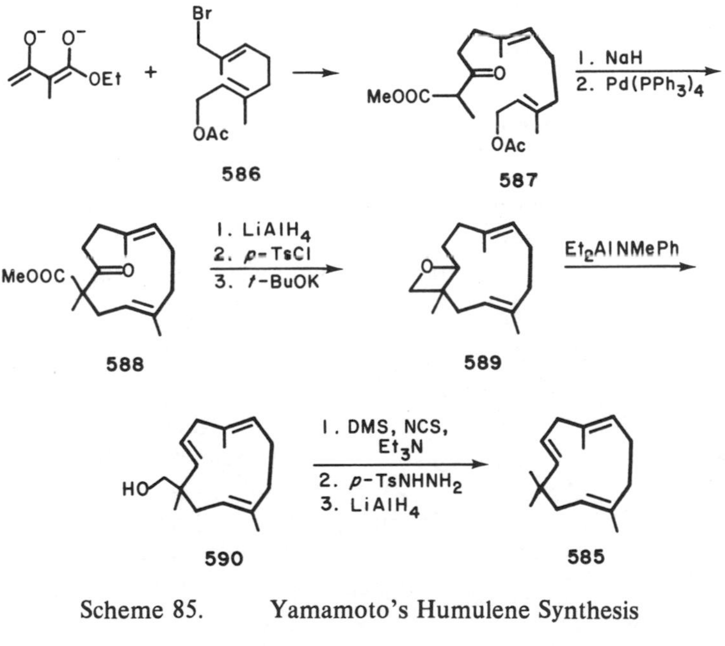

Scheme 85.   Yamamoto's Humulene Synthesis

4.   BICARBOCYCLIC   SESQUITERPENES; HYDRONAPHTHALENES

The largest single group of sesquiterpenes from the standpoint of syntheses which have been reported, is the hydronaphthalene group. In this section, we consider this group of compounds, further subdivided into four major classes: eudesmanes (**1**), cadinanes (**2**), drimanes (**3**), and eremophilanes (**4**). Finally, we consider a few syntheses of

miscellaneous hydronaphthalenes containing an additional cyclopropane ring.

A. Eudesmanes

(1) Occidol, Emmotin H, Rishitinol, and Platphyllide

The first group of eudesmanes which we now consider is the rearranged ring-A aromatic compounds occidol (**5**), emmotin H (**6**) and rishitinol (**7**) and the desmethyl aromatic eudesmane platphyllide (**8**).

The simplest of the group is occidol, in that there is no stereochemical problem and no real skeletal challenge. Three different syntheses, all by T.-L. Ho, have been reported (Scheme 1).[155-157] The first synthesis begins with 5,8-dimethyl-α-tetralone (**9**), requires three steps, and affords occidol in 45% overall yield. The second synthesis, published a year later, begins with 3,6-dimethylphthalic anhydride, requires four steps, and provides occidol in only 7% overall yield. The key step is Diels-Alder cycloaddition of the o-quinodimethane derived from α,α'-dibromoprehnitene and methyl acrylate. A year later, Ho reported

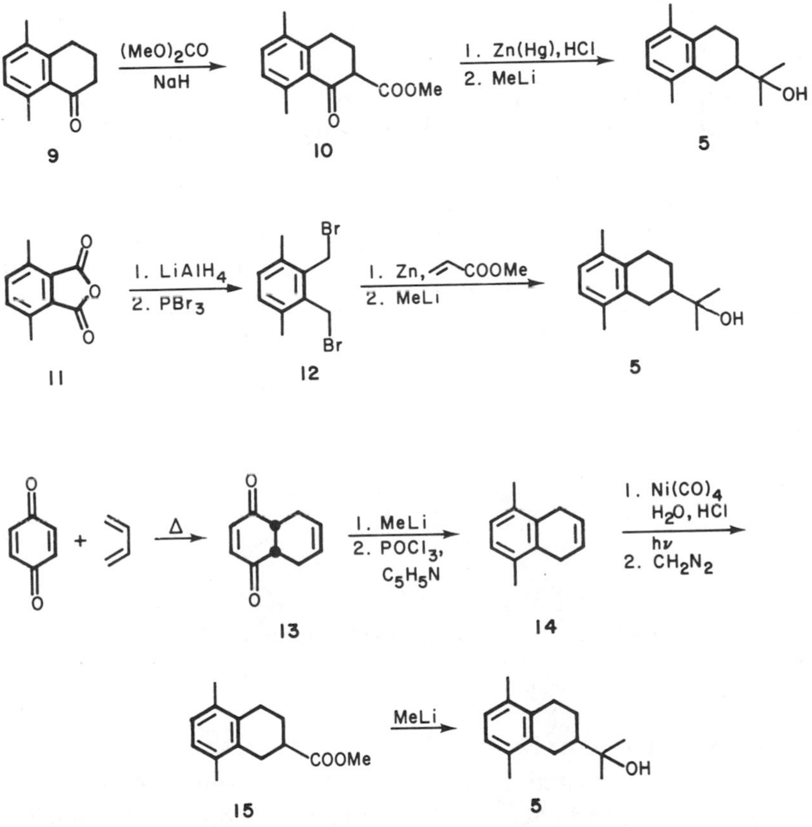

Scheme 1.   The T.-L. Ho Occidol Syntheses

his third synthesis, which begins with the Diels-Alder reaction between benzoquinone and butadiene. This synthesis requires six steps, and gives occidol in about 14% overall yield.

Emmotin-H is interesting for its novel 1,2-naphtho quinone structure. Krishna Rao has recently reported the straightforward synthesis summarized in Scheme 2.[158] Friedel-Crafts cyclization of diacid **16** provides a mixture of keto acids **17** and **18** from which **17** may be crystallized in 63% yield. Oxidation of the derived methyl ester with selenium dioxide gives *o*-quinone **19**, which is reduced and acetylated to obtain **20**. Treatment of triester **20** with excess methylmagnesium bromide, followed by air oxidation, gives emmotin-H (**6**).

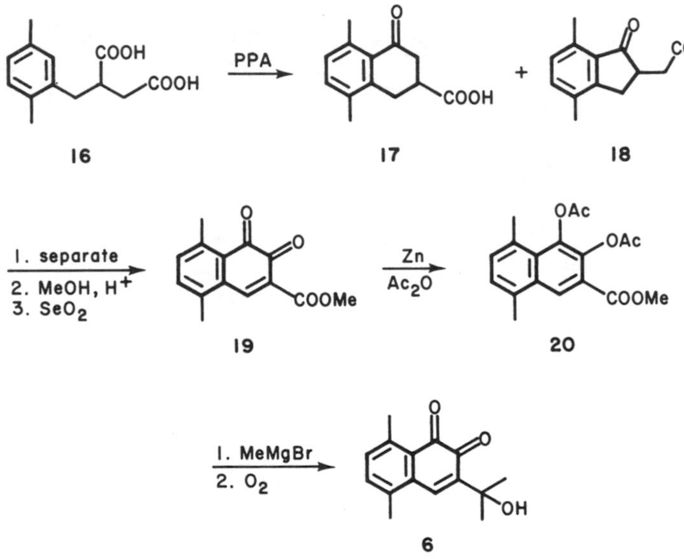

Scheme 2.   Krishna Rao's Emmotin-H Synthesis

Tomiyama and co-workers have synthesized rishitinol, an interesting hydroxy occidol isolated from fungus-infected white potatoes.[159] The synthesis is outlined in Scheme 3. Successive alkylation of malonic ester with benzylic chloride **21** and ethyl chloroacetate affords a triester (**23**),

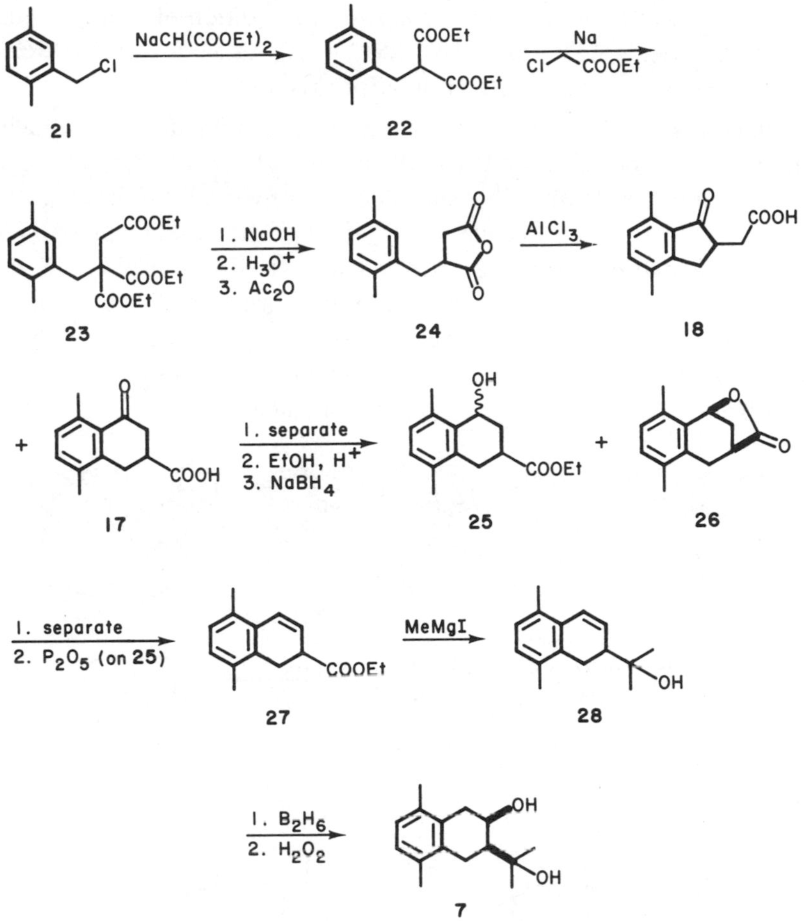

Scheme 3.   Tomiyama's Synthesis of Rishitinol

which is converted into anhydride **24**. Friedel-Crafts cyclization affords tetralone **17** and indanone **18** in a ratio of 2:1, exactly as was later found by Krishna Rao for cyclization of the succinic acid **16** (see Scheme 2). The tetralone **17** is separated by fractional crystallization, esterified and reduced to obtain a mixture of hydroxy ester **25** and lactone **26**. Dehydration of the former compound, followed by addition of methylmagnesium iodide, yields dehydrooccidol (**28**), which undergoes regioselective and stereoselective hydroboration to afford rishitinol.

Bohlmann's synthesis of the nor-eudesmane platphyllide is outlined in Scheme 4.[160] It begins with 1-naphthoic acid, which is hydrogenated and converted into amide **30**. This substance is converted into lactone **31** by oxidation with lead tetraacetate (22% yield). After methanolysis of the lactone, the hydroxy group is oxidized and the isopropanol side chain introduced by Mukaiyama's titanium tetrachloride mediated aldol condensation. After reduction of the keto ester and reoxidation of the primary alcohol, lactones **37** and **38** are obtained in a ratio of 2:1. The minor isomer is dehydrated to obtain platphyllide.

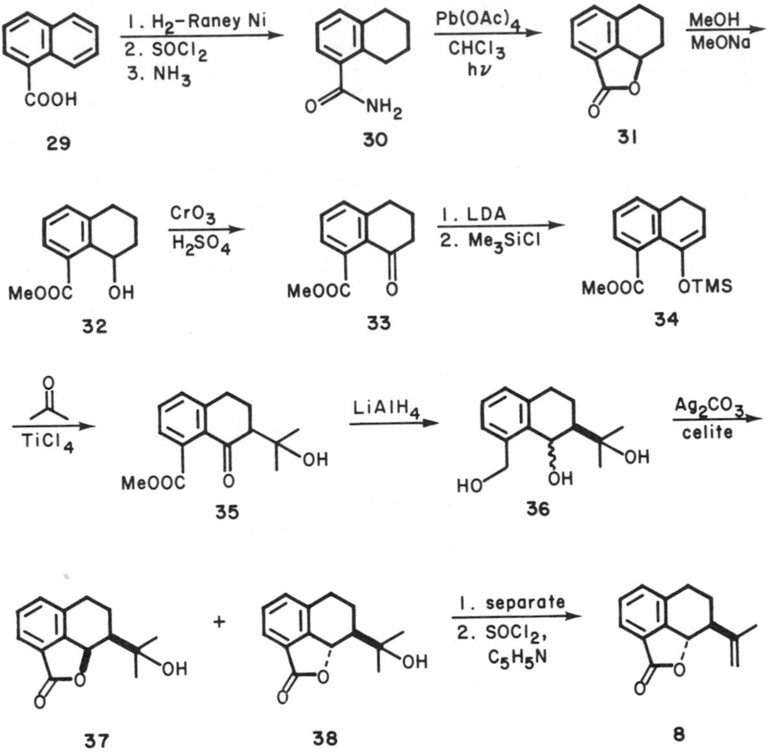

Scheme 4.   Bohlmann's Synthesis of Platphyllide

(2) α-Cyperone, β-Cyperone, β-Eudesmol,
*   and β-Selinene*

α-Cyperone (**39**) has a stereochemical complexity relating the isopropenyl substituent to that of the angular methyl group. β-Cyperone (**40**)
is even simpler, as there is only one stereocenter. β-Eudesmol (**41**) and
β-selinene (**42**) have three stereocenters and present somewhat more
complex problems of synthesis. Extensive early work was directed at
these simple eudesmanes and is reported in Volume 2 of this series.
The new syntheses have involved novel approaches to the carbon skeleton.

Although it only contains two stereocenters, α-cyperone (**39**) presents
a stereochemical problem since simple Robinson annelation of 2,5-disub-
stituted cyclohexanones affords mainly the diastereomeric octalone having the two substituents both axial:[161]

Caine and Gupton provided an interesting solution to this problem,
shown in Scheme 5.[162] Michael addition of (−)-2-carone (**43**) to ethyl
vinyl ketone affords diketone **44**. Stereoselectivity is good because of
the *gem*-dimethyl group which hinders attack from the top face of **43**.
Treatment of **44** with ethanolic HCl causes ring opening of the cyclopropyl ketone and aldol cyclization. The three-step synthesis provides **39** in
46% overall yield and is much more efficient than the earlier Pinder synthesis.[161]

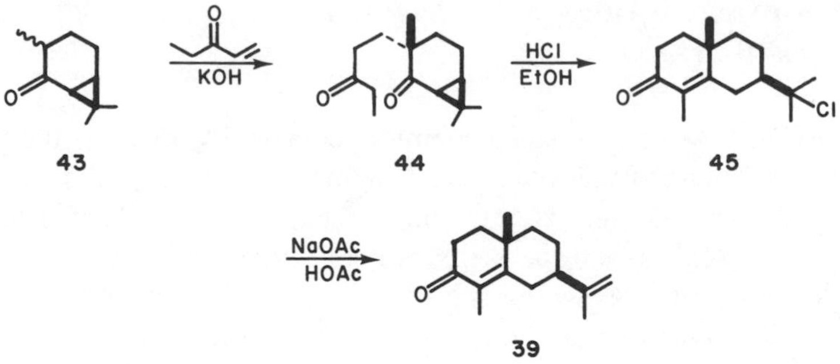

Scheme 5.   Caine-Gupton Synthesis of α-Cyperone

Gammill and Bryson have reported the synthesis of β-cyperone outlined in Scheme 6.[163] The synthesis begins with enone 47, available in
two steps from dihydroresorcinol (46). Enone 47 is alkylated with
dibromide 48, a homolog of the familiar Wichterle reagent, to obtain
enone 49, which is treated with isopropylmagnesium bromide under
copper catalysis, to obtain 50. Hydrolysis of the vinyl bromide and aldol
cyclization occur concomitantly upon treatment of 50 with perchloric acid
in refluxing formic acid; β-cyperone is obtained in about 18% overall
yield for the five-step synthesis beginning with 46.

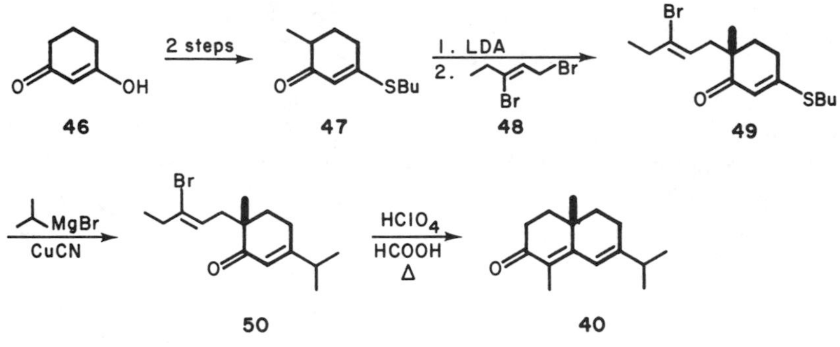

Scheme 6.   Gammill-Bryson Synthesis of β-Cyperone

Huffman and Mole have reported a synthesis of β-eudesmol which does not involve Robinson annelation as the method of constructing the decalin nucleus (Scheme 7).[164] The synthesis begins with keto acid **52**,

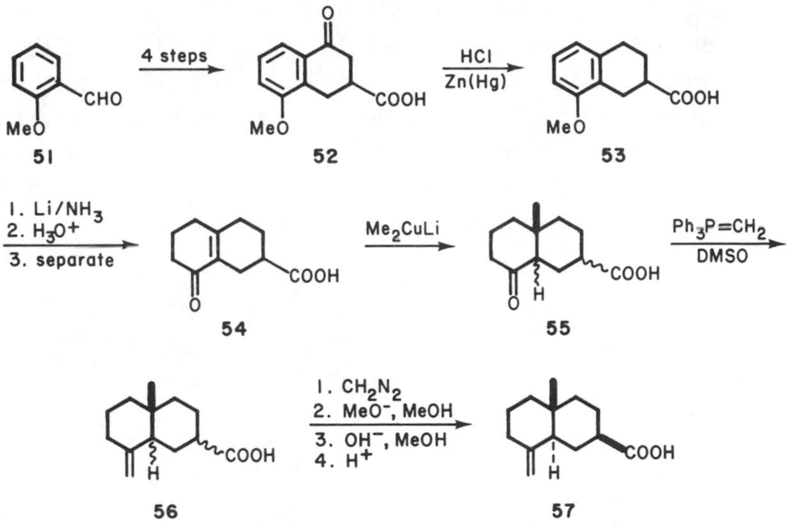

Scheme 7.   Huffman-Mole Synthesis of β-Eudesmol

available in four steps (22% yield) from o-anisaldehyde. After Clemmensen reduction, the methoxy acid **53** is subjected to Birch reduction to obtain a complex mixture of acids. The desired unsaturated keto acid **54** is obtained in 25% yield after chromatography. Addition of lithium dimethylcuprate to **54** gives a mixture of three keto acids which is subjected to Wittig methylenation to obtain acids **56**. This mixture of acids is esterified with diazomethane (mineral acid causes isomerization of the double bond) and the resulting mixture of methyl esters is equilibrated by sodium methoxide in methanol. Saponification affords methylene acid **57**, which had previously been converted into β-eudesmol. The overall yield for the 11 steps from aldehyde **51** to acid **57** is only 1.4%, even though the authors advertise the synthesis as short and efficient.[165]

Carlson and Zey have reported a similar approach to the eudesmanes (Scheme 8).[166] Stobbe condensation of di-t-butyl glutarate and keto ester **58** affords **59**, which is deprotected and dehydrated to obtain **60**. Cyclization affords keto acid **54**. The overall yield for the three-step synthesis of **54** (35% from keto ester **58**) is superior to that employed by Huffman and Mole (Scheme 7, 5% from aldehyde **51**). Addition of lithium dimethylcuprate to enone **61** provides a 48:24:17:11 mixture of the four diastereomers of **62**, with the desired stereoisomer being the least abundant. Base-catalyzed equilibration changes the initial ratio to 14:10:18:58. Saponification and chromatography gives pure keto acid **63**, which has previously been converted into β-eudesmol.[167]

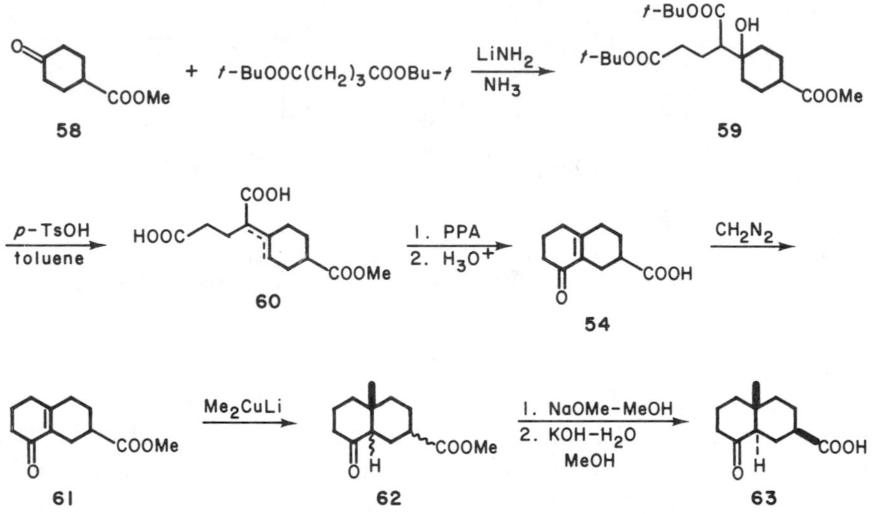

Scheme 8.   Carlson-Zey Synthesis of β-Eudesmol

Wolinsky and co-workers have synthesized β-selinene as outlined in Scheme 9.[168] Unsaturated chloride **65**, obtained in two steps from limonene, is acylated with vinylacetyl chloride (**66**) to obtain a mixture of enone **67** (34%) and enone **68** (33%). After separation of the mixture, enone **67** is hydrogenated, and dehydrochlorinated to obtain ketal

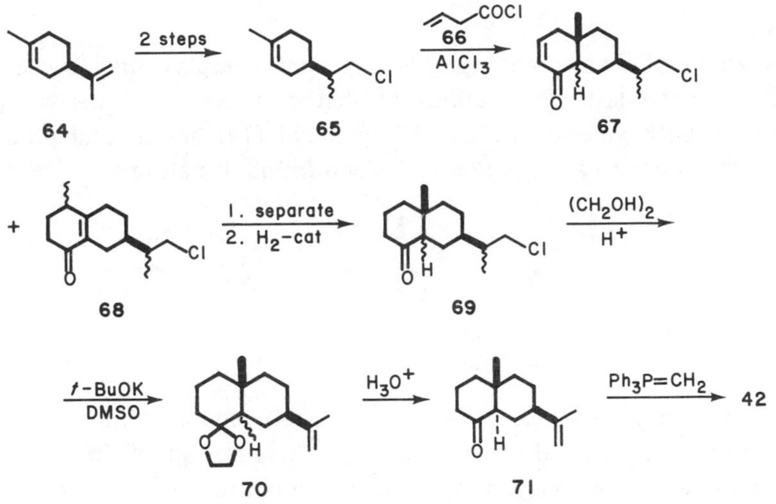

Scheme 9.   Wolinsky's Synthesis of β-Selinene

70. Hydrolysis followed by Wittig methylenation affords β-selinene (**42**). Stereochemically, ketone **71** is obtained as a 9:1 mixture with its

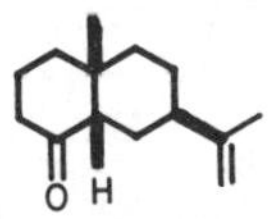

diastereomer **72.** The relative stereochemistry of the side chain and the angular methyl center is set in the cyclization leading to **67**:

The other enone (**68**), which is suggested to be formed by carbonium ion rearrangements, was converted into occidol.

(3) Juneol, 10-Epijuneol, and 4-Epiaubergenone

This group of eudesmanes is slightly more complex than those considered in the last two sections in that each has four stereocenters. Although both juneol (**73**) and 10-epijuneol (**74**) are natural products, 4-epiaubergenone (**75**) has not itself been found in nature.

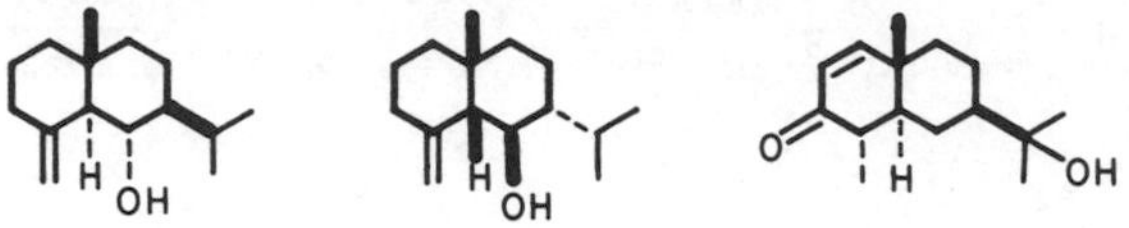

Schwartz and co-workers have synthesized juneol from methyl farnesate (**76**) as outlined in Scheme 10.[169] Cyclization of **76** is brought about using Lucas' reagent, to obtain a mixture consisting mainly of chloro ester **77**. Treatment of this mixture first with aqueous $ZnCl_2$, then with $ZnCl_2$ in refluxing benzene affords a mixture of diastereomeric esters **78**, accompanied by 16% of a double bond isomer

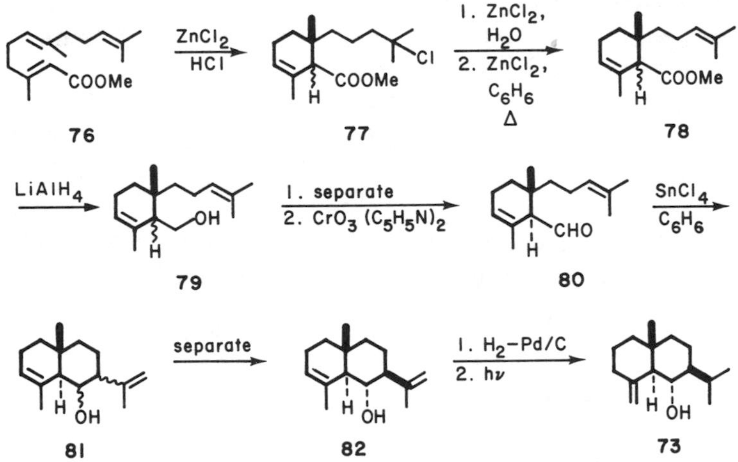

Scheme 10.   Schwartz Synthesis of Juneol

and 17% of a bicyclic isomer. The mixture of isomers **78** is reduced to obtain a 1:1 mixture of diastereomeric alcohols **79**. After chromatographic separation, each isomer is oxidized to the corresponding aldehyde. Cyclization of the two aldehydes is brought about by treatment with stannic chloride in benzene. Each isomer undergoes ring closure with retention of stereochemistry at the formyl group. However, each yields a mixture of stereoisomers. The desired isomer **82** is obtained in 18% yield by chromatography of the cyclization product from isomer **80**. Selective hydrogenation of the isopropenyl double bond and photoisomerization of the endocyclic double bond yields ($\pm$)-juneol. Although this synthesis produces the desired natural product, it leaves a great deal of room for improvement in selectivity.

Wender and Lechleiter have carried out an elegant synthesis of 10-epijuneol which is outlined in Scheme 11.[170] Photoaddition of methyl cyclobutenecarboxylate (**83**) to piperitone (**84**) provides adduct **85**. Deoxygenation by the Bamford-Stevens reaction gives alkene **86**, which is reduced, hydroborated, and oxidized to obtain diol **87**. This substance is selectively phosphorylated on the primary hydroxy group and the resulting monophosphate reduced by lithium naphthalenide to obtain 10-epijuneol (**74**). The synthesis is short, efficient, and completely stereoselective.

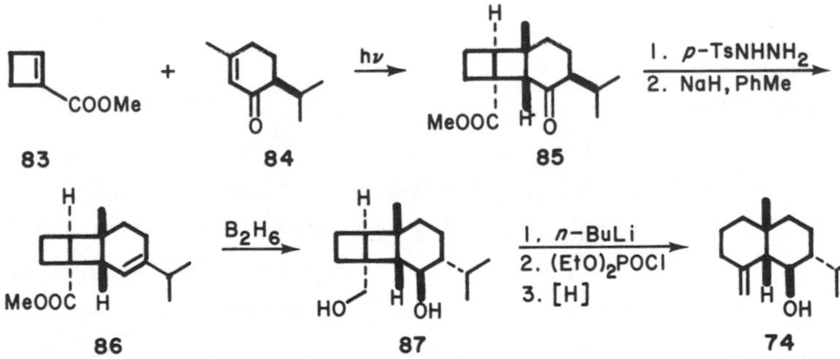

Scheme 11.   Wender-Lechleiter Synthesis of 10-Epijuneol

R. B. Kelly and co-workers at the University of New Brunswick synthesized 4-epiaubergenone, an isomer of the natural product, as outlined in Scheme 12.[171] Ketone 88 is subjected to a straightforward sequence of steps to obtain racemic dihydrocarissone (91), which is brominated and dehydrobrominated to obtain racemic enone 75. The optically active form of this material is obtained by base-catalyzed equilibrium of natural aubergenone.

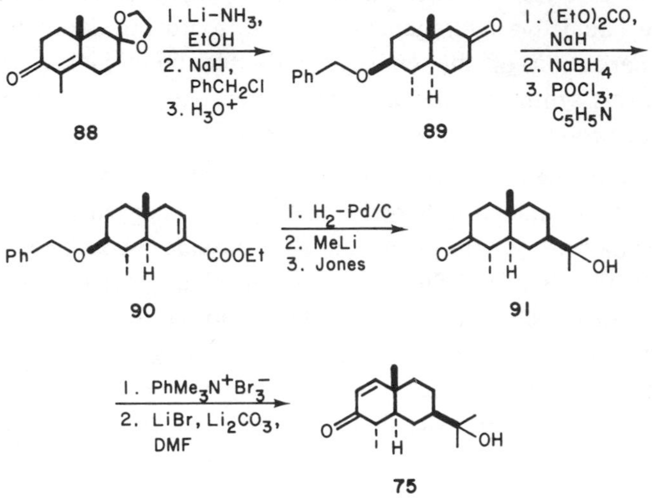

88   89   90   91   75

Scheme 12.   Kelly Synthesis of 4-Epiaubergenone

(4) Cuauhtemone

Cuauhtemone (98) is a sesquiterpene growth inhibitor which is isolated from a Mexican medicinal shrub. It has been synthesized by Goldsmith and Sakano as shown in Scheme 13.[172] Monoketal 88 (see also Scheme 12) is reduced and the resulting enolate is deoxygenated by Ireland's method to obtain unsaturated ketal 92. Hydrolysis and acylation with diethyl carbonate affords a β-keto ester (93) which is hydroxylated to obtain a 3:1 mixture of diastereomeric diols, with 94 the major isomer.

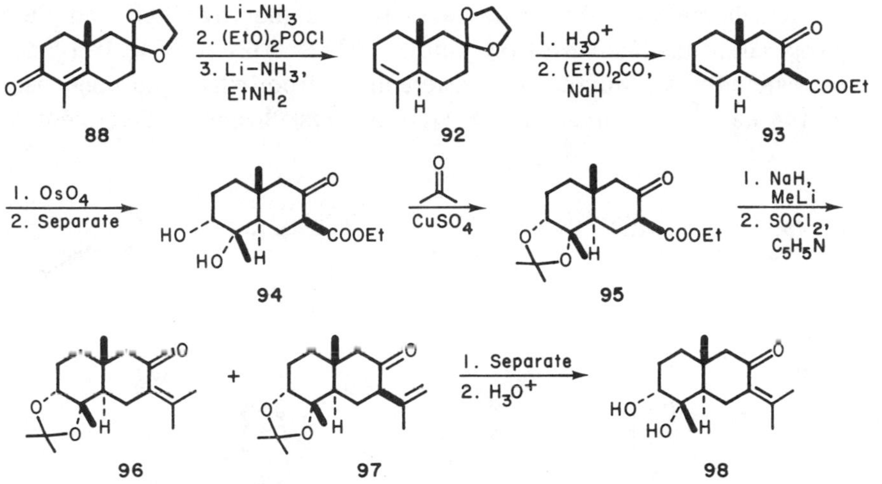

Scheme 13.   Goldsmith Synthesis of Cuauhtemone

The enone system is introduced by addition of methyllithium to the salt of the protected β-keto ester **95**, followed by dehydration of the resulting β-hydroxy ketone.[173] Isomers **96** and **97** are produced in a 2:1 ratio. Deprotection of **96** affords cuauhtemone (**98**).

(5) β-Agarofuran, Norketoagarofuran, and Evuncifer Ether

β-Agarofuran (**99**), norketoagarofuran (**100**), and evuncifer ether (**101**) are related structurally in that each is a cyclic ether.

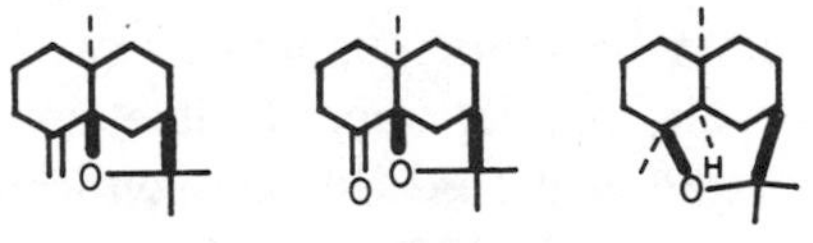

Büchi and Wuest have reported the synthesis of β-agarofuran summarized in Scheme 14. Dihydrocavone (102) is first hydrated and then hydrogenated to obtain hydroxy ketone 103.[174] Alkylation with 5-iodo-1-pentyne affords a mixture of diastereomers from which the major isomer 104 may be obtained in 41% yield by crystallization. Treatment of

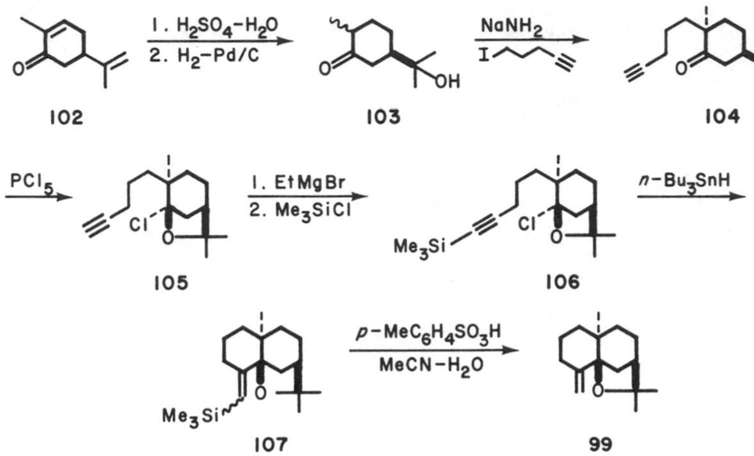

Scheme 14.   Büchi and Wuest Synthesis of β-Agarofuran

104 with PCl_5 provides the chloro ether 105. After silylation of the terminal acetylene, the A-ring is closed by free radical cyclization. Desilylation affords β-agarofuran (99). This imaginative synthesis is one of the rare examples of the use of free radical chemistry in natural product synthesis.

In Volume 2 of this series, we reported a synthesis of norketoagarofuran by Heathcock and Kelly.[175] Although the synthesis was brief, the final steps turned out to be capricious and irreproducible. Kelly subsequently reported an alternative, reproducible route (Scheme 15).[176] Treatment of unsaturated acid 108 with peroxyformic acid yields a mixture of diol monoformates which is saponified to obtain dihydroxy acid

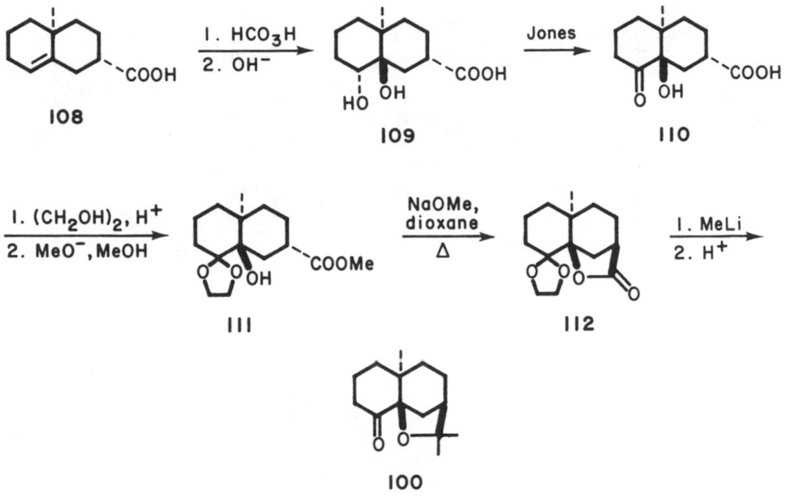

Scheme 15.   Kelly's Synthesis of Norketoagarofuran

109. Jones oxidation gives a ketol (**110**) which is ketalized and the resulting ethylene glycol ester trans-esterified to obtain **111**. Treatment of this hydroxy ester with sodium methoxide in refluxing dioxane using a Soxhlet extractor containing calcium hydride causes isomerization and lactonization. Lactone **112** is treated with methyllithium and the resulting dihydroxy ketal hydrolyzed with acid to obtain norketoagarofuran (**100**).

Evuncifer ether (**101**), a defense secretion of the West African soldier termite, has been prepared by Baker, Evans and McDowell as summarized in Scheme 16.[177] Hydroxy ketone **103**[178] is annelated using 1-chloro-3-pentanone to obtain enone **114**. Deoxygenation of this material affords epi-γ-eudesmol (**115**), which is oxymercurated to obtain a 3:1 mixture of **101** and isodihydroagarofuran (**116**).

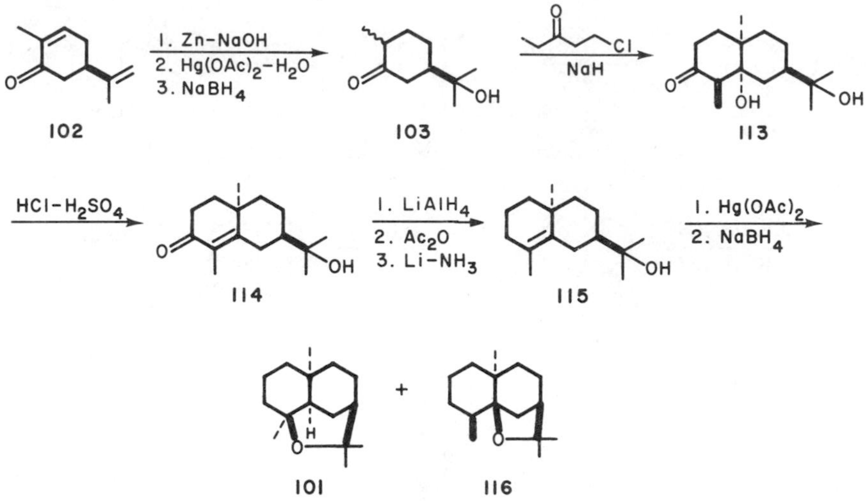

Scheme 16.   Baker-Evans-McDowell Synthesis of Evuncifer Ether

(6) Rishitin and Glutinosone

Rishitin (**117**) and glutinosone (**118**) are nor-eudesmanes having four asymmetric centers. Both have been synthesized by Masamune and co-workers at Hokkaido University in Sapporo.

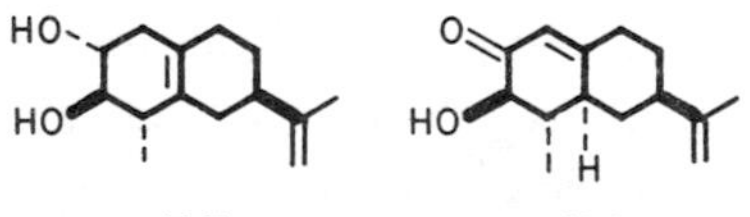

The rishitin synthesis[179] uses santonin as a relay (Scheme 17). Although long (19 steps), the conversion of santonin into rishitin proceeds with good stereocontrol and embodies some exceedingly clever methodology. Removal of the angular methyl group with establishment of the C_9-C_{10} double bond is accomplished by a modified Barton reaction (**122-123**) followed by decyanation of the allyl nitrile **124**. The latter transformation is rarely used in synthesis.

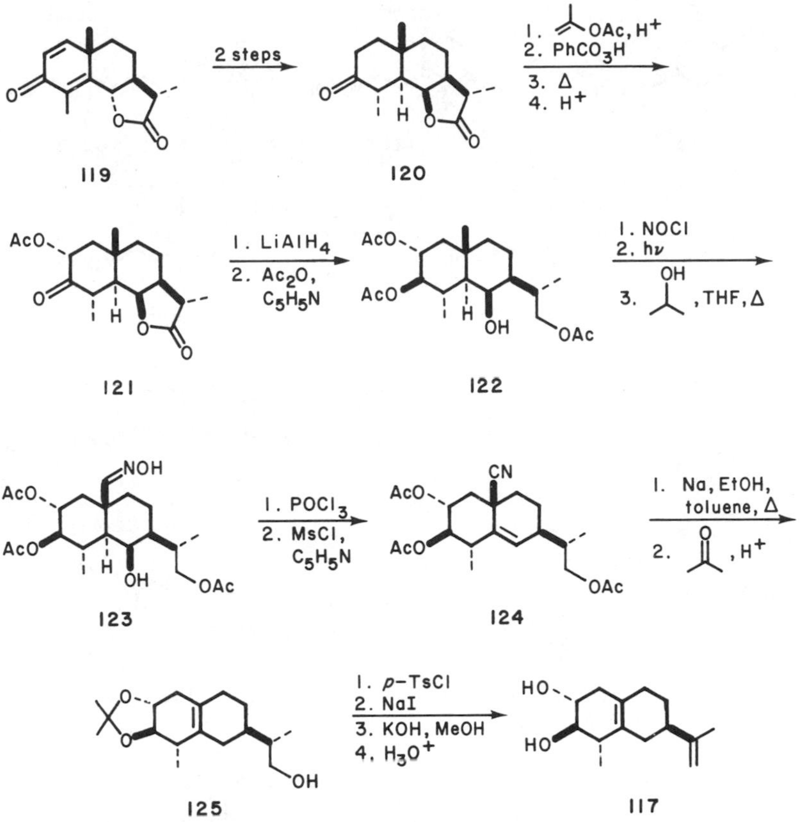

Scheme 17.   Hokkaido Synthesis of Rishitin

The Hokkaido synthesis of glutinosone[180] is summarized in Scheme 18. The synthesis begins with Robinson annelation on keto diester **126** to obtain octalone **127**. The C_2 oxygen is introduced by epoxidation of the 2,3-enol acetate, as in the rishitin synthesis. However, in the present case, the acetoxy ketone **130** is equilibrated with its isomer **131** by heating with tetramethylammonium acetate; an approximately equimolar mixture of the two isomers is obtained. Treatment of the triester **132** with a large excess of methylenetriphenylphosphorane affords a mixture

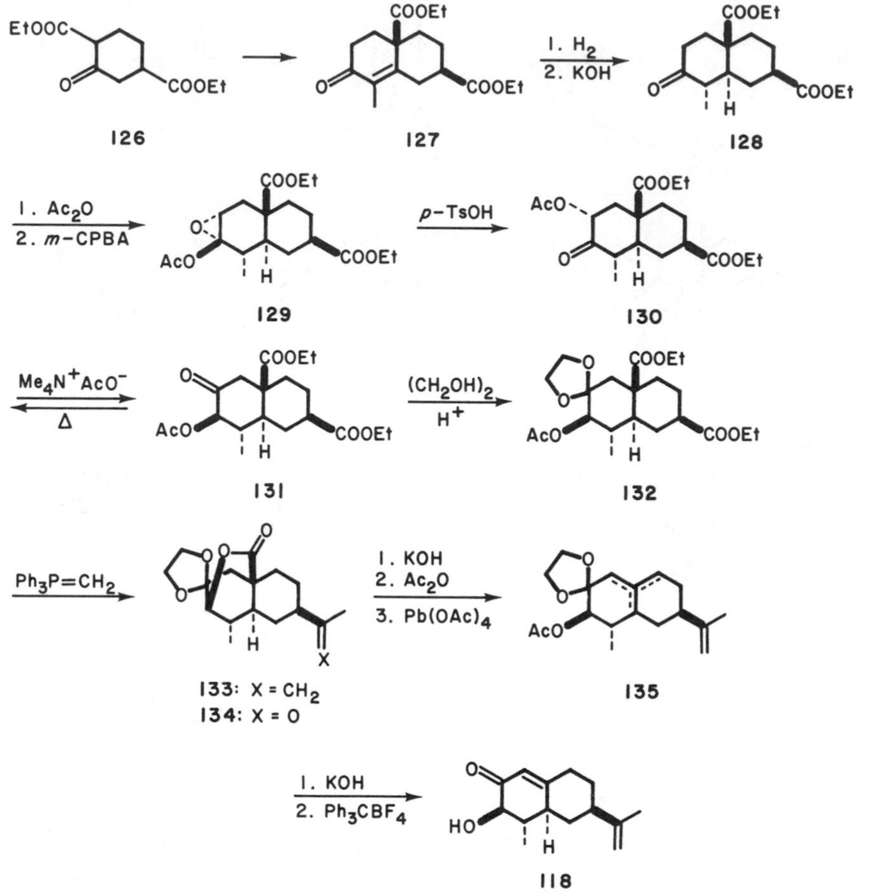

Scheme 18.   Hokkaido Synthesis of Glutinosone

of lactones **133** (22%) and **134** (50%). Treatment of the latter compound with methylenetriphenylphosphorane affords more of the desired isopropenyl compound **133** (88%). Hydrolysis of **133**, followed by lead tetraacetate decarboxylation gives a mixture of unsaturated compounds (**135**). Cleavage of the ketal and conjugation of the enone system provides glutinosone (**118**) along with large quantities of phenolic by-products. The overall yield of glutinosone in this 14-step synthesis is 2.8%.

(7) Occidentalol

(+)-Occidentalol (136) is of interest as a rare eudesmane in which the angular methyl center has the α configuration. Its synthesis was first

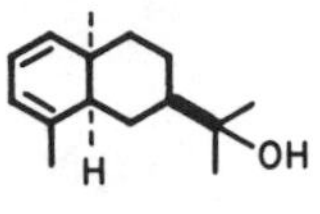

136

achieved by Hortmann (Scheme 19).[181,182] The Hortmann synthesis begins with (+)-carvone (137), which is converted into ester 138 in a 10-step sequence.[183] The A-ring diene system is introduced by straightforward reactions, giving the trans-fused diene 140. Photoisomerization

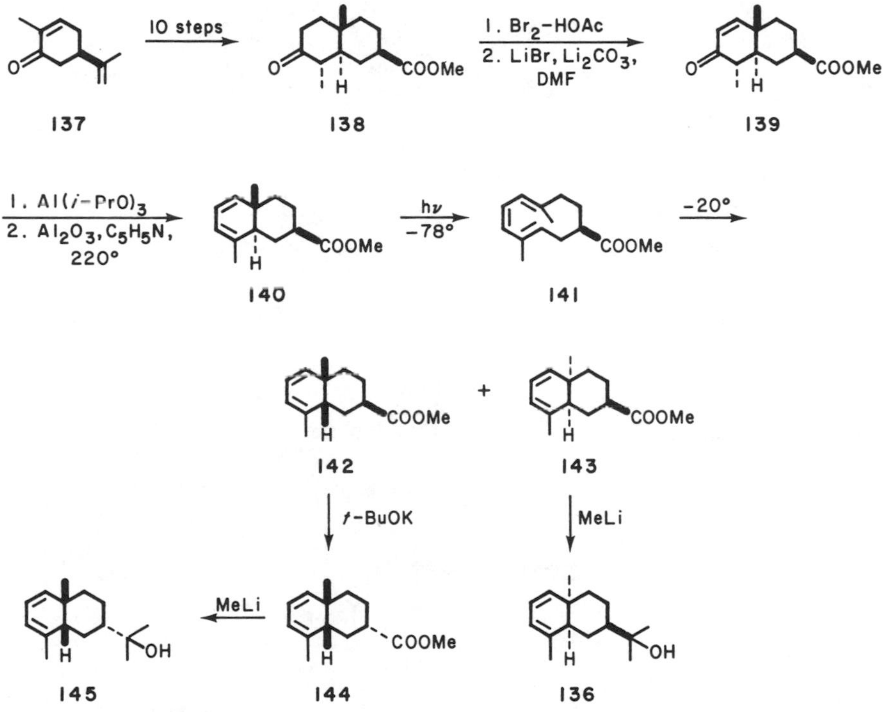

Scheme 19.   Hortmann Synthesis of Occidentalol

at -78°C affords the *trans,cis,trans*-cyclodecatriene **141**, which undergoes disrotatory closure to a mixture of cis-fused dienes **142** and **143**. Treatment of the latter with methyllithium provides (+)-occidentalol (**136**). Base-catalyzed epimerization of isomer **142** gives a 25:1 mixture of isomers **144** and **142**. Isomer **144** reacts with methyllithium to afford (−)-occidentalol (**145**).

Antipode **145** has also been prepared by Ando and co-workers in a relay synthesis from santonin.[184] The Ando synthesis is summarized in Scheme 20. The synthesis begins with the vinyl decalone **146**, which had

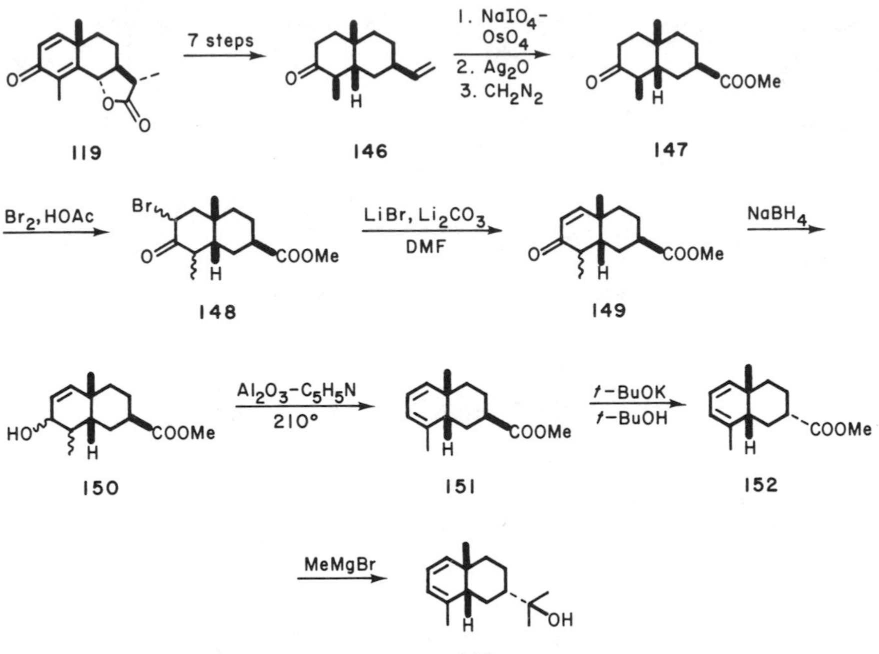

Scheme 20.   Ando Synthesis of (-)-Occidentalol

previously been prepared from santonin in seven steps.[185] The steps used to convert keto ester **147** into (−)-occidentalol are very similar to those employed by Hortmann.

Shorter syntheses of (+)-occidentalol were reported in 1972 by Deslongchamps and co-workers in Sherbrooke[186] and by Amano and Heathcock in Berkeley.[187] The latter synthesis was reported in Volume 2 of this series.[188] The Deslongchamps synthesis is summarized in Scheme 21. The synthesis proceeds via epicarrisone (**153**), which is hydrogenated to establish the cis-fused decalone system. The diene system is introduced by a precedented series of steps. The most interesting transformation is selective dehydration of the allylic alcohol in **157**. In fact, this transformation gives occidentalol in 62% yield, along with 26% of recovered **157**.

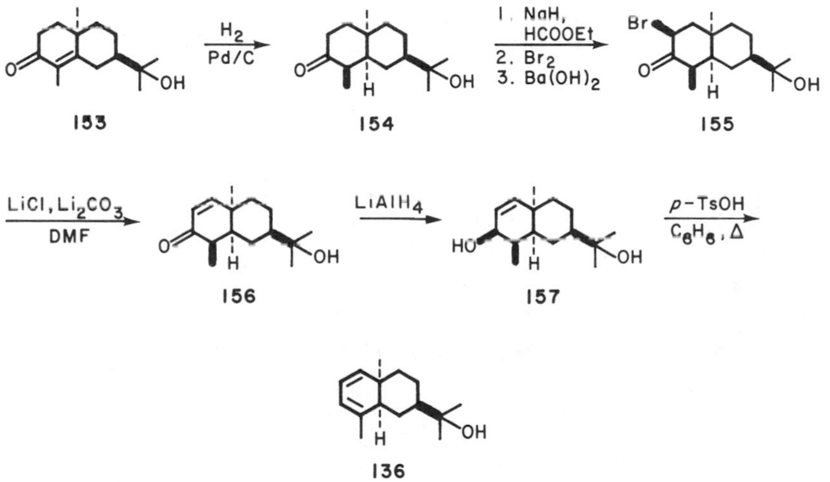

Scheme 21.   Deslongchamps Synthesis of (±)-Occidentalol

A synthesis of (±)-occidentalol has been reported by Watt and Corey (Scheme 22).[189] The synthesis begins with an unusual "inverse-electron demand" Diels-Alder reaction between pyrone 158 and alkene 159. The initial adduct extrudes CO_2, providing diene 160 in 25-40% yield. The ester group is reduced to methyl by a sequence previously developed in Corey's laboratory and the side chain is introduced by a Wadsworth-Emmons reaction. Hydrolysis of the enol thioether 164 gives a 3:1 mixture of epimeric methyl ketones. Addition of methyllithium to the major isomer yields racemic occidentalol.

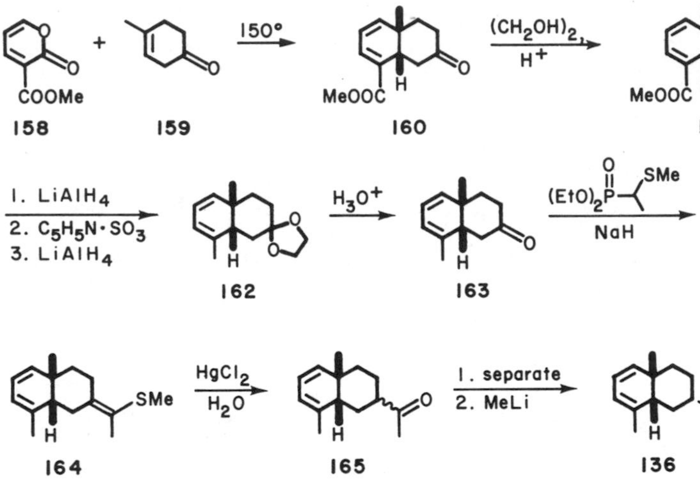

Scheme 22.   Watt-Corey (±)-Occidentalol Synthesis

The final occidentalol synthesis, which also yields the racemate, is that of Marshall and Wuts (Scheme 23).[190] Alkylation of the dianion resulting from Birch reduction of *m*-toluic acid affords diene acid **167**, which is converted into ester **170** by a sequence of straightforward steps. Conversion of **170** to malonate **171** is surprisingly difficult, and may only be accomplished using lithium 2,2,6,6-tetramethylpiperidide (LTMP) as the base. Unsaturated aldehyde **172** undergoes smooth cyclization in a remarkably stereospecific reaction, affording diene acylal **173** as the only isomer. Straightforward conversion of **173** provides racemic occidentalol.

The six published occidentalol syntheses are compared in Table 1.

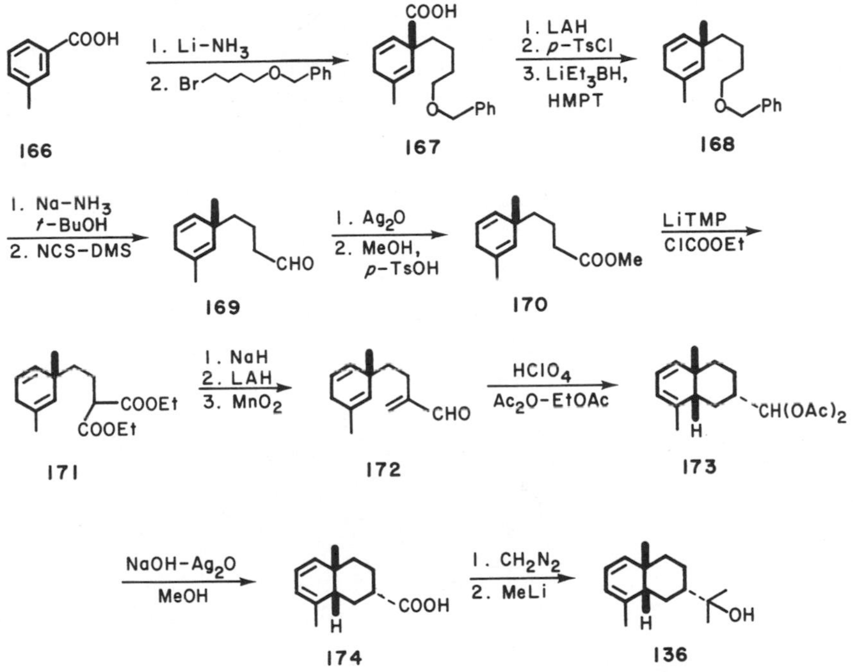

Scheme 23. Marshall-Wuts Synthesis of (±)-Occidentalol

Table 1. Comparison of Occidentalol Syntheses

Authors	Starting Material	Product	Steps	Overall Yield
Hortman, et al.		(+)-occidentalol	17	0.37%
		(−)-occidentalol	18	0.13%
Ando, et al.		(−)-occidentalol	16	a
Deslongchamps, et al.		(+)-occidentalol	10	5.8%
Heathcock and Amano		(+)-occidentalol	8	2.3%
Watt and Corey		(±)-occidentalol	9	1.5-2.4%
Marshall and Wuts		(±)-occidentalol	15	15%

a. Yields not provided in paper.

*(8) Santonin, Yomogin, Tuberiferine, Alantolactone,
     Isotelekin, Dihydrocallitrisin, and Frullanolide*

The eudesmanolides have been particularly inspiring with regard to synthesis.[191] During the period of this review, syntheses of α-santonin (**119**), yomogin (**175**), tuberiferine (**176**), alantolactone (**177**), isotelekin (**178**), dihydrocallitrisin (**179**), and frullanolide (**180**) have been accomplished.

119	175	176
177	178	179

180

Marshal and Wuts have carried out the synthesis of α-santonin which is summarized in Scheme 24.[192] The synthesis begins with methylation of the dianion produced by Birch reduction of *m*-toluic acid to obtain dienoic acid **181**, which is converted into bromide **184** by standard transformations. Alkylation of ethyl thioethoxymethyl sulfoxide with **184** affords a thioacetal monosulfoxide which is hydrolyzed with concomitant cyclization to dienol **185**. Alkylation of the derived ketone occurs from the β-face, affording keto ester **186** with good stereoselectivity. Reduction of **186** with lithium aluminum hydride affords an equimolar mixture of epimeric diols. After chromatographic separation, the desired isomer is oxidized by silver carbonate on Celite to obtain the trans-fused

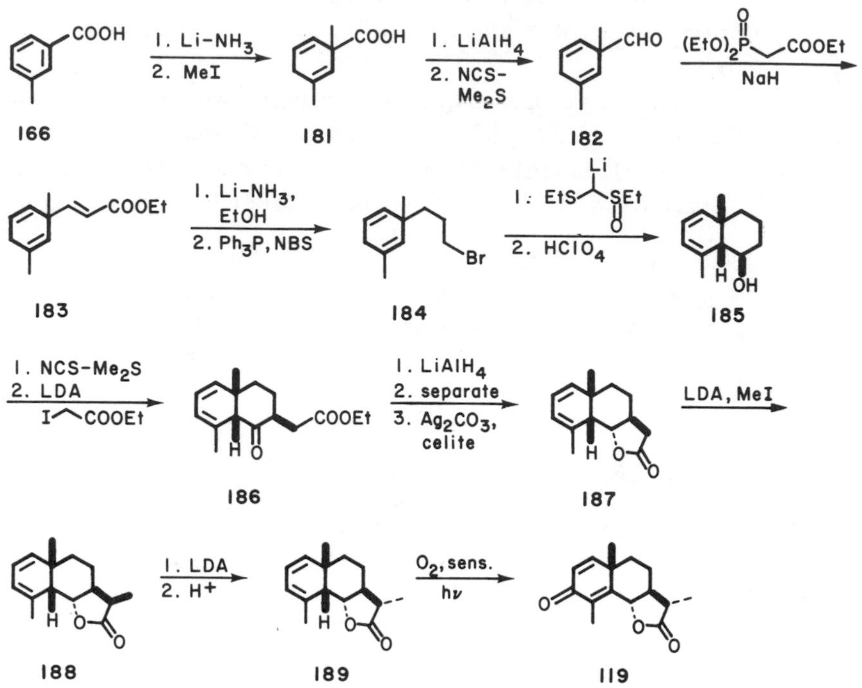

Scheme 24.   Marshall-Wuts Synthesis of (±)-α-Santonin

lactone **187**. Although methylation occurs stereospecifically from the β-face, if the enolate of the methylated lactone is formed and quenched, the α-methyl epimer **189** is produced in good yield. Singlet-oxygen oxidation of **189** affords α-santonin.

Caine and Hasenhuettl synthesized yomogin as is summarized in Scheme 25.[193] The octalindione **191**, available by a straightforward six step synthesis from diester **190**, is alkylated as its enamine to obtain diketo ester **192**. Selective reduction of the saturated ketone affords the cis-fused lactone **193**, which is methylenated by Grieco's procedure. The final double bond is smoothly introduced by DDQ oxidation of enone **194**.

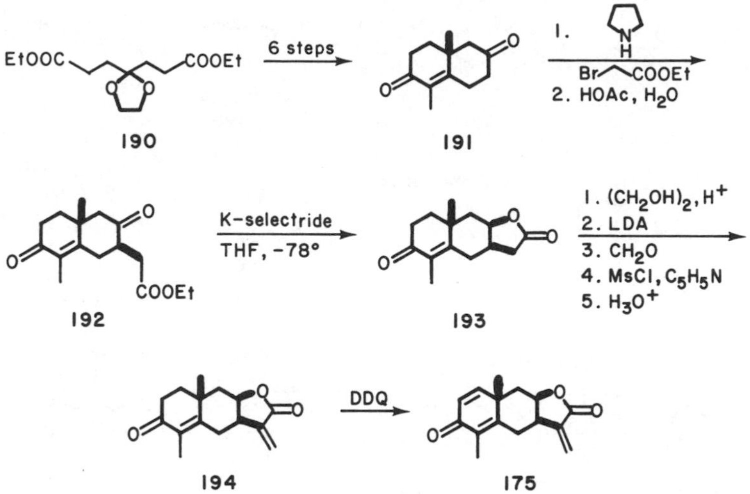

Scheme 25. Caine-Hasenhuettl Synthesis of (±)-Yomogin

Tuberiferine was synthesized by Grieco and Nishizawa as outlined in Scheme 26.[194] Enone **195**, obtained in 85% yield by acid-catalyzed Robinson annelation of 2-methylcyclohexanone, is ketalized by reaction with ethylene glycol. Since **195** is known to give a 1:1 mixture of the 4,5 and 5,6 isomers upon ketalization,[195] the low isolated yield obtained by Grieco and Nishizawa (57%) is understandable. Alkene **196** is converted into trans-fused decalone **198** by the method developed by Marshall in his β-eudesmol synthesis.[196] The side-chain destined to become the butenolide ring is introduced by alkylation of **198** with methyl bromoacetate; the desired stereochemistry is achieved by equilibration of the acetic ester group to the equatorial position. The stereochemistry of the oxygenated carbon is established as equatorial by lithium-ammonia reduction. The final transformations, methylenation of the lactone and introduction of the ring-A unsaturation, are accomplished using selenoxide methodology.

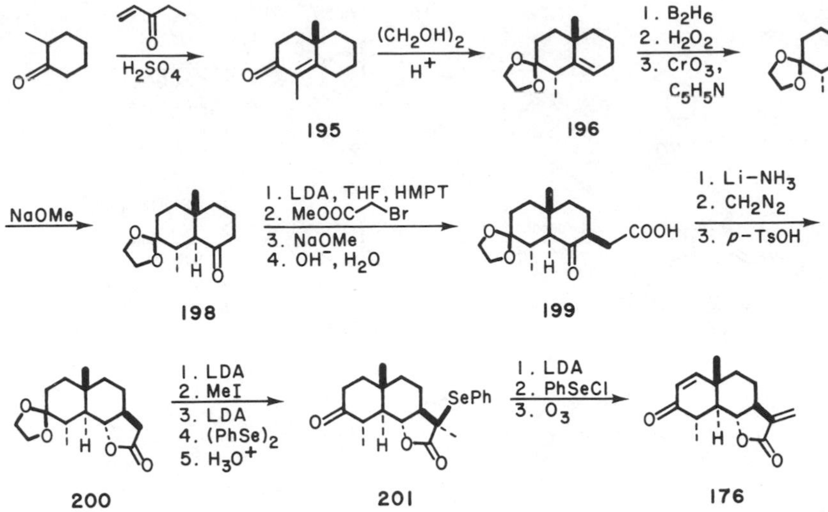

195   **196**   **197**

198   **199**

200   **201**   **176**

Scheme 26.   Grieco-Nishizawa Synthesis of (±)-Tuberiferine

Posner and Loomis have developed a synthesis of enones **205** and **206** as shown in Scheme 27.[197] Compound **205** has been converted into alantolactone (**177**) by nine further transformations.[198] The real utility of the Posner-Loomis approach is for the synthesis of the furanoeudesmane attractylon (**207**). Enone **206** has been converted into **207** by a straightforward three-step sequence.[199] Thus, the Posner-Loomis synthesis of

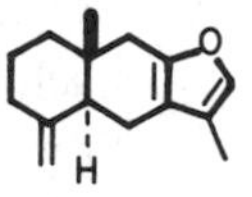

207

207 from **202** requires only eight steps, compared with sixteen by the Minato-Nagasaki route.[199]

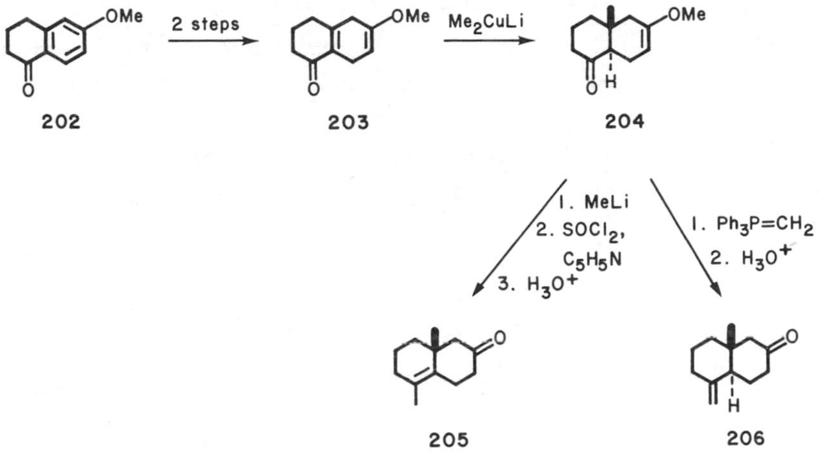

Scheme 27.   Posner-Loomis Approach to Eudesmanes

Miller's synthesis of racemic isotelekin is summarized in Scheme 28.[200] The synthesis is laid out along well-precedented classical lines and proceeds with high stereoselectivity and affords excellent yields at each step. Noteworthy features are the unexpected trans-esterification which occurs in conversion of **214** to **215** and the use of the Mannich method for methylenation of the lactone.

Schultz and Godfrey carried out the interesting synthesis of dihydro-callitrisin which is shown in Scheme 29.[201a,b] Enol ether **217** is condensed with butenolide **218** and the resulting product brominated to obtain **219** (52% overall yield). Reductive removal of the bromine provides the deconjugated diene **220**, which is reduced by sodium cyanoborohydride to obtain the cis-fused lactone **221**. Hydroboration occurs predominantly from the β-face as usual.[202] Oxidation of the secondary alcohol and epimerization of the cis to the trans decalone affords **223**. The remaining carbons are introduced by Wittig methylenation and methylation of **224**. The stereochemistry of the lactonic methyl group is

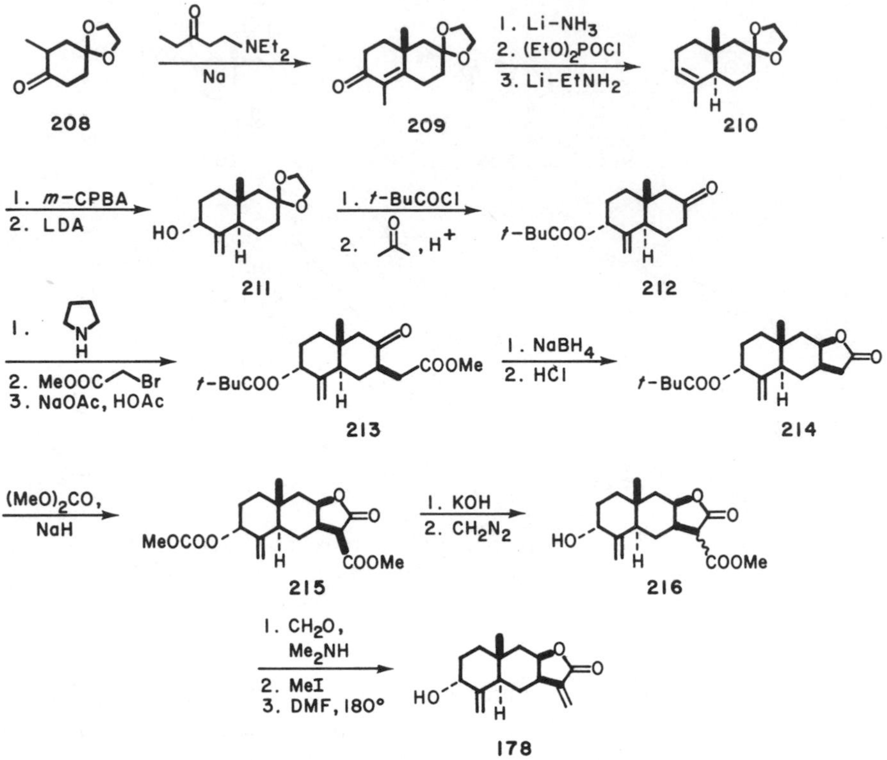

Scheme 28.   Miller-Behare Synthesis of (±)-Isotelekin

established in the decarbomethoxylation step, which is carried out by heating with cyanide ion in DMF. Intermediate **219** has also been converted into (±)-7,8-epialantolactone (**225**), an isomer of the natural product **177**.

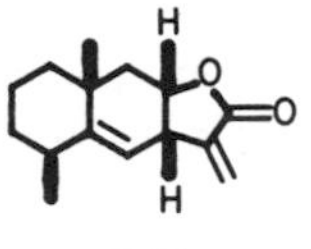

225

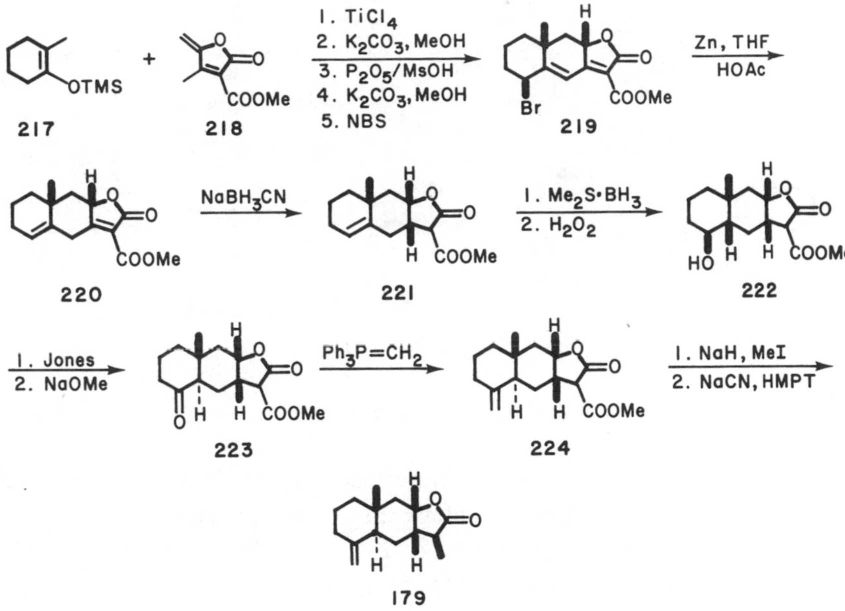

Scheme 29.   Schultz-Godfrey Synthesis of (±)-Dihydrocallitrisin

Frullanolide has been synthesized by two groups. The first synthesis was by Still and Schneider (Scheme 30).[203] Octalone **227** is prepared from 2,6-dimethylcyclohexanone by a route involving acid-catalyzed Michael addition and base-catalyzed aldolization. The two-step process is necessary because other direct annelation procedures fail. Epoxidation occurs from the β-face to give an epoxy ketone which is rearranged by the Wharton procedure to obtain the cis-fused octalol **228**. The elements of the methylenebutyrolactone moiety are introduced by an inventive application of Ireland's enolate Claisen rearrangement. Iodolactonization, followed by elimination of HI, completes this original synthesis.

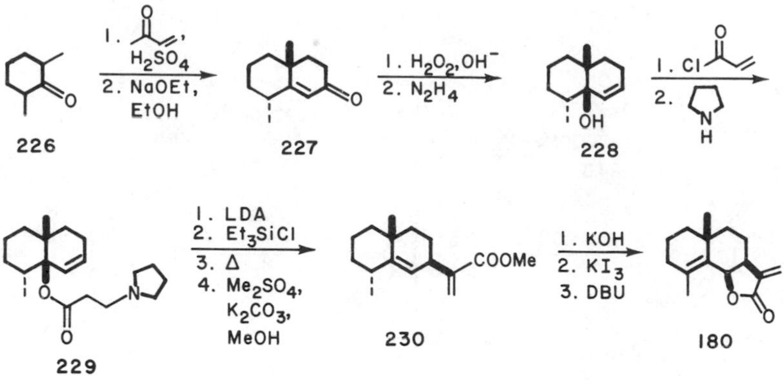

Scheme 30.   Still-Schneider Frullanolide Synthesis

Yoshikoshi and co-workers synthesized frullanolide as shown in Scheme 31.[204] The synthesis employs the novel annelation reagent 232, which is used to convert β-keto ester 231 into adduct 233 (33% yield). The main problem with reagent 232 is that it does not work with simple ketones. Thus, in this case, an extra six steps are required to convert the angular methoxycarbonyl group to a methyl group. The overall yield

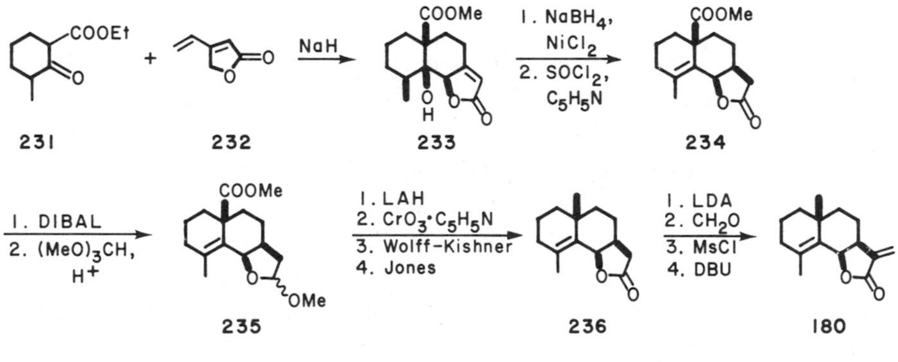

Scheme 31.   Yoshikoshi's Synthesis of Frullanolide

of frullanolide is 2.6% for the 12 steps from **232**. This is to be compared with 8.3% for the 11 steps of the Still synthesis, which begins with commercially available material.

B. Cadinanes

(1) Aromatic Cadinanes

A number of aromatic cadinanes (cadalenes and calamenes) have been synthesized. Compounds **237-241**, constituents of elm wood, have been prepared by Krishna Rao and co-workers.[205,206] Mansonone D (**242**) and mansonone G (**243**), interesting natural o-quinones, have also been prepared by Krishna Rao,[207,208] as has the structurally related furan pyrocurzerenone (**244**).[209] Lacinilene C methyl ether (**245**), a compound which may be responsible for byssinosis, a debilitating disease prevalent in the textile industry, has been prepared by McCormick and co-workers.[210]

237: R = Me   **239**: R = Me
238: R = CHO   **240**: R = CHO   **241**   **242**

243   **244**   **245**

Krishna Rao's synthesis of the phenolic calamenes **237** and **238** is summarized in Scheme 32.[205] A noteworthy feature of this synthesis is cyclization of **247** to **248**, in which it appears that a secondary, rather than a tertiary, carbonium ion is involved. The reason for this unusual behavior cannot be formation of a six- rather than a seven-membered

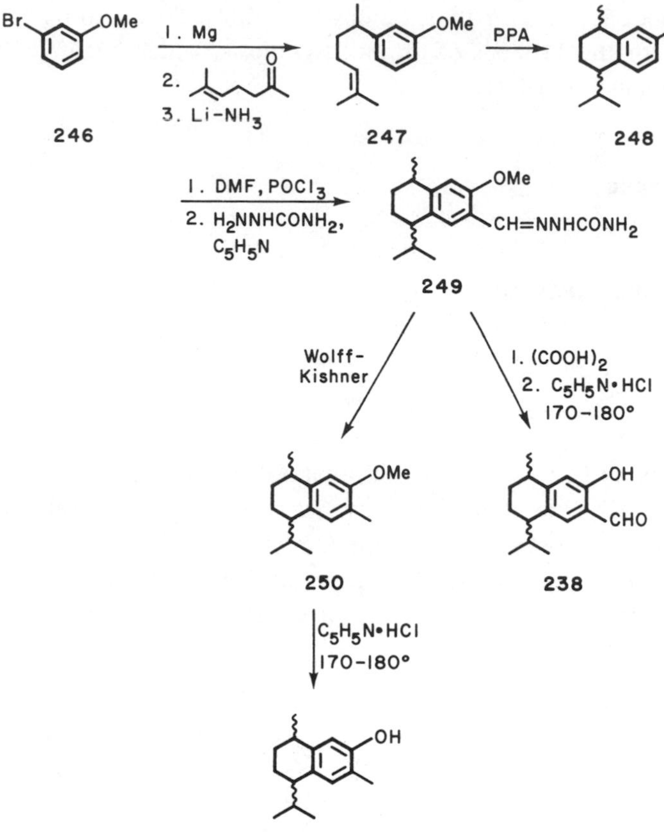

237

Scheme 32. Alexander-Krishna Rao Synthesis of Phenolic Calamenes

ring, since *ipso*-cyclization of the tertiary carbonium ion could lead to the reasonable spirocyclic ion **251**, which would probably rearrange to a mixture of **252** and **253**. Dehydrogenation of **249** is effected by heating with

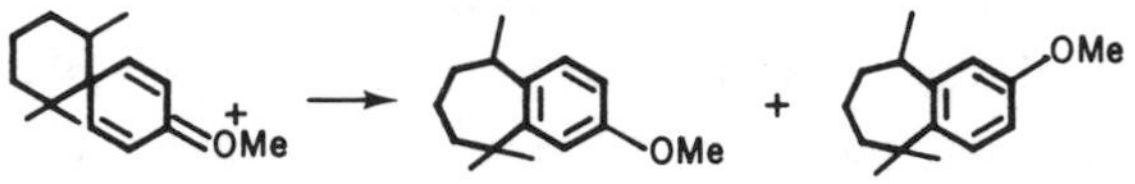

251 252 253

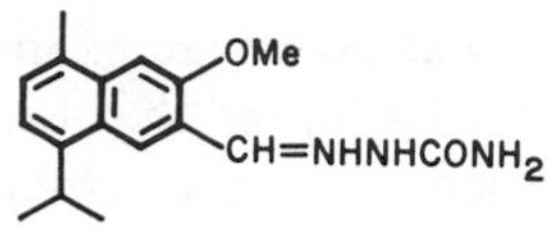

254

2,3-dichloro-5,6-dicyanoquinone to afford the cadalene **254** in 55% yield. This material is converted into cadalenes **239** and **240** by methods similar to those used to convert **249** into **237** and **238**. Cadalene **241** was prepared from the known tetralin **255**[211] as shown in Scheme 33. The conversion of tetralin **255** to β-tetralone **257** is carried out using a method developed by Stork for his early diterpene synthesis work. The most noteworthy step in the synthesis is conversion of ketone **257** to the hydroxynaphthoic acid **258**, which occurs in unspecified yield upon treatment of **257** with dimethyl oxalate and sodium methoxide.

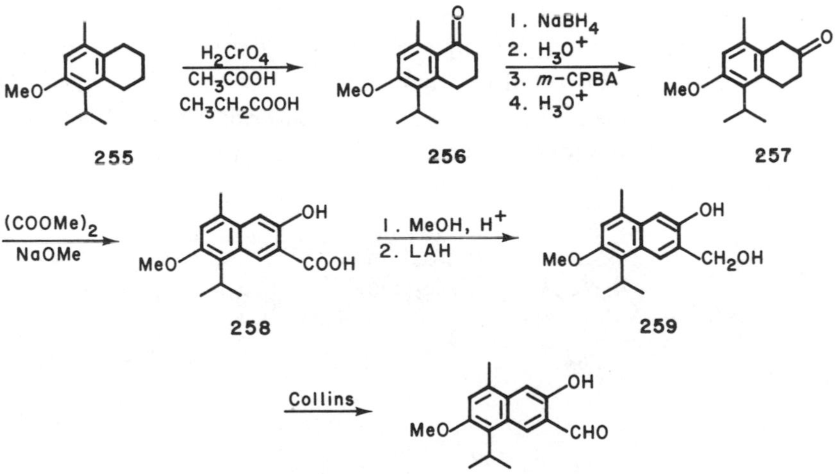

Scheme 33.   Viswanatha-Kirshna Rao Synthesis
of 3-Methoxy-7-hydroxycadalenal

Krishna Rao synthesized pyrocurzerenone (**244**) as shown in Scheme 34.[209] Friedel-Crafts cyclization of the readily available acid **260** occurs with concomitant deisopropylation to provide tetralone **261**. The furan

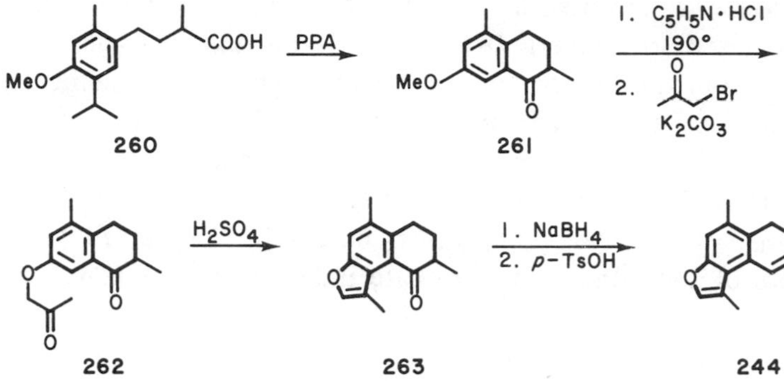

Scheme 34.   Viswanatha-Krishna Rao Synthesis of Pyrocurzerenone

ring is introduced in good yield by cyclization of the aryloxy ketone **262**. Pyrocurzerenone is converted into mansonone D by catalytic hydrogenation, followed by successive oxidations with chromic acid and selenium dioxide.[207] A similar sequence may be used to convert the cadalene **265** into mansonone G.[208]

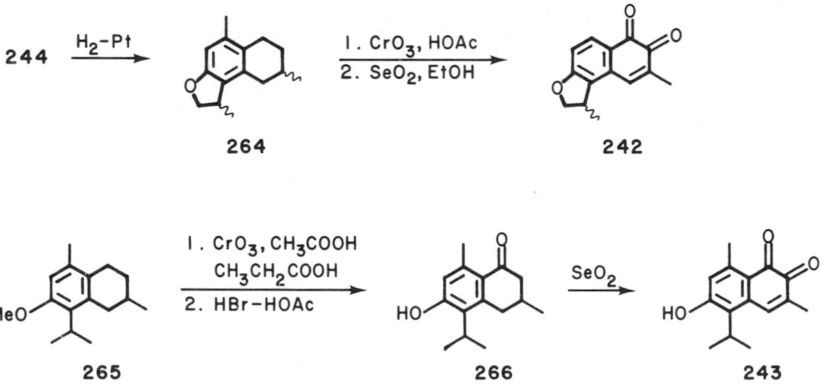

McCormick's synthesis of lacinilene C methyl ether (Scheme 35)[210] begins with keto ester **267**, available in quantity by Friedel-Crafts acylation of *o*-cresol methyl ether. The most noteworthy feature of the synthesis is the method of introducing the unusual functionality of ring A. The three-step sequence **270→245** occurs in 27% overall yield.

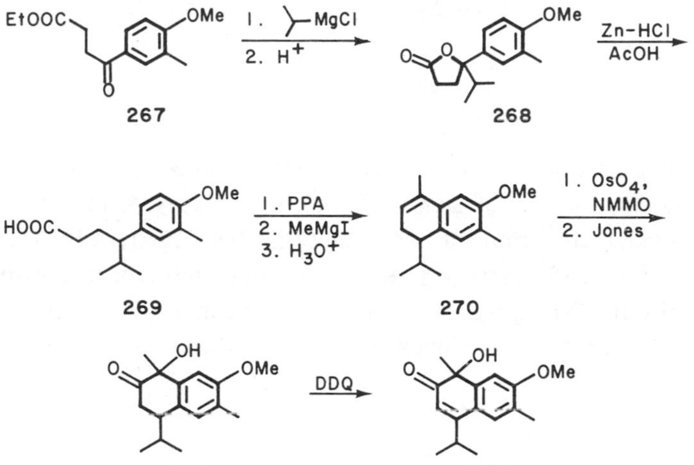

Scheme 35.   McCormick's Synthesis of Lacinilene C Methyl Ether

(2) ε-Cadinene, γ₂-Cadinene, α-Amorphene,
  Zonarene, and Epizonarene

At the time of the writing of Volume 2 in this series, little synthetic work had been directed at the nonaromatic cadinanes. In the intervening 10 years, a considerable amount of research in this area has appeared. Syntheses of the hydrocarbons ε-cadinene (**272**), γ₂-cadinene (**273**), α-amorphene (**274**), zonarene (**275**), and epizonarene (**276**) are discussed in this section.

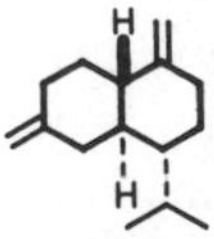

272

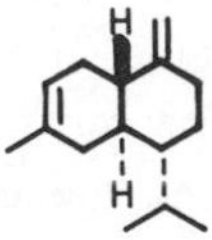

273

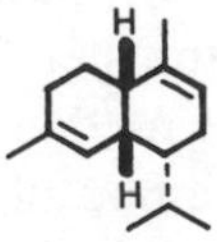

274

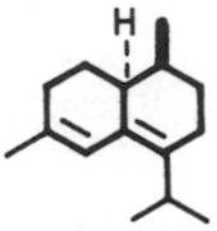

275

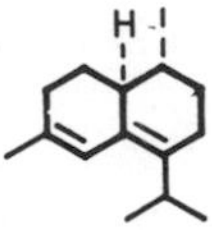

276

Vig's synthesis of ϵ-cadinene is summarized in Scheme 36.[212] The initial Diels-Alder adduct of diene **278** and methyl vinyl ketone has the acetyl group axial, the result of endo addition. Base-catalyzed epimerization affords isomer **280**, in which the acetyl substituent occupies the equatorial position. Wittig methylenation provides diene **281**, which is selectively hydrogenated and hydroborated, oxidized, and hydrolyzed to

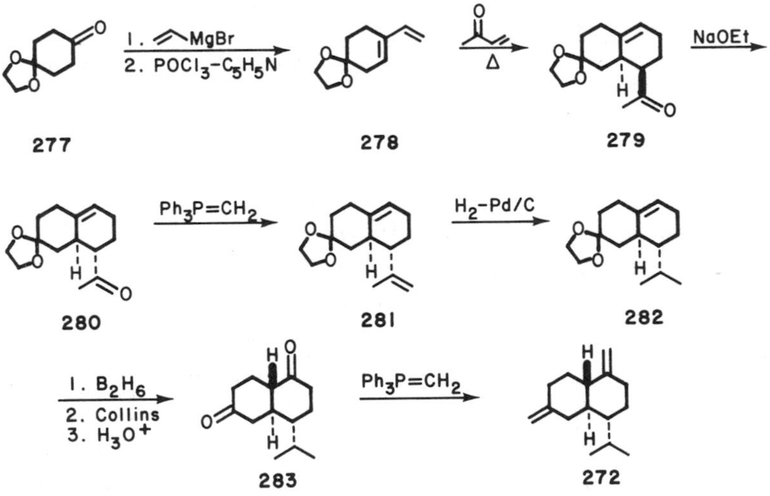

Scheme 36. (±)-ϵ-Cadinene: Vig Synthesis

obtain dione **283**, the racemic version of an intermediate previously employed by Burk and Soffer in a synthesis of $(+)$-ϵ-cadinene.[213] For comparison, the Burk-Soffer synthesis is reproduced in Scheme 37. In addition to providing $(+)$-ϵ-cadinene, a simple modification provides $(-)$-γ_2-cadinene (**273**).

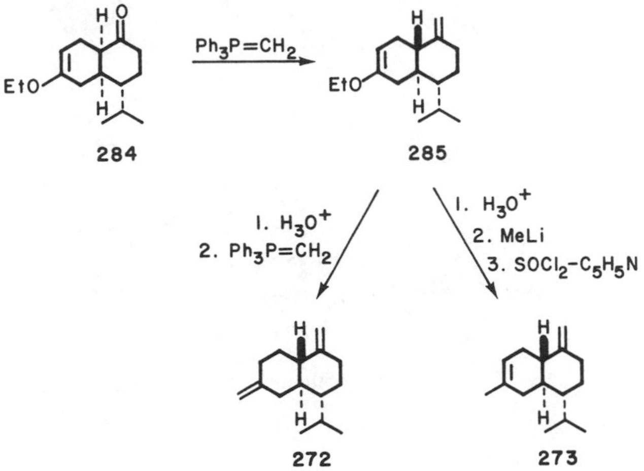

Scheme 37. $(\pm)$-ϵ-Cadinene and $(-)$-γ_2-Cadinene: Burk-Soffer Synthesis

Like the cadinenes just discussed, α-amorphene (**274**) presents the challenge of three stereocenters—two at the ring junction and the third at the isopropyl substituent. Unlike the cadinenes, α-amorphene possesses a cis ring fusion. In principal, α-amorphene could be prepared from the Burk-Soffer intermediate **284**. However, it was found that epimerization of the ring junction position occurs under the conditions of the Wittig reaction, thus leading into the trans-fused cadinene series. Remarkably, Burk and Soffer have also reported that **284** undergoes epimerization *even in its reaction with methyllithium*.[214] Clearly this reaction invites further study.

The only synthesis of an amorphene to date is the Mirrington-Gregson synthesis of α-amorphene, summarized in Scheme 38.[215] Diels-Alder

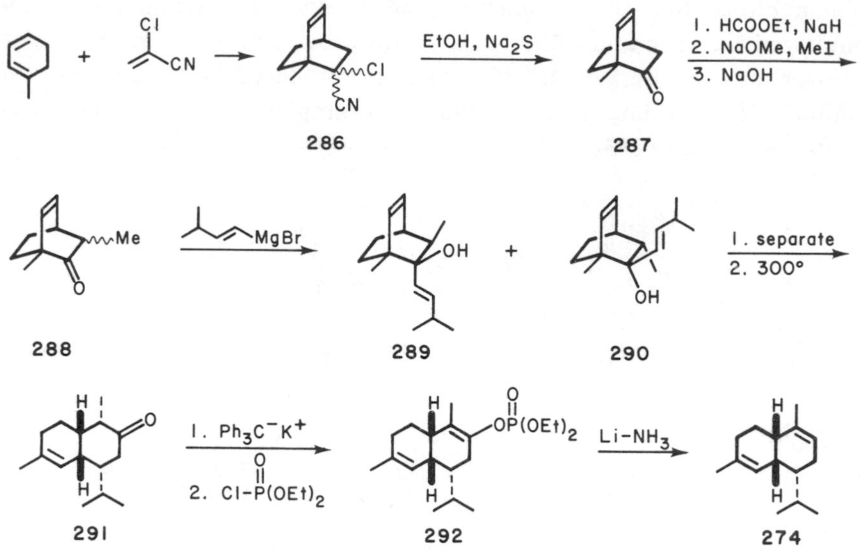

Scheme 38.   (±)-α-Amorphene: Mirrington-Gregson Synthesis

addition of 1-methyl-1,3-cyclohexadiene and α-chloroacrylonitrile provides an adduct (286), which is hydrolyzed by Evans' procedure to bi-
cyclooctenone 287. Methylation, via the formyl ketone, affords an
epimeric mixture (288) which reacts with the Grignard reagent derived
from (E)-1-bromo-3-methyl-1-butene 'to give a mixture of tertiary
alcohols 289 and 290 in a ratio of 3:1. The alcohols are separated and
the minor isomer pyrolyzed at 300°C to obtain the cis-fused octalone
291. Deoxygenation of the more highly substituted enol phosphate provides (±)-α-amorphene. This synthesis provides a very nice example of
the solution of a difficult problem by the judicious application of two
modern methods—the oxy-Cope rearrangement and the Ireland deoxygenation sequence. It is unfortunate that the synthesis is marred by the
lack of stereoselectivity in the methylation step.

Zonarene (275) and epizonarene (276) are simpler targets than the
cadinenes and α-amorphene, since there are only two stereocenters. Of
the two, zonarene is the more challenging target, since the methyl group

is axial. The only synthesis to date is one due to Yamamura (Scheme 39).[216] The synthesis starts with dienone **293**, previously prepared by Yamamura and co-workers by modification of α-santonin (**119**).[216b]

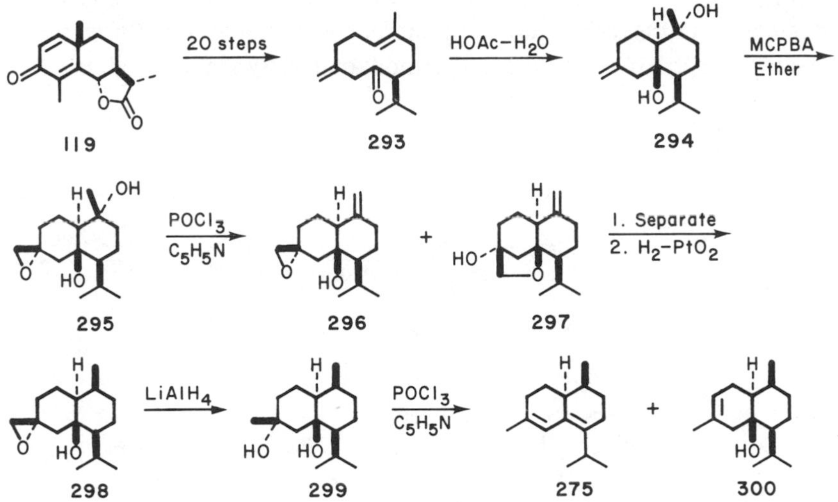

Scheme 39.   Yamamura's Synthesis of (±)-Zonarene

Acid-catalyzed cyclization affords diol **294**, which is epoxidized by *m*-chloroperoxybenzoic acid in ether to provide epoxy diol **295**. Selective dehydration of one of the two tertiary alcohols is achieved by POCl$_3$-pyridine, but some isomerization occurs, leading to a mixture of unsaturated hydroxy ethers **296** and **297** in a ratio of 40:27. The isomers are separated and **296** is hydrogenated to obtain hydroxy ether **298**, which is reduced to diol **299**. Dehydration of this substance also gives rise to a mixture—(±)-zonarene (**275**) being accompanied by unsaturated alcohol **300**. Although Yamamura's synthesis will be recorded as the first successful synthesis of zonarene, it leaves much room for improvement. The synthesis requires over 25 steps *from santonin* and requires separation of isomers at several stages.

A short synthesis of (+)-epizonarene (276) from (−)-menthone (301) is summarized in Scheme 40.[217] A noteworthy feature of this synthesis is the Mannich alkylation of 301, which occurs at the methylene, rather than the methine position. Annelation of ring A is accomplished by reaction of enone 302 with acetoacetic ester. Enone 303 is produced under equilibrating conditions; the stereochemistry at the three stereocenters corresponds to the more thermodynamically stable product.

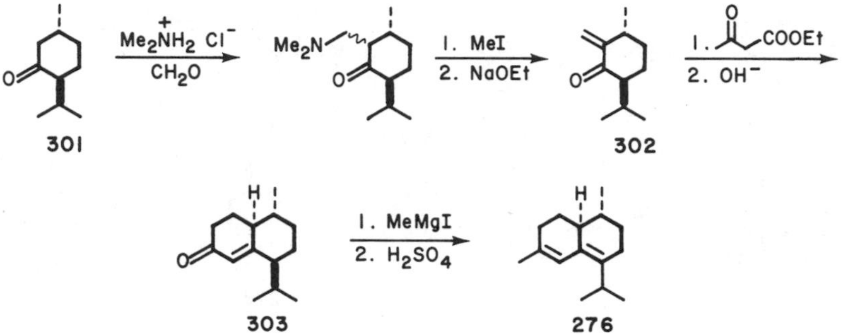

Scheme 40.   Kulkarni's Synthesis of (+)-Epizonarene

(3) α -Cadinol and Torreyol

α-Cadinol (304a) and torreyol (304b) present the additional synthetic challenge of the tertiary alcohol group—a fourth stereocenter compared to the cadinenes and α-amorphene.

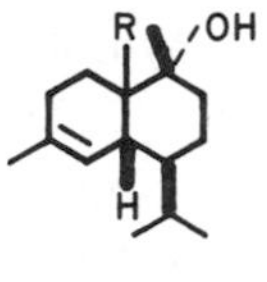

304

a: R = α−H
b: R = β−H

A synthesis of (±)-α-cadinol, by Caine and Frobese, is outlined in Scheme 41.[218] Enone **306**, obtained as a 3:2 mixture of stereoisomers, is dehydrogenated to trienone **307** which is reduced by transfer hydrogenation to stereochemically pure **306** in which the angular methyl and

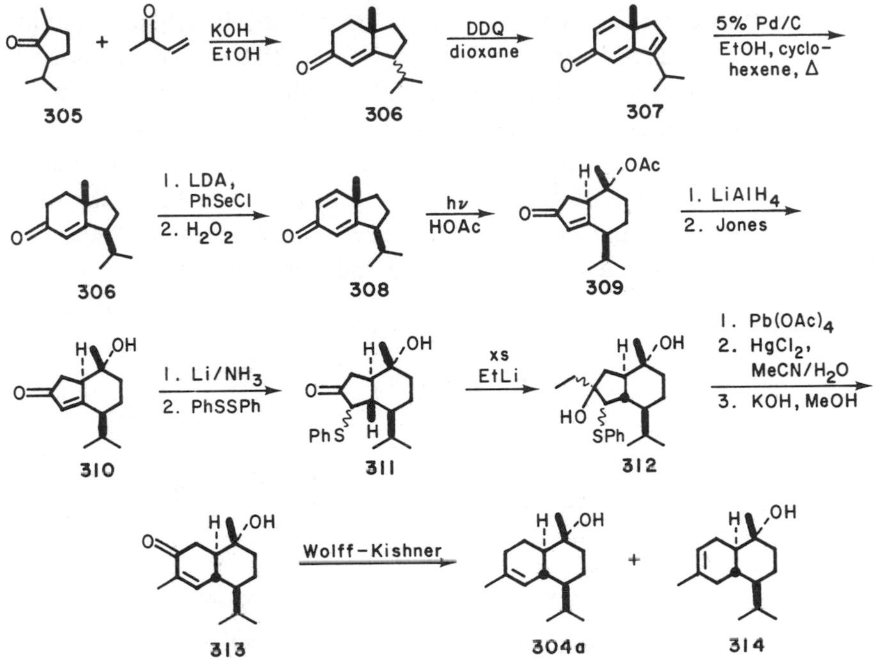

Scheme 41.   Caine-Frobese Synthesis of (±)-α-Cadinol

isopropyl groups are cis. This compound is converted to the cross-conjugated cyclohexadienone **308**. Photochemical rearrangement of the latter provides acetoxy enone **309**, which is transformed into hydroxy enone **310**. Lithium ammonia reduction of **310**, followed by sulfenylation of the resulting enolate, gives a 3:1 mixture of epimeric sulfides **311** in 76% yield. The mixture is treated with excess ethyllithium and the resulting mixture of diols (**312**) is cleaved by lead tetraacetate. After hydrolysis, aldol condensation and dehydration gives 3-oxo-α-cadinol (313). Wolff-Kishner reduction of this compound gives (±)-α-cadinol (304a) and its isomer 314 in a ratio of 2:1.

Taber and Gunn reported the synthesis of torreyol which is summarized in Scheme 42.[219] Aldehyde 314, prepared by alkylation of the piperidine enamine of isovaleraldehyde with ethyl acrylate, is condensed with the Wittig reagent derived from methylallyltriphenylphosphonium chloride to give exclusively the E-stereoisomer 315. Ester 315 is converted to an aldehyde, which reacts with vinylmagnesium bromide to give a diastereomeric mixture of trienols (316). Oxidation of this material at 0 °C gives trienone 317, which spontaneously cyclizes to a 9:1 mixture of diastereomeric octalones 318 and 319. Treatment of the mixture with methyllithium provides a mixture of (±)-torreyol (304b, 64% based on the 318 in the starting material) and stereoisomers.

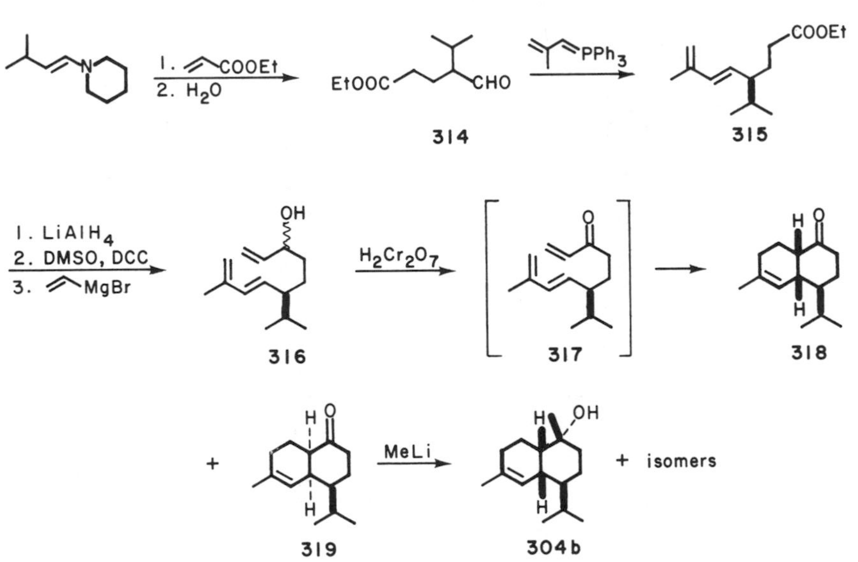

Scheme 42.   Taber-Gunn Synthesis of (±)-Torreyol

C. Drimanes

(1) Driman-8-ol, Driman-8,11-diol, and Drim-8-en-7-one

Driman-8-ol (**320**), driman-8,11-diol (**321**), and drim-8-en-7-one (**322**) are constituents of Greek tobacco. Enzell and co-workers have prepared

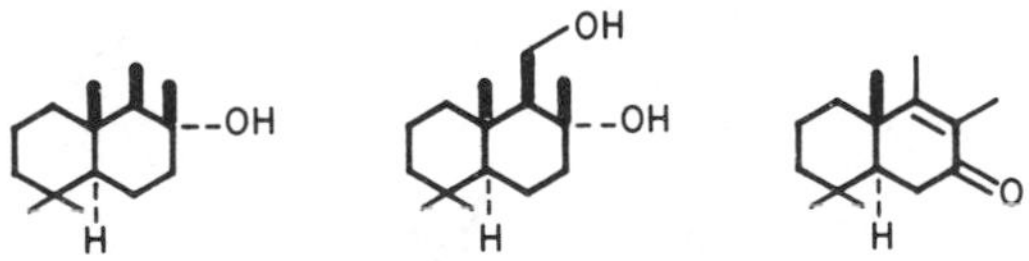

all three of these compounds[220,221] from drim-7-en-11-ol (**323**, bicy-clofarnesol), which has been prepared previously.[222] Scheme 43 outlines the synthesis of alcohol **320** and diol **321**. Bicyclofarnesyl acetate under-goes stereoselective epoxidation from its α face and epoxy acetate **324** reacts with lithium aluminum hydride to give diols **321** and **325** as a 1:9 mixture. Deoxygenation of the minor isomer is accomplished by reduc-tion of its monomesylate to obtain alcohol **320**. For synthesis of enone

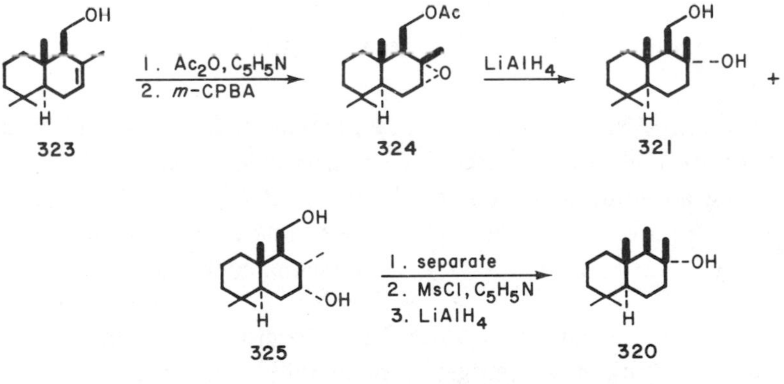

Scheme 43.   Enzell's Synthesis of Driman-8-ol and Driman-8,11-diol

322 (Scheme 44) epoxy acetate 324 is hydrolyzed and the resulting alcohol oxidized using Brown's two-phase conditions to obtain β,γ-epoxy aldehyde 327. Base-catalyzed elimination, followed by reduction of the

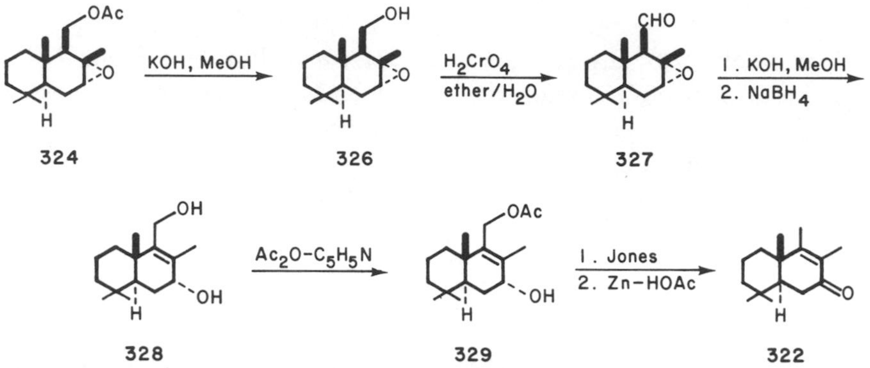

Scheme 44.   Enzell's Synthesis of Drim-8-en-7-one

aldehyde function affords diol 328. Careful acetylation yields a mixture of products, of which monoacetate 329 is the major product. Oxidation and reductive removal of the acetoxy group provides drim-8-en-7-one (322).

(2) Confertifolin, Isodrimenin, Cinnamolide, Drimenin, Futronolide, Polygodial, Isotadeonal, and Warburganal

Confertifolin (330), isodrimenin (331), cinnamolide (332), and drimenin (333) are simple drimanic lactones. Futronolide (334) is related to 331 by having an additional 7α hydroxy group. Polygodial (335) and isotadeonal (336) are dialdehydes having the drimane and epidrimane stereochemistry. Warburganal (337) is an oxygenated relative of polygodial. The discovery of warburganal, which has potent antifeedant activity against the African army worm, KB cytotoxicity, and antimicrobial properties, has sparked a resurgence of activity in synthesis of drimanic sesquiterpenes.

330   331   332   333

334   335   336   337

Because of its importance in several of the syntheses discussed in the following pages, we reiterate a basic reaction in the drimane field—the acid-catalyzed cyclization of farnesic acid to give the bicyclofarnesic acids. This reaction, which was examined extensively by Schinz, Stork, and Eschenmoser[223] produces predominantly drimanic acid (338), with 8-epi-drimanic acid (339) being a minor product; typical yields are 35% and 5%, respectively.

338   339

The first syntheses of confertifolin and isodrimenin were carried out by Wenkert.[224] In connection with his project to synthesize warburganal, Oishi and his co-workers developed the synthesis outlined in Scheme 45.[225] The synthesis begins with dihydro-β-ionone, which is converted by successive treatment with methyl orthoformate and pyridinium bromide in methanol into the keto acetal 340. Compound 340 is subjected to a Darzens condensation and the intermediate epoxide is treated with acid to obtain a furan ester (342). Friedel-Crafts cyclization of this substance provides a tricyclic ester, which is saponified to 343. It is noteworthy that the cyclization is stereospecific, affording only the trans-fused isomer. After decarboxylation, furan 344 is oxidized by lead

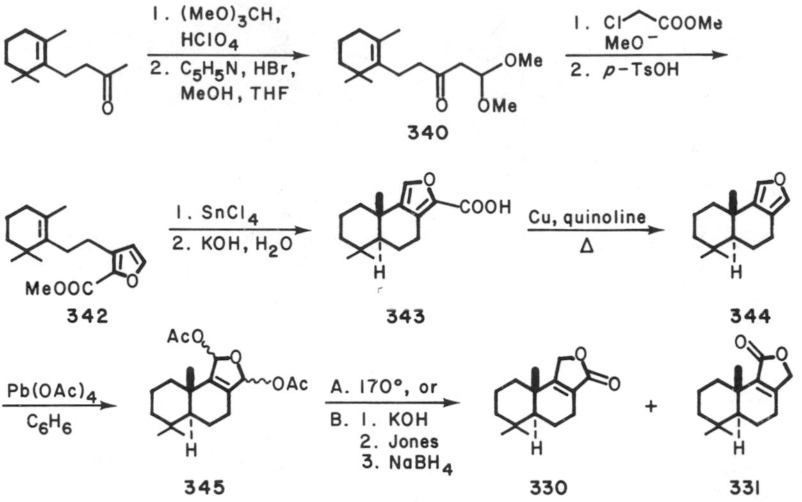

Scheme 45.   (±)-Confertifolin and (±)-Isodrimenin: Oishi Synthesis

tetraacetate to obtain a crystalline diacetate (**345**, undetermined stereo-chemistry). Thermolysis of this compound yields (±)-confertifolin (**330**, 74%) and (±)-isodrimenin (**331**, 17%). To obtain isodrimenin as the major product, the diacetate is first saponified and the resulting diol oxidized to (±)-winterin (**346**). Reduction of (±)-winterin by sodium

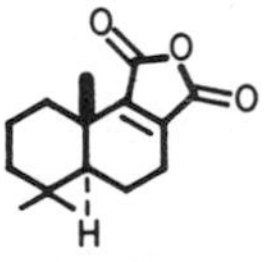

346

borohydride in THF produces (±)-isodrimenin (81%) along with 14% of (±)-confertifolin. The Oishi synthesis is reasonably short (10-12 steps from β-ionone) and produces **330** and **331** in reasonable yields (about 13% overall). Furthermore, the reactions employed appear to be adapt-able to large scale applications.

The group headed by Kitahara at Sendai has developed syntheses of (±)-drimenin, (±)-cinnamolide, and (±)-polygodial. Although part of that work was reported in Volume 2 of this series,[226] the full synthesis is repeated in Scheme 46, since the applications to (±)-cinnamolide and (±)-polygodial have appeared since 1970.[227,228] The Sendai group begins with cyclization of monocyclofarnesic acid to give mainly drimanic acid, which is esterified to obtain methyl cyclofarnesate (347). Oxidation of 347 with singlet oxygen produces allylic alcohol 348 in 33% yield. Treatment of 348 with aqueous sulfuric acid provides (±)-drimenin (333) in a direct four-step route from 346 (15% overall yield). Reduction of 333 gives a diol which is smoothly oxidized by manganese diox-

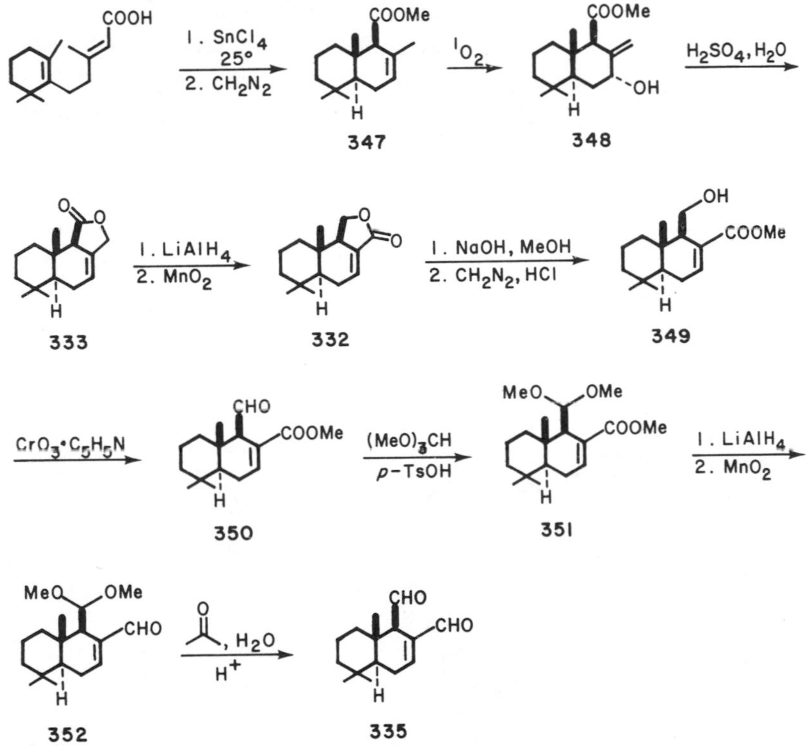

Scheme 46.   (±)-Drimenin, (±)-Cinnamolide, and (±)-Polygodial: Sendai Syntheses

ide to (±)-cinnamolide (**332**). Saponification of the latter compound gives a sodium salt which is treated with ethereal diazomethane and then slowly acidified with ethereal HCl to obtain hydroxy ester **349** admixed with the starting lactone **332**. The mixture is oxidized and the resulting aldehyde converted into an acetal (**351**). At this stage **351** is separated from unreacted lactone **332** by chromatography. Crystalline **351** is obtained in 30% overall yield based on (±)-cinnamolide. Reduction of the ester function, followed by manganese dioxide oxidation provides the monoacetal **352**, which is hydrolyzed with oxalic acid in aqueous acetone to obtain (±)-polygodial (**335**).

The Sendai group has also used methyl bicyclofarnesate in a synthesis of futronolide (**334**), as summarized in Scheme 47.[229] The singlet oxygen oxidation of **347** is followed by isomerization of the double bond and oxidation of the secondary alcohol to obtain unsaturated keto ester **353**. Ketalization, allylic bromination, and hydrolysis provide keto lactone **355**. Reduction of the ketone affords entirely the equatorial alcohol

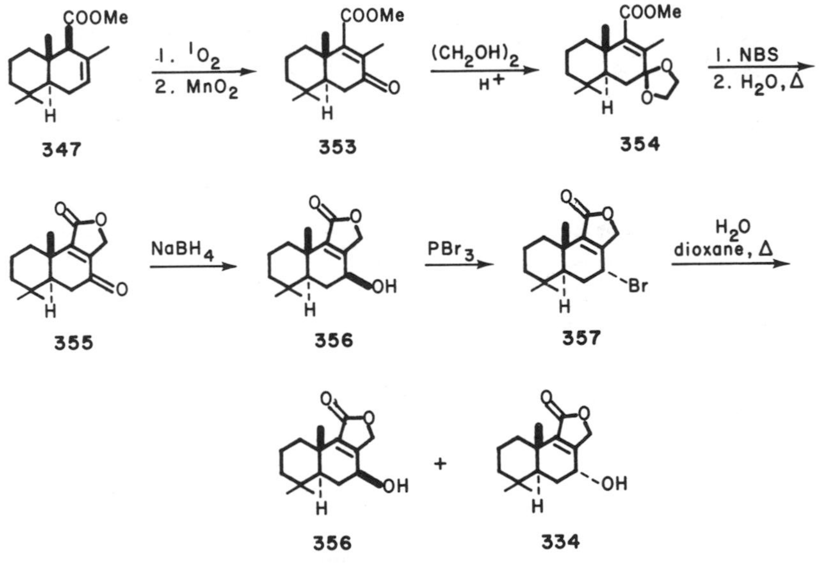

Scheme 47.   (±)-Futronolide: Kitahara's Synthesis

356, which is epimeric with (±)-futronolide at C-7. However, the crystalline bromide **357** undergoes solvolysis in aqueous dioxane to provide a 1:1 mixture of **356** and (±)-futronolide (**334**), which may be separated by repeated crystallization.

The first successful synthesis of warburganal to be published was that of Ohsuka and Matsukawa of Osaka City University.[230] The synthesis (Scheme 48) begins with methyl 9-epibicyclofarnesate (**358**). It is important to recall that cyclization of either farnesic acid (or its esters) or monocyclofarnesic acid (or its esters) affords this stereoisomeric bicyclic acid as the minor product, usually in yields on the order of 5%. However, the first step of the Ohsuka-Matsukawa synthesis, allylic oxidation with selenium dioxide, does not work with methyl bicyclofarnesate. Thus, the synthesis must be carried out starting with the less abundant epimer. Reduction of acetal ester **359**, followed by Collins oxidation, provides an acetal (**360**) which is hydrolyzed by aqueous acid to obtain (±)-isotadeonal (**336**). Alternatively, the enolate derived from **360** is

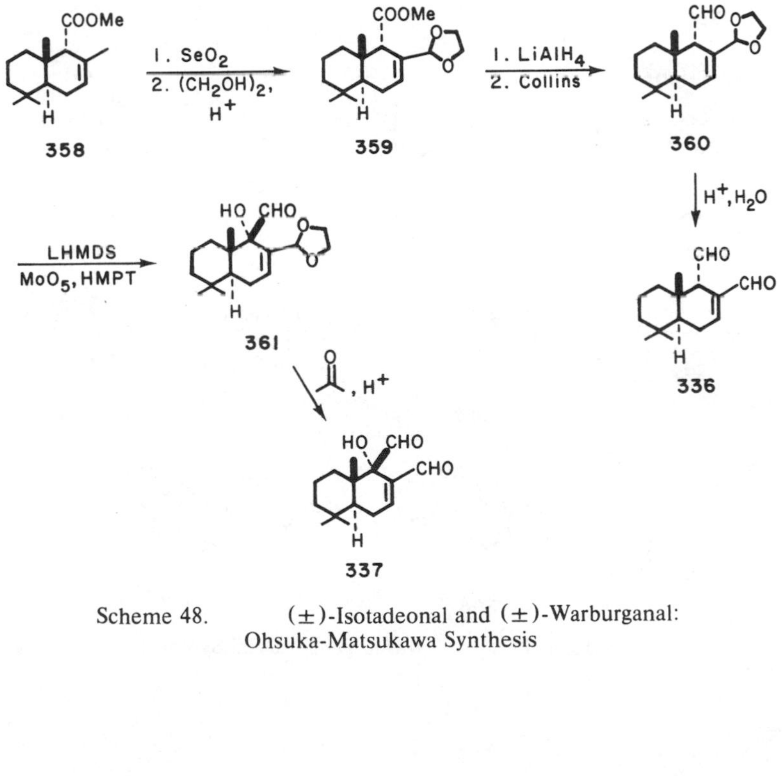

Scheme 48.   (±)-Isotadeonal and (±)-Warburganal:
Ohsuka-Matsukawa Synthesis

oxidized with MoO_5 to obtain the hydroxylated aldehyde **361** in 24% yield. The oxidation is stereospecific, the bulky oxidant apparently being directed to the α-face of the molecule by the angular methyl group. Hydrolysis of **361** provides ($\pm$)-warburganal (**337**). Although this synthesis is short (six steps) and produces warburganal in reasonable yield (9% overall), it is marred by the scarcity of the starting material. However, if the problems associated with the allylic oxidation of the more abundant **347** could be solved, this approach to warburganal would be very attractive, especially in light of Nakanishi's experience with the hydroxylation step (see Scheme 50 below).

Oishi and co-workers have synthesized ($\pm$)-warburganal beginning with ($\pm$)-isodrimenin (**331**) as shown in Scheme 49.[231] Allylic oxidation

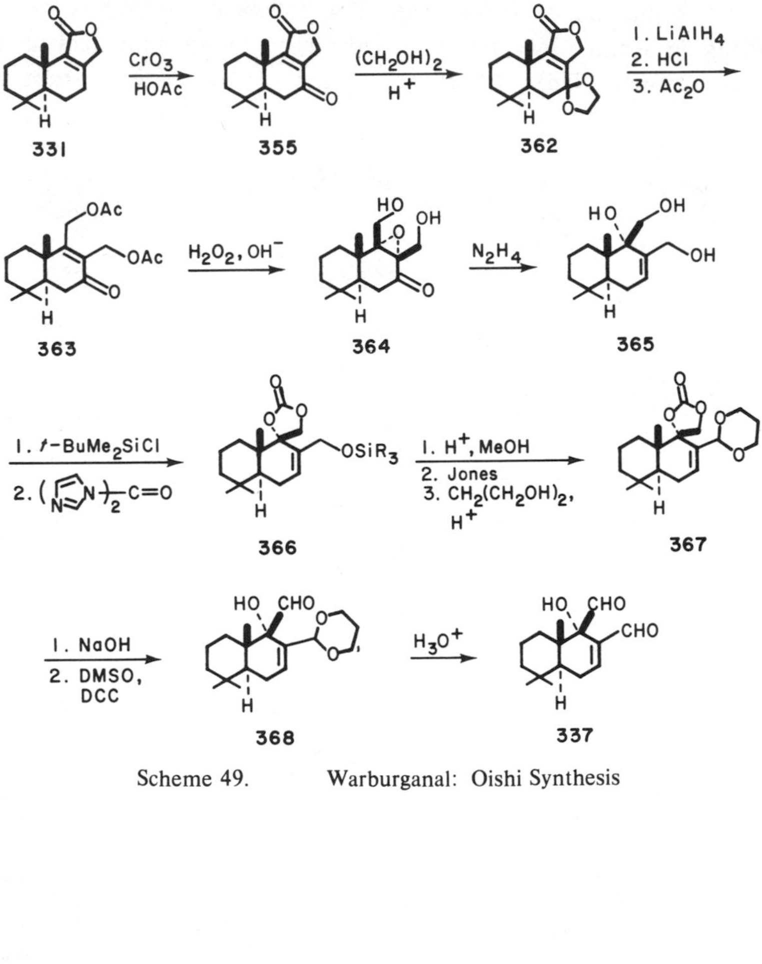

Scheme 49.   Warburganal: Oishi Synthesis

affords **355**, which had been prepared earlier by the Kitahara group (see Scheme 47). Intermediate **355** is converted by several standard steps into enone **363**, which is epoxidized using alkaline hydrogen peroxide to obtain epoxy ketone **364**. Wolff-Kishner reduction occurs with concomitant elimination to afford unsaturated triol **365**, which only needs to have its two primary alcohol functions oxidized to obtain warburganal. Unfortunately, such a simple solution does not succeed, and it is necessary to put triol **365** through an intricate series of protection, deprotection, and oxidation steps to achieve the desired goal. In any event, the eight-step sequence works, and ($\pm$)-warburganal (**337**) is produced in 38% overall yield from epoxide **364**. The major loss seems to come in the silylation of **365**, which is not completely selective. In all, the Oishi synthesis requires 25 steps from dihydro-β-ionone (see Scheme 44) and produces ($\pm$)-warburganal in an overall yield of about 2.6%.

Nakanishi's synthesis of warburganal (Scheme 50)[232] begins with a Diels-Alder reaction between diene **369** and dimethylacetylenedicarboxylate, a reaction employed earlier by Brieger in a synthesis of ($\pm$)-winterin.[233] Allylic oxidation of the adduct **370**, obtained in 83% yield, affords dienone **371**. Catalytic hydrogenation of **371** provides the trans-fused decalone **372**. The overall yield of **372** is 76% from diene **369**. The keto function is removed and the necessary C_7-C_8 double bond is introduced by a straightforward method and the resulting diester **373** is reduced to obtain diol **374** (which may be oxidized to polygodial, **335**). Again, an intricate sequence of protection, deprotection, and oxidation steps leads to the monoprotected polygodial (**377**), which is hydroxylated by the Vedejs procedure to obtain the axial alcohol. Hydrolysis of the acetal provides ($\pm$)-warburganal in 85% yield for the last two steps. It is not clear why the Nakanishi hydroxylation is so superior to that reported by Ohsuka and Matsukawa (see Scheme 48), which proceeds in only 24%. Different acetals and different bases were employed by the two groups. Overall, the Nakanishi synthesis provides ($\pm$)-warburganal in 16 steps (15.7% yield) from diene **369**. Of the three syntheses of this interesting compound, this is clearly the most efficient at this stage. However, note that application of Nakanishi's hydroxylation conditions in the Ohsuka-Matsukawa route would raise the overall yield of that synthesis to 32%.

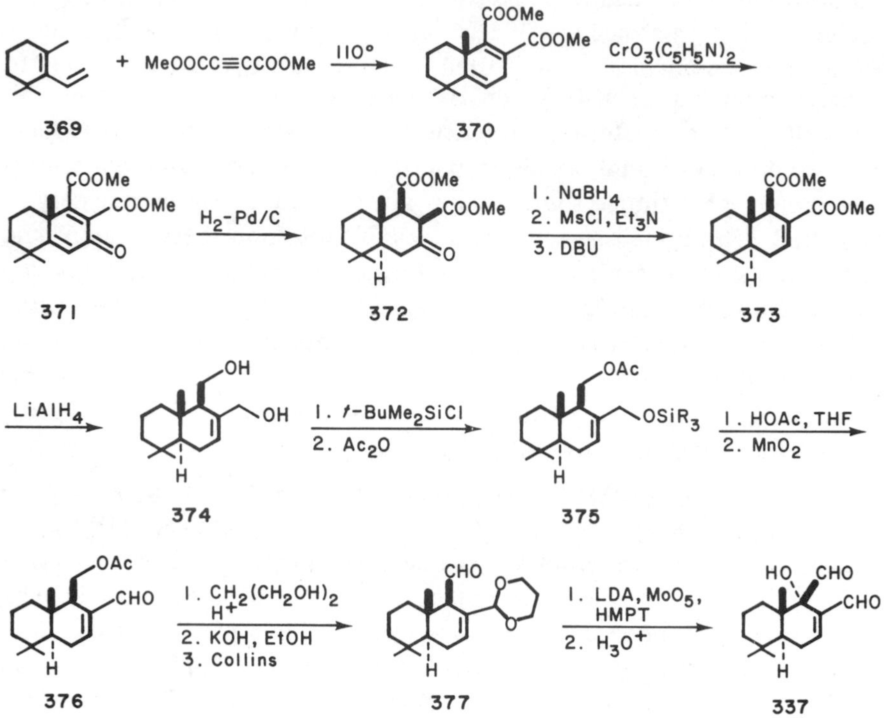

Scheme 50.   (±)-Warburganal: Nakanishi's Synthesis

(3) Pallescensin-A

Pallescensin A (**379**) does not fit into any of the normal sesquiterpene skeletal classes. It has been suggested that it may arise from the acyclic precursor **378**.[234] Because of its structural similarity to the drimane class, and because the key reaction employed for its synthesis is one of the characteristic reactions leading into the drimane family, we consider pallescensin-A at this point. Nasipuri and Das constructed the sesquiterpene in a straightforward manner[234] by coupling geranyl chloride (**380**) with 3-furylmethylmagnesium chloride to obtain diene **378** (Scheme 51). Treatment of this material with boron trifluoride affords (±)-palles-

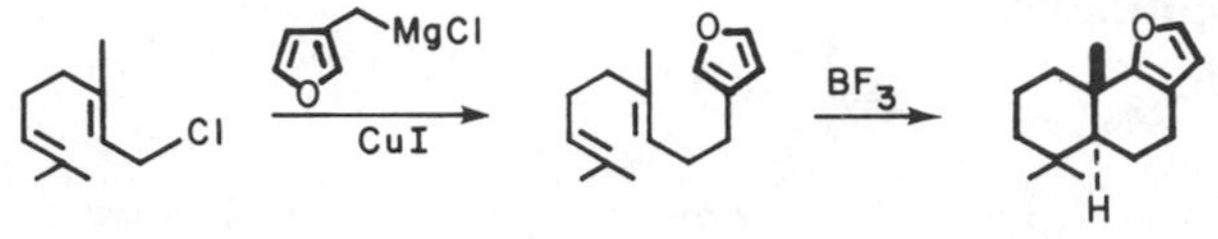

Scheme 51.   Nasipuri-Das Synthesis of (±)-Pallescensin A

censin-A in 84% yield. The cyclization step is stereospecific; only the trans-fused diastereomer is produced.

A related cyclization was employed by Matsumoto and Usui in a synthesis of (+)-pallescensin-A (Scheme 52).[235] The cyclohexene derivative 383 is prepared in a 2-step sequence beginning with R-(−)-α-cyclocitral (381). Cyclization of 383 is effected by aluminum chloride. Unfortunately, (+)-pallescensin-A is produced as the minor component of a 2:1 mixture, in 10% yield.

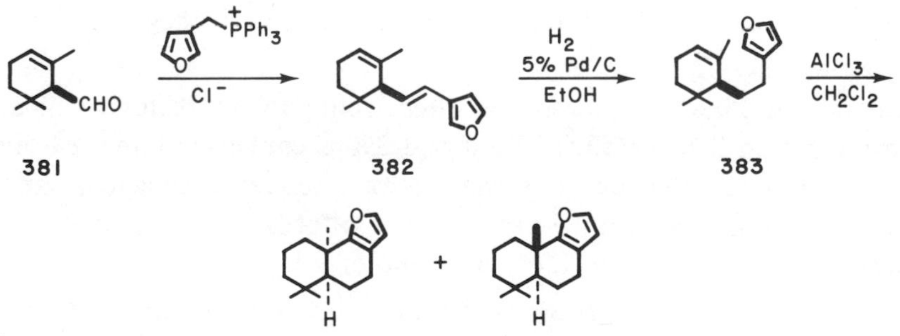

Scheme 52.   Matsumoto-Usui Synthesis of (±)-Pallescensin A

D. Eremophilanes

(1) Valencene, Nootkatone, 7-Epinootkatone, Isonootkatone,
    and Dihydronootkatone

The eremophilane family of sesquiterpenes has received a great deal of attention during the past 10 years. In this section, we review the work on valencene (385), nootkatone (386), 7-epinootkatone (387), and isonootkatone (388). In addition, we describe a synthesis of dihydronootkatone (389). Although this substance is not a natural product, its synthesis is styled after the probable biogenesis of the eremophilanes.

Full papers reporting Marshall's syntheses of (±)-nootkatone (386)[236] and (±)-isonootkatone (388)[237] have appeared. Since these syntheses have been reported in Volume 2 of this series,[238] they will not be repeated.

Odom and Pinder explored the direct route to nootkatone which is summarized in Scheme 53.[239] Keto acid 390 is converted into (±)-dihydrosylvecarvone (394) by the straightforward sequence depicted. Annelation of ketone 394 with 3-penten-2-one affords (±)-7-epinootkatone (387) and (±)-nootkatone (386) in a ratio of 9:1.

Based on earlier observations by Piers,[240] Takagi and co-workers improved the ratio of nootkatone and 7-epinootkatone to 1:1 by converting ketone 394 into the 6-n-butylthiomethylene derivative before Robinson annelation (Scheme 54).[241] Actually, at the time of Takaga's paper,

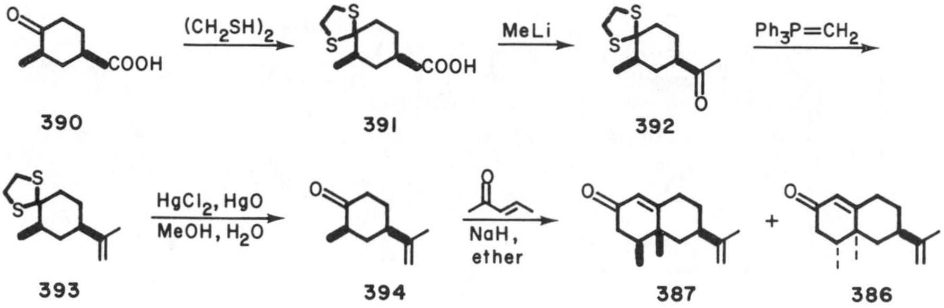

Scheme 53.   Odom-Pinder Synthesis of (±)-7-Epinootkatone

and (±)-Nootkatone

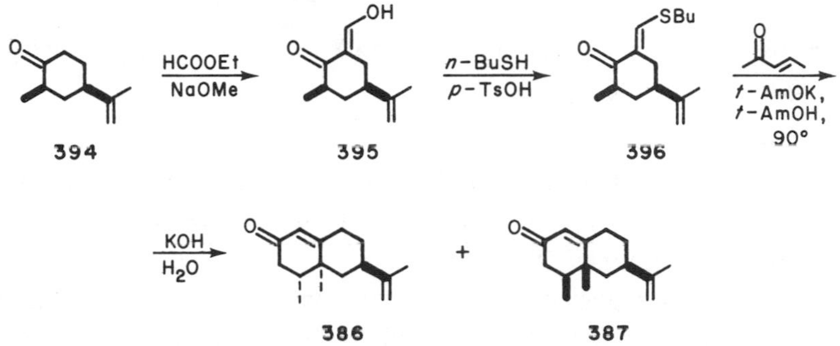

Scheme 54.   Takagi's Synthesis of (+)-Nootkatone

Pinder[242] had already provided a superior solution to the problem of stereochemistry at C_7. The problem was solved (Scheme 55) by simply delaying introduction of the C_7 side chain until after the octalin system has already been formed, so that the desired equatorial disposition of the C_7 substituent can be established by equilibration. The required starting material is monoketal **397**, which is transformed into **398**. The carbonyl group is removed by standard steps, affording ketal **399**. After hydrolysis, the acetonyl group is introduced by a modified Wadsworth-

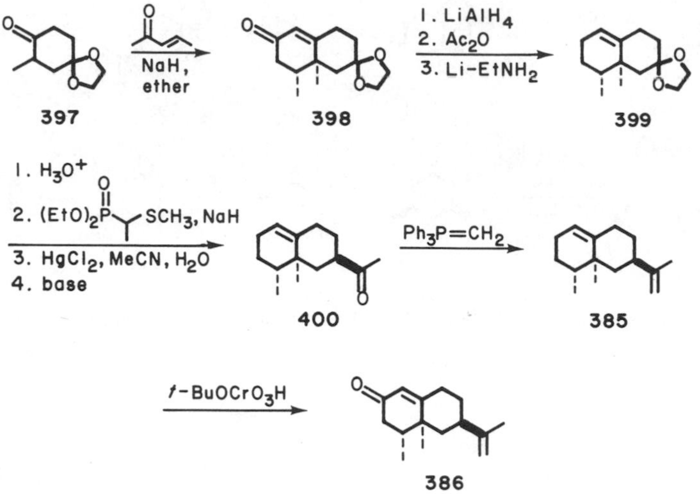

Scheme 55.   Pinder's Synthesis of (±)-Valencene and (±)-Nootkatone

Emmons reaction and ketone **400** is then subjected to Wittig methylenation to obtain (±)-valencene. Allylic oxidation restores the enone function, giving (±)-nootkatone (**386**).

In 1973, Dastur reported the imaginative synthesis of (±)-nootkatone which is outlined in Scheme 56.[243] Birch reduction of 3,4,5-trimethylanisole affords diene **401**, which is heated with methyl acrylate in the presence of dichloromaleic anhydride to provide adduct **402**. Although products arising both from exo and endo transition states are formed, addition appears to occur only trans to the secondary methyl in the active diene, **403**. Thus, although the product is a mixture of epimers at the

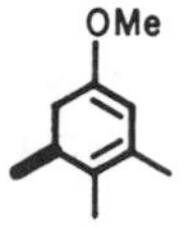

403

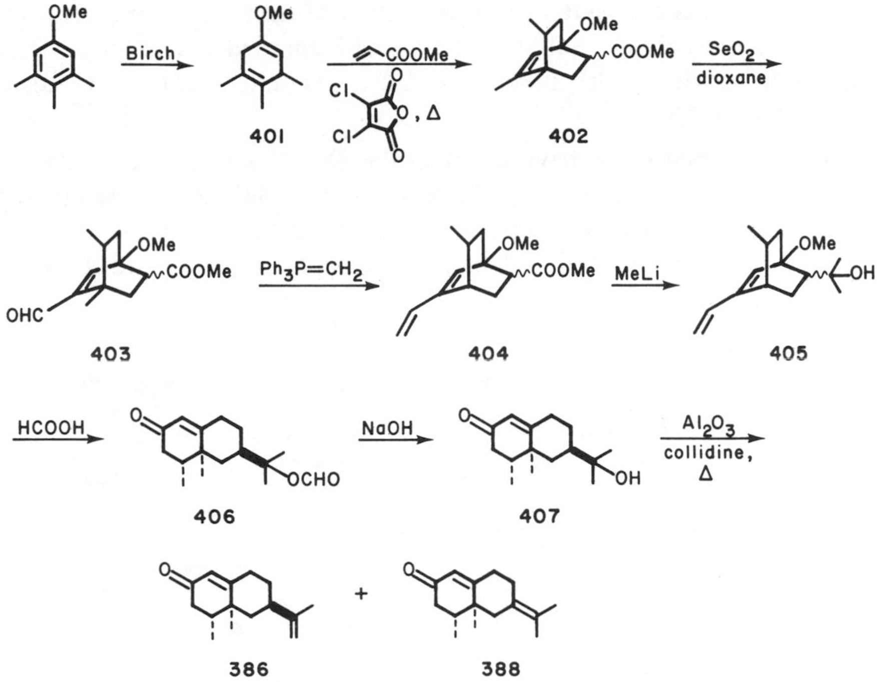

Scheme 56.   Dastur's Synthesis of (±)-Nootkatone

methoxycarbonyl position, the secondary methyl group is completely syn to the double bond. This is important, as this relative stereochemistry governs the relative stereochemistry of the vicinal methyl groups. The allylic methyl is functionalized by selenium dioxide oxidation and the resulting aldehyde is methylenated to obtain diene **404**. After addition of methyllithium, tertiary alcohol **405** is solvolyzed in formic acid to obtain formate **406**. The conversion of **405** may be formulated as:

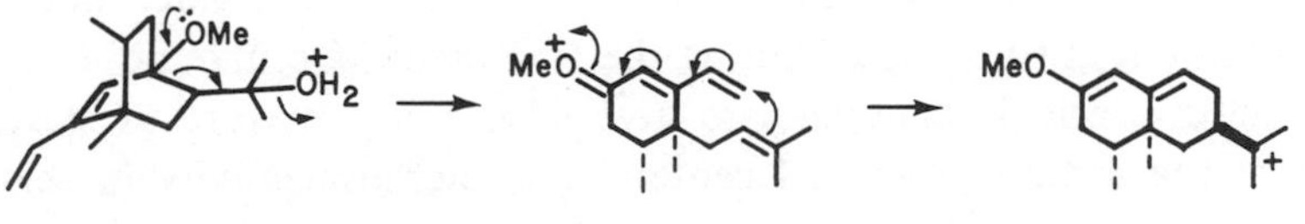

The synthesis is completed by saponification of the formate and dehydration of the tertiary alcohol by heating it with alumina and collidine. The final step provides a 3:1 mixture of (±)-nootkatone (386) and (±)-iso-nootkatone (388).

Heathcock and Clark have modified the Dastur synthesis as shown in Scheme 57.[244] The enol ether of dihydroorcinol (408) is alkylated by the Stork-Danheiser procedure first with methyl iodide and then with prenyl bromide. The product is a 4:1 mixture of diastereomers of which 409 is

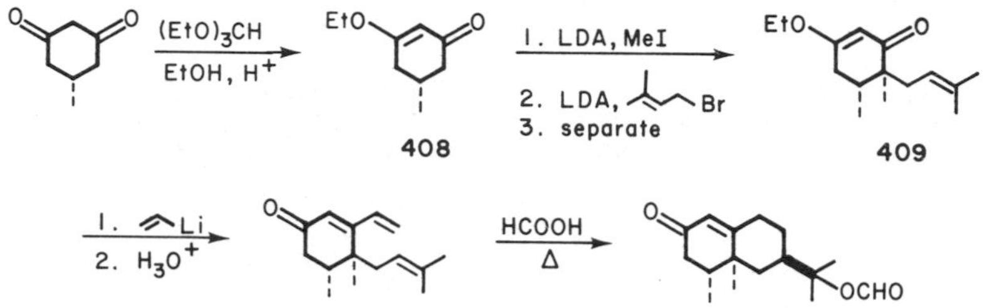

Scheme 57.   Heathcock-Clark Modification
of the Dastur Nootkatone Synthesis

the major isomer. This isomer is separated chromatographically and treated with vinyllithium to obtain, after hydrolysis, trienone 410. Formolysis of 410 affords the Dastur enone formate (406).

Hiyama and co-workers have developed an interesting stereospecific synthesis of (±)-nootkatone from keto ester 412, which they prepare by Diels-Alder addition of methyl acrylate to diene 411 (Scheme 58).[245] This material is condensed with the dilithium salt of 3-butyn-2-ol to obtain 413 as a diastereomeric mixture. The methyl ester undergoes saponification at some point during the treatment. Compound 413 reacts with sulfuric acid in methanol to give 414, a 3:2 mixture of stereoisomers at the methoxycarbonyl center. The stereoselectivity in this ring

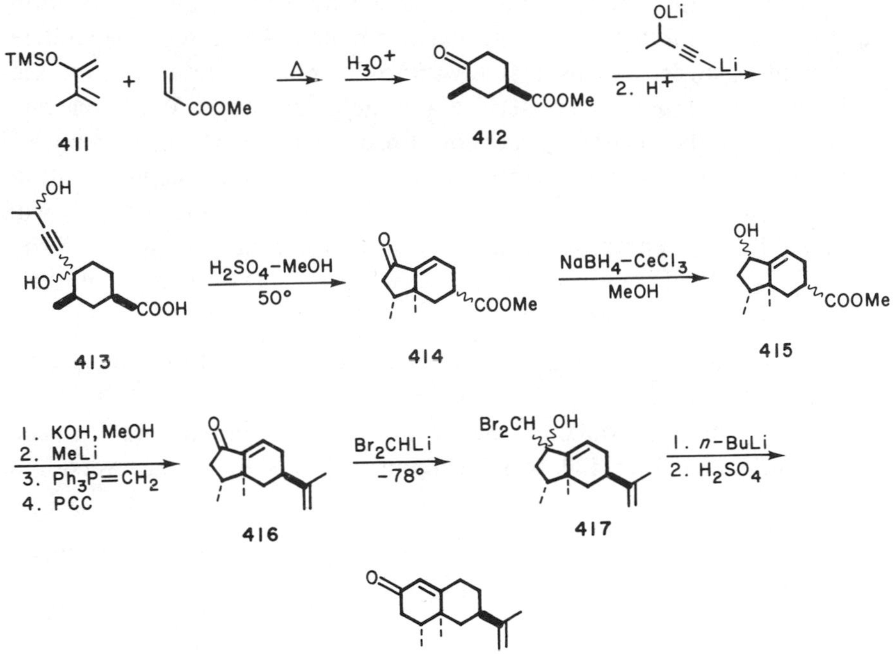

411 412 413 414 415 416 417 386

Scheme 58. Hiyama's Synthesis of (±)-Nootkatone

closure is interesting and may be regarded as a conrotatory closure of the protonated dienone:

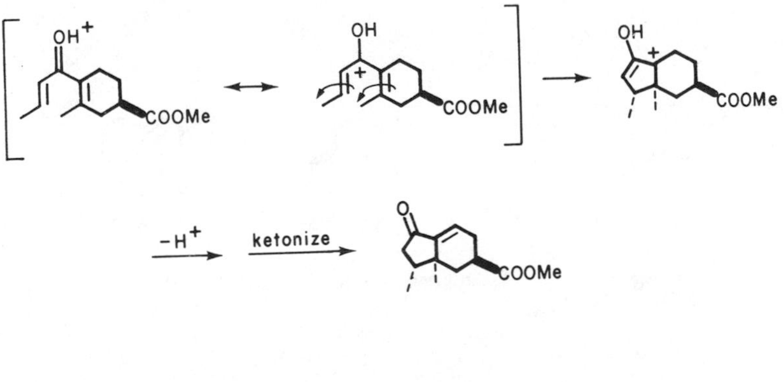

Reduction of the enone is accomplished with $NaBH_4$-$CeCl_3$, to afford alcohol **415**, a mixture of several diastereomers. The ester is saponified, the resulting hydroxy acid treated with excess methyllithium and the acetyl group thus produced epimerized and methylenated by Wittig reaction. Oxidation of the secondary alcohol affords dienone **416** in isomerically pure form. The five-membered ring is expanded by addition of dibromomethyllithium, and conversion of the resulting adduct with *n*-butyllithium. The product is a β,γ-unsaturated ketone, which is conjugated with acid to obtain ($\pm$)-nootkatone. The overall synthesis requires 11 steps and proceeds in 17% yield.

A final nootkatone synthesis, which produces the natural enantiomer, is that of Yoshikoshi and his co-workers (Scheme 59).[246] Nopinone (**419**), conveniently obtained by ozonolysis of β-pinene (**418**) is converted into silyl ether **420**, which is employed in Mukaiyama's version of

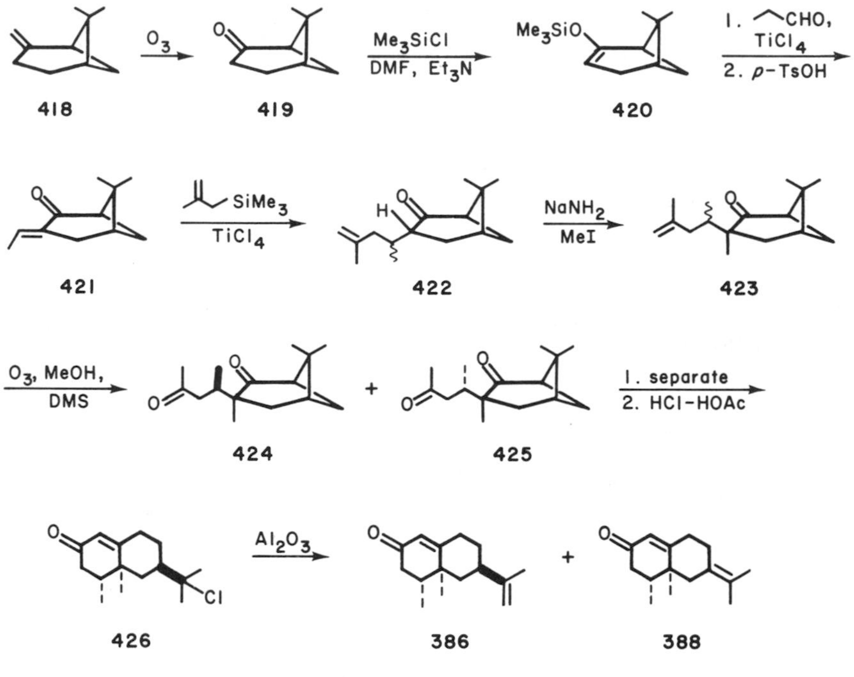

Scheme 59.   Yoshikoshi's Synthesis of ($\pm$)-Nootkatone

the aldol condensation to obtain enone **421**. Application of the Sakurai reaction to this enone affords diastereomeric ketones **422** in a ratio of 3:1. The ketone is then methylated via its sodium enolate; alkylation occurs exclusively anti to the *gem*-dimethyl bridge, providing isomers **423**. Ozonolysis gives diones **424** and **425** in a 3:1 ratio. The isomers are separated and the major isomer cyclized. The cyclobutane ring suffers concomitant ring opening to give **426**. Dehydrochlorination provides (+)-nootkatone (**386**) and (−)-isonootkatone (**388**) in a ratio of about 9:1. The synthesis provides **386** in eight steps, 13% overall yield.

The final synthesis we discuss in this section is that of (+)-dihydronootkatone by Caine and Graham (Scheme 60).[247] Annelation of (+)-carone with methyl vinyl ketone affords enone **428** exclusively. Stereoselectivity is assured in this annelation by the steric hindrance provided by the *gem*-dimethyl groups. Treatment of **428** with HCl in ethanol yields chloro enone **429**, which is dehydrochlorinated to obtain dienone **430**. Compound **430** is transformed into the methylated dienone **431** by a straightforward sequence and **431** is subjected to photoisomerization to obtain tricyclic dienone **432**. Catalytic hydrogenation

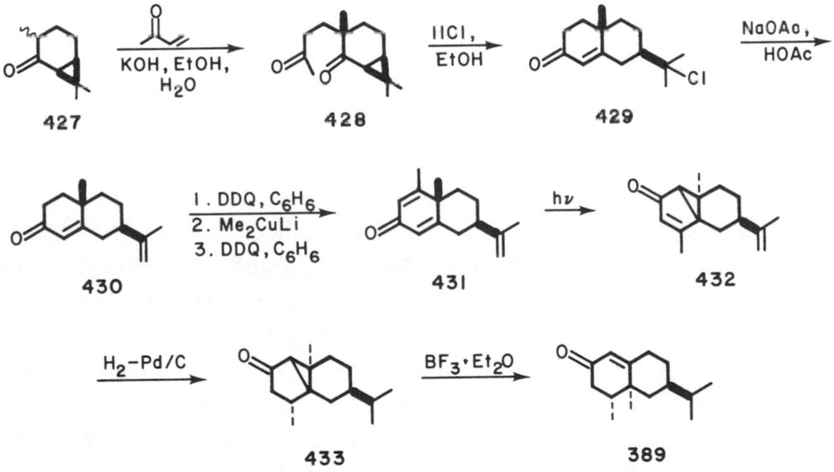

Scheme 60.   Caine-Graham Synthesis of (+)-Dihydronootkatone

reduces both double bonds. Stereoselectivity in reduction of the cyclopentenone double bond results from adsorption of the molecule on the
less hindered exo-face of the bicyclo[3.1.0]hexane system comprising
rings A and B of 432. Acid-catalyzed rearrangement of 433 provides
(+)-dihydronootkatone (389). The final step may be formulated as follows:

A similar angular methyl migration may be involved in the biogenesis of
the eremophilanes.

(2) Fukinone and Dehydrofukinone

Syntheses of fukinone (434) and dehydrofukinone (435) were reported
in Volume 2 of this series.[248] Full details on Piers' synthesis of

434   435

(±)-fukinone have subsequently been published.[249] Shortly thereafter,
Torrence and Pinder reported a similar synthesis of (±)-fukinone, which
is summarized in Scheme 61.[250] The method employed by Torrence and
Pinder to reach enone 441 is essentially identical to that used by Piers
and Smillie. A somewhat different method was used to introduce the
isopropylidine unit. Whereas the Piers route from 441 to (±)-fukinone
requires nine steps, the Pinder approach accomplishes the same task in
only five steps. In all, the Torrence-Pinder synthesis requires 13 steps
and proceeds in about 10% overall yield.

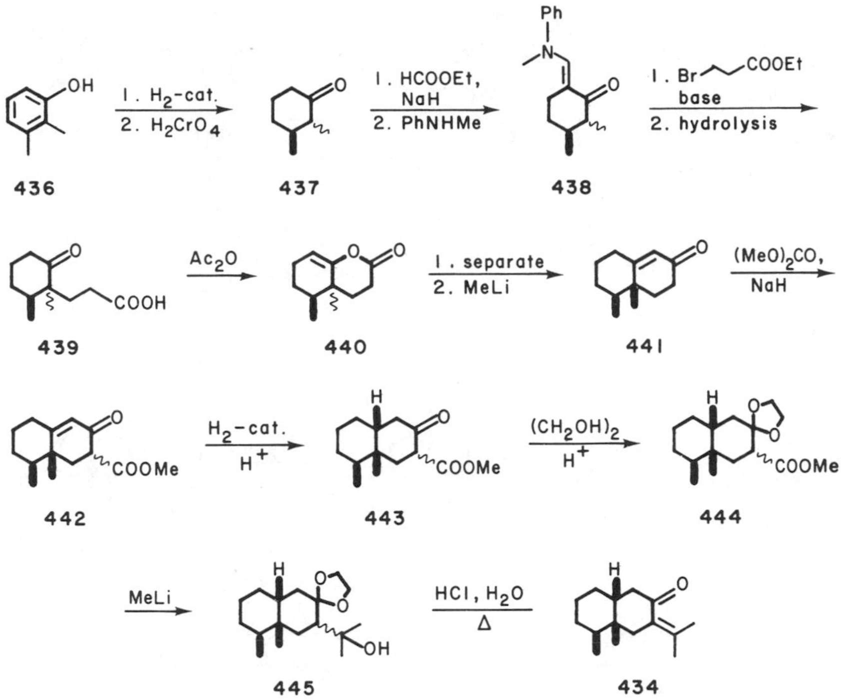

Scheme 61.   Torrence-Pinder Synthesis of (±)-Fukinone

About the same time, Marshall and Cohen reported their synthesis of (±)-fukinone, which is outlined in Scheme 62.[251] o-Anisaldehyde is elaborated into cyclohexenone **449**, which, after purification as its 1,4 adduct with piperidine, is treated with methyllithium to obtain allylic alcohol **450**. Formolysis of **450**, after the manner of Johnson, results in the bicyclic formate **451**, which is transformed into unsaturated acetate **452**. After allylic oxidation to enone **453**, the second methyl group is introduced by cuprate addition to obtain keto ester **454**. Stereoselectivity in this addition is a result of the folded nature of the cis-octalin system. In such molecules, reagents invariably attack trigonal carbons adjacent to the bridgeheads from the convex face. After removal of the keto function, the acetate is hydrolyzed, the resulting alcohol is oxidized, and the

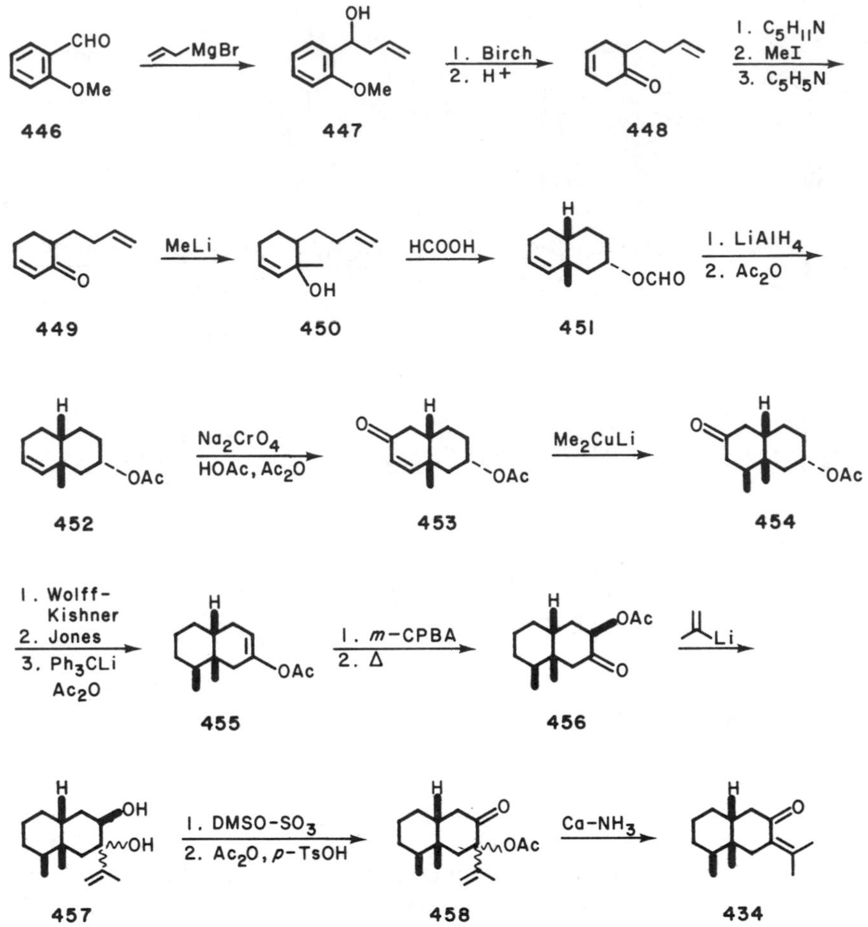

Scheme 62.   Marshall-Cohen Synthesis of (±)-Fukinone

ketone so produced is converted into enol acetate **455**. The ensuing transformations serve to transport the carbonyl group to C_3 while introducing the necessary isopropylidine unit at C_2. This interesting construction was employed because more direct methods were unsuccessful. For example, the following sequence fails in the last step, as unsaturated alcohol **463** cannot be oxidized to (±)-fukinone:

459   460   Ph₃CLi / MeI   461   m-CPBA

462   Na-NH₃   463   434

In all, the Marshall-Cohen synthesis requires 19 steps from o-anisaldehyde and produces **434** in only 0.25% yield.

In 1979, Torii and co-workers reported a second synthesis of (±)-dehydrofukinone, which is outlined in Scheme 63.[252] Diels-Alder reaction of unsaturated keto ester **464** and butadiene provides octalone **465** as a mixture of cis and trans ring junction isomers. The mixture may be equilibrated by base to the pure trans isomer **466**. The cis disposition of the vicinal methyl groups results from a steric bias in the Diels-Alder reaction due to the secondary methyl group in the dienophile. The ketone is ketalized and the ester function is reduced and converted into a mesylate. After deprotection of the ketone, base treatment of the resulting keto mesylate **468** results in the formation of tricyclic ketone **469**. Lithium-ammonia reduction of the latter compound gives unsaturated ketone **470**, in which the angular methoxycarbonyl group has finally been reduced to a methyl. Eight further steps are required to convert this compound into enone **441**, an intermediate employed in both the Piers and Pinder (±)-fukinone syntheses. The Torii synthesis provides a vivid illustration of a trap into which synthetic chemists are prone to fall. It often happens that a worker discovers an interesting new method and, in an honest attempt to demonstrate the utility of the method, proceeds to apply it to solve a problem which has already been solved in a more satisfactory manner by some other, simpler method. In the present case, the Piers-Pinder synthesis of enone **441** requires only eight steps from a readily available aromatic precursor and provides the compound in >30% overall yield. The Torii procedure necessitates 15 steps from a less accessible starting material, and gives **441** in less than

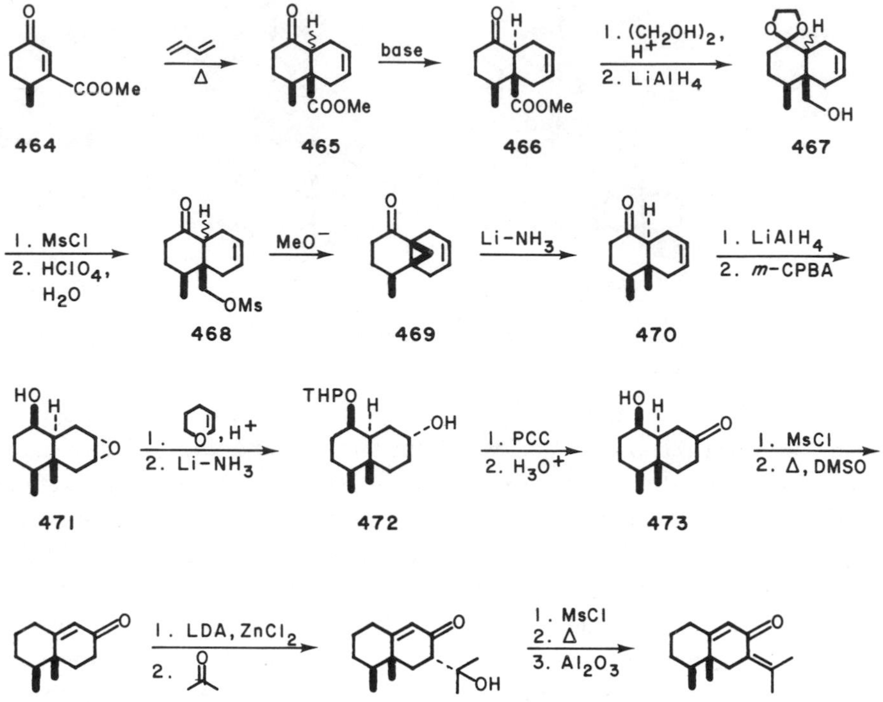

Scheme 63.   Torii's Synthesis of (±)-Dehydrofukinone

10% overall yield. However, the Torii synthesis does make one important contribution to methodology in this class of compounds. The introduction of the isopropylidine function by directed aldol condensation and
dehydration appears to be the most direct method employed to date.

(3) Isopetasol, 3-Epiisopetasol, and Warburgiadione

Isopetasol (475), 3-epiisopetasol (476), and warburgiadione (477) are
more highly oxidized fukinone derivatives. All three compounds have

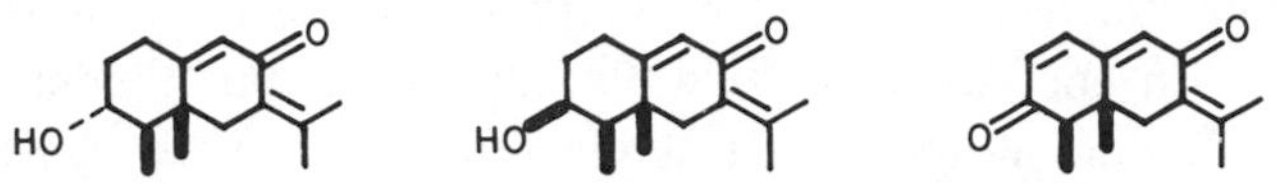

475   476   477

been prepared by Yamakawa and co-workers as shown in Scheme 64.[253] Dione **479** is obtained from 2,3-dimethylphenol via 2,3-dimethylbenzoquinone. Michael addition of **479** to methyl vinyl ketone gives a mixture of adducts **480, 481,** and **482** in a ratio of 20:20:1. After chromatographic separation, aldols **481** and **482** are treated with acid to obtain enone **483**, along with a small amount of the bicyclo[3.3.1]nonane isomers **484**. Enone **483** may be prepared from **479** in this manner in about 25% yield. After selective ketalization of the unconjugated carbonyl, the

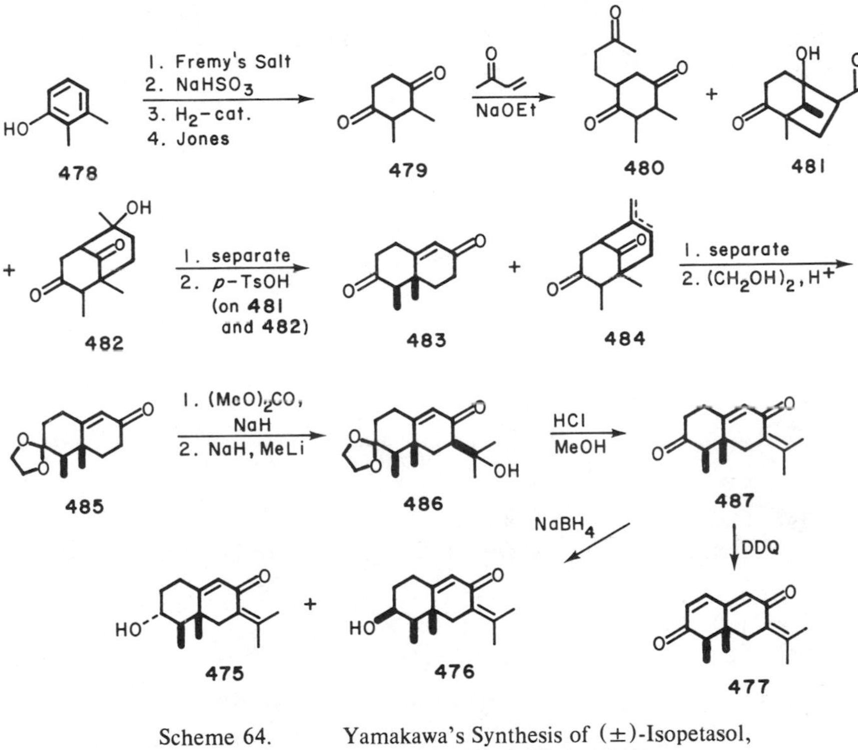

478   479   480   481

+   482   483   484

485   486   487

475   476   477

Scheme 64.   Yamakawa's Synthesis of (±)-Isopetasol,
(±)-3-Epiisopetasol and (±)-Warburgiadione

isopropylidine unit is introduced by methodology which is now more or less standard in the fukinone field (cf. the Piers and Pinder fukinone syntheses). Enedione **487** is selectively reduced to give (±)-isopetasol (**475**) and (±)-3-epiisopetasol (**476**) in a ratio of 1:5. Alternatively, **487** may be dehydrogenated with dichlorodicyanobenzoquinone to obtain (±)-warburgiadione (**477**).

Torii and co-workers have published the synthesis of (±)-isopetasol, which is outlined in Scheme 65.[254] The synthesis is similar to the Torii dehydrofukinone synthesis, in that the decalin nucleus is reached by a Diels-Alder reaction and also because a cyclopropyl ketone plays a key role. In this case, however, cyclopropyl ketone **493** is cleaved reduc-

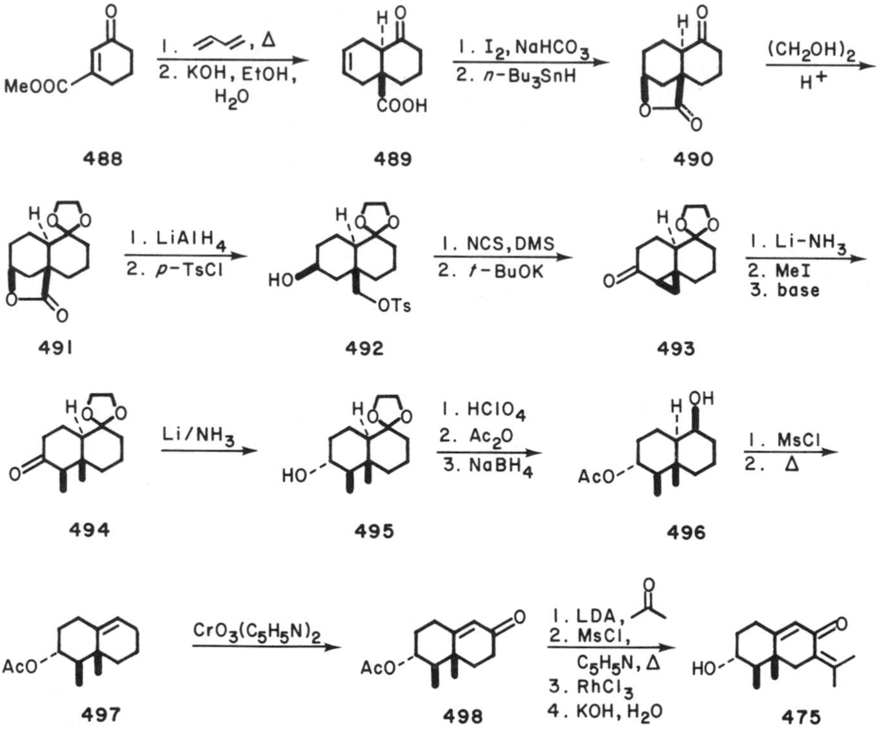

Scheme 65.   Torii's Synthesis of (±)-Isopetasol

tively and the resulting enolate methylated to obtain **494**, an intermediate with the required vicinal methyl groups. Although the stereochemistry resulting from the reductive methylation is trans, the C_2-methyl center is easily corrected by base equilibration to the more stable cis isomer, in which the secondary methyl group is equatorial. Lithium in ammonia reduction of ketone **494** affords the thermodynamically favored (equatorial) alcohol **495**. The remaining steps are unexceptional and ($\pm$)-isopetasol is obtained in 19 steps in an overall yield of about 4%.

(4) Eremophilone

Eremophilone (**499**) has stood as a special symbol to the natural products chemist of the perversity of nature. It was isolated by Simonson and co-workers in 1932. Simonson proposed an erroneous structure, ruling out the correct structure because of the isoprene rule. Later, how-

499

ever, on the basis of a suggestion by Robinson, Simonson reconsidered and eventually advanced the correct structure, which was subsequently confirmed by Grant's X-ray analysis of a derivative. At the time of the writing of Volume 2 in this series, no successful syntheses of eremophilone had been achieved, even though it was the oldest and most well-known member of the group. The principle synthetic challenge is controlling the stereochemistry of the isopropenyl group, which must be axial. Since 1970, there have been three total syntheses, and a fourth synthesis which leads to a key intermediate in one of the others.

The first successful synthesis was reported by Ziegler and Wender in 1974 (Scheme 66).[255] The synthesis begins by conjugate addition of lithium divinylcuprate to 3,4-dimethylcyclohexenone. The cis disposition of the vicinal methyl groups is assured by the well-established preference

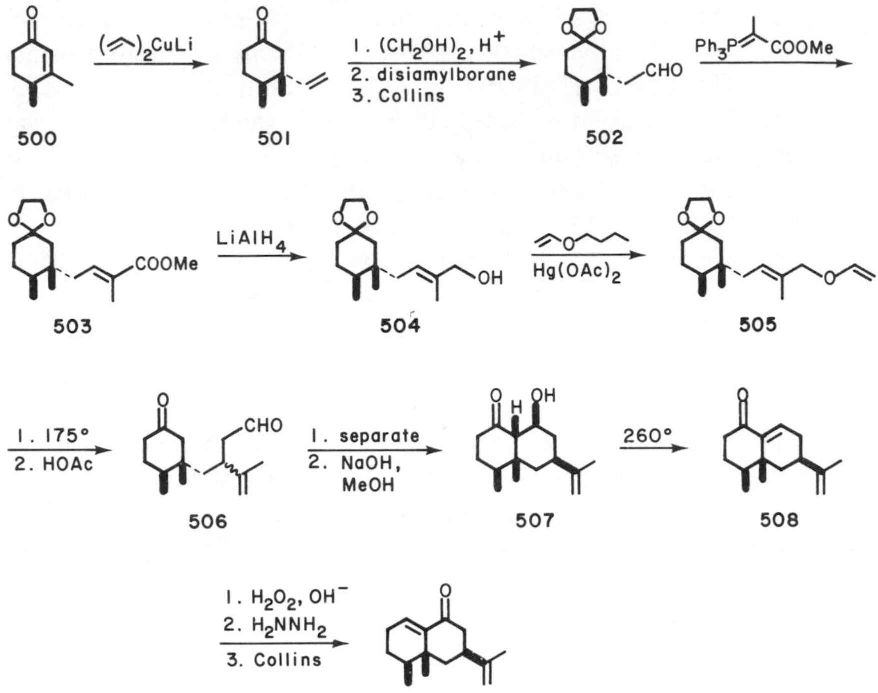

Scheme 66.   (±)-Eremophilone: Ziegler-Wender Synthesis

for such conjugate additions to occur trans to a substituent at C_4. Unsaturated ketone 501 is elaborated by straightforward methods into alcohol 504, which is subjected to Claisen rearrangement using the butyl vinyl ether technique. The ingenious plan is marred by the fact that the Claisen rearrangement is essentially stereorandom; the two diastereomers are produced in a ratio of 55:45. The desired isomer is the minor product of the reaction. After separation, the two diastereomers may be independently converted into (±)-eremophilone (499) and its C_7-epimer.

Subsequently, Ziegler and his co-workers developed an alternative synthesis which, while not totally stereoselective, provides (±)-eremophilone with about 70% control of stereochemistry of C_7.[256] This synthesis

(Scheme 67) begins with the same unsaturated ketone (**501**) used in the first eremophilone synthesis. It is epoxidized by the three-step sequence shown (direct epoxidation of **501** fails) and epoxy ketone **510** is cyclized by treatment with potassium *t*-butoxide. As it turns out, it is in this step

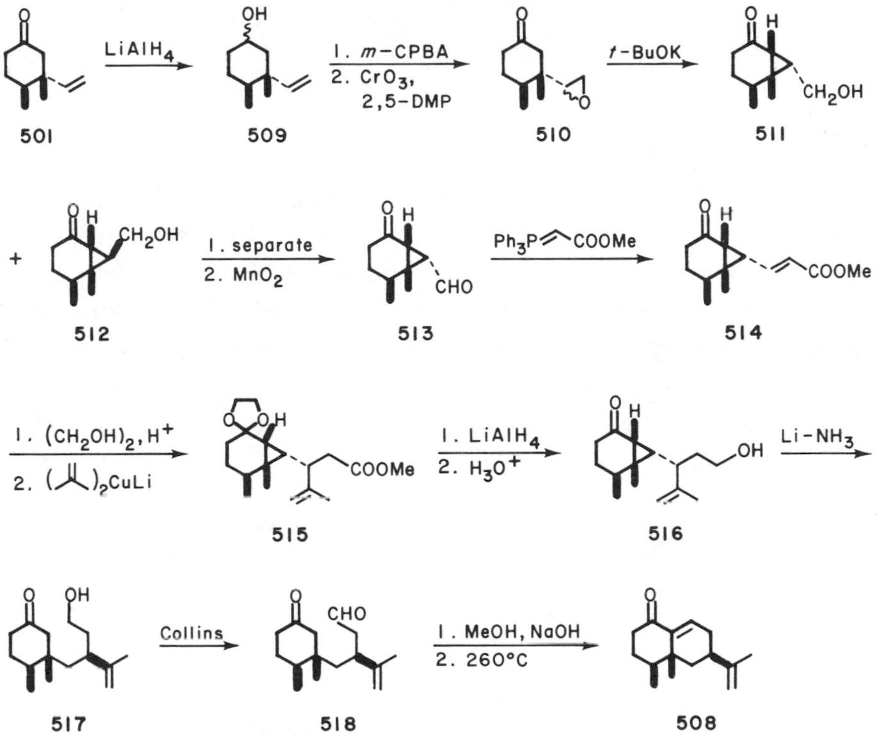

Scheme 67.　　(±)-Eremophilone: Ziegler's Stereoselective Variant.

that the eventual stereochemistry of C_7 is effectively established. Endo and exo compounds **511** and **512** are produced in a ratio of about 70:30. The isomers are separated and the major isomer **511** is oxidized to aldehyde **513**, which is converted by Wittig reaction to unsaturated ester **514**. After protection of the ketone group, the isopropenyl unit is added by way of the cuprate reagent; compound **515** is the sole product of the reaction. The stereochemical outcome of this reaction is understandable in terms of the following depiction of the *endo*-cyclopropylacrylate:

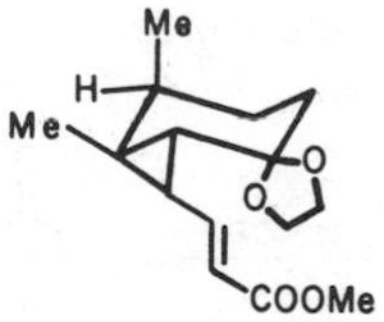

Attack of the organometallic reagent from the front face of the double bond is severely hindered by the dioxolane ring. Standard transformations serve to convert **515** into keto alcohol **516**, which is ring-opened by lithium in ammonia to give keto alcohol **517**. Oxidation, ring closure, and dehydration provide enone **508**, an intermediate in the first Ziegler synthesis. The minor cyclopropyl carbinol **512**, when submitted to the same sequence of steps, provides enone **508** and its C_7-epimer in a ratio of 15:85.

In 1975, McMurry and co-workers reported a completely stereoselective synthesis of (±)-eremophilone (Scheme 68).[257] Ketone **394**,

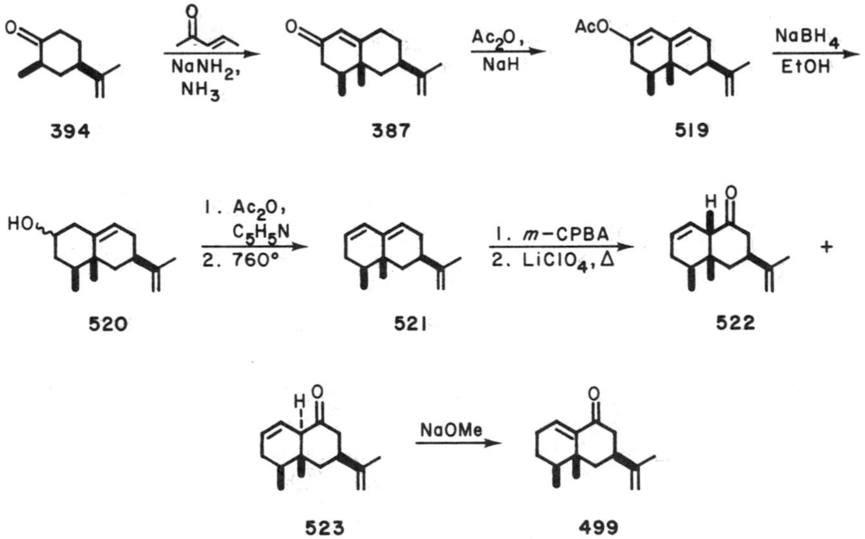

Scheme 68.   (±)-Eremophilone: McMurry's Synthesis

prepared from β-pinene in five steps by a procedure introduced by Van der Gen, is condensed with 3-penten-2-one to give 7-epinootkatone **387** (see Scheme 53). Enone **387** is converted into diene **521**, which is epoxidized and the mixture of epoxides rearranged with lithium perchlorate to give the cis-fused enone **522** (itself a minor natural product) and the trans isomer **523** in a ratio of 3:1. Conjugation of the mixture provides ($\pm$)-eremophilone (**499**). Although (+)-β-pinene is employed as starting material, racemization occurs in the course of its conversion to ketone **394**. The complete synthesis from β-pinene requires 13 steps and gives eremophilone in about 3% overall yield.

Ficini and Touzin synthesized ($\pm$)-eremophilone as shown in Scheme 69.[258] Keto ester **524** reacts with excess methyllithium to give keto alcohol **525** after hydrolysis. The tertiary alcohol is protected and the elements of the second ring are added in the form of an organometallic reagent. In this step, the necessary cis disposition between the angular methyl and the isopropenyl groups is established with high selectivity. Since the protecting groups on the two alcohols are the same, four additional steps are required to reach keto alcohol **528**, even though the overall change amounts to no more than hydrolysis of one of the two ether groups. The primary hydroxy is oxidized and the keto aldehyde is aldolized to obtain **530**, a mixture of diastereomers at the secondary hydroxy position. Parallel studies had shown that catalytic hydrogenation of the exocyclic methylene group, either at this stage or after dehydration of the aldol, would probably be stereorandom. Consequently, it is necessary to arrange the functionality in the molecule so that this double bond can be reduced chemically. This requires that the ketone function be protected. Ficini and Touzin chose to reduce it temporarily to an alcohol. Of course, this choice in turn requires that the existing secondary alcohol be protected in order to differentiate it from the secondary alcohol to be produced by reduction of the ketone. For this purpose the ethoxyethyl group is again called into service. After the proper manipulations, the reduction of **531** turns out to be stereoselective, as anticipated. The keto function is regenerated, the protecting groups removed, and diol **533** is heated with alumina at 280°C to obtain a mixture of ($\pm$)-eremophilone (**499**, 25% yield) and the deconjugated enone **522**.

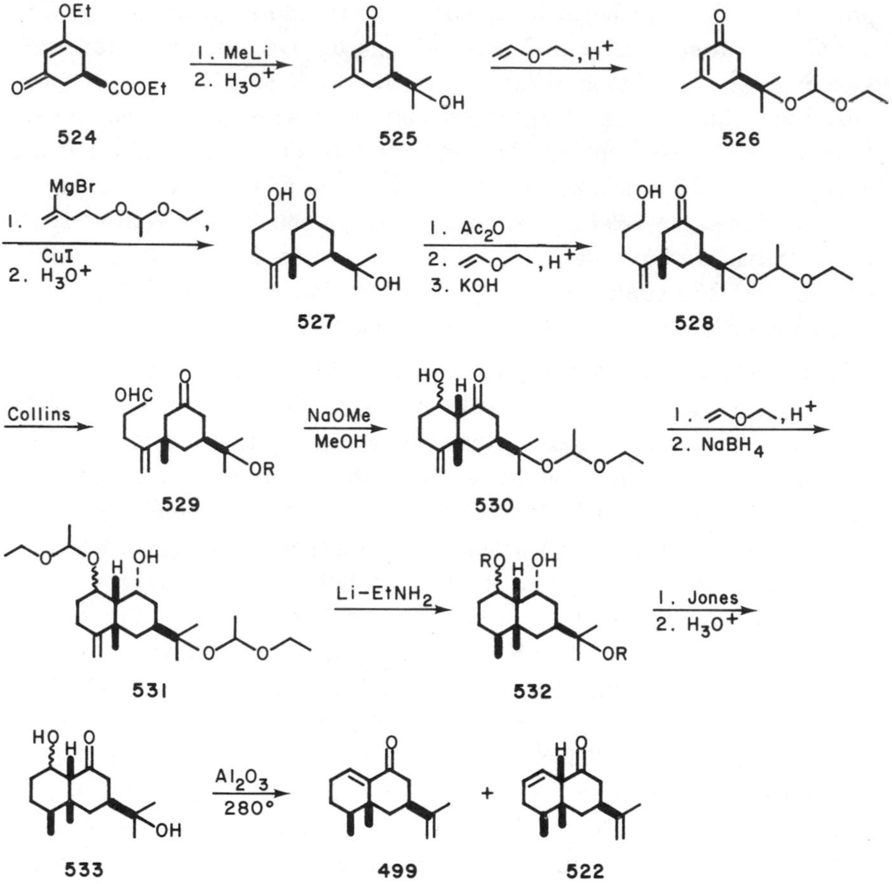

Scheme 69.   (±)-Eremophilone: Ficini-Touzin Synthesis

This synthesis provides a graphic illustration of the extent to which organic chemists have come to rely upon protecting groups. Of the sixteen steps in the synthesis, seven are taken up with introducing or removing hydroxy protecting groups.

The final synthesis of eremophilone was reported in 1979 by the Firmenich group of Näf, Decorzant and Thommen (Scheme 70).[259] Enone

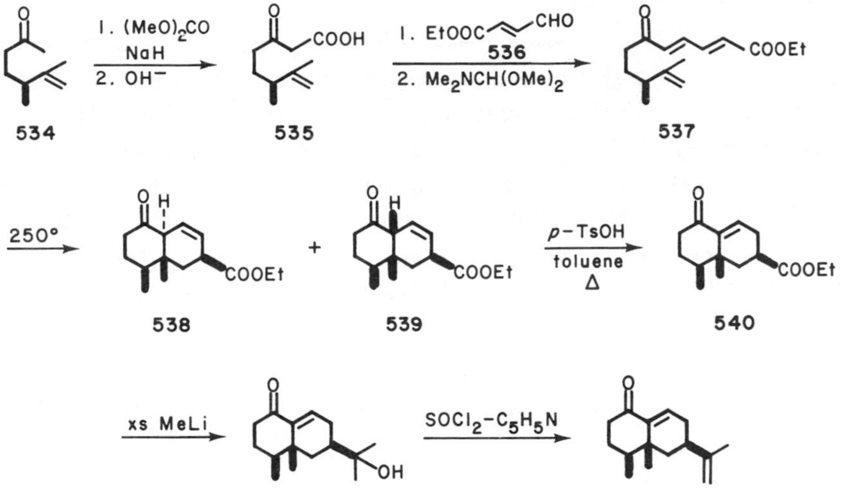

Scheme 70.   (±)-Eremophilone: Firmenich Synthesis

534 is carboxylated and the resulting β-keto acid **535** is condensed with aldehydo ester **536** to give, after decarboxylation, intermediate **537**. When this material is heated at 250°C, intramolecular Diels-Alder addition occurs, providing a mixture of trans- and cis-fused adducts **538** and **539**. It is suggested that isomer **538** is the kinetic product and that some epimerization occurs under the conditions of the cycloaddition. The mixture is conjugated by treatment with acid to obtain enone ester **540**, which reacts with excess methyllithium to give keto alcohol **541**. It is remarkable that the ketone carbonyl does not react under these conditions. Dehydration of **541** provides the Ziegler-Wender enone **508**. The relative configurations of the three chiral centers are established in the intramolecular Diels-Alder reaction, which is proposed to proceed by way of the following transition state:

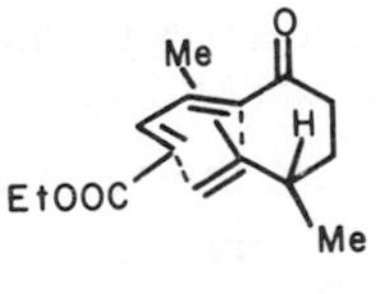

(5) Furanoeremophilanes

There exists a large family of eremophilanes in which the three-carbon side chain is incorporated into a furan or butenolide ring. Those that have been synthesized in the past 10 years are furanoeremophilane (542), furanoligularone (543), eremophilenolide (544), tetrahydro-ligularenolide (545), 9,10-dehydrofuranoeremophilane (546), ligularone (547), isoligularone (548), euryopsonol (549), epi-euryopsonol (550), 3β-hydroxyfuranoeremophilane (551), and 3β-furanoligularanol (552).

542   543   544   545

546   547   548   549

550   551   552

A common synthetic problem in this group is establishing the cis stereochemistry of the vicinal methyl groups. In addition, the ring junction stereochemistry must be controlled—usually to produce the cis isomer. The additional stereocenter in butenolides 544 and 545 actually presents no problem, since it is defined in the thermodynamically more stable configuration and may be equilibrated via the hydroxyfuran tautomer. Compounds 547-550 present a problem in regioselectivity, in placing the oxo function at the proper site. As will be seen, this problem has not been solved, and the syntheses to date give mixtures of regioisomers which must be separated. Finally, compounds 549-552 present the problem of the additional chiral center at the secondary hydroxy group. This problem has also not been solved satisfactorily.

The first reported synthesis of one the compounds in this group was Piers' synthesis of (±)-eremophilenolide (**544**).[260] The synthesis was later extended to produce (±)-tetrahydroligularenolide (**545**) as well.[261] Enone **441**, used previously in the Piers and Pinder fukinone syntheses[262] is converted into β-keto ester **556** by the rather circuitous sequence indicated in Scheme 71. Torrence and Pinder report that

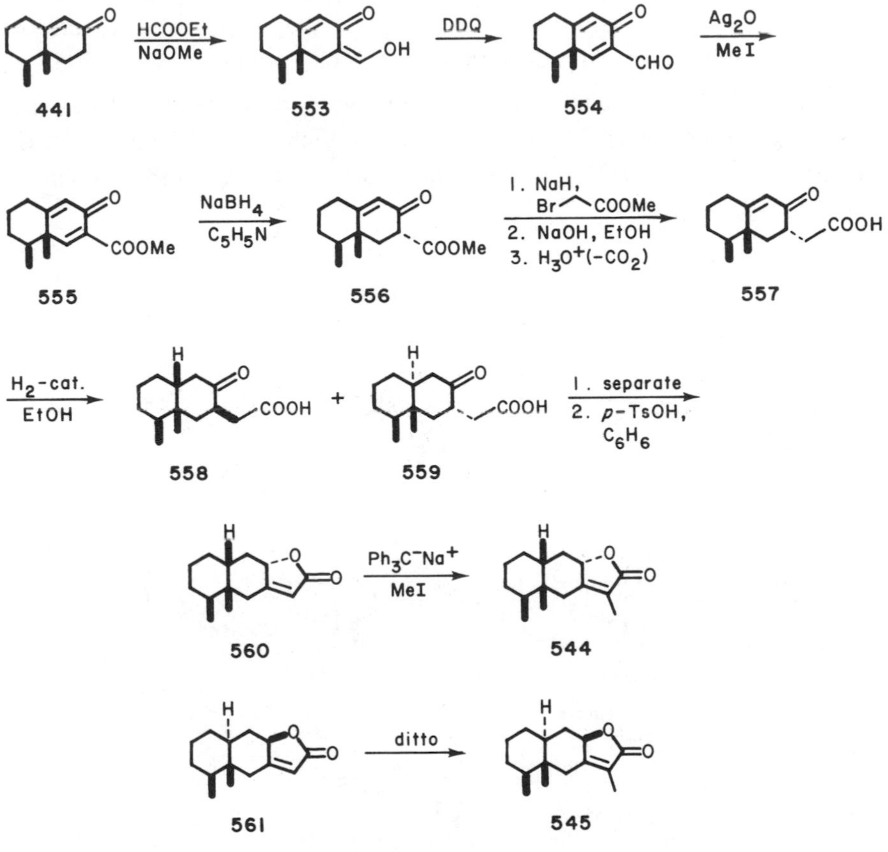

Scheme 71.   Piers Synthesis of (±)-Eremophilenolide and (±)-Tetrahydroligularenolide

enone **441** may be directly carbomethoxylated to give a β-keto ester corresponding to **556**. However, the Piers sample is a solid, mp 108°C, while the specimen obtained by Torrence and Pinder is reported to be an oil. Because of reports that similar enones undergo direct car-bomethoxylation both at C_1 and at C_3, the purity of Pinder's product is questionable and Piers and co-workers chose the more complex route. Compound **556** is alkylated and the resulting keto diester is hydrolyzed and decarboxylated to give enone acid **557**. Catalytic hydrogenation pro-vides a mixture of diastereomeric keto acids **558** and **559**. The ratio of **558** to **559** depends upon the catalyst employed. With Pd/C, the ratio of **558 to 559** is 3:2; with Rh/C, the ratio is 2:3. The two isomers are separated and converted independently into butenolides **544** and **545**.

In 1973, Takahashi and Tatee reported a second synthesis of (±)-tetra-hydroligularenolide (**545**).[263] The synthesis (Scheme 72) begins with the known keto ether **562**, which is obtained from dihydroresorcinol in six

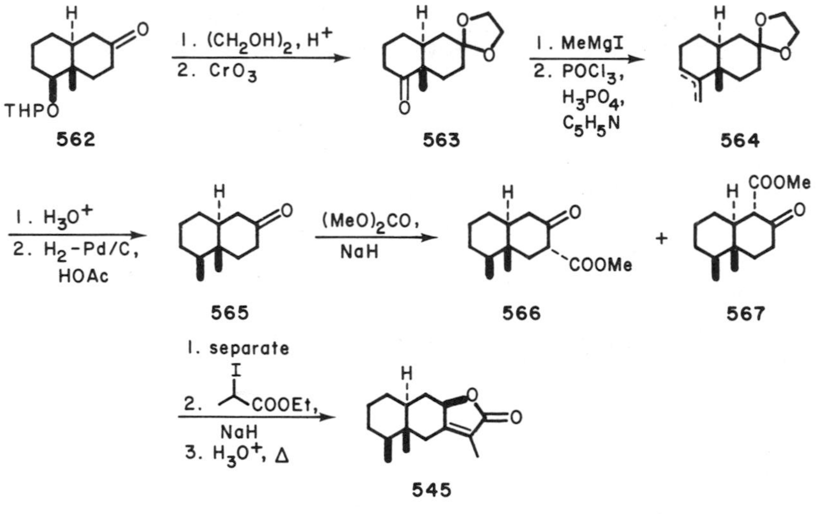

Scheme 72.   Takahashi's Synthesis of (±)-Tetrahydroligularenolide

steps. Compound **562** is converted into keto ketal **563**, which is methylated with methyllithium and the resulting alcohol is dehydrated to obtain a mixture of alkenes **564**. Catalytic hydrogenation appears to be highly stereoselective, with ketone **565** being the only stereoisomer produced. The overall yield from **562** to **565** is 16%. Methoxycarbonylation of **565** yields a mixture of β-keto esters, which is separated chromatographically to obtain 75% of **566** and 12% of **567**. Alkylation of the former isomer with ethyl α-iodopropionate gives, after hydrolysis and decarboxylation, ($\pm$)-tetrahydroligularenolide (**545**). The alkylation proceeds in only 30% yield, the principal by-product being the O-alkylated isomer.

Kitahara and his co-workers reported the synthesis of ($\pm$)-eremophilenolide (**544**) and ($\pm$)-furanoeremophilane (**542**) which is summarized in Scheme 73.[264] Octalone **568**, obtained by the Diels-Alder reaction of 2-methylcyclohexenone and butadiene, reacts with methyllithium to produce a 2:1 mixture of diastereomers **569** and **570**. When the mixture is subjected to oxymercuration conditions, only the major isomer reacts, leading to ether **571**. Treatment of **571** with acetyl chloride in the presence of mercuric acetate as catalyst yields a chloro acetate, which, upon treatment with base, is converted into unsaturated alcohol **572**. After oxidation of the secondary alcohol and protection of the resulting ketone, the double bond is hydroborated and the resulting ketal alcohol is oxidized to **574**. Since hydroboration occurs from the convex face of the *cis*-octalin system, the stereochemistry of the C_1 methyl group is adjusted by epimerization with base. As shown by the following projection formulas, the interaction labeled *a* must be more destabilizing than the one labeled *b*.

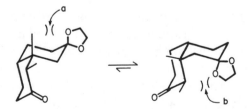

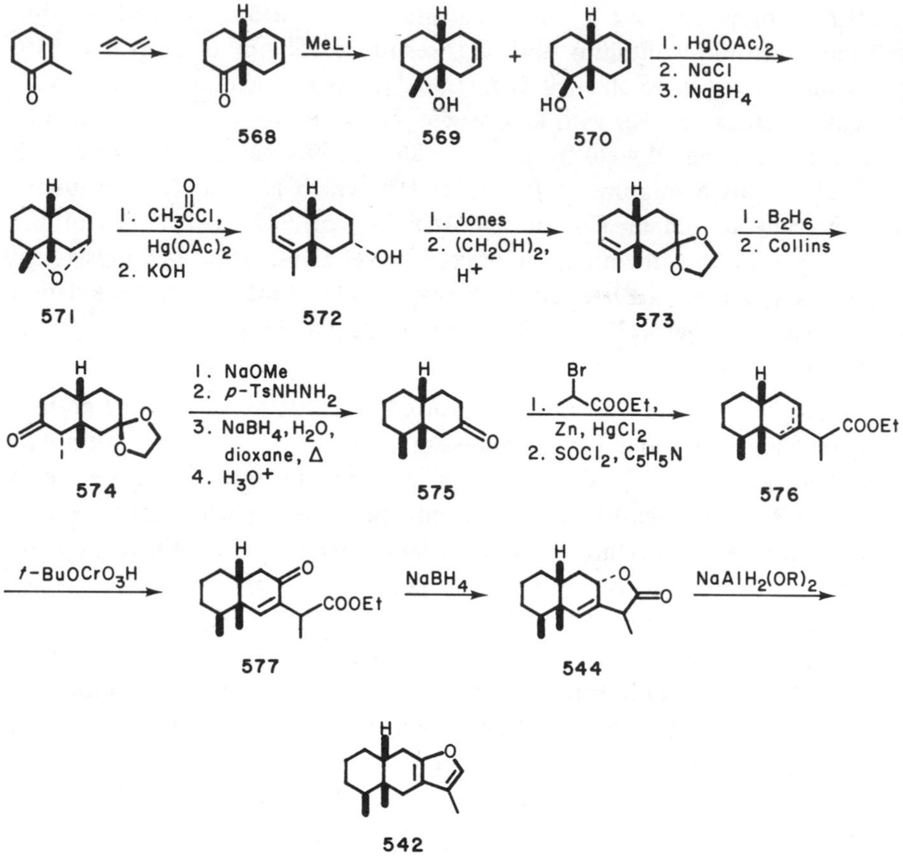

568   **569**   **570**

571   **572**   **573**

574   **575**   **576**

577   **544**

542

Scheme 73.   Kitahara's Synthesis (±)-Eremophilenolide
and (±)-Furanoeremophilane

The carbonyl group is removed by the Bamford-Stevens method. After
deprotection, the three-carbon side chain is introduced by a Reformatsky
reaction. After dehydration, a mixture of alkenes (**576**) is produced.
Allylic oxidation with *t*-butyl chromate occurs regiospecifically, giving
enone **577**, which is reduced by sodium borohydride to obtain
(±)-eremophilenolide (**544**). Further reduction of the lactone, followed
by dehydration, provides (±)-furanoeremophilane (**542**).

In 1976, Bohlmann and co-workers reported syntheses of (±)-ligular-one (547), (±)-euryopsonol (549) and (±)-epieuryopsonol (550).[265] The synthesis of ligularone (Scheme 74) begins with p-cresol, which is converted by a nine-step sequence into furanoquinone 584. Diels-Alder

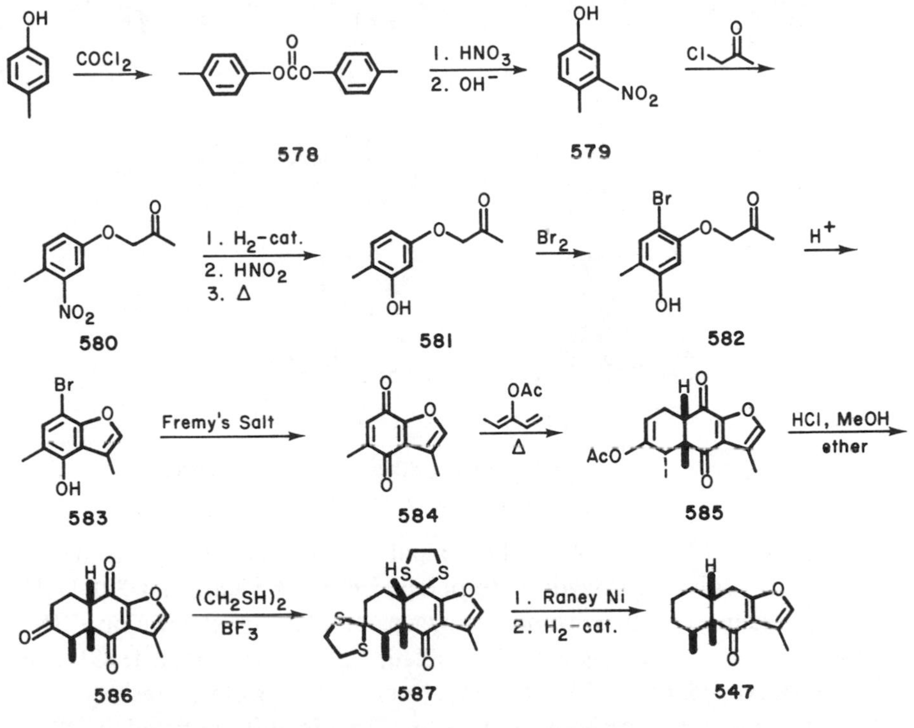

Scheme 74.   Bohlmann's Synthesis of (±)-Ligularone

reaction of this material with 3-acetoxy-1,3-pentadiene provides adduct 585. Upon hydrolysis, the secondary methyl center undergoes epimeri-zation, giving the key intermediate 586. This compound reacts with excess ethanedithiol to give the bis-dithioketal 587 in 58% yield; the more hindered carbonyl does not react. Desulfurization of 587 with Raney nickel gives (±)-ligularone.

For the synthesis of $(\pm)$-euryopsonol (**549**) and $(\pm)$-epieuryopsonol (**550**) (see Scheme 75), the key trione **586** is carefully reduced with $NaAlH_2(OCH_2CH_2OCH_3)_2$ to obtain a mixture of alcohols, which is acetylated. Fractional crystallization of the mixture provides acetates **588**

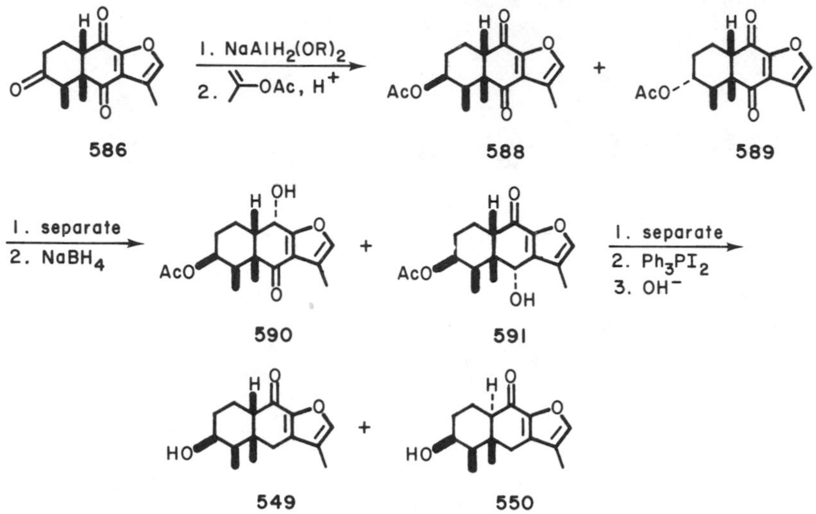

Scheme 75.   Bohlmann's Synthesis of $(\pm)$-Euryopsonol and $(\pm)$-Epieuryopsonol

(25% yield) and **589** (13% yield) in pure form. The major isomer is reduced by sodium borohydride to obtain **590** (63% yield) and **591** (26% yield). Again, the two isomers are separated by crystallization. The minor isomer is deoxygenated to obtain $(\pm)$-euryopsonol (**549**) and $(\pm)$-epieuryopsonol (**550**) in 36% and 11% yields, respectively. Although the three carbonyl groups in **586** are differentiated in their reactivity toward various reagents, the difference in reactivity is not sufficient for the synthesis of **549** in a highly selective manner.

Yamakawa and Satoh have also explored the use of triketone **586** for the synthesis of various furanoeremophilanes.[266] As shown in Scheme 76, they prepared **586** in a similar manner to that employed by Bohlmann and co-workers. In agreement with Bohlmann's observation, Yamakawa found the C_3 carbonyl to be more reactive toward ethylene glycol; the crystalline 3-ketal is obtained in 88% yield. However, no

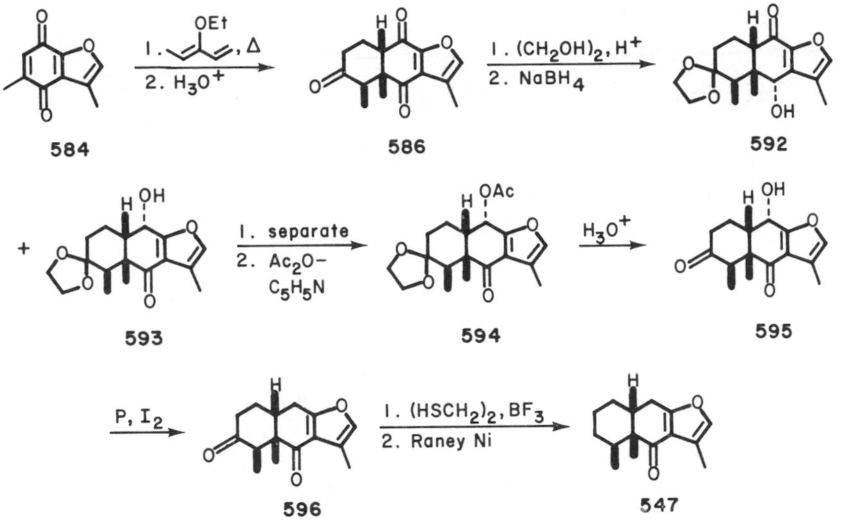

Scheme 76.   Yamakawa's Synthesis of (±)-Ligularone

regioselectivity is observed in reduction of this monoketal, isomers **592** and **593** being produced in equal amounts. The isomers are separated and isomer **593** is converted into dione **596** by removal of the hydroxy group. For the synthesis of (±)-ligularone (**547**), advantage is again taken of the greater reactivity of the C_3-carbonyl.

Yamakawa and Satoh have also converted triketone **586** into alcohols **551** and **552**, as shown in Scheme 77. Hydrolysis of isomer **592** (see Scheme 76) gives diketo alcohol **597** in 75% yield. Deoxygenation of this substance gives isomeric diones **598** (20% yield), **599** (20% yield), and **600** (30% yield). Reduction of **598** by sodium borohydride occurs with good stereoselectivity, providing keto alcohol **601** in 81% yield, along with 7% of its C_3-epimer. Further reduction of **601** gives diol **602**, which is converted to a diacetate and hydrogenolyzed to obtain (±)-3β-hydroxyfuranoeremophilane (**551**). Similar treatment of isomer **599** gives keto alcohol **603** (93% yield), which is further reduced and hydrogenolyzed to obtain (±)-3β-furanoligularanol (**522**). Diones **598** and **599** have also been converted into compounds **542**, **543**, and **546**.

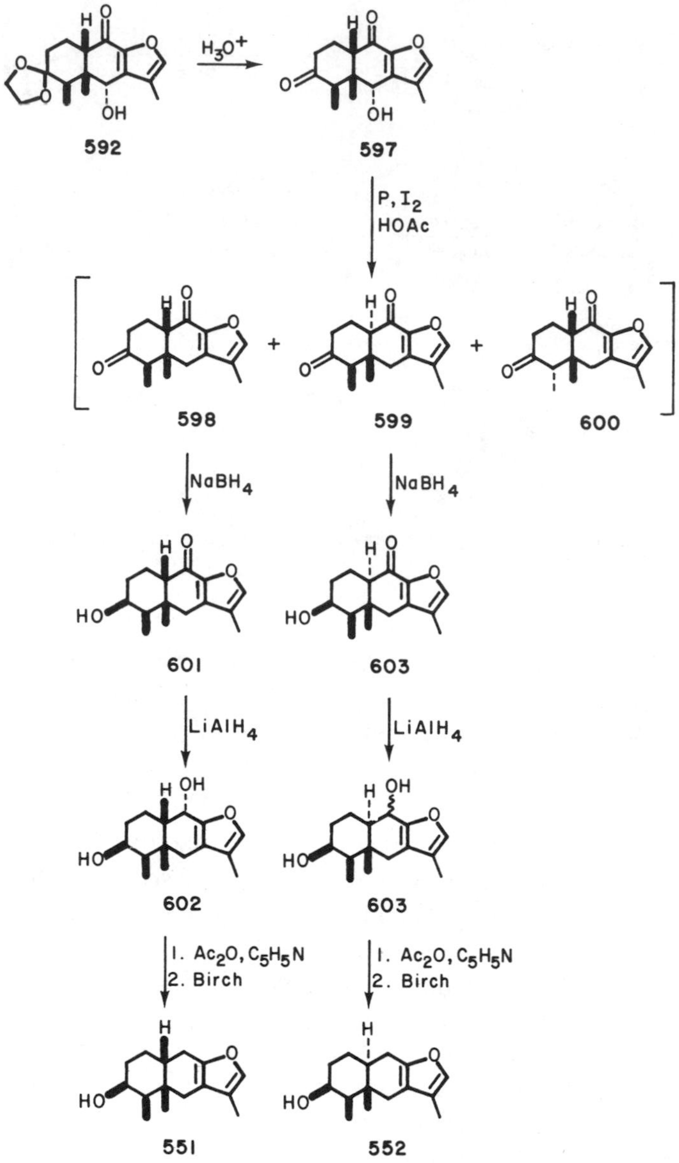

Scheme 77.   Yamakawa's Synthesis of (±)-3β-Hydroxyfuranoeremophilane and (±)-3β-Furanoligularanol

The final synthesis in this area is that due to Yoshikoshi and his co-workers, which produces (±)-ligularone and (±)-isoligularone (Scheme 78).[267] Enedione **604**, conveniently prepared by Diels-Alder reaction of 2-methyl-cyclohexenone and the Danishefsky diene is ketalized at the nonconjugated carbonyl and treated with lithium dimethylcuprate to obtain keto ketal **605**. As in the Marshall fukinone synthesis, the cis-stereochemistry of the vicinal methyl groups results from the folded nature of the octalone. After removal of the exposed carbonyl group, the other carbonyl is deprotected and converted into enone **607**. This

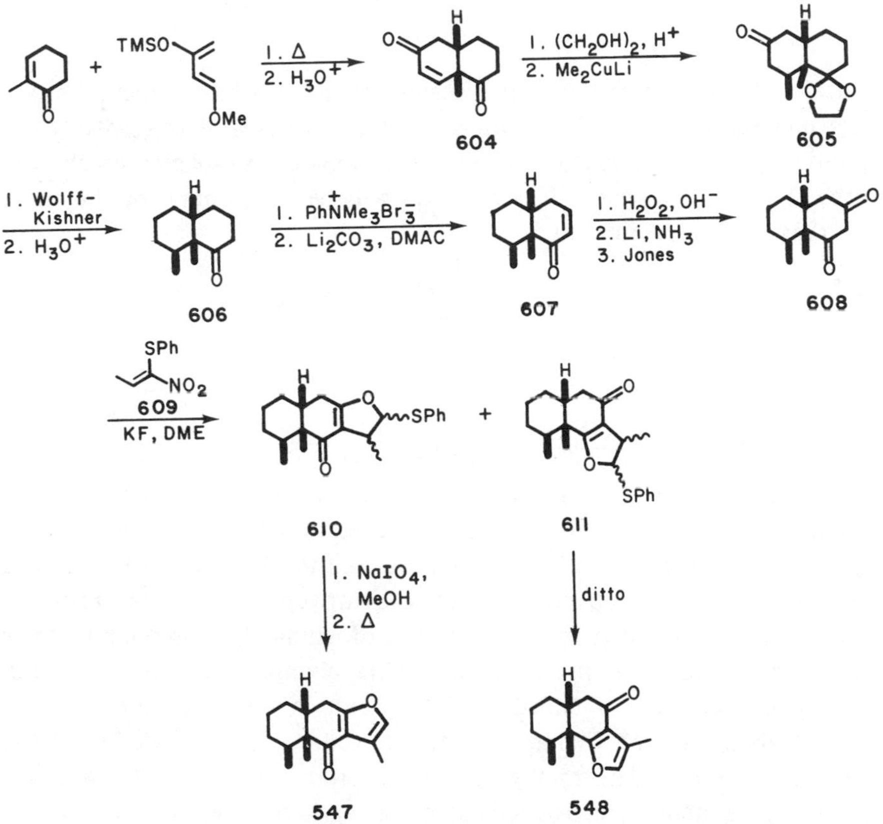

Scheme 78.   Yoshikoshi's Synthesis of (±)-Ligularone and (±)-Isoligularone

material is subjected to alkaline epoxidation, and the mixture of diastereomeric epoxides is reduced with lithium in ammonia. The resulting mixture of diols is oxidized to dione **608**. Michael reaction of **608** with unsaturated nitro compound **609** provides **610** and **611**, each a mixture of diastereomers, in a ratio of 1:2. Oxidation of **610**, followed by thermolysis of the resulting sulfoxide gives (±)-ligularone (**547**) in 47% yield. Similar treatment of isomer **611** provides (±)-isoligularone (**548**) in 54% yield.

(6) Cacalol

Cacalol (**612**) is an example of a sesquiterpene in which a methyl group has undergone a net 1,3-migration. Three syntheses of this novel compound have been reported. In spite of its apparent simplicity, cacalol has proven to be a formidable target. As will be seen, none of the three syntheses are very efficient.

612

The first published synthesis (Scheme 79) came from Inouye, Uchida, and Kakisawa of Tokyo Kyoiku University.[268] 2-Methoxy-5-methylphenol is subjected to Friedel-Crafts reaction with γ-valerolactone to give an acid which is methylated and cyclized to obtain ketone **614**. The carbonyl group is removed by catalytic hydrogenolysis and the resulting tetralin **615** is acylated to obtain **616**. Unfortunately, compound **616** is the minor product of this reaction, being obtained in only 20% yield. The major product, obtained in 53% yield, is the ester produced by demethylation of the C_1-methoxy group, followed by acetylation of the resulting phenol. The methyl ethers are removed with BBr_3 and the resulting bisphenol is converted to the bis-aryloxyacetic ester **617**, which is hydrolyzed and the resulting diacid is cyclized by treatment with acetic anhydride. Removal of the other carboxymethyl group is accomplished

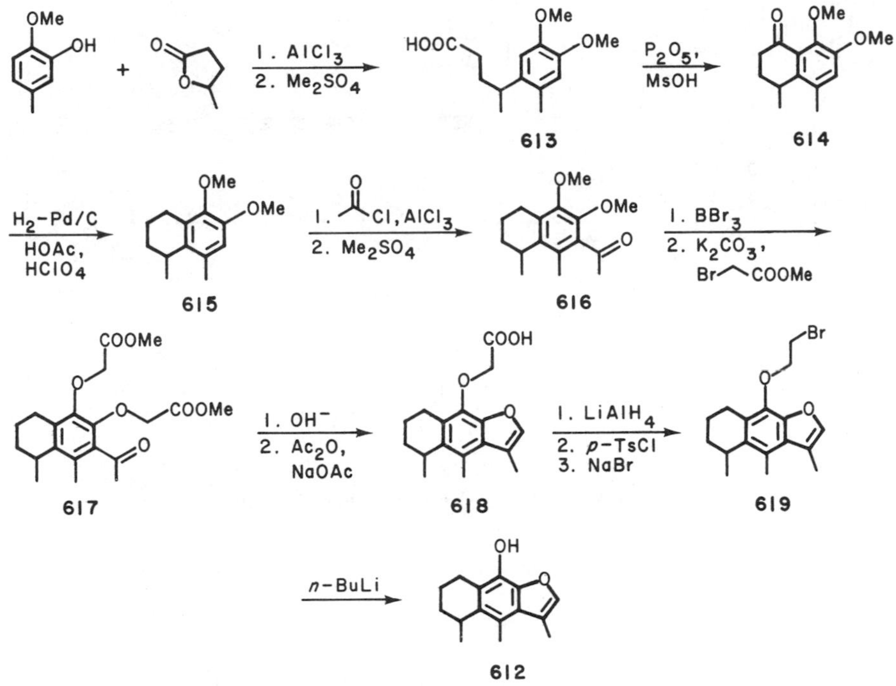

Scheme 79.   (±)-Cacalol: Tokyo Synthesis

by conversion into bromide **619**, which fragments upon being metallated, thus giving (±)-cacalol (**612**).

Walls' synthesis (Scheme 80)[269] also begins with a Friedel-Crafts reaction, but with 2-methoxy-4-methylphenol. Use of BF$_3$ allows both bonds to be formed, resulting in tetralone **620**, albeit in only 13% yield. The elements of the furan ring are introduced by Claisen rearrangement of crotyl ether **621**. Ozonolysis of the resulting rearrangement product gives hemiacetal **623**, which undergoes dehydration upon being heated with phosphoric acid. Clemmensen reduction of **624** removes the carbonyl group and provides (±)-cacalol (**612**).

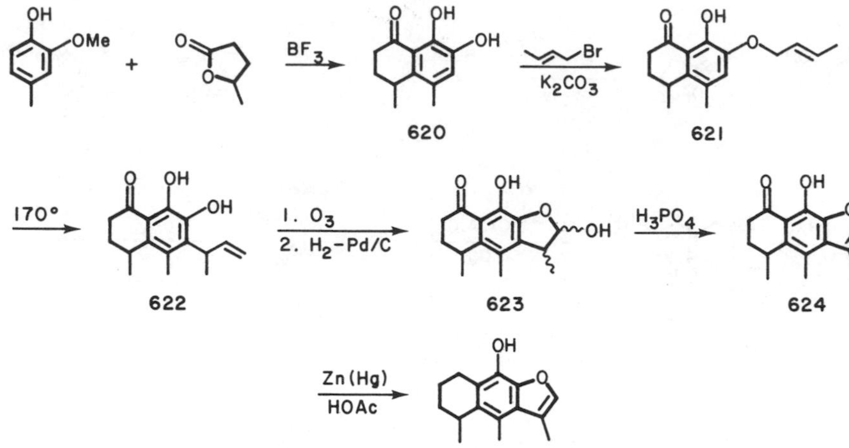

Scheme 80.   (±)-Cacalol: Walls' Synthesis

Huffman's synthesis (Scheme 81)[270] utilizes *p*-cresol for the Friedel-Crafts cyclization (which proceeds in only 26% yield), thus necessitating the subsequent introduction of the other ring oxygen. After Clemmensen reduction of the tetralone carbonyl, tetrahydronaphthol **626** is oxidized at C_2 by acylation and Bayer-Villiger oxidation. The eventual furan ring is added by alkylation of phenol **629** with α-chloroacetoacetic ester. However, compound **630** is obtained in only 14% yield. Although Walls found that a compound analogous to **630**, but lacking the ethoxycarbonyl group, fails to cyclize, Huffman was successful in cyclizing **630**. After the appropriate further manipulations, (±)-cacalol (**612**) is produced. Of the three syntheses, the latter appears to be the least efficient, requiring nine steps and producing (±)-cacalol in only 0.17% overall yield. The Walls synthesis requires six steps and gives the final product in 1.5% yield. The Tokyo synthesis requires 13 steps. Although yields are not reported for most of the steps, one of the key steps, Friedel-Crafts acylation of **615**, occurs in only 20% yield.

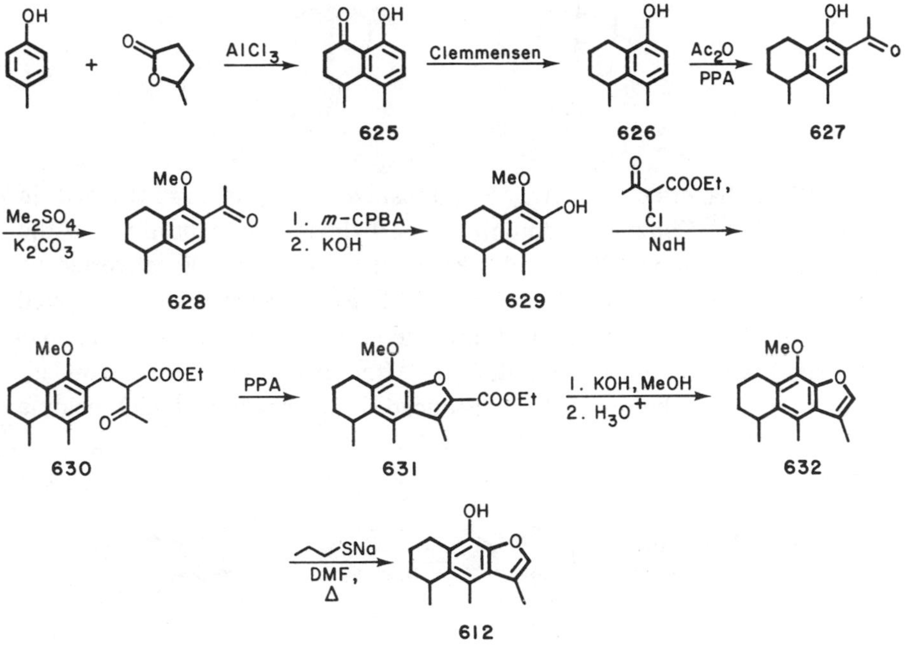

Scheme 81.   (±)-Cacalol: Huffman's Synthesis

E. Miscellaneous Hydronaphthalenes

(1) Valeranone and Valerane

Valeranone (633) is a novel sesquiterpene which, although isoprenoid, does not fit the "head-to-tail isoprene rule." It has attracted considerable synthetic attention, and several synthetic approaches were reported in Volume 2 of this series. In this section, we will discuss one additional synthesis of 633, as well as two papers which deal with the synthesis of the parent hydrocarbon, valerane (634).

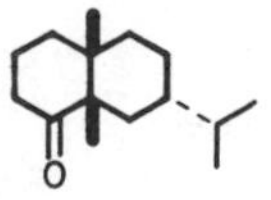

633 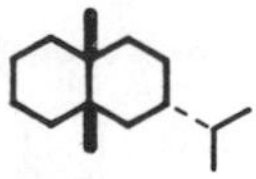   634

In 1973, Banerjee and Angadi published the synthesis outlined in Scheme 82.[271] Conjugate addition of methylmagnesium bromide to the well-known octalone 635 provides keto ester 636, which is converted by sodium hydride and diethyl carbonate into 637. The ketone is removed by desulfurization of the derived dithioketal and the resulting diester 638 is treated with excess methyllithium. After oxidation of the secondary alcohol, hydroxy ketone 639 is produced. Dehydration of 639 yields a

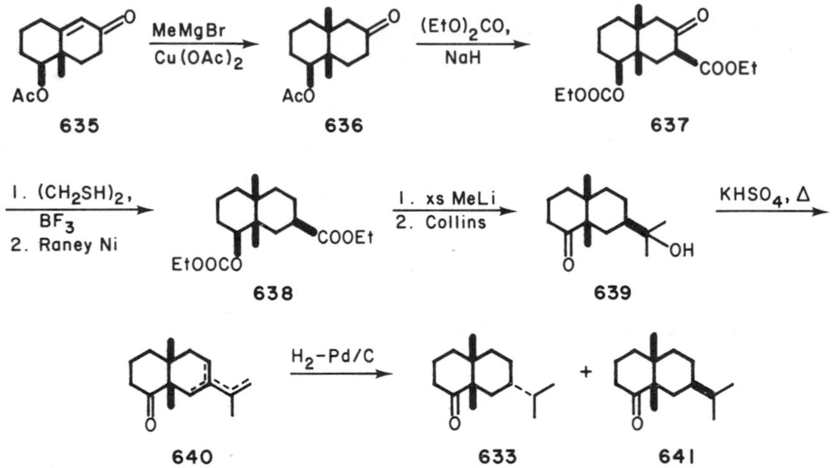

Scheme 82.   Banerjee's Synthesis of (±)-Valeranone

mixture of four unsaturated ketones (640), which is hydrogenated to obtain 20% (±)-valeranone (633) and 80% of its diastereomer 641. In light of this result, the author's claim of a "stereoselective total synthesis of (±)-valeranone..." seems specious.

Posner and his co-workers reported the interesting synthesis of valerane outlined in Scheme 83.[272] Enol ether **642** is converted into cyclohexenone **643** by a well-precedented method. This material is

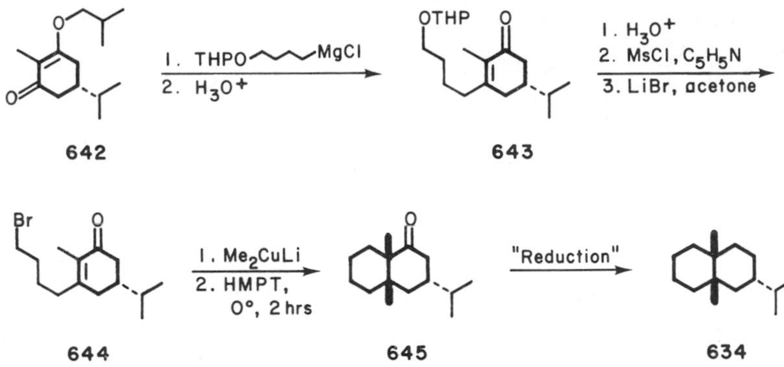

Scheme 83.   Posner's Synthesis of ($\pm$)-Valerane

transformed into enone bromide **644**, which is subjected to conjugate addition with lithium dimethylcuprate. The resulting enolate is treated with hexamethylphosphoric triamide whereupon internal alkylation occurs to produce decalone **645** in 25% yield. Removal of the carbonyl group provides ($\pm$)-valerane (**634**). Although this synthesis is brief, it is not clear how it would be applied to valeranone itself, the only important compound of the valerane class.

Baldwin and Gawley explored the interesting synthesis outlined in Scheme 84.[273] Photoaddition of 1,2-dimethylcyclohexene and β-keto aldehyde **646** produces keto aldehyde **647** in quantitative yield. Acid-catalyzed aldolization gives enone **648**, which is hydrogenated to decalone **649**. Wittig reaction of **649** gives alkene **650**, which is hydrogenated to a mixture of 40% ($\pm$)-valerane (**634**) and 60% of its diastereomer (**674**). Attempts to significantly alter this ratio by hydrogenating other alkene isomers were not successful. Thus, Baldwin confirms Banerjee's finding that catalytic hydrogenation of an alkene is *not* the way to establish the correct valerane stereochemistry.

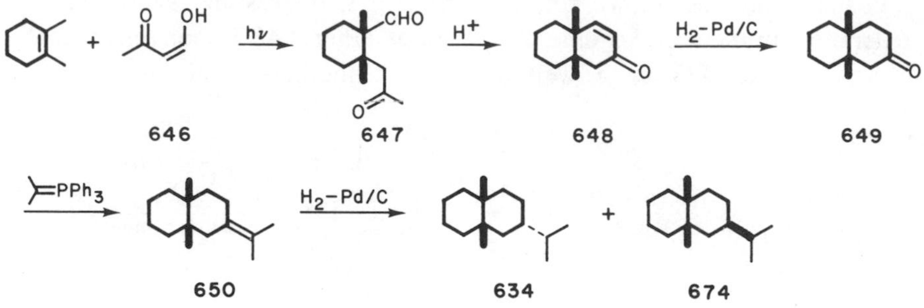
hν
CHO
H+
H2-Pd/C
646
647
648
649
PPh3
H2-Pd/C
650
634
674

Scheme 84. Baldwin's Synthesis of (±)-Valerane

(2) Khusitene and β-Gorgonene

Khusitene (**675**) is a nor-cadinane which occurs in Indian vetiver oil. The nonisoprenoid β-gorgonene (**676**) may be a rearranged eudesmane. In this section, we discuss two syntheses of the former and one of the latter.

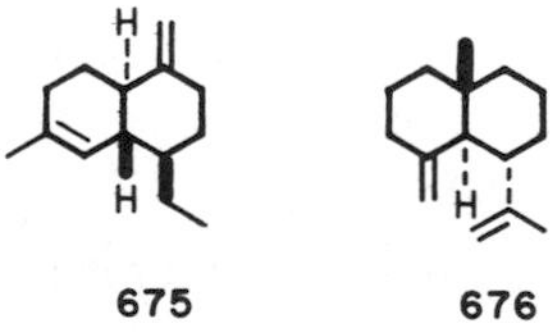
H
H
H
675
676

The first synthesis of (±)-khusitene came from the laboratories of O. P. Vig at the Punjab University in Chandigark. The Vig synthesis is summarized in Scheme 85.[274] Keto aldehyde **676**, obtained from the piperidine enamine of butanal and ethyl acrylate, is condensed with methallyltriphenylphosphorane to obtain diene ester **677**. This material undergoes a Diels-Alder reaction with ethyl acrylate to produce adduct **678**. After Dieckman cyclization, decarboxylation, and Wittig methylenation, (±)-khusitene is obtained.

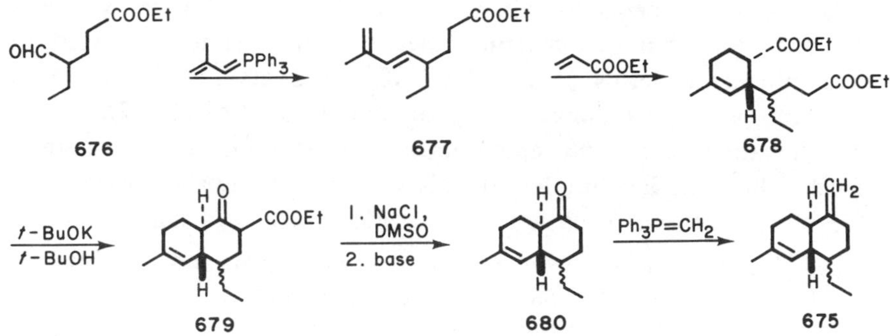

Scheme 85.   Vig's First (±)-Khusitene Synthesis

In 1977 Vig reported the alternative synthesis outlined in Scheme 86.[275] Diene aldehyde **686**, prepared by the straightforward route depicted, is condensed with vinylmagnesium bromide to obtain an alcohol (**687**) which is oxidized by manganese dioxide. The resulting trienone

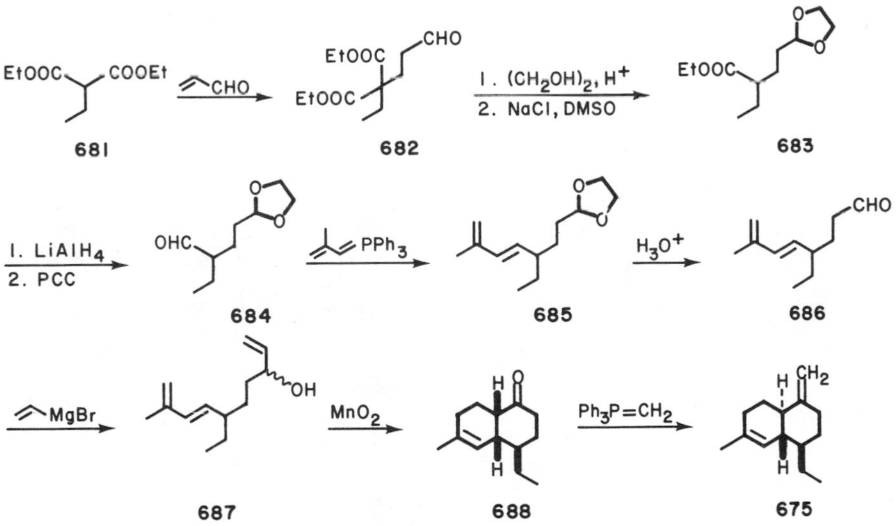

Scheme 86.   Vig's Second (±)-Khusitene Synthesis

undergoes spontaneous intramolecular Diels-Alder addition, giving an octalone. Although the question of stereochemistry is not addressed, the predominant isomer is presumably **688**, based on Taber's experience with the analogous isopropyl compound (see Scheme 42). Wittig methylenation occurs with epimerization to the *trans*-octalin system, as usual. Thus, the final product of this synthesis is probably diastereomer **675**, although the stereostructure of natural khusitene is still unknown.

(±)-β-Gorgonene has been prepared by Boeckman and co-workers as shown in Scheme 87.[276] The known decalone **692** is prepared by well-established reactions. Chlorination of **692**, followed by dehydro-chlorination with hot quinoline affords enone **693**. Conjugate addition of isopropenylmagnesium bromide provides a 7:2:1 mixture of ketones **694**, **695**, and **696**. The major isomer, **694**, is isolated in 45% yield. Methylenation of **694** by the Peterson method provides (±)-β-gorgonene (**676**).

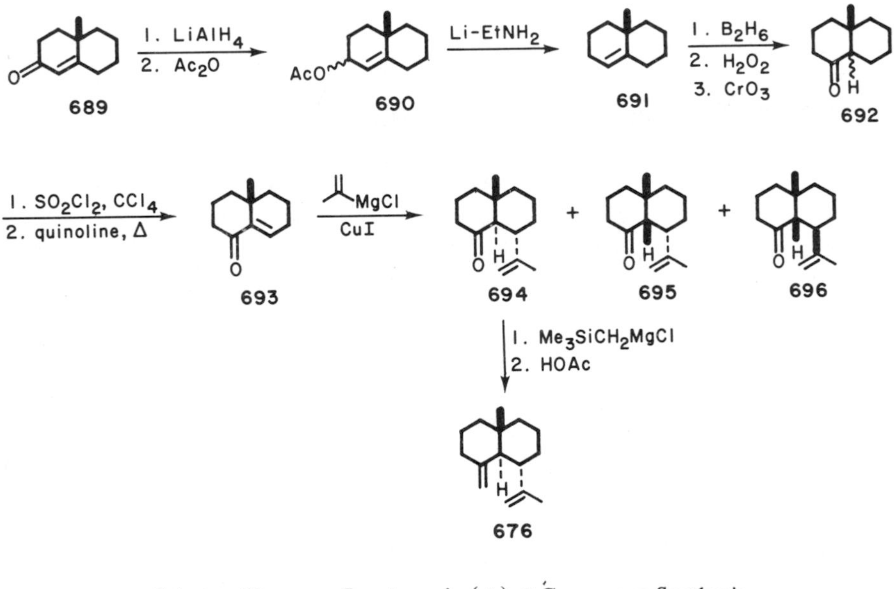

Scheme 87.   Boeckman's (±)-β-Gorgonene Synthesis

F. Hydronaphthalenes Containing an Additional Cyclopropane Ring

In this section, we discuss syntheses of β,β-cycloeudesmol (697), β,α-cycloeudesmol (698), α,β-cycloeudesmol (699), mayurone (700), thujopsene (701), and thujopsadiene (702).

697   698   699

700   701   702

There are four diastereomers corresponding to the cycloeudesmol structure. In 1974, an antibiotic substance was isolated from a marine alga by Fenical and Sims and proposed to have the cycloeudesmol structure. However, the relative stereochemistry of the natural product was not determined. This proposal prompted several groups to synthesize various isomers. As it turned out, none of the synthesized isomers 697-699 are identical with the natural product. Hence, natural cycloeudesmol is suspected to have one of the α,α-structures 703.

703

Moss and co-workers synthesized the β,β and β,α isomers 697 and 698 as shown in Scheme 88.[277] For synthesis of the β,β diastereomer, the known[278] unsaturated ketal 704 is employed. It is known that this compound undergoes addition reactions preponderantly from the top face of the double bond.[278] Thus, reaction with dichlorocarbene under conditions of phase transfer catalysis gives a dichlorocyclopropane which is reduced by sodium in ammonia to tricyclic ketal 705. After ketal hydrolysis, the carbonyl group is reduced to give predominantly alcohol 706. Replacement of the hydroxy group, by reaction of the tosylate with

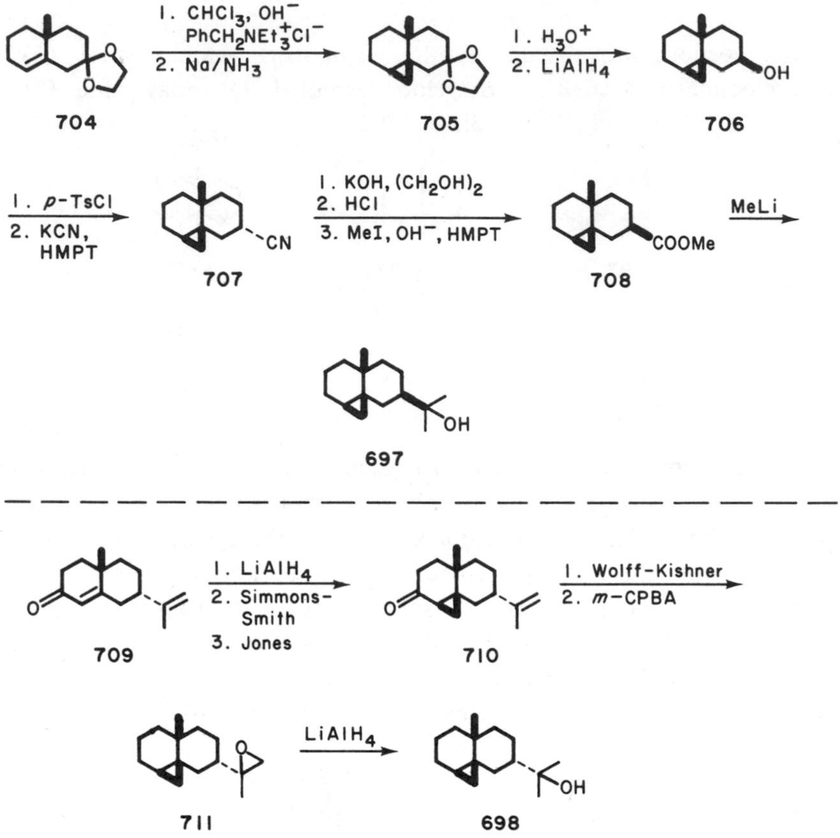

704   **705**   **706**

707   **708**

697

- -

709   **710**

711   **698**

Scheme 88.   Moss Synthesis of Cycloeudesmol Isomers

cyanide ion, is accompanied by some loss of stereochemistry at C_3. However, this is of no consequence as nitrile **707** undergoes ready epimerization upon hydrolysis, yielding after esterification, methyl ester **708**. Reaction of this compound with methyllithium gives $(\pm)$-β,β-cycloeudesmol (**697**).

For synthesis of the β,α diastereomer, Moss and co-workers employed octalone **709**, readily prepared from dihydrocarvone and methyl vinyl ketone. It is well-known that dienone **709** has the isopropenyl and angular methyl groups trans.[279] Reduction of the carbonyl gives the equatorial allylic alcohol which undergoes assisted Simmons-Smith cyclopropanation to provide, after oxidation, the cyclopropyl ketone **710**. The

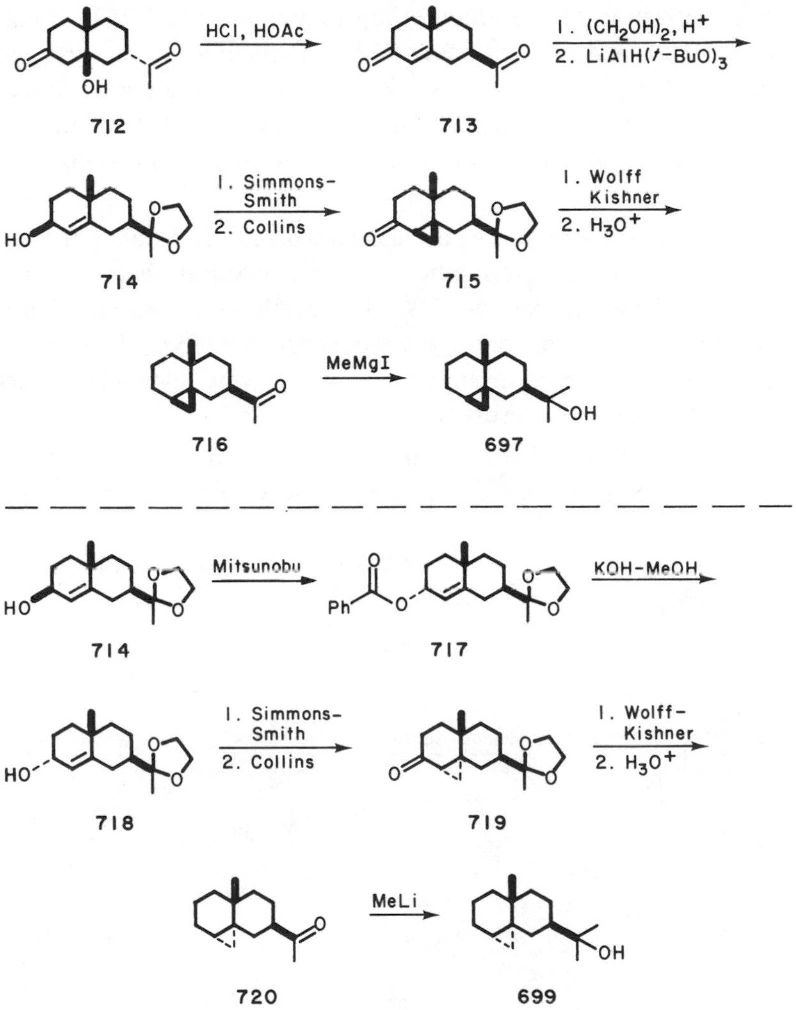

Scheme 89.   Ando Synthesis of Cycloeudesmol Isomers

carbonyl group is removed and the isopropenyl group is hydrated to obtain $(\pm)$-β,α-cycloeudesmol (**698**).

Ando and co-workers prepared all three isomers **697-699**.[280] As shown in Scheme 89, the synthesis of the β,β diastereomer is different from that employed by Moss. Compound **712**[281] is dehydrated with subsequent epimerization of the acetyl group to give enedione **713**. Ketalization of the saturated carbonyl, followed by reduction of the enone gives the equatorial allylic alcohol **714**. The remaining steps are similar to those used by Moss for the β,α isomer. The β,β stereochemistry is obtained by the assistance of the β-hydroxy in the Simmons-Smith reaction.

To make the α,β isomer, Ando and co-workers first invert the allylic alcohol by the Mitsunobu method. Simmons-Smith reaction on **718** gives the α-cyclopropyl ketone **719**. The synthesis is completed in the normal manner to give $(\pm)$-α,β-cycloeudesmol (**699**). The Japanese group prepared the β,α diastereomer **698** by the identical sequence employed by Moss and co-workers.

Caine and co-workers prepared the β,β isomer by the efficient route summarized in Scheme 90.[282] Chloroenone **429** (see Scheme 60) is sol-

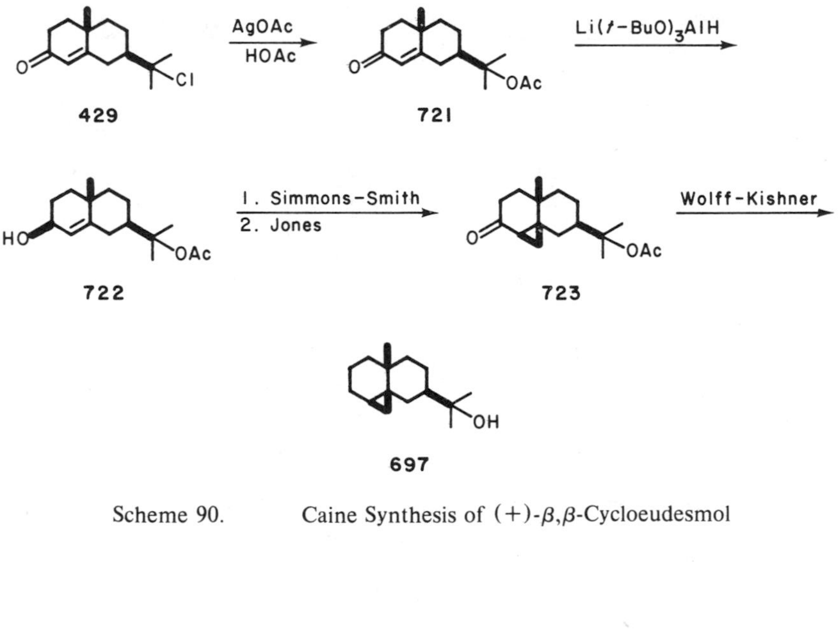

Scheme 90.   Caine Synthesis of $(+)$-β,β-Cycloeudesmol

volyzed to obtain acetate **721**. The remaining steps in the synthesis are essentially the same as are used in the other syntheses discussed.*

McMurry and Blaszczak have synthesized (±)-mayurone (**700**) as outlined in Scheme 91.[283] Aldehyde **724** (an intermediate in Büchi's thujopsene synthesis)[284] is converted into an ester, which is alkylated with allyl

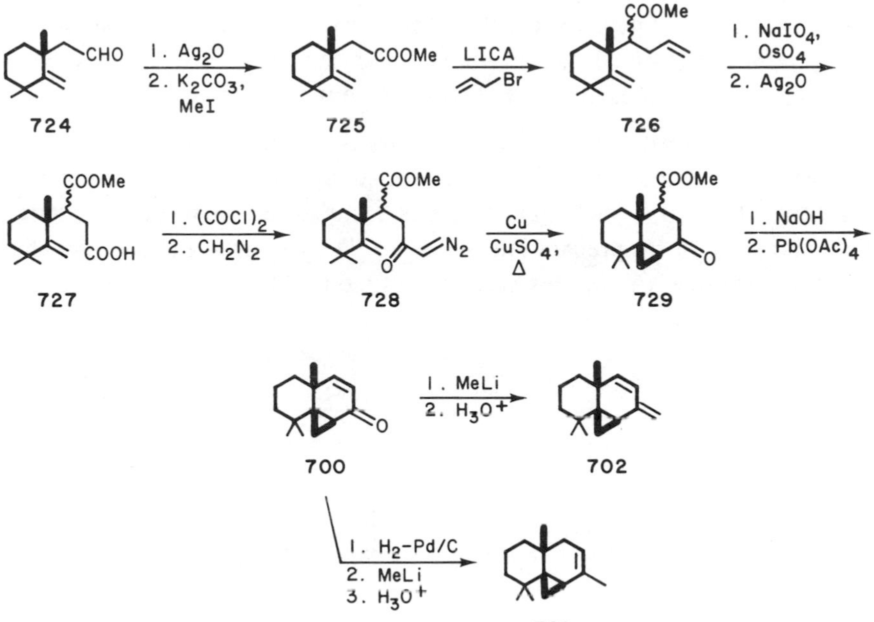

Scheme 91.   McMurry's Synthesis of (±)-Mayurone,
(±)-Thujopsene and (±)-Thujopsadiene

---

*In 1981, Ando and co-workers synthesized the fourth stereoisomer (α, α) and found that it is also not identical with the natural product [M. Ando, S. Sayama, and K. Takese, *Chem. Lett.*, 377 (1981)]. The structure is now known to be:

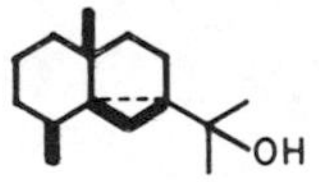

[T. Suzuki, A. Furusaki, H. Kikuchi, E. Kurosawa, and C. Katayama, *Tetrahedron Lett.*, **22**, 3423 (1981)].

bromide. The allyl group is cleaved to an acid (727), which is converted into a diazoketone (728). Copper-catalyzed decomposition of 728 gives a carbene which inserts in the double bond to give 729. This step is based on Büchi's earlier observation[284] that such intramolecular carbene insertions provide the correct thujopsene stereochemistry. The novel transformation in the McMurry synthesis is the oxidative decarboxylation of γ-keto ester 729, which is brought about by treating the acid with lead tetraacetate. The ester function thus serves as a latent double bond. The product (±)-mayurone (700) is converted into (±)-thujopsadiene (702) and (±)-thujopsene (701) by literature procedures.

Smith and co-workers have also prepared (±)-mayurone (700), as summarized in Scheme 92.[285] The Smith synthesis is a showcase for diazoketone chemistry, such intermediates being used on three occasions. The first use is in preparation of ester 725 (an intermediate in the McMurry synthesis). In this case, diazoketone 734 is rearranged under

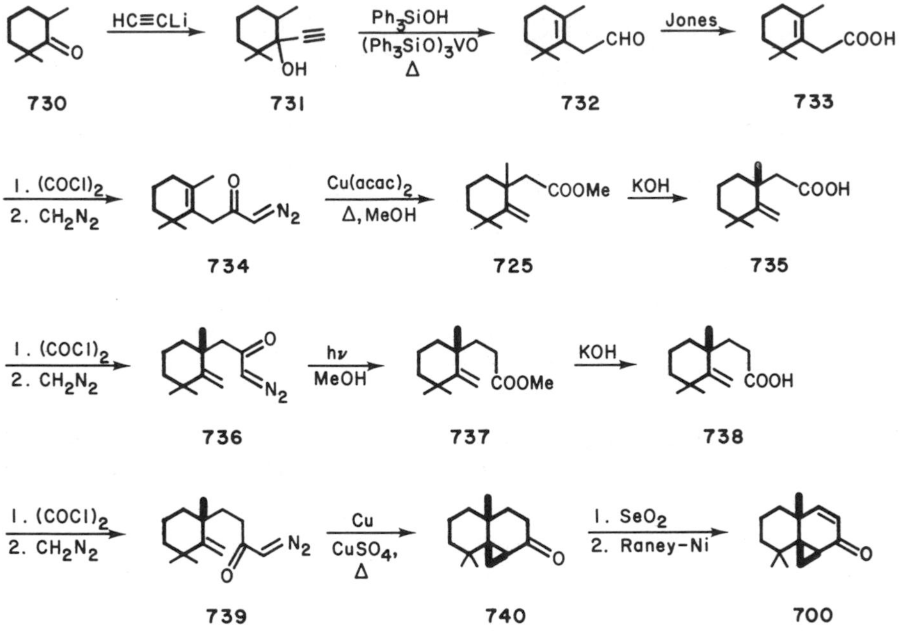

Scheme 92.   Smith's Synthesis of (+-)-Mayurone

the influence of $Cu(acac)_2$, presumably via the intermediate bicyclo-pentanone **741**. In the second use, diazoketone **736** is subjected to photochemical Wolff rearrangement as a method for homologation of the

734   741   742

side chain. Finally, diazo ketone **739** is subjected to copper-catalyzed decomposition, leading to cyclopropyl ketone **740**, as in the McMurry synthesis.

It is interesting to compare the efficiency of the four syntheses which have been reported in this group. Since the earlier workers did not prepare mayurone, we compare on the basis of thujopsene. The 1963 Dauben synthesis of thujopsene requires seven steps from octalone **743**

743   701

and gives thujopsene in 7.3% yield. Büchi's 1964 synthesis requires nine steps from cyclocitral (**744**) and proceeds in 0.8% overall yield. The

744   701

McMurry synthesis is 16 steps from cyclocitral and gives the final product in 32% yield. Finally, the Smith synthesis also requires 16 steps and

744   701

gives (±)-thujopsene in 5.7% overall yield.

730   701

5. OTHER BICARBOCYCLIC SESQUITERPENES

A. Isolated Rings

(1) Taylorine and Hypacrone

Taylorine (**756**) was synthesized from $(\pm)$-Δ^3-carene (**745**) by Nakayama and co-workers as is outlined in Scheme 93.[286] The steps to ester aldehyde **747** proceed without incident. Unfortunately, the Wittig reaction leading to **748** is complicated by the formation also of the acyclic isomer **749**. The two isomers are obtained in yields of 38% and 17%, respectively, after "repeated chromatography". Intermediate **748** is elaborated as shown in the scheme. The cyclpentenone ring is added by the six-step sequence **751**→**754** in 16% overall yield. The synthesis is completed by a low yield (13%) Wittig reaction on **754**, followed by oxidation of the side-chain hydroxy to obtain $(-)$-taylorine. Although all yields are not published, the overall yield for this 15-step conversion of carene into taylorine is no greater than 0.5%.

Hypacrone (**763**), a novel seco-illudoid from the fern *Hypolepis punctata* Mett., was synthesized by Hayashi and co-workers by the brief synthesis summarized in Scheme 94.[287] Unsaturated acid **757**, conveniently prepared by alkylation of the dianion of isobutyric acid, is cyclized by way of its acyl chloride to cyclopentenone **758**. This material is converted into the dienol ether **759**, which is condensed by the method of Mukaiyama with diketone **760**. Adduct **761** is obtained in 30% yield after chromatography on silica gel. Dehydration of **761** occurs upon

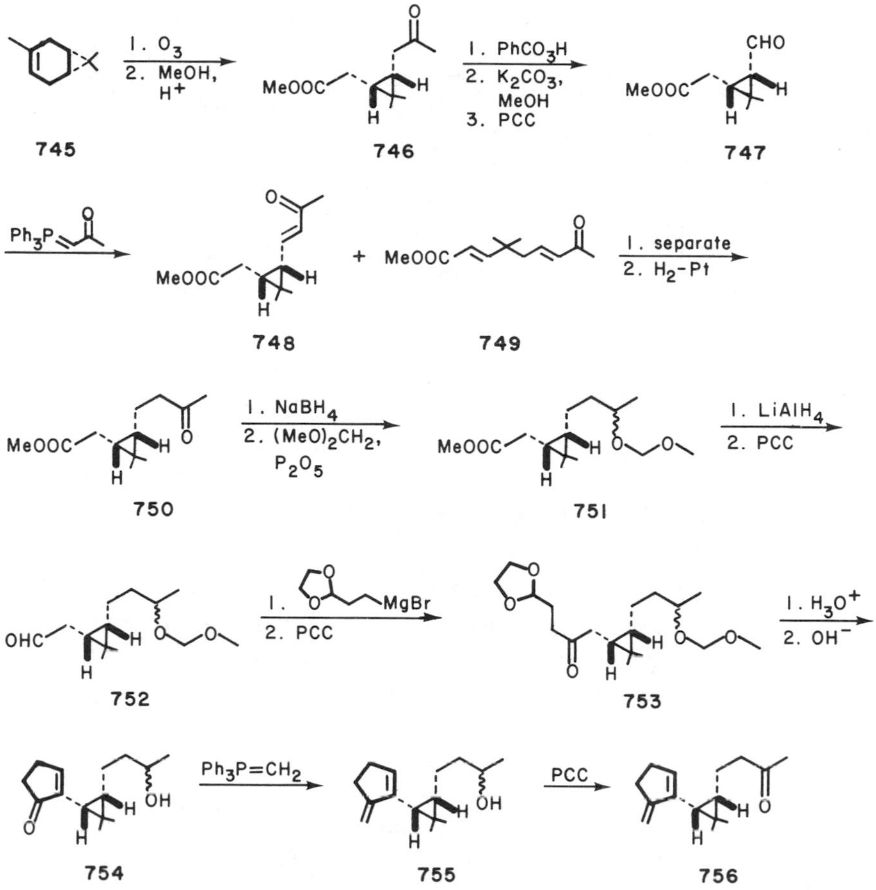

Scheme 93.   Nakayama's Synthesis of (-)-Taylorine

preparative gas chromatography at 190°C to give a dienone in 60% yield. Not unexpectedly, the more stable *E* diastereomer **762** is the sole produce of the reaction. Sensitized photolysis of **762** produces a photostationary state (1:1) of **762** and hypacrone (**763**), the *Z* diastereomer. At the time of publication[287] isomers **762** and **763** had not been separated, and the production of hypacrone was inferred only from the ¹H-NMR spectrum of the mixture.

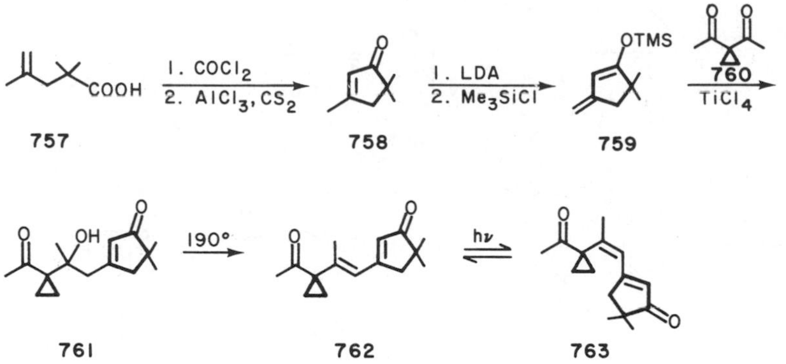

Scheme 94.   Hayashi's Synthesis of Hypacrone

(2) Cuparene, α-Cuparenone, and β-Cuparenone

During the period of this review, the cuparanes cuparene (764), α-cuparenone (765) and β-cuparenone (766) have been synthesized. The synthetic challenge is the same in each case—creation of the adjacent quaternary centers. A variety of methods have been used to solve this problem.

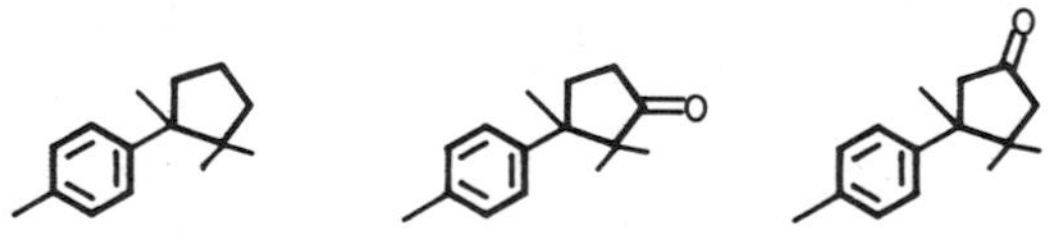

Vig and co-workers have reported a synthesis of (±)-cuparene (Scheme 95).[288] The synthesis begins with addition of the p-tolyl Grignard reagent to β-keto ester 767 to give an alcohol which is allegedly dehydrated to 768. However, their report should be taken with caution, since 768 is described as a "conjugated ester" and ascribed a carbonyl absorption of 1710 cm⁻¹. The remaining methyl group is said to be introduced by coupling of the tertiary, benzylic bromide 769 with lithium dimethylcuprate.

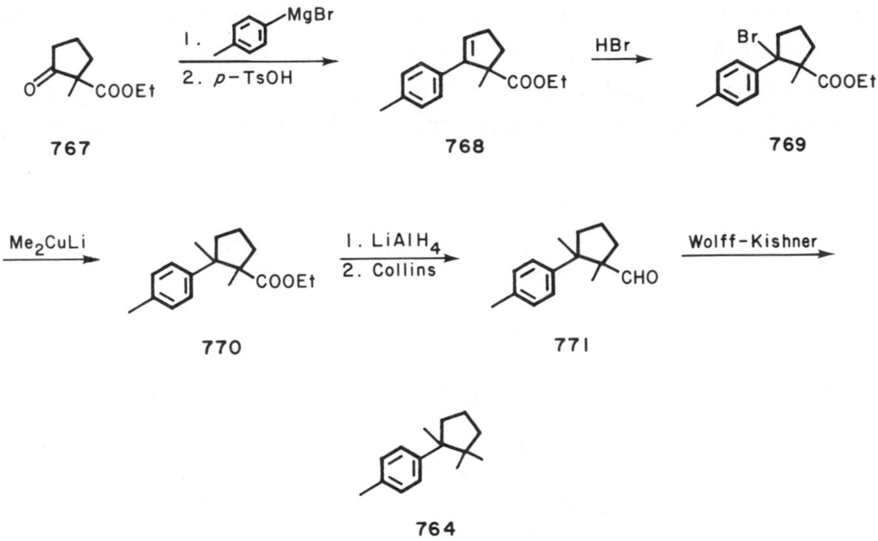

Scheme 95.   Vig's Synthesis of (±)-Cuparene

Bird, Yeong, and Hudec, in England, have prepared (±)-cuparene as shown in Scheme 96.[289] Cyclohexene 772, obtained from 2,2-dimethyl-cyclohexanone and *p*-tolyllithium, is epoxidized to obtain 773. Upon treatment with BF_3 etherate, 773 isomerizes to a mixture of aldehyde 774 (22%) and ketone 775 (25%). Wolff-Kishner reduction of 774,

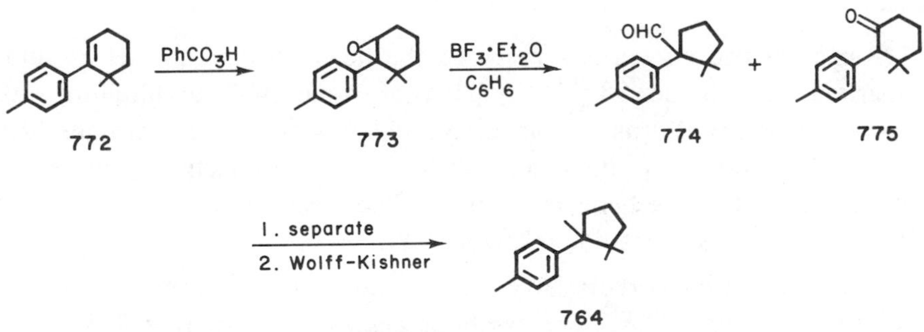

Scheme 96.   Bird-Yeong-Hudec Synthesis of (±)-Cuparene

obtained by preparative thin layer chromatography, provides (±)-cuparene.

De Mayo and Suau synthesized (±)-cuparene by the inventive route outlined in Scheme 97, which incorporates the photocyclization of thione

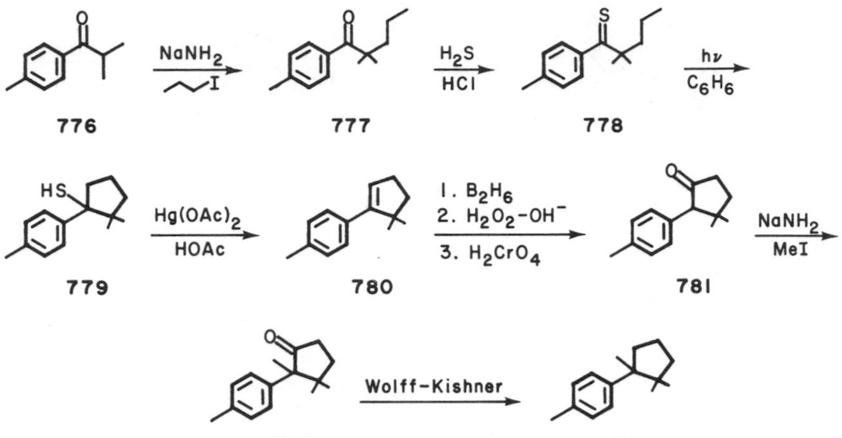

776 **777** **778** **779** **780** **781** **782** **764**

Scheme 97. De Mayo's Synthesis of (±)-Cuparene

778 as a key step.[290] The crude benzylic thiol **779** is treated with mercuric acetate to accomplish the elimination of hydrogen sulfide. The remainder of the synthesis is straightforward, proceeding through cyclopentanone **781**, which is methylated to establish the second quaternary center.

Kametani and co-workers prepared the de Mayo ketone **781** as summarized in Scheme 98.[291] Coupling of 5-methyl-2-furyllithium with *p*-xylyl bromide affords a furan (**783**), which is hydrolyzed to dione **784**. The aldol closure of the latter substance is regioselective, due to the greater acidity of the benzylic protons. The second methyl is introduced by cuprate addition, leading to ketone **781**.

Wenkert and co-workers have synthesized (±)-α-cuparenone (**765**) as shown in Scheme 99.[292] The synthesis begins with aldehyde **786**, which is obtained in 40% yield by the chromyl chloride oxidation of *p*-cymene.

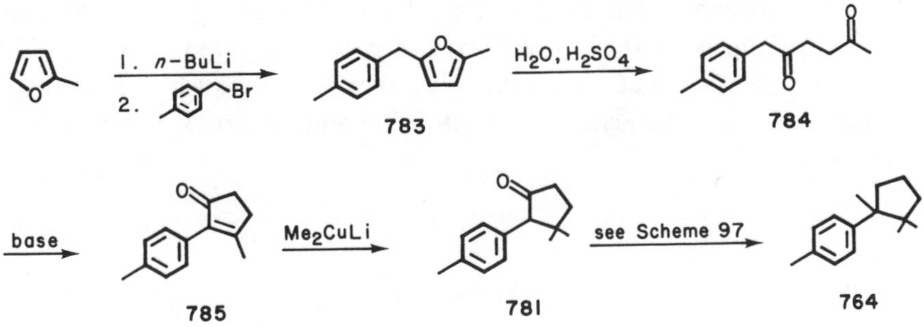

Scheme 98. Kametani's Synthesis of (±)-Cuparene

Decomposition of diazoacetone in the presence of enol ether **788** provides a mixture of diastereomeric cyclopropyl ketones (**789**), which is ring-opened by treatment with HCl. The resulting keto aldehyde is aldolized to give enone **790**, which is methylated to obtain **791**. Saturation of the double bond affords (±)-α-cuparenone.

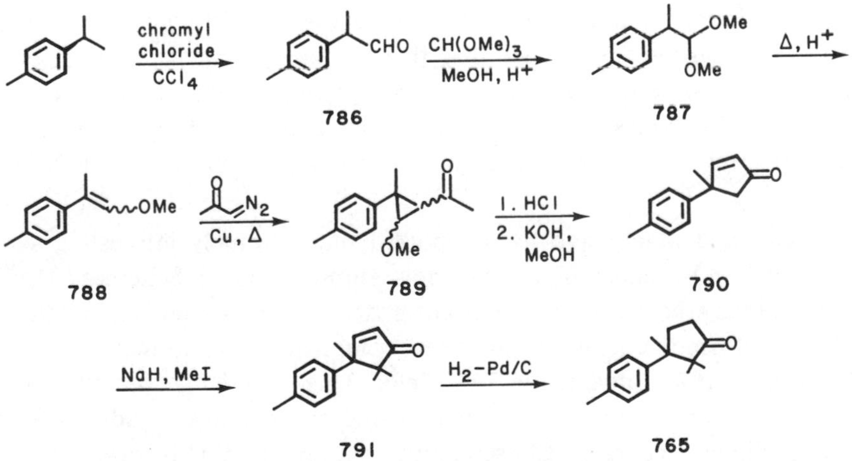

Scheme 99. Wenkert's Synthesis of (±)-α-Cuparenone

($\pm$)-β-Cuparenone has been prepared by Mane and Krishna Rao as outlined in Scheme 100.[293] The Grignard reagent derived from chloride **792** undergoes conjugate addition with the Knoevenagel adduct from acetone and ethyl cyanoacetate. Hydrolysis and decarboxylation of the

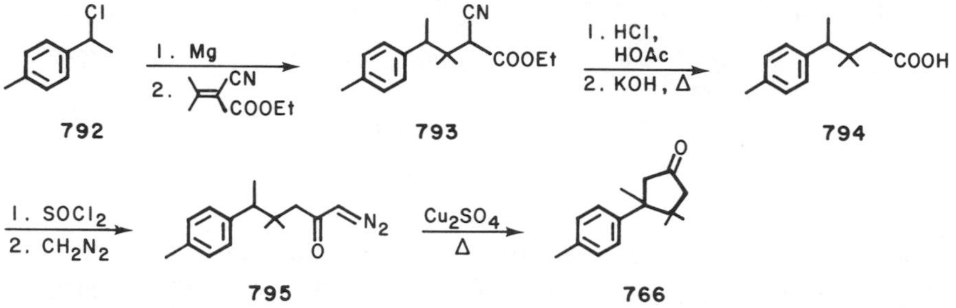

Scheme 100.   Krishna-Rao's Synthesis of ($\pm$)-β-Cuparenone

crude adduct (**793**) provides acid **794** in 50% overall yield. The carbene derived from diazoketone **795** undergoes insertion into the tertiary, benzylic C-H bond to give ($\pm$)-β-cuparenone (**766**) in 39% yield, based on acid **794**. It should be noted that this reaction is one of only a small number of examples of the insertion of an α-keto carbene into a σ-bond. The synthetic β-cuparenone is accompanied by several by-products, no doubt resulting from other reactions of the intermediate carbene.

Casares and Maldonado have reported the extremely interesting synthesis of ($\pm$)-β-cuparenone, which is summarized in Scheme 101.[294] The synthesis begins with an extraordinary reaction—conjugate addition of the preformed enolate of protected cyanohydrin **796** to mesityl oxide. The adduct **797** is obtained in 70% yield. This material is converted into the known enone **799**, which fails to undergo conjugate addition with lithium dimethylcuprate. Consequently, the final methyl is introduced in an indirect manner, via cyclopropyl ketone **802**. The reductive cleavage

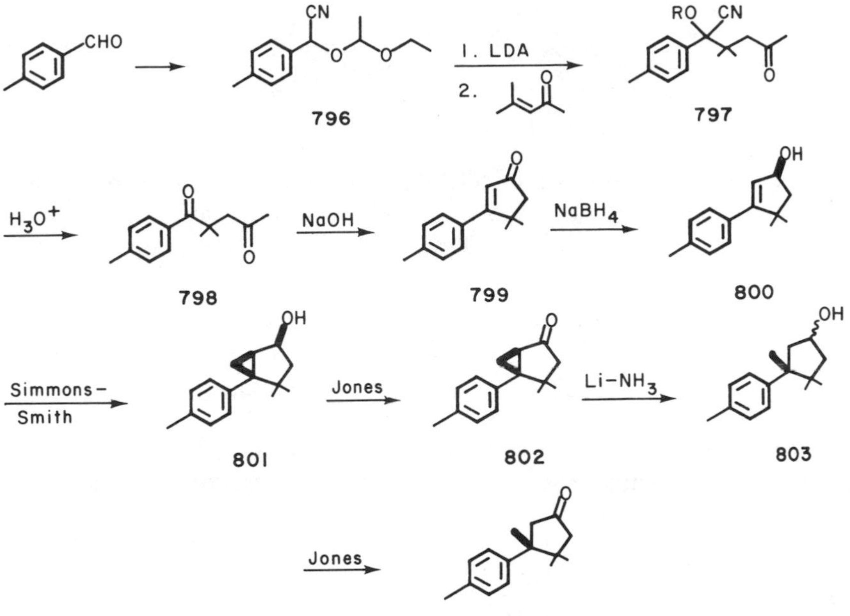

Scheme 101.   Casares-Maldonado Synthesis of (±)-β-Cuparenone

of **802** provides a striking demonstration of the importance of orbital overlap in this reaction; the peripheral bond is cleaved even though cleavage of the ring fusion bond would lead to a benzylic radical.

(3) Laurene and Aplysin

Laurene (**804**) and aplysin (**805**) are related to the simpler cuparanes, but have suffered a methyl migration. Both molecules present the problem of relative stereochemistry. In laurene, the methyls must be made trans, the less stable relative configuration. In aplysin, the secondary methyl is endo with respect to the bicyclo[3.3.0]octane system, also the less stable configuration.

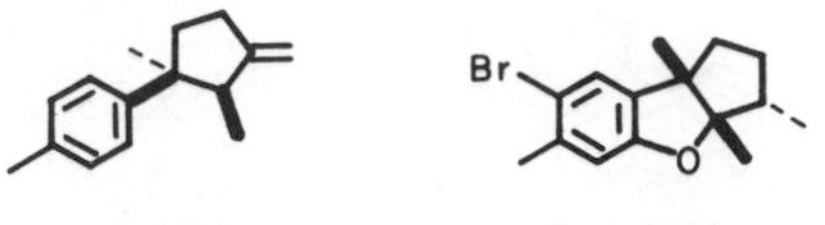

804   805

An attempted synthesis of **804** by Irie and co-workers went astray when oxidation of alcohol **806** with Sarett's reagent gave the epimerized ketone **807**.[295] Subsequently, McMurry and von Beroldingen modified

806   807

the basic Irie approach, as summarized in Scheme 102.[296] The requisite relative stereochemistry is established by hydroboration of cyclopentene

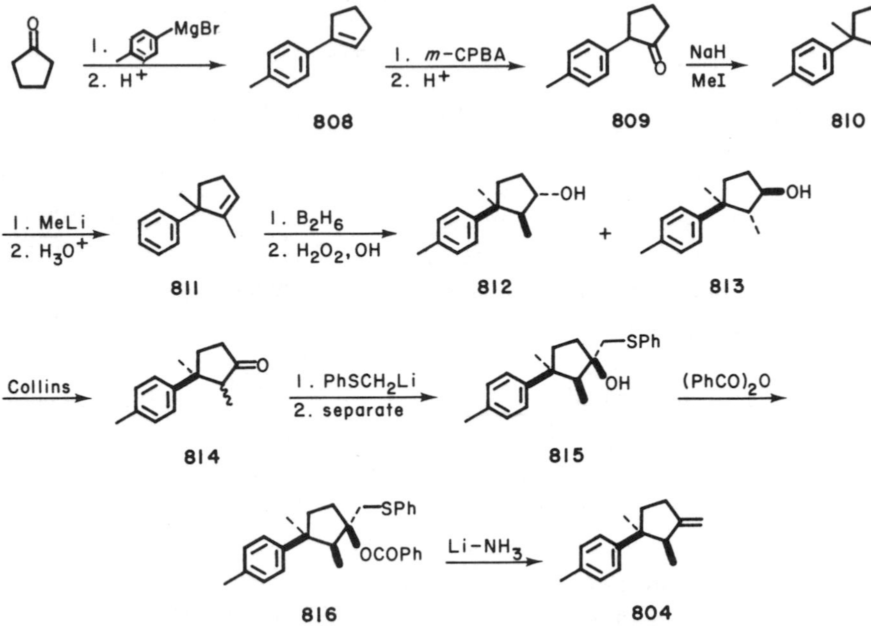

Scheme 102.   McMurry's Synthesis of (±)-Laurene

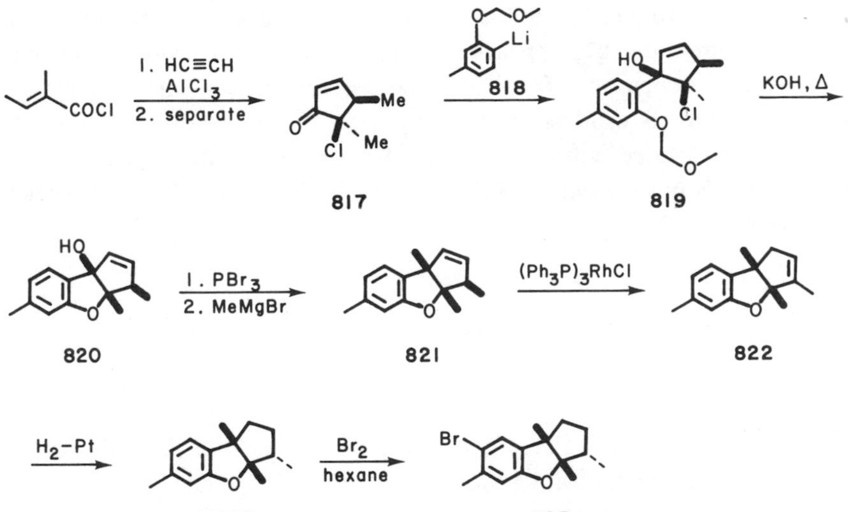

Scheme 103.   Ronald's Synthesis of (±)-Aplysin

811; a 4:1 mixture of diastereomers **812** and **813** is produced. McMurry found that, unlike the Sarett procedure (compare **806→807**), Collins oxidation of the hydroxy group is not accompanied by epimerization of the methyl group; a 4:1 mixture of ketones (**814**) is obtained. Although simple Wittig methylenation of this mixture causes extensive epimerization, ketones **814** react with phenylthiomethyllithium without isomerization. The diastereomers are separated at this point and the major isomer is eliminated by Coates' procedure to obtain (±)-laurene (**804**).

Ronald has synthesized (±)-aplysin by the novel route shown in Scheme 103.[297] Tigloyl chloride reacts with acetylene and aluminum chloride to give a 3:2 mixture of the chlorocyclopentenone **817** and its diastereomer. After separation of these isomers, by silica gel chromatography, compound **817** is condensed with the aryllithium reagent **818** to obtain chlorohydrin **819**. Treatment of this material with hot KOH affords allylic alcohol **820**, which is treated with phosphorus tribromide and the crude product immediately added to excess methylmagnesium bromide. Compound **821** appears to be the only isomer produced in this remarkable sequence, in 20% yield. The correct sense of chirality at the final stereocenter is established by isomerization of the double bond to the more substituted position, followed by catalytic hydrogenation to give (±)-debromoaplysin (**823**). Bromination of **823** provides (±)-aplysin.

(4) Trichodiene, Norketotrichodiene, 12,13-Epoxytrichothec-9-ene, Trichodermin, and Trichodermol

The trichothecane antibiotics are a group of oxygenated sesquiterpenes produced as metabolites of various fungi. They have elicited considerable interest due to their cytotoxic and phytotoxic effects. Syntheses have been recorded for trichodiene (**824**, a presumed biosynthetic intermediate), norketotrichodiene (**825**), 12,13-epoxytrichothec-9-ene (**826**), trichodermin (**827a**), and trichodermol (**827b**).

824

825

826

827a R = Ac
827b R = H

The first publication in the area was the Colvin-Raphael synthesis of (±)-trichodermin, which sets important precedents for further efforts in the group.[298a,b] As shown in Scheme 104, the synthesis begins with preparation of the key lactone **832** from *p*-cresol methyl ether. Two

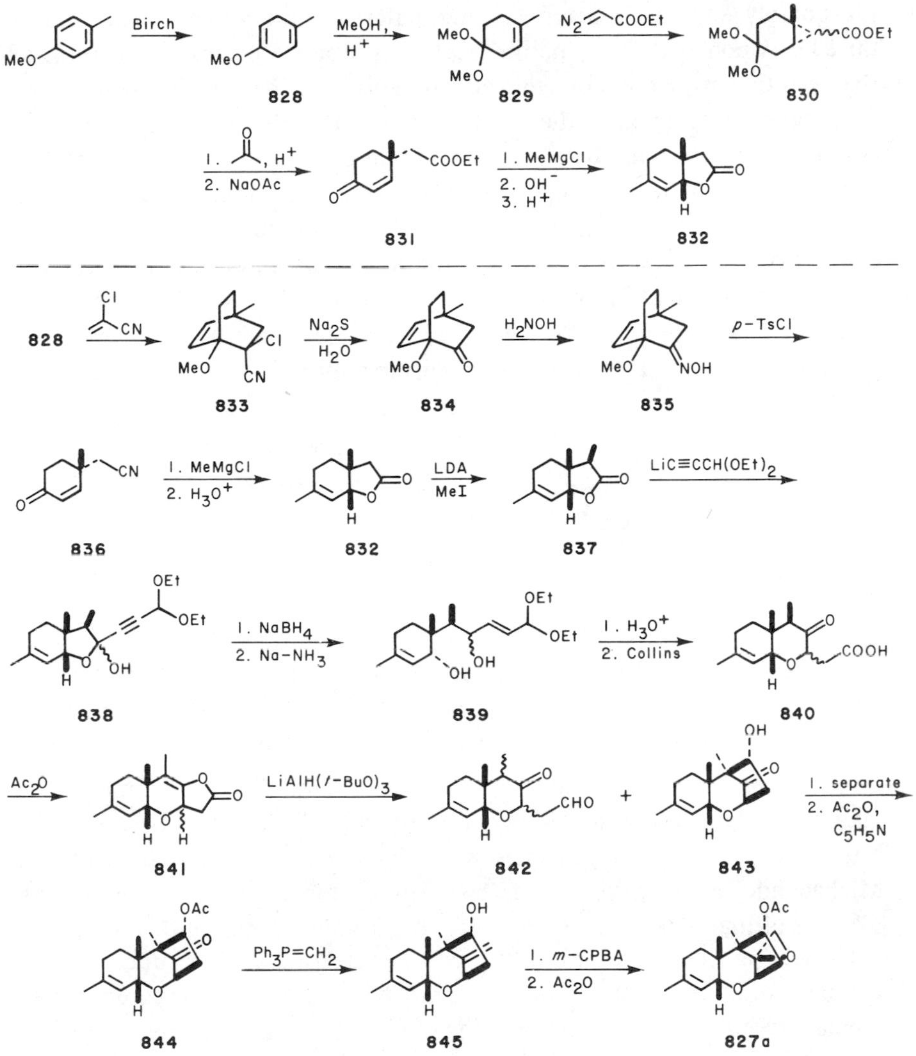

Scheme 104.   Colvin-Raphael Synthesis of (±)-Trichodermin

routes to this important intermediate were developed, one proceeding by way of cyclopropane **830**, and the other by way of the Diels-Alder adduct **833**. Methylation of **832** provides a single diastereomeric product, presumably **837**. This lactone is manipulated, via hemiketal **838**, to acetal **839**. Upon hydrolysis of the acetal, cyclization occurs. Oxidation of the resulting hydroxy aldehyde affords keto acid **840**, which reacts with acetic anhydride to give the enol lactone **841**. Reduction of the latter compound gives a mixture of keto aldehyde **842** (52%) and aldol **843** (7%). Unfortunately, compound **842** cannot be induced to cyclize to provide more of aldol **843**. The reason for this failure is not clear, since the various diastereomers of **842** should be in equilibrium under the reaction conditions. However, it is not certain that **843** even arises by aldol cyclization. It is at least conceivable that it may form by "alkoxide-assisted [1,3]sigmatropic rearrangement":

The Colvin-Raphael synthesis is completed by methylenation of the acetate of **843**, followed by epoxidation of alcohol **845**. Good selectivity is observed in the epoxidation, presumably because the proximate hydroxy group guides the reagent to the exocyclic double bond.

In 1974, Fujimoto and co-workers reported a synthesis of (±)-12,13-epoxytrichothec-9-ene, a metabolite identical to trichodermin but lacking the acetoxy function.[299] The synthesis begins with the Michael addition of β-keto ester **846** onto crotonaldehyde (Scheme 105). The resulting aldehyde is converted into an acetal and the enone carbonyl is reduced by the Meerwein-Pondorf method. Upon acidification, the cis-fused dihydropyran **848** is obtained, apparently as a single stereoisomer. Although the stereochemistry at the secondary methyl center is unimportant, as it is destined to be lost at a later stage, it is noteworthy that a single isomer is produced in the Michael reaction. The stereochemistry is the same as that observed by Marshall in a

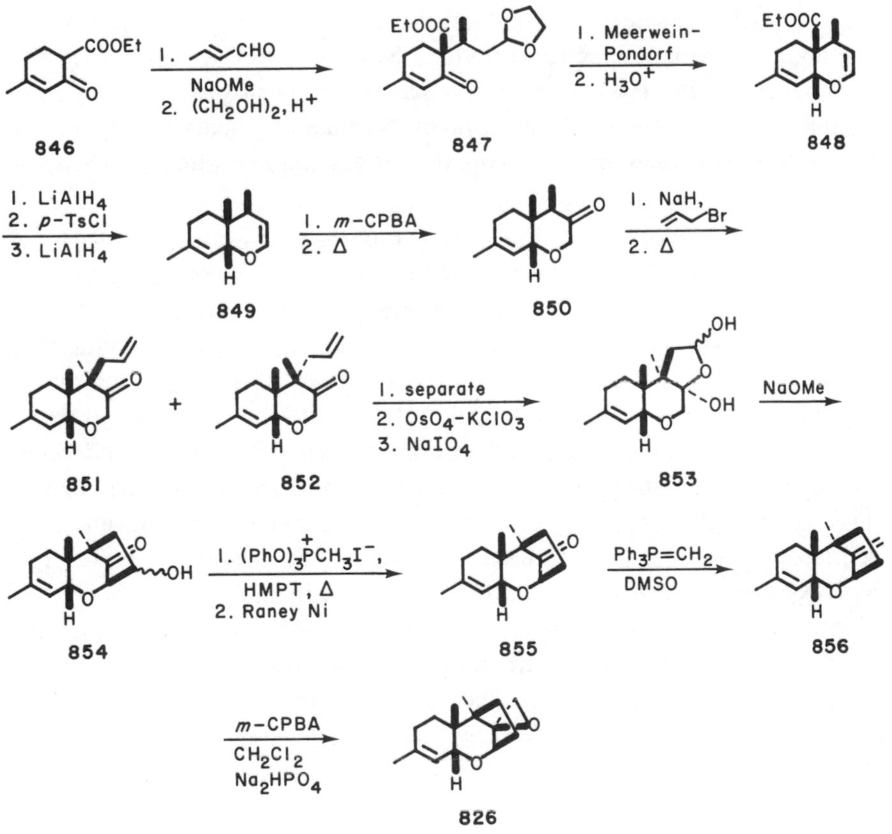

Scheme 105.   Fujimoto's Synthesis of (±)-12,13-Epoxytrichothec-9-ene

related reaction.[300] Epoxidation of **849** occurs on the more nucleophilic enol ether double bond. Pyrolysis of the resulting mixture of diastereomeric epoxides provides ketone **850**. Allylation occurs, rather unexpectedly, on oxygen. However, Claisen rearrangement rectifies the situation and affords a 2:1 mixture of diastereomers **851** and **852**. The isomers are separated by silica gel chromatography and the allyl double bond of the major isomer is cleaved. Treatment of the resulting product (**853**) under basic conditions affords tricyclic aldol **854** in 90% yield. The aldol is deoxygenated and the ketone is methylenated by the Wittig method. Epoxidation of **856** provides **826** (30% yield), along with 40% of recovered **856**. No comment is made regarding oxidation of the other double bond. However, later work by Kamikawa suggests that the two double bonds of **856** undergo oxidation at comparable rates (see Scheme 106).

In 1976, Masuoka and Kamikawa reported a second synthesis of (±)-12,13-epoxytrichothec-9-ene (Scheme 106).[301] Photoaddition of ketal **857** onto 3-methyl-2-cyclopentenone provides a mixture of adducts, from which **858** may be obtained in 16% yield. Acetolysis of adduct **858** yields **859**, in which the relative stereochemistry of the two asymmetric carbons has been neatly arranged. Dione **859** is manipulated by a straightforward sequence of steps into β-keto ester **861**, which undergoes Mannich condensation at C_5 of the cyclopentanone to provide **862** in 37% yield. After removal of the ethoxycarbonyl group, the enone is reduced to a mixture of diastereomeric allylic alcohols (**864**). The hydroxy group is acetylated and the dithioketal hydrolyzed to obtain **865**. Upon removal of the acetate, a mixture of **866** (60% yield) and **867** (12% yield) is obtained. The mixture is separated and the hydroxy group in the major isomer inverted to obtain **868**. Reaction of the latter with methylmagnesium iodide, followed by acid, yields diene **856**. Oxidation of **856** gives a 1:1 mixture of (±)-12,13-epoxytrichothec-9-ene (**826**) and **870**.[302]

In 1980, Still and Tsai reported the outstanding piece of synthetic craftsmanship which we summarize in Scheme 107.[303] Dienyl ether **871** reacts with benzoquinone to give the endo adduct **872**, which is regiospecifically ring-contracted to cyclopentenone **874** by alkaline epoxidation, followed by Favorskii rearrangement. A second stereoselective epoxidation, followed by lithium-ammonia reduction provides triol **875**, in which the relative stereochemistry of the eventual secondary hydroxy and the two methyl-bearing centers is correctly established. Selective acetylation of the primary hydroxy gives acetate **876**, which is reductively cleaved by photolysis in aqueous HMPT. The next four steps serve to rearrange the protecting groups and set up the desired 1,3-diol system for fragmentation (**878→879**), which unmasks the trichothecane carbon

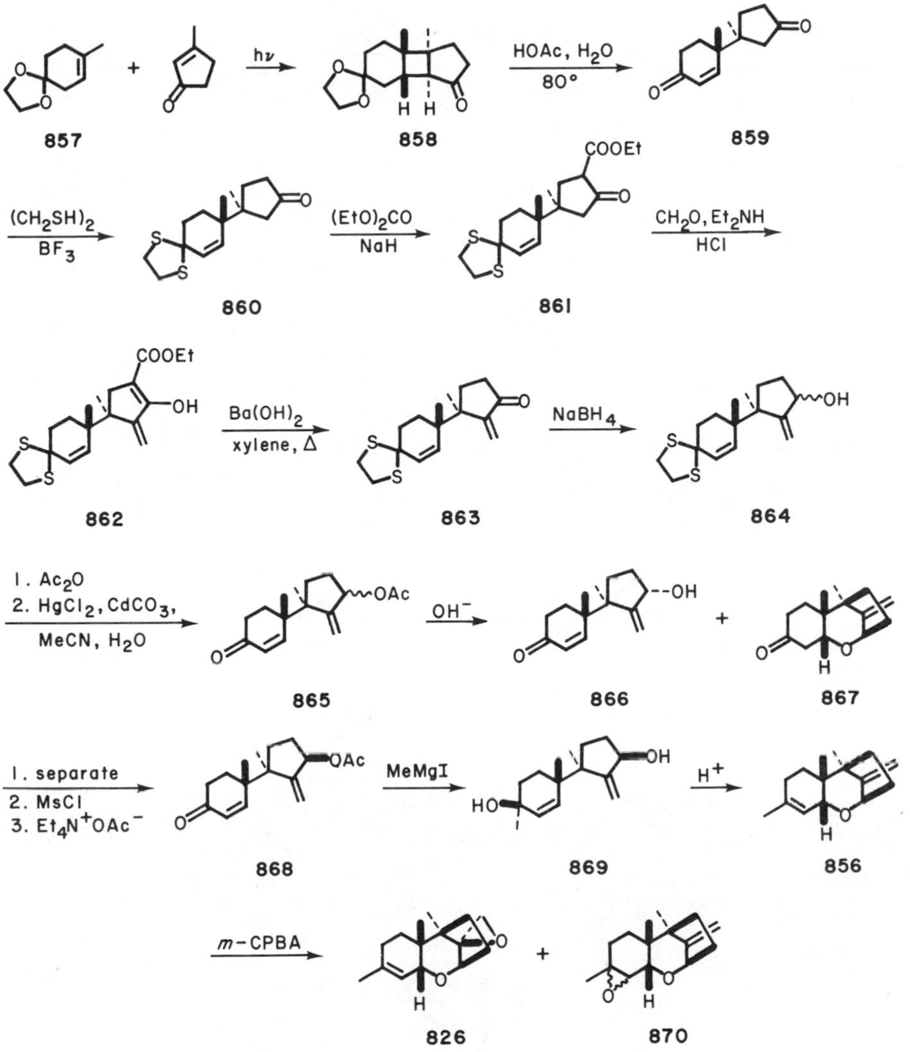

Scheme 106.   Masuoka-Kamikawa Synthesis
of (±)-12,13-Epoxytrichothec-9-ene

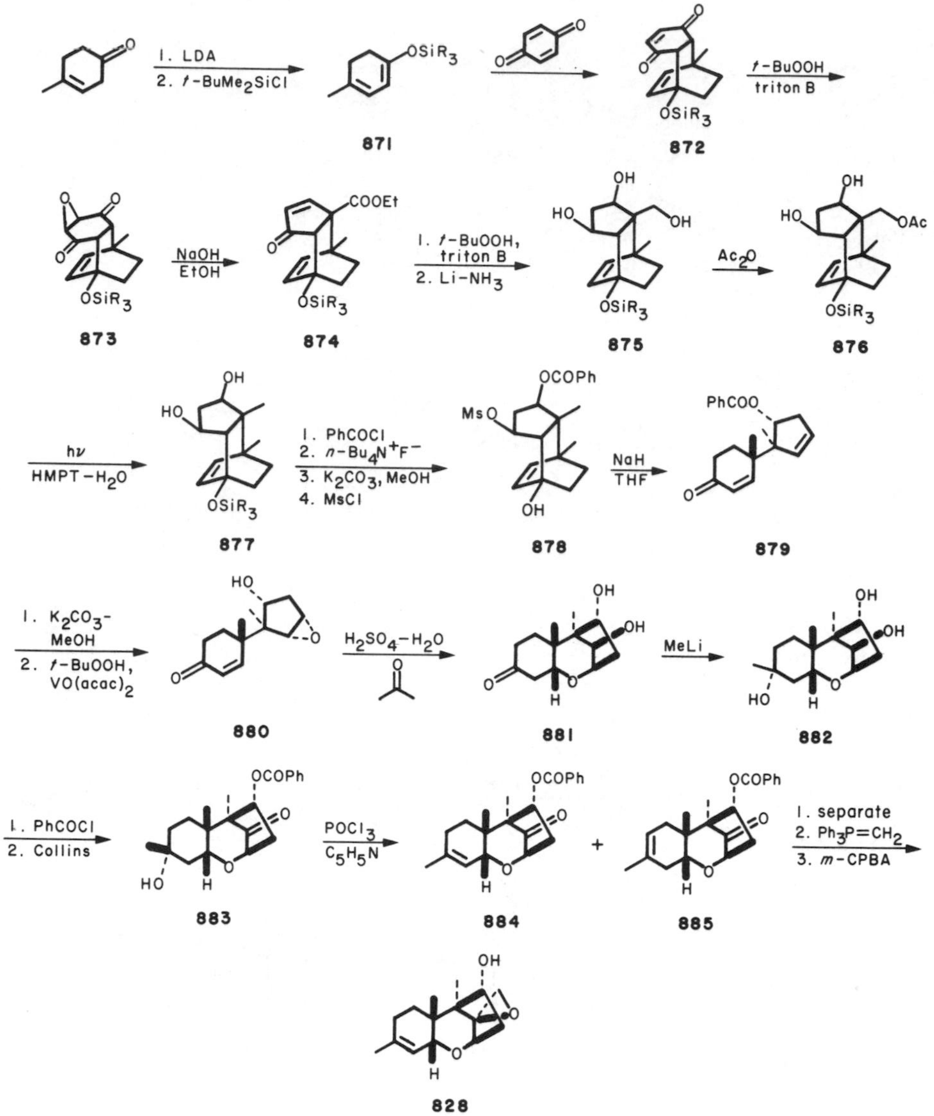

Scheme 107.   Still's Synthesis of (±)-Trichodermol

skeleton. The introduction of the oxygen bridge between the two carbo-
cycles now remains to be done. This is accomplished by stereoselective
epoxidation of the alcohol derived by hydrolysis of benzoate **879**. The
resulting epoxide (**880**) undergoes acid-catalyzed ring-opening to a triol
which undergoes intramolecular Michael addition to give **881**. The
specificity of the ring-closure is interesting as two other tricyclic ethers
(**886** and **887**) might also have formed by addition of the other two
hydroxy groups to the enone. However, each of these compounds is of

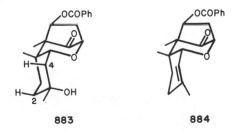

necessity a trans-fused bicyclo[3.3.0]octane system, which would be considerably less stable than the bridged system in the actual product. The
synthesis is completed by addition of methyllithium to the carbonyl
group, protection of the less hindered secondary hydroxy group, and oxidation of the other. Dehydration of the tertiary alcohol gives a 7:1 mixture of the desired alkene **884** and its isomer **885**. The reason for this
fortunate selectivity is not clear. It may simply be that the acidity of the
C_4-hydrogen is slightly greater than that of the C_2-hydrogen because of
the inductive effect of the proximate oxygen. The synthesis is completed

in the same manner as by Colvin and Raphael in their earlier trichodermin synthesis. In all, the Still synthesis requires 21 steps from dienyl
ether **871** and proceeds in an overall yield of only 0.5%. For comparison, the Colvin-Raphael synthesis (Scheme 104) requires 17-19 steps

and gives (±)-trichodermol in an overall yield of 0.006%. Thus, although the Still synthesis represents an advance, it leaves considerable room for improvement in the area.

In 1976 there appeared three papers reporting syntheses of trichodiene and the related ketone, norketotrichodiene. Welch and co-workers reported two syntheses of diene **824**.[304a,b] In the first (Scheme 108), the Colvin-Raphael lactone **832**, which was prepared by Welch and Wong in

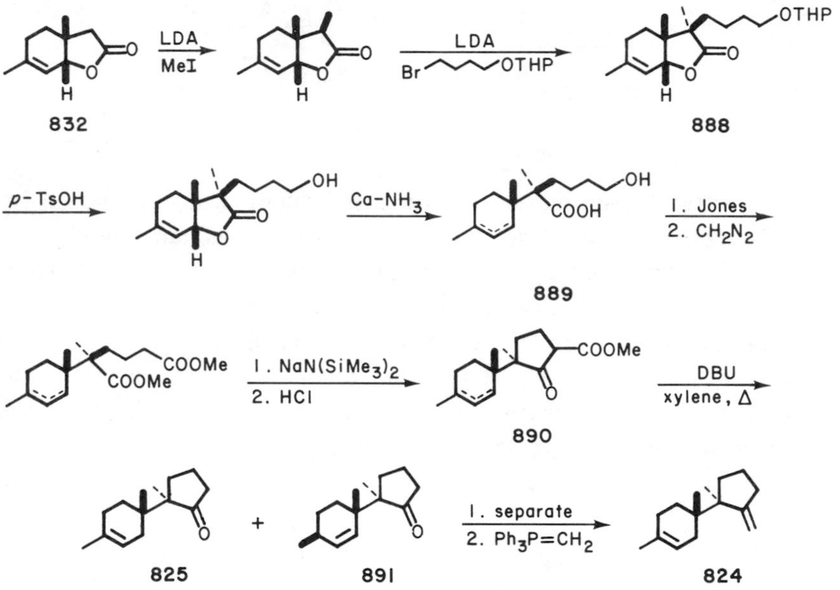

Scheme 108.   Welch's First Synthesis of (±)-Trichodiene

an alternate way,[305a,b] is alkylated first with methyl iodide and then with the tetrahydropyranyl ether of 4-bromo-1-butanol to obtain an adduct (**888**) as a 96:4 mixture of diastereomers. After removal of the protecting group, the allylic oxygen is eliminated by reduction with calcium in ammonia. Unfortunately, as it develops, this reduction gives two isomers, of which the desired is the minor component. Straightforward elaboration of hydroxy acid **889** eventually gives a 1:2 mixture of

(±)-norketotrichodiene (**825**) and its isomer **891**. Wittig methylenation of the former compound yields (±)-trichodiene (**824**). In an alternative approach (Scheme 109), the cyclohexene double bond is introduced by lithium-ammonia reduction of a diene (**895**), which gives a 4:1 mixture of ketones **825** and **897**, this time favoring the right one.

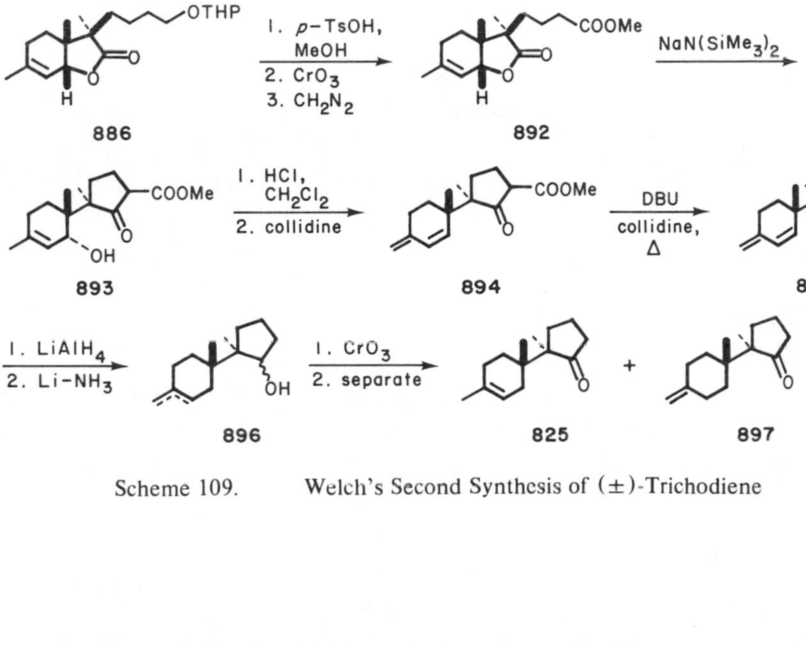

Scheme 109.   Welch's Second Synthesis of (±)-Trichodiene

Yamakawa and co-workers have prepared (±)-norketotrichodiene (**825**) as shown in Scheme 110.[306] Photocycloaddition of enones **898** and **899** yields a mixture of numerous products from which adducts **900** and **901** may be isolated. Reverse aldol reaction of **900** yields **902**, which is tosylated and reduced to hydroxy ketone **904**. Dehydration provides **825**. In spite of the poor yield in the first step, this synthesis provides ketone **825** in a very direct manner and is clearly competitive with the Welch synthesis.

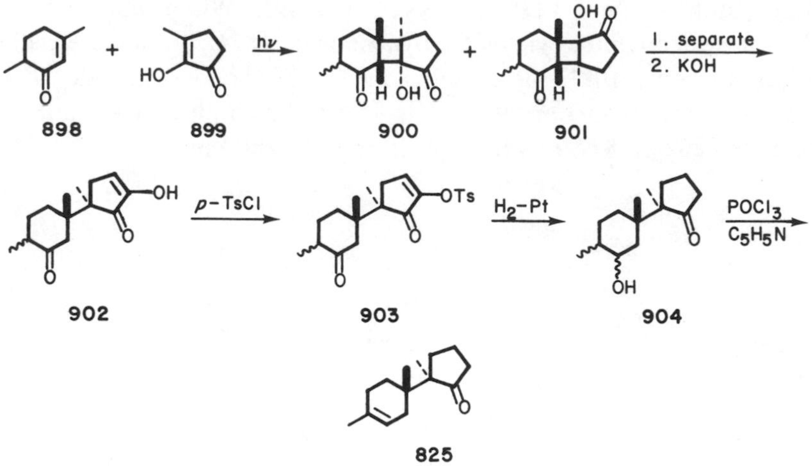

Scheme 110.   Yamakawa's Synthesis of (±)-Norketotrichodiene

(5) Debromolaurinterol Acetate

In 1973, Mirrington and co-workers reported a synthesis of (±)-debro-molaurinterol acetate (913), a possible precursor to a number of racemic sesquiterpenes including laurinterol (915).[307] The early part of the syn-

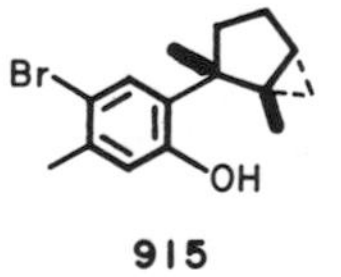

915

thesis (Scheme 111), preparation of ketone 907 from the anisole deriva-tive 905, is almost identical to work published by Hirata and co-workers in 1969.[308] Dehydration of the alcohol resulting from methylation of 907 gives an inseparable mixture of alkenes 908 and 909. The ensuing Simmons-Smith methylenation of the mixture gives a complex mixture of products including both (±)-laurinterol methyl ether (910) and its diastereomer 911 in a 1:3 ratio. Again, the isomers turn out to be

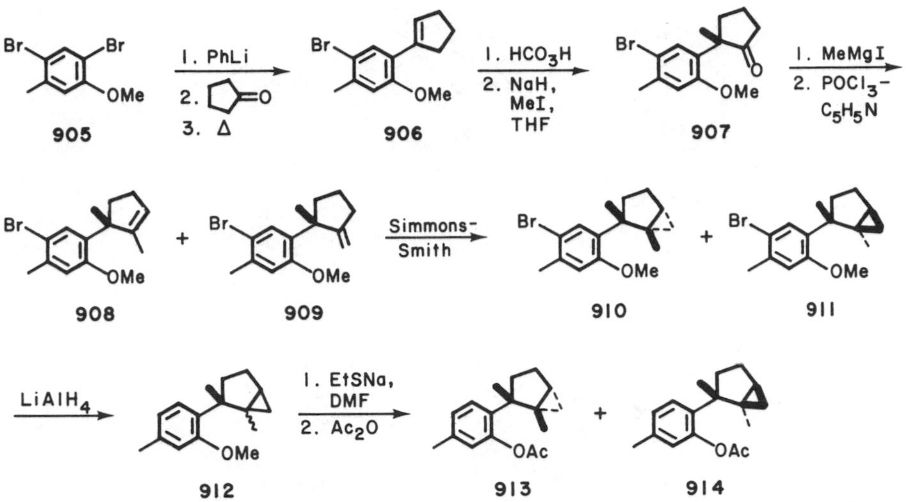

Scheme 111.   Mirrington's Synthesis of (±)-Debromolaurinterol Acetate

inseparable, so the bromine and the methyl are removed, and the phenol acetylated to give a mixture of **913** and **914**, which can be separated by preparative gas chromatography. The enantiomerically homogeneous version of **913** has been converted into (−)-laurinterol previously.

B. Bridged Systems

(1) Camphorenone, Epicamphorenone, α-Santalene, α-Santalol, β-Santalene, epi-β-Santalene, β-Santalol, and Sesquifenchene

A number of sesquiterpenes have the bicyclo[2.2.1]heptane or closely related tricyclo[2.2.1.0]heptane skeleton. Those which have been synthesized in this period are camphorenone (**916**), epicamphorenone (**917**), α-santalene (**918**), α-santalol (**919**), β-santalene (**920**), epi-β-santalene (**921**), β-santalol (**922**) and sesquifenchene (**923**).

916   917   918

919   920   921

922   923

Money and co-workers, at the University of British Columbia, in Vancouver, have synthesized (±)-camphorenone and (±)-epicamphorenone as shown in Scheme 112.[309] Racemic dihydrocarvone (924), after ketalization, is metallated and allowed to react with ethylene oxide to obtain alcohol 926. After replacement of the hydroxy group with chlorine, the ketal is removed and the ketone converted into an enol acetate. Cyclization of 929 is brought about by boron trifluoride in wet dichloromethane to yield a mixture of chloro ketones 930 and 931 in 55-60% yield (ratio not specified). After protection of the carbonyl, the remaining three carbons of the side chain are introduced by reaction of the derived Wittig reagent with acetone. The final product is a mixture of (±)-camphorenone (916) and (±)-epicamphorenone (917), which is separated by preparative glpc. Compound 917 was subsequently converted into (±)-β-santalene (920), (±)-α-santalene (918), (±)-copacamphor (934) and (±)-ylangocamphor (935). Compound 918 was converted into (±)-epi-β-santalene (921).[310] Since these syntheses involve

934   935

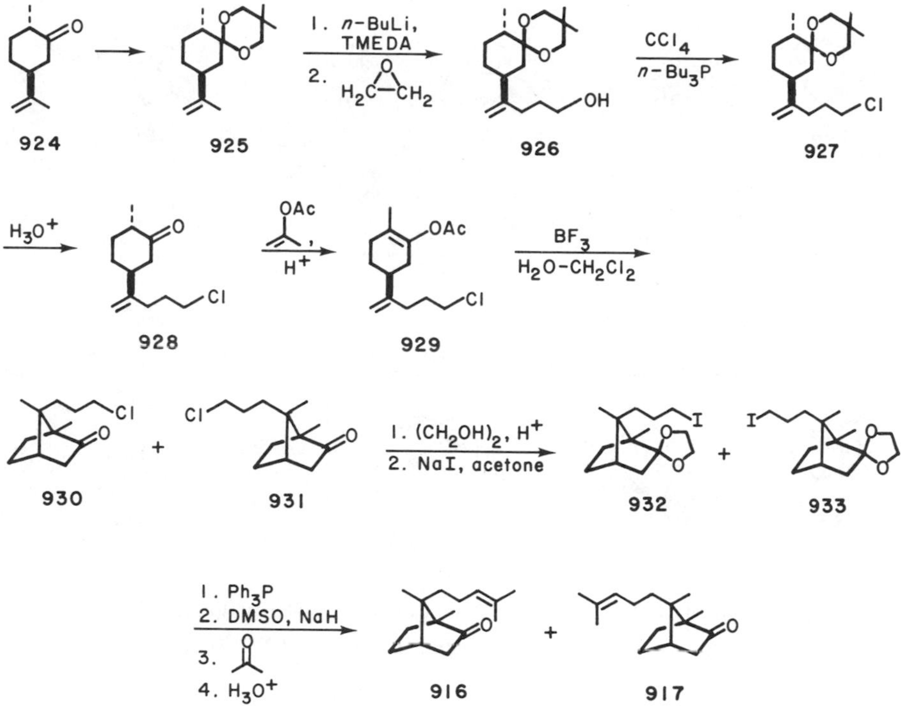

924   925   926   927

928   929

930   +   931   932   +   933

916   +   917

Scheme 112.   Money's Synthesis of (±)-Camphorenone

essentially the conversion of one sesquiterpene to another, the reader is referred to the original source for details.

In 1973, Money and co-workers reported a synthesis beginning with (+)-9-bromocamphor (**936**, Scheme 113) which affords (+)-epicamphorenone (**917**), and (+)-epi-β-santalene (**921**).[311] A parallel sequence beginning with (−)-8-iodocamphor (**939**) provides (−)-camphorenone (**916**) and (−)-β-santalene (**920**). As will be seen from the scheme, the key transformation is coupling of the neopentyl iodide in each case with the π-allylnickel reagent derived from the reaction of nickel carbonyl with prenyl bromide. Note that the Wagner-Meerwein rearrangement of the tosylate derived from camphorenone leads to the epi-β-santalene

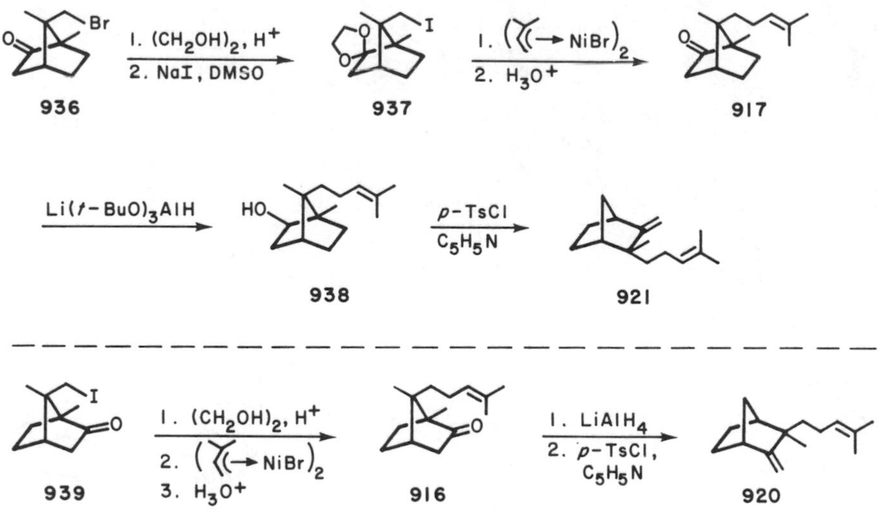

Scheme 113.   Money's Synthesis of (-)-Camphorenone,
(+)-Epicamphorenone, (-)-β-Santalene, and (+)-Epi-β-santalene

stereochemistry while the analogous tosylate derived from the epicamphorenone gives rise to normal β-santalene stereochemistry.

In 1970, Corey and co-workers had reported a synthesis of (+)-α-santalol (**919**, Scheme 114) beginning with (−)-π-bromotricyclene (**940**).[312] The key to the synthesis is reduction of propargylic alcohol **943** by diisobutylaluminum hydride, followed by iodination of the intermediate vinyl alane. Since the alkyne reduction occurs by an anti mechanism under these conditions, the resulting vinyl iodide has the *Z* stereochemistry (**944**). This is unfortunate, for if it had the *E* stereochemistry only one more operation would be necessary to obtain α-santalol.[313] As it is, the CH$_2$OH group must be converted into CH$_3$ and the I into CH$_2$OH. These manipulations require an additional five steps. The final product is (+)-α-santalol (**919**).

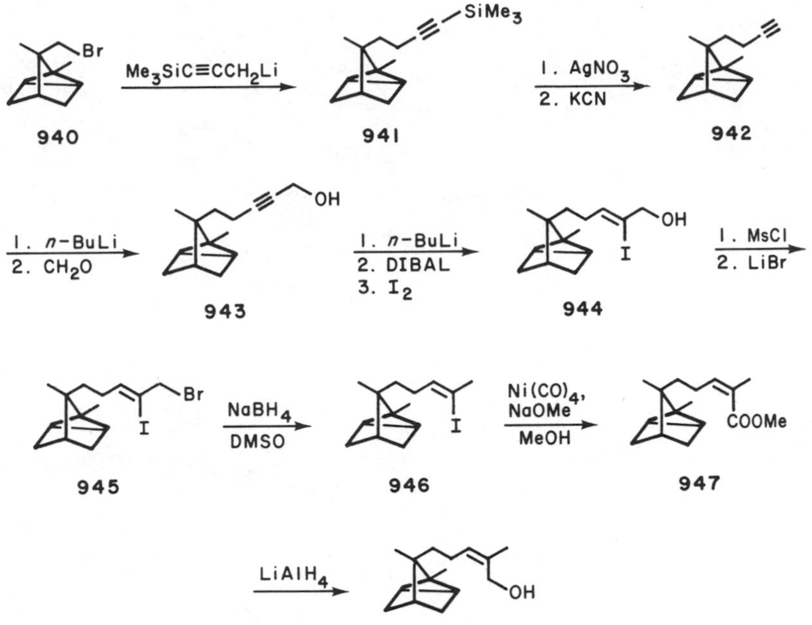

Scheme 114.   Corey's Synthesis of (+)-α-Santalol

Julia and Ward have also employed (−)-π-bromotricyclene in syntheses of (+)-α-santalene and (+)-α-santalol (Scheme 115).[314] In this case the side chain is attached by displacement of the neopentyl bromide with sodium benzenesulfinate in DMF to obtain a sulfone (941) which is alkylated either with prenyl bromide to prepare α-santalene or with the dichloride 944 to prepare α-santalol. In each case the benzenesulfenate group is removed by reduction of an aryl-alkyl sulfone (942 or 946) with sodium amalgam.

Takagi and co-workers have reported the conversion of (−)-π-bromotricyclene into (+)-α-santalol by a direct coupling using the π-allylnickel

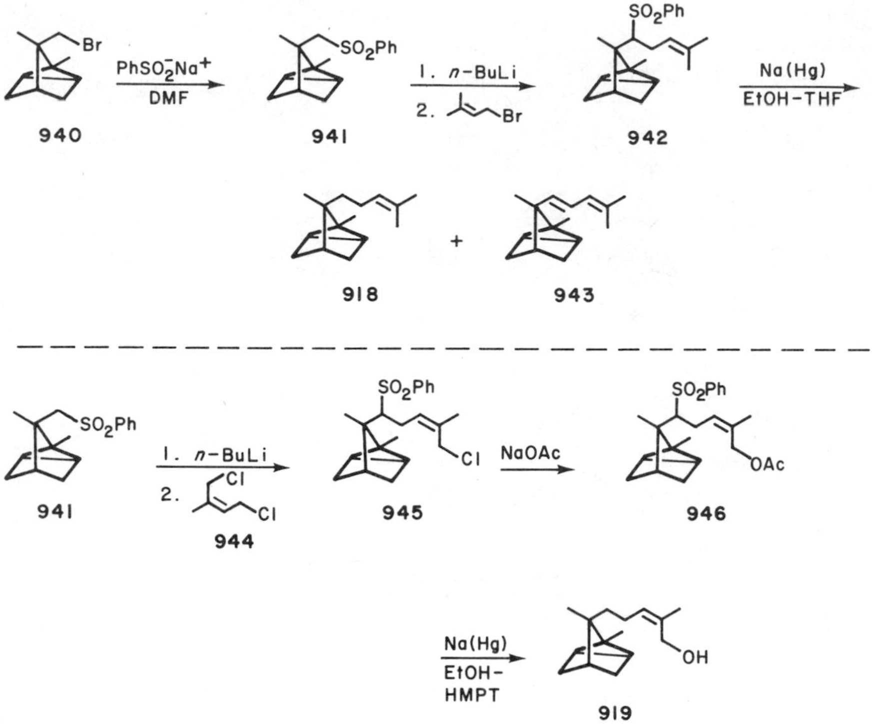

Scheme 115.   Julia's Synthesis of $(+)$-α-Santalene and $(+)$-α-Santalol

derivative **947** as the source of the side chain (Scheme 116).[315] The coupling reaction is carried out in HMPT and affords ethers **948** and **949** in 45% yield. Unfortunately, the desired stereoisomer is the minor component of a 3:2 mixture. Debenzylation of the mixture gives a mixture of $(+)$-α-santalol **(919)** and its *E* stereoisomer **(950)** which may be separated by silica-gel chromatography.

In 1978, Monti and Larsen at the University of Texas published the very interesting synthesis of $(\pm)$-α-santalene which is shown in Scheme 117.[316] Unlike the other syntheses discussed so far, this synthesis does not begin with the tricyclene nucleus, but rather with the unsaturated

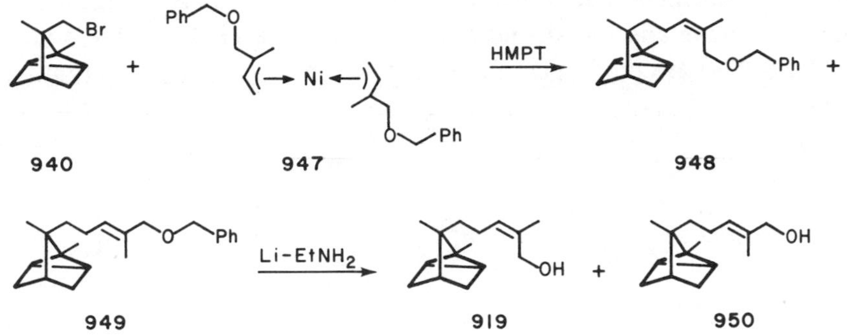

Scheme 116.   Takagi's Synthesis of (+)-α-Santalol

acid **951**. Intramolecular acylation provides the bicyclo[3.2.1]octenone **952**[317] which undergoes smooth photoisomerization to tricyclic ketone **953** (71% yield). Photochemical Wolff rearrangement of diazo ketone **954** provides the ring-contracted acid **955**, which is alkylated as its dianion with 5-iodo-2-methyl-2-pentene to obtain acid **956**. Conversion of the carboxy group to methyl completes the synthesis of (±)-α-santalene.

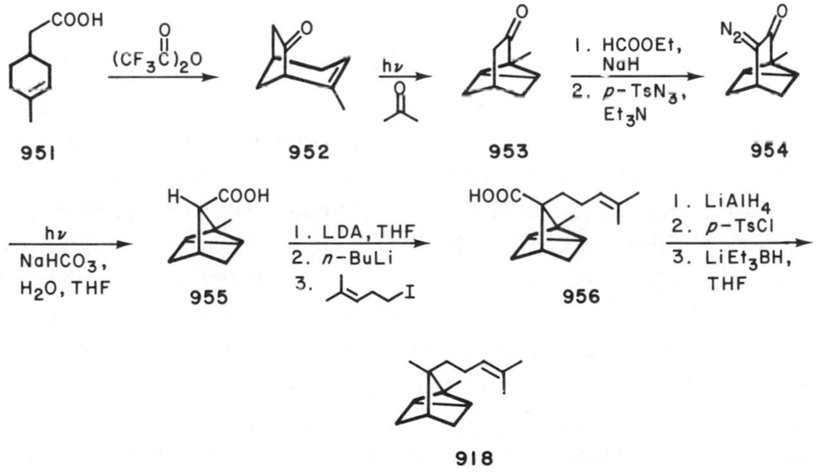

Scheme 117.   Monti-Larsen Synthesis of (±)-α-Santalene

In 1979, Bertrand and co-workers reported the Diels-Alder reaction of cyclopentadiene and the allenic ester **957** (Scheme 118).[318] Two adducts (**958** and **959**) are produced. Reduction of the more reactive norbornene

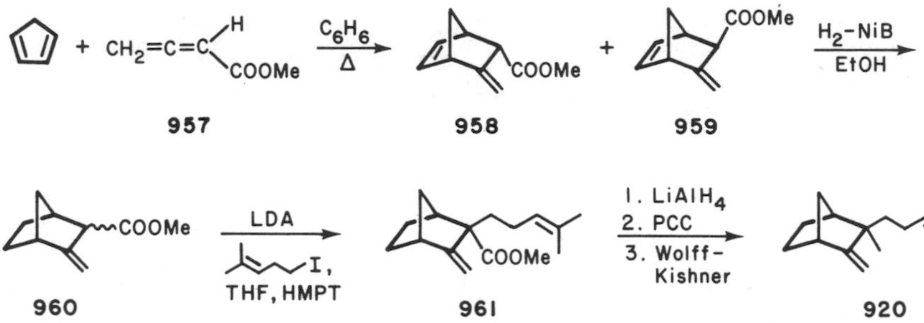

Scheme 118.   Bertrand's Synthesis of (±)-β-Santalene

double bond using Brown's nickel boride catalyst provides a mixture of esters (**960**) which is alkylated with 5-iodo-2-methyl-2-pentene. As in Corey's earlier β-santalene work[319] the alkylation is highly stereoselective and the exo isomer **961** is the sole alkylation product. Conversion of the methoxycarbonyl group to methyl provides (±)-β-santalene (**920**).

Christenson and Willis, of the Fritzsche, Dodge and Olcott Company, have reported a synthesis of β-santalene and β-santalol beginning with racemic camphene (Scheme 119).[320] Epoxidation with peroxyacetic acid yields two epoxides (**963** and **964**) in 87% yield. Treatment of the mixture with the dilithio derivative of acetic acid, followed by acidification yields a mixture of exo and endo spiro lactones **965** and **966**. When this mixture of lactones is stirred in concentrated sulfuric acid, rearrangement occurs to give a mixture of two lactones in a 5:1 ratio. The major lactone (**967**) is isolated in 45% yield by silica gel chromatography. After reduction to the aldehyde oxidation state, the isopropylidine unit is added by Wittig reaction to obtain alcohol **969**. Dehydration of this

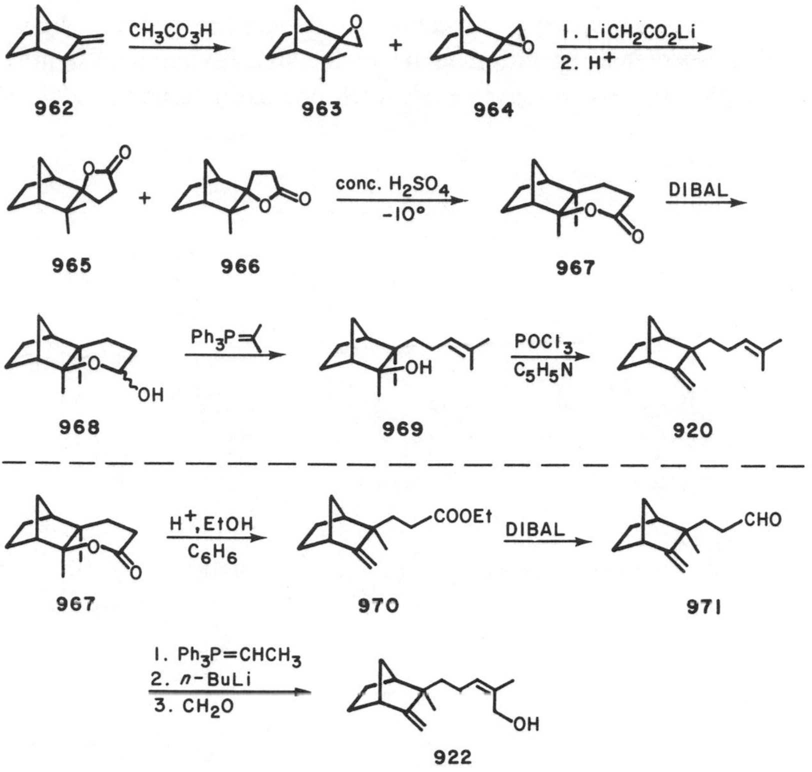

Scheme 119.   Christenson-Willis Synthesis
of (±)-β-Santalene and (±)-β-Santalol

material yields (±)-β-santalene. For preparation of β-santalol, lactone
967 is treated with *p*-toluenesulfonic acid and ethanol in refluxing benzene, whereupon unsaturated ester **970** is produced. The side-chain construction is completed as in the earlier Corey-Yamamoto synthesis of
(+)-α-santalol.[321]

In 1979, Baumann and Hoffmann, of BASF in Ludwigshafen, reported a synthesis of (±)-β-santalol (Scheme 120).[322] The synthesis begins with a Diels-Alder reaction between cyclopentadiene and (Z)-4-chloro-2-methyl-2-pentenal, an intermediate in the BASF vitamin A synthesis. Surprisingly, the reaction gives exclusively the exo addition product 972,

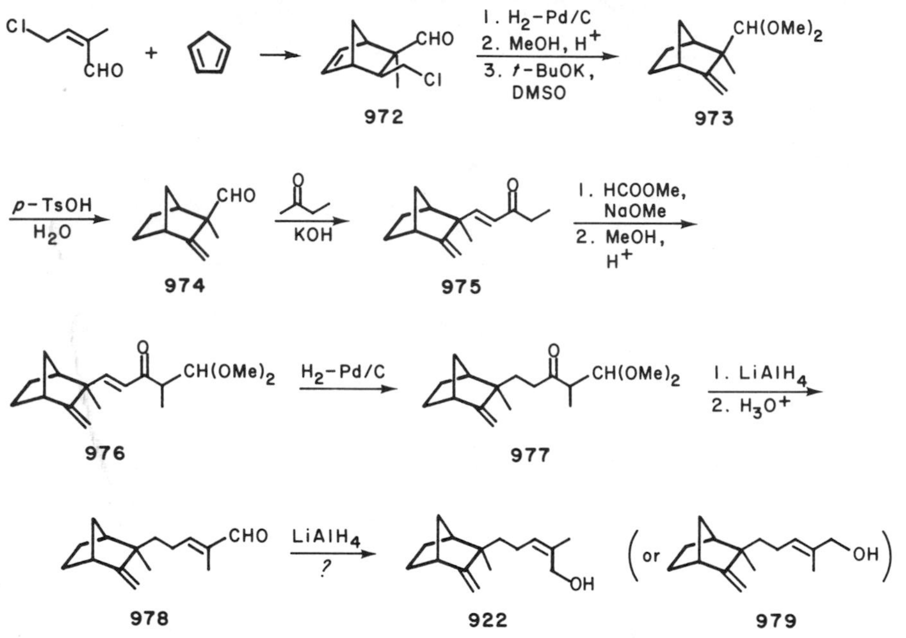

Scheme 120.   Baumann-Hoffmann Synthesis of (±)-β-Santalol

which is obtained in greater than 80% yield. After hydrogenation of the double bond, the aldehyde is protected and the chloride eliminated. After regeneration of the aldehyde function, intermediate 974 is obtained. Mixed aldol condensation of aldehyde 974 with 2-butanone gives enone 975 in 84% yield. The remaining carbon is added by formylation; after acetalization and hydrogenation of the double bond, compound **977** is obtained. Reduction of the carbonyl group gives a hydroxy acetal which is treated with acid to obtain an α,β-unsaturated aldehyde. This substance is depicted by Baumann and Hoffmann as the E stereoisomer **978**, which is a reasonable assignment based on analogy to simpler examples.[323] Inexplicably, the product of reduction of this aldehyde is depicted as $(\pm)$-β-santalol **(922)**. However, the spectra reported for the product seems to be more consistent with the spectral properties of **922** than with the E stereoisomer **979**. In particular, the ^{1}H-NMR CH$_2$OH resonance of **922** occurs at 4.06 ppm while that of **979** occurs at 3.85 ppm.[320] Baumann and Hoffmann report that their alcohol has this resonance at 4.03 ppm. Until this confusion is clarified, the Baumann-Hoffmann synthesis should be considered with caution.

In 1963, Bhattacharyya isolated a new sesquiterpene from *Valeriana Waalichi* and assigned it a bergamotene structure. After all four bergamotene racemates had been synthesized, it was clear that Bhattacharyya's hydrocarbon has a different structure and Corey and co-workers proposed the sesquifenchene structure **923**.[324] In 1972, Bessiere-Chretien and Grison confirmed this assignment by the synthesis outlined in Scheme 121.[325] Ether **980**[326] undergoes skeletal rearrangement upon treatment with boron trifluoride etherate in acetic anhydride to provide acetoxy alcohol **981**. The derived *endo*-norbornyl tosylate undergoes Wagner-Meerwein rearrangement upon solvolysis. After

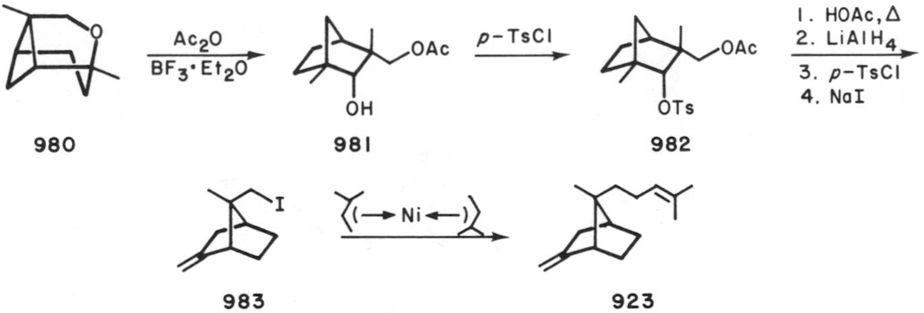

Scheme 121.   Bessiere-Chretien-Grison Synthesis of $(+)$-Sesquifenchene

reduction of the ester, the primary alcohol is iodinated to obtain **983**. The prenyl side chain is attached using the π-allyl nickel procedure to obtain (+)-sesquifenchene.

Three years later, Grieco and his co-workers reported two additional syntheses of ($\pm$)-sesquifenchene.[327a,b] In the first (Scheme 122), nor-bornadiene is converted in two steps following a procedure developed by Corey in connection with prostaglandin synthesis into the tricyclic keto acid **984**. Three further steps suffice to convert this compound into the bromo ester **985**, which is converted into the enolate ion and then methylated to obtain **986** and **987**. It turns out that the bromine is more effective than the ketal oxygen at hindering the approach of the electrophile to the planar enolate and the **986 to 987** ratio is 15:1. The major isomer is debrominated and the methoxycarbonyl group transformed into CH_2I as shown. The remaining carbons of the side chain are introduced by Julia's method (see Scheme 115) to obtain ($\pm$)-sesquifenchene.

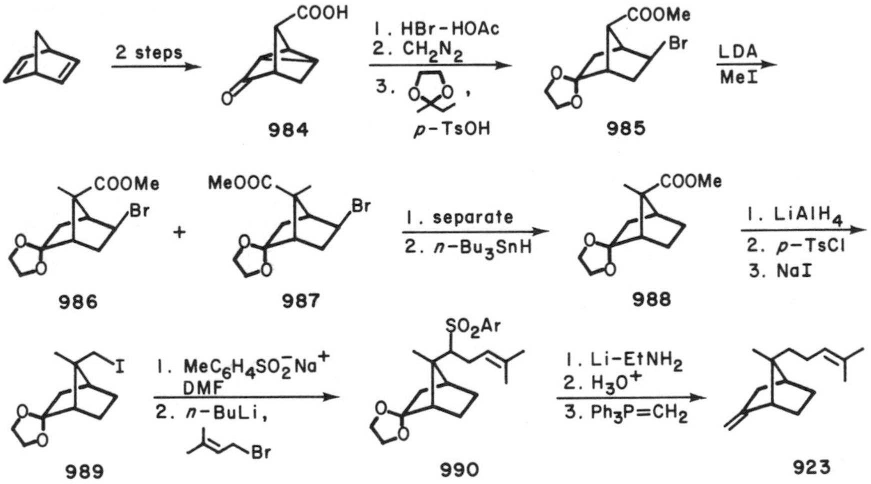

Scheme 122.   Grieco's First Synthesis of ($\pm$)-Sesquifenchene

In his second synthesis (Scheme 123) Grieco begins with *endo*-dicyclopentadiene, which is selectively hydrogenated using the Brown nickel boride catalyst to obtain **991**. Hydroboration of this alkene, followed by direct chromic acid oxidation of the resulting alkylborane provides a 3:2 mixture of isomeric ketones **992** and **993**. These are easily separated, since **993** forms a bisulfite addition product while **992** does not. Baeyer-Villiger oxidation gives a lactone (**994**), which is reduced to an aldehyde and condensed with isopropylidenetriphenylphosphorane to obtain unsaturated alcohol **995**. After oxidation, the resulting ketone is methylated to obtain ketone **996**. (This ketone had previously been prepared in two steps from norbornanone.)[319] Reduction occurs to give

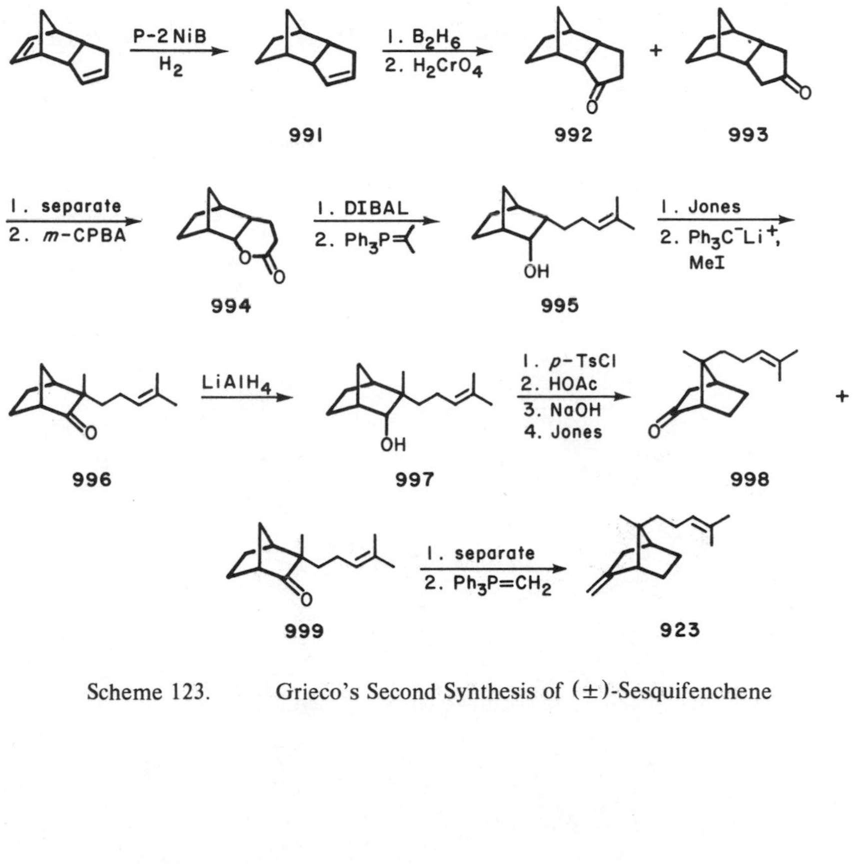

Scheme 123.   Grieco's Second Synthesis of (±)-Sesquifenchene

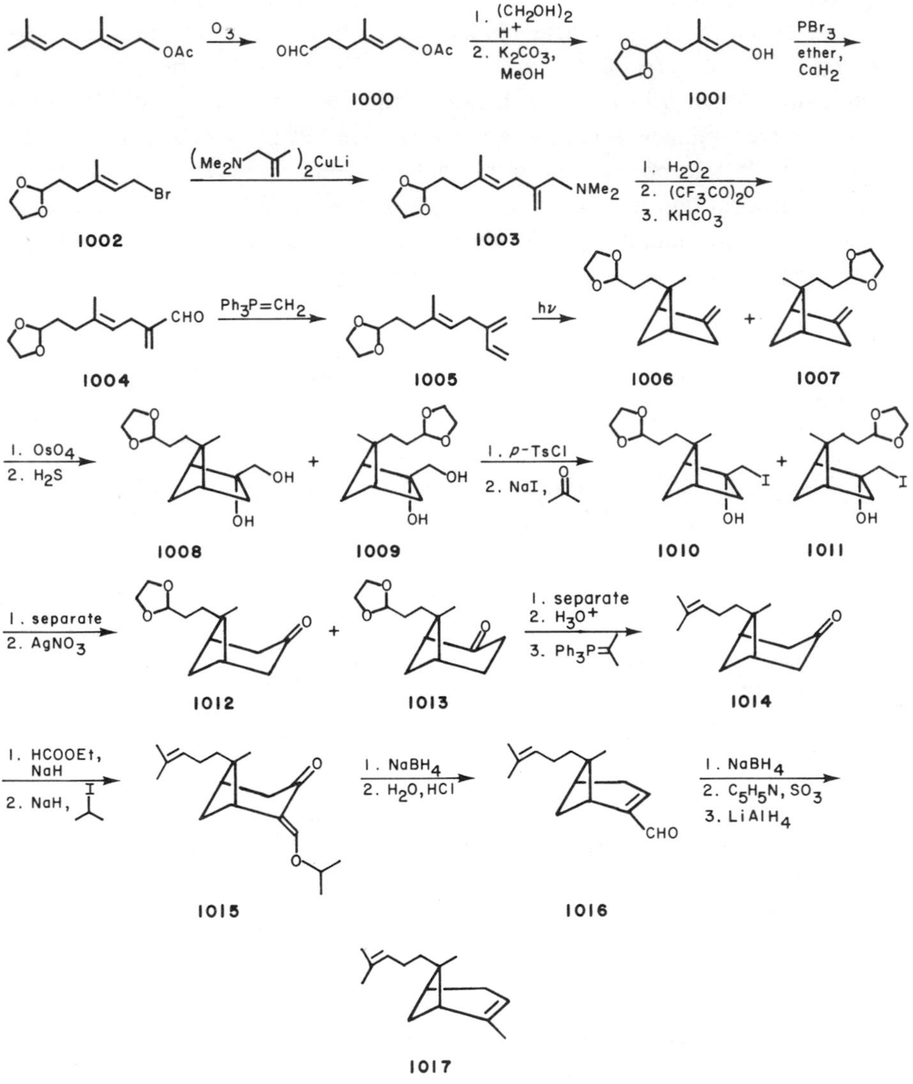

Scheme 124.   Corey's Synthesis of $(\pm)$-α-*trans*-Bergamotene

the endo alcohol **997**, which is rearranged as its tosylate to obtain a mixture of acetates. Conversion of these compounds into ketones gives a 7:3 mixture of ketones **998** and **999**. Methylenation of the major isomer gives (±)-sesquifenchene (**923**). It is not clear that either Grieco synthesis represents an improvement over the Bessiere-Grison synthesis.

(2) α-trans-Bergamotene

In 1975, Corey, Cane, and Libit reported a synthesis of (±)-α-*trans*-bergamotene (**1017**).[324] The synthesis begins with geranyl acetate, which is ozonized to acetoxy aldehyde **1000** (Scheme 124). Straightforward manipulations provide the allylic bromide **1002** which is coupled with a vinyl cuprate reagent to obtain allylamine **1003**. Oxidation of this compound by the modified Polonovski method gives an α,β-unsaturated aldehyde (**1004**), which is converted by Wittig reaction into triene **1005**. Photochemical isomerization of this compound gives bicyclo[2.2.1]-hexane derivatives **1006** and **1007** in a ratio of 5:3. The mixture is hydroxylated, converted into monotosylates, and iodinated to obtain **1010** and **1011**, which are separated by silica gel chromatography. Rearrangement of the major isomer **1010** gives isomeric ketones **1012** and **1013** in yields of 55% and 22%, respectively. Again, the isomers are separated by chromatography and the major isomer converted into unsaturated ketone **1014**. A series of six straightforward steps completes the synthesis of (±)-α-*trans*-bergamotene (**1017**). Although the synthesis outlined in Scheme 124 was the first reported synthesis of a *trans*-bergamotene, it left much room for improvement, since it requires over 20 steps and requires isomer separations at two stages.

An improvement was recorded by Larsen and Monti in their 1977 report of the synthesis outlined in Scheme 125.[317] Bicyclic enone **952** is hydroxylated and the resulting secondary-tertiary diol is sequentially mesylated and then converted into the trimethylsilyl ether **1019**. The enolate ion undergoes intermolecular alkylation faster than intramolecular alkylation at -40°C, to give ketone **1020**, which is cyclized to the

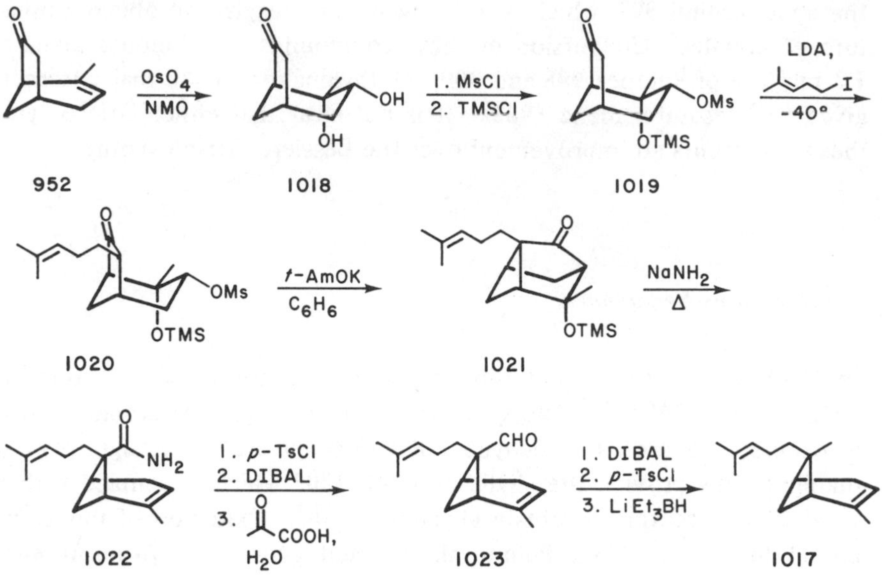

Scheme 125.   Monti-Larsen Synthesis of (±)-α-*trans*-Bergamotene

tricyclic ketone **1021**. Haller-Bauer fragmentation provides **1022**, which has the correct α-*trans*-bergamotene skeleton. Unfortunately, the amide is very unreactive and a further six steps are required to convert it into a methyl group. The Monti synthesis is an outstanding contribution, even with this minor flaw.

C. Spirocyclic Systems

(1) Spirovetivanes

In the time since β-vetivone (**1024**) and its congeners were recognized to have the spiro[4.5]decane rather than the hydroazulene skeleton,[328] the members of this class have become a favorite target for synthesis. In

Volume 2 of this series, we documented three syntheses of vetivane sesquiterpenes. In this volume, we report nine more syntheses of **1024** (a compound which is rapidly becoming the "most synthesized" sesquiterpene), in addition to syntheses of agarospirol (**1025**), hinesol (**1026**), β-vetispirene (**1027**), α-vetispirene (**1028**), solavetivone (**1029**), and anhydro-β-rotundol (**1030**). In addition, we call attention to the full paper reporting the Marshall-Brady synthesis of (±)-hinesol (**1026**)[329] which was discussed earlier.[330]

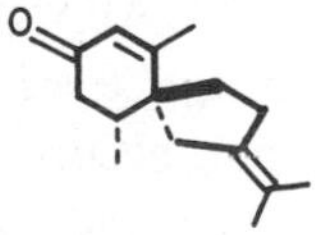
1024

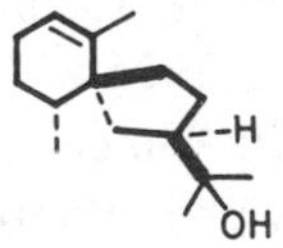
1025

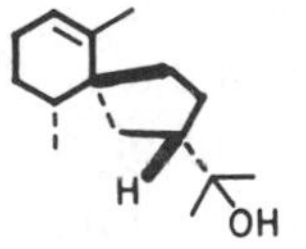
1026

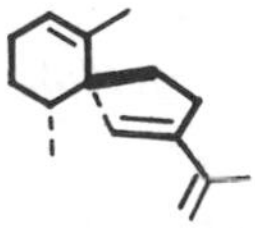
1027

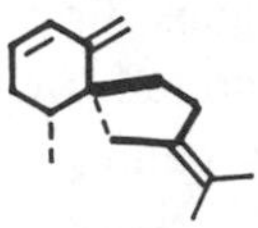
1028

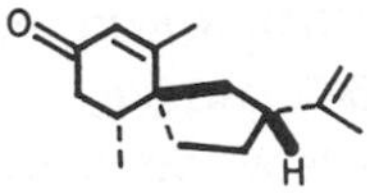
1029

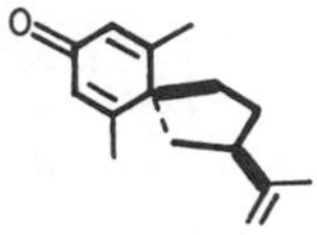
1030

In 1973, Stork, Danheiser, and Ganem published an astoundingly simple synthesis of (±)-β-vetivone (Scheme 126).[331] The ethyl enol ether

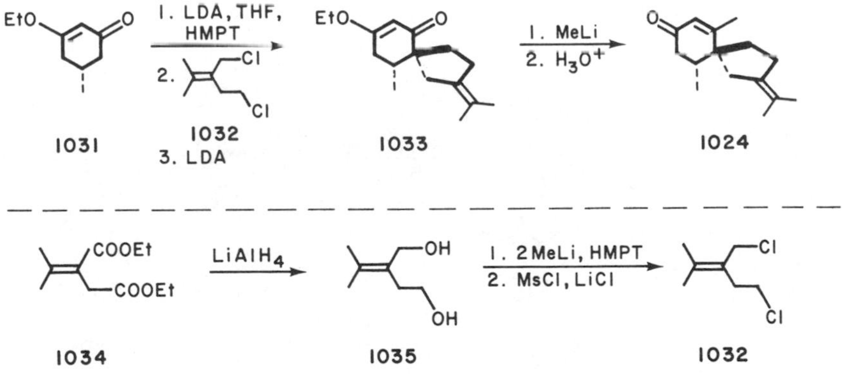

Scheme 126. Stork's Synthesis of (±)-β-Vetivone

of dihydroorcinol (1031) is deprotonated by LDA in THF containing one equivalent of HMPT. Dichloride 1032 is added, followed by more LDA. Spiroannelated enol ether 1033 is obtained as one stereoisomer in 45% yield. Addition of methyllithium and acid hydrolysis gives (±)-β-vetinone (1024) in an overall yield of 27% for the three steps. The requisite dihalide is prepared from diethyl isopropylidinesuccinate as shown in the scheme.

Also published in 1973 was a synthesis by McCurry and Singh leading to (±)-β-vetivone and (±)-α-vetispirene (Scheme 127).[332] Alkylation of the pyrrolidine enamine of aldehyde 1036 with crotyl bromide affords 1037 as a 1:1 mixture of diastereomers. Since the alkylation occurs first on nitrogen, and the resulting N-crotylenammonium ion then undergoes [3.3]sigmatropic rearrangement, only the terminal alkenes are produced. The diastereomeric mixture is converted by a well-precedented sequence to unsaturated aldehyde 1040 which undergoes Prins cyclization to a mixture of 1041 and 1042, which is separated by chromatography. Moffatt oxidation of 1041 is accompanied by conjugation of the double bond, leading to (±)-β-vetivone (1024). Alternatively, the mesylate of 1041 may be eliminated to obtain (±)-α-vetispirene (1028). In an earlier publication, McCurry and co-workers had synthesized alcohol 1043 and studied its formolysis. Predictably, alkylation of the allylic cation occurs on the face of the cyclohexane double bond trans to the methyl group, leading to (±)-10-epi-β-vetivone (1044) in 40% yield.

1043   1044

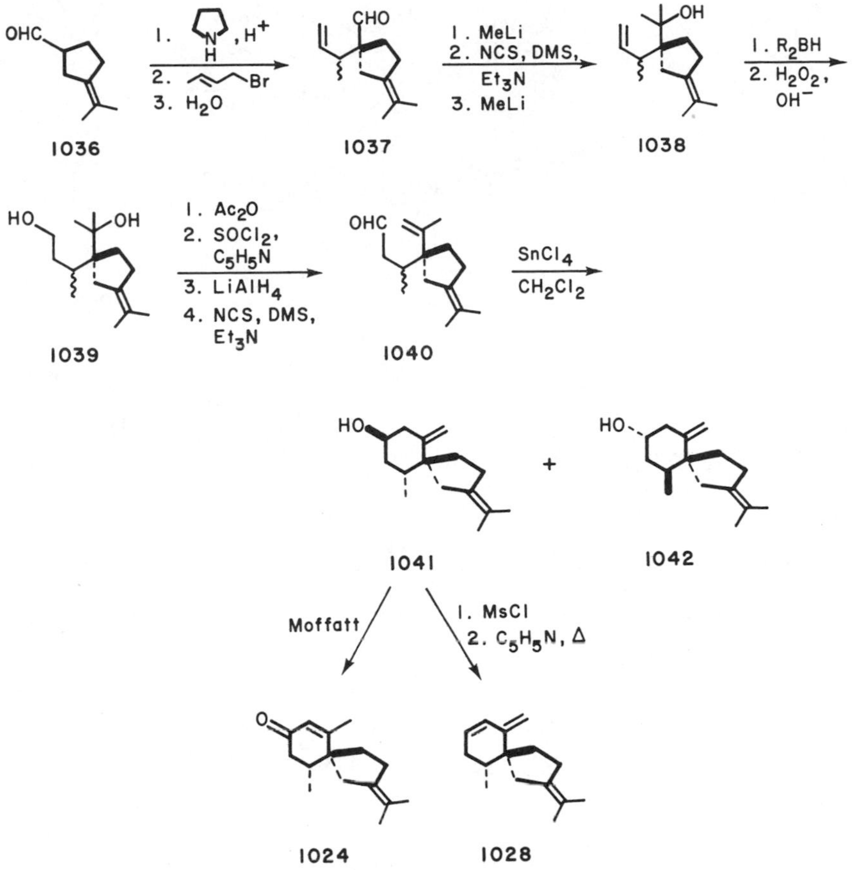

Scheme 127.   McCurry's Synthesis of $(\pm)$-β-Vetivone
and $(\pm)$-α-Vetispirene

Yamada and co-workers have developed the rather long, but highly
stereoselective synthesis of $(\pm)$-β-vetivone which we summarize in
Scheme 128.[333] The starting point aldehyde is **1045** (obtained in six steps
from *m*-cresol methyl ether). After homologation, the resulting

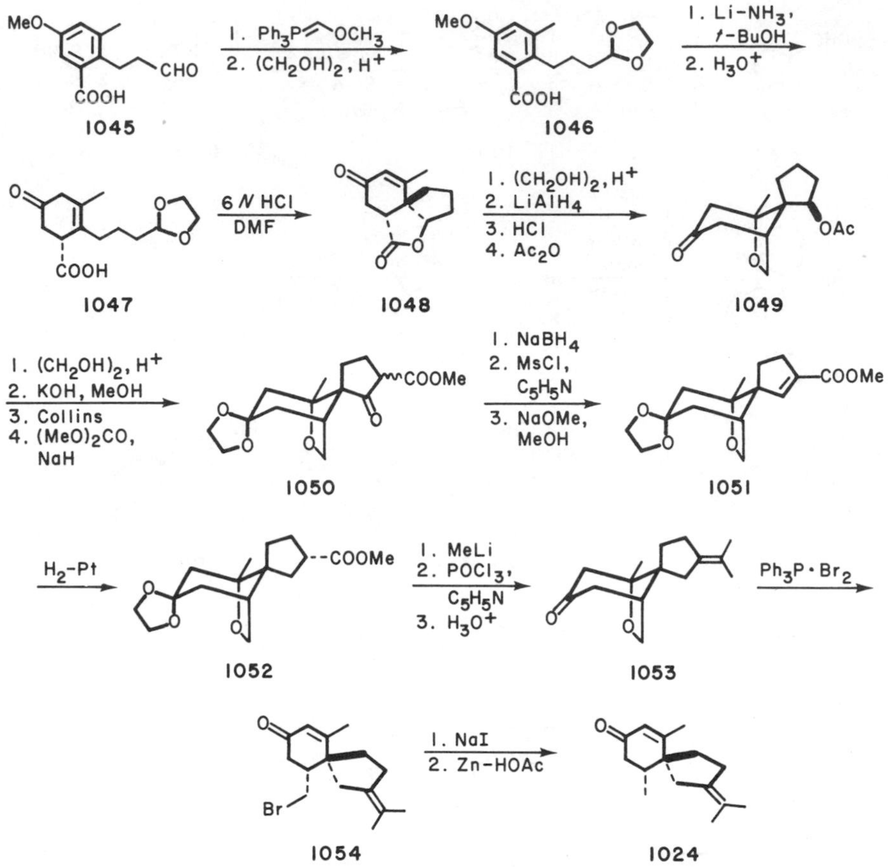

Scheme 128.   Yamada's Synthesis of (±)-β-Vetivone

aldehyde is protected, and the acetal acid **1046** is reduced by Birch's procedure to obtain compound **1047**. Hydrolysis of the acetal gives an aldehyde which undergoes aldol condensation at the γ-carbon of the α,β-unsaturated ketone to generate the [5.4]spirodecane system. Studies on simple model compounds had shown that this condensation is reversible, and that the initial vinylogous aldol is prone to rearrange to other products. However, in this case the carboxy group traps the initial product, leading to lactone **1048**. Protection of the ketone carbonyl, reduction of the lactone, and regeneration of the enone system is followed by spontaneous cyclization to give a tricyclic alcohol, which is acetylated to obtain keto ester **1049**. Precedented transformation of this material into α,β-unsaturated ester **1051** is followed by catalytic hydrogenation. The hydrogenation is highly stereoselective, the sole product being stereoisomer **1052**. After conversion of the methoxycarbonyl group to isopropylidene, the ether bridge is eliminated by reaction with triphenylphosphine dibromide, giving **1054**, which is readily transformed into $(\pm)$-β-vetivone **(1024)**.

In a companion paper[334] Yamada and co-workers describe conversion of intermediates from the β-vetivone synthesis into $(\pm)$-hinesol **(1026)**, $(\pm)$-α-vetispirene **(1027)**, and $(\pm)$-β-vetispirene **(1028)**. These conversions, which are all rather routine, are summarized in Scheme 129.

In 1977, Yamada reported utilization of compound **1049** for synthesis of $(\pm)$-solavetivone **(1029)**, a stress metabolite from infected potato tubers.[335] A key step in this synthesis is conjugate addition of isopropenylmagnesium bromide to enone **1060**. An approximate equimolar mixture of diastereomers is obtained. The isomers were readily separated by chromatography and the complete stereostructure of one isomer was elucidated by X-ray analysis. This isomer was used to complete the synthesis outlined in Scheme 130, thus establishing the stereostructure of the natural product. The synthesis was complicated by the necessity of protecting the isopropenyl double bond, which otherwise was found to undergo extensive isomerization.

In 1974, Caine and Chu reported another approach to the β-vetivone problem (Scheme 131).[336] $(\pm)$-Nootkatone **(386)** is converted into the cross-conjugated dienone **1065**, and the isopropenyl double bond is isomerized by conversion into the tertiary bromide, which is eliminated under E_2C conditions to obtain **1066**. Successive photoisomerizations, first in dioxane, then in aqueous acetic acid, lead to trienone **1068**, which is selectively hydrogenated to a 7:3 mixture of $(\pm)$-β-vetivone **(1024)** and its diastereomer **1044**.

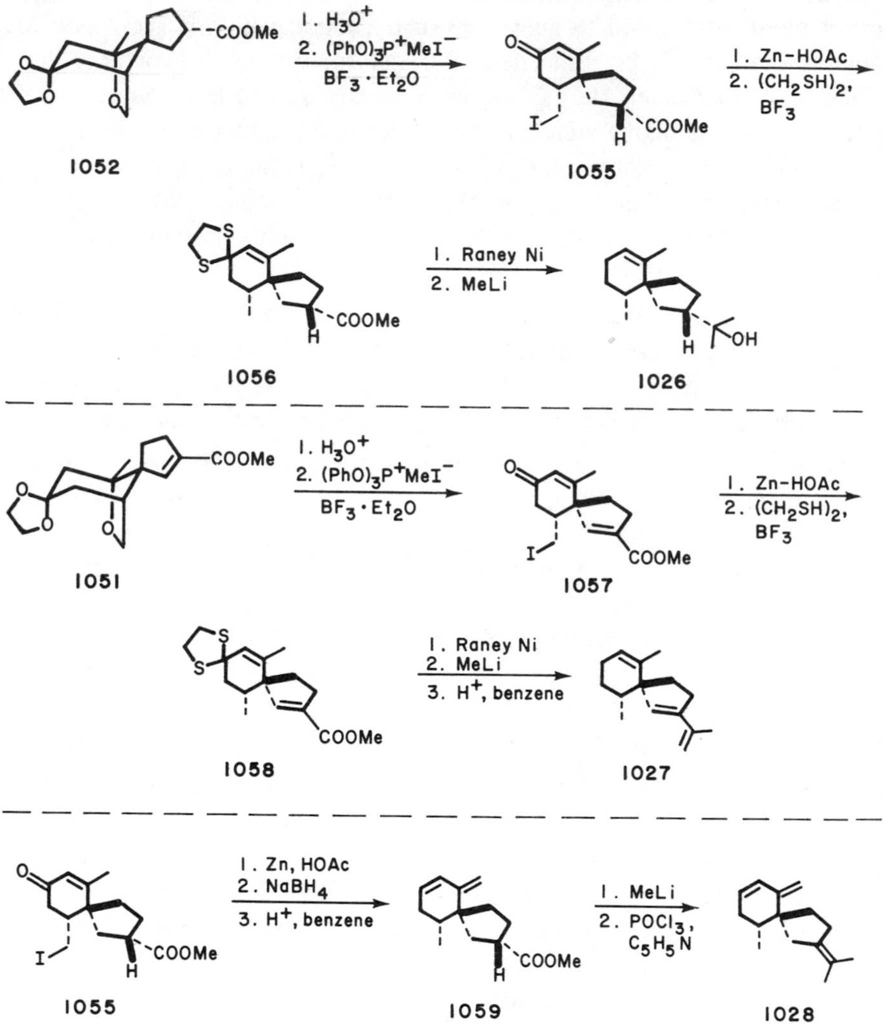

Scheme 129.   Yamada's Synthesis of (±)-Hinesol, (±)-α-Vetispirene, and (±)-β-Vetispirene

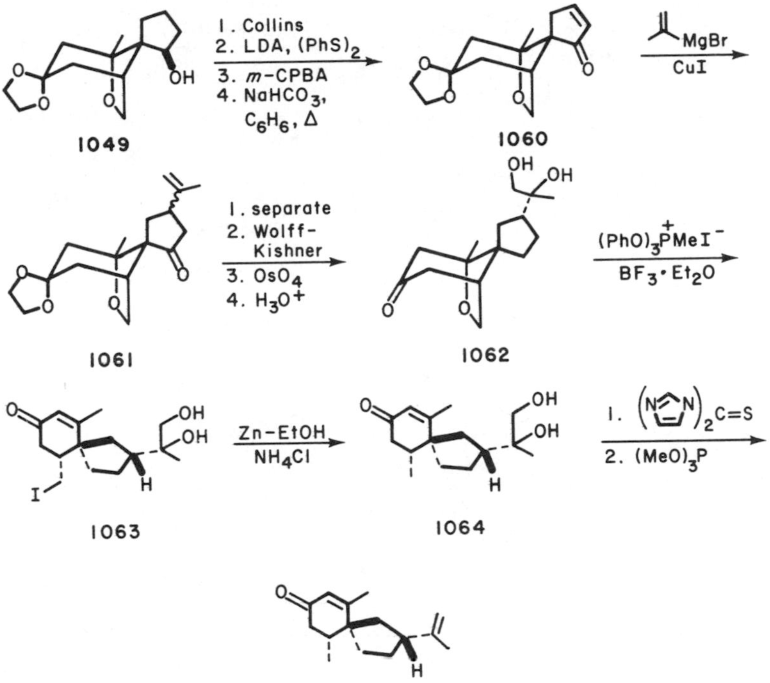

1049

1. Collins
2. LDA, (PhS)$_2$
3. m-CPBA
4. NaHCO$_3$,
   C$_6$H$_6$, Δ

1060

$\xrightarrow{\text{MgBr}}$
CuI

1061

1. separate
2. Wolff-
   Kishner
3. OsO$_4$
4. H$_3$O$^+$

1062

(PhO)$_3\overset{+}{\text{P}}$MeI$^-$
BF$_3$·Et$_2$O

1063

Zn–EtOH
NH$_4$Cl

1064

1. $\left(\overset{N\diagup\diagdown N}{\diagdown\diagup}\right)_2$C=S
2. (MeO)$_3$P

1029

Scheme 130.   Yamada's Synthesis of (±)-Solavetivone

In the same year, Pesaro and co-workers reported the synthesis summarized in Scheme 132.[337] The synthesis begins with (±)-α-terpineol (1069), which is ozonized to obtain hemiacetal 1070. This compound cyclizes upon treatment with acid to provide the acetylcyclopentene 1071. Catalytic hydrogenation of the less substituted double bond gives 1072, which is elaborated into unsaturated nitrile 1074. The derived acyl halide undergoes intramolecular acylation to give ketone 1075 (see also Scheme 127, 1040→1041). The cross-conjugated dienone 1076 undergoes conjugate addition when treated with several methylcopper reagents to give (±)-β-vetivone and its diastereomer. Although isomer 1024 is invariably the major isomer, the highest selectivity (1024 to 1044 = 5:1) is obtained using [MeCuBr]$^-$[Li(i-Bu$_2$NH)$_2$]$^+$.

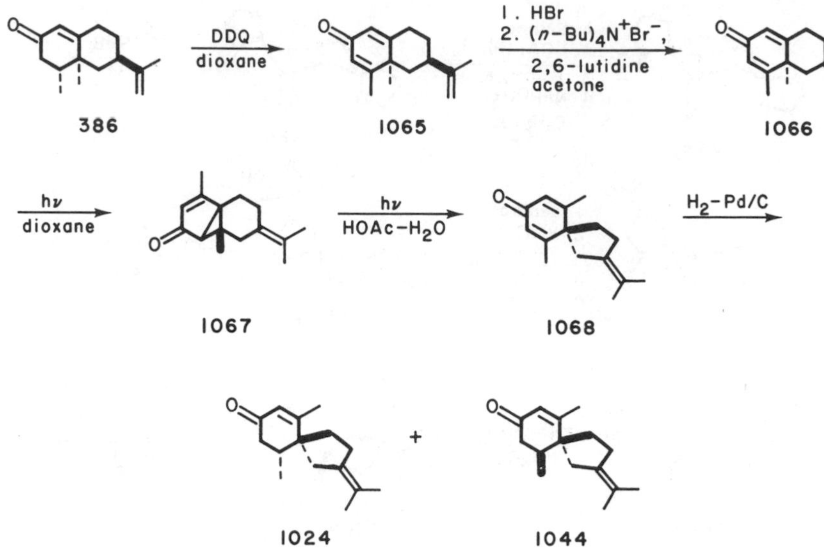

Scheme 131.   Caine's Synthesis of (±)-β-Vetivone from (±)-Nootkatone

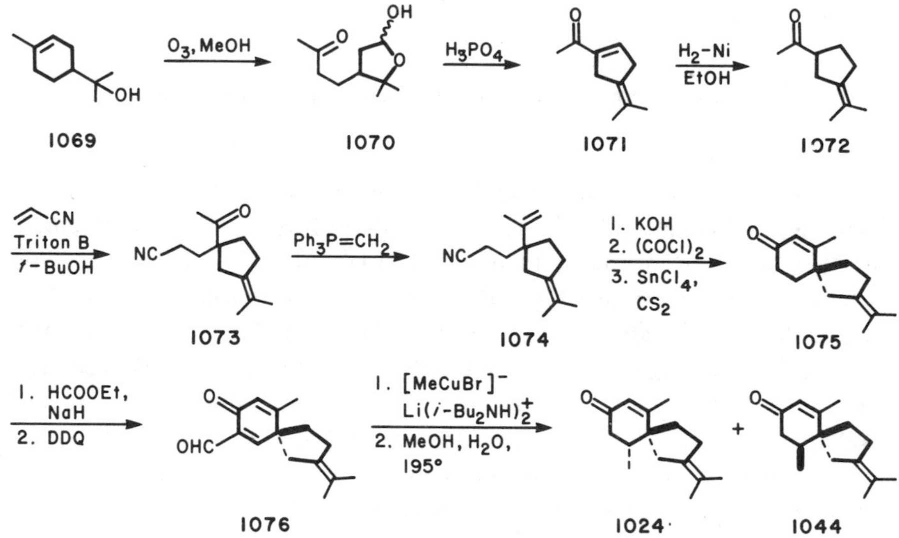

Scheme 132.   Pesaro's Synthesis of (±)-β-Vetivone

Trost and co-workers have studied the applicability of their cyclobutanone spiroannelation procedure for the synthesis of spirovetivanes as
outlined in Scheme 133.[338] Treatment of 2,6-dimethylcyclohexenone
(1077) with 1-lithiocyclopropyl phenyl sulfide gives an oxaspiropentane

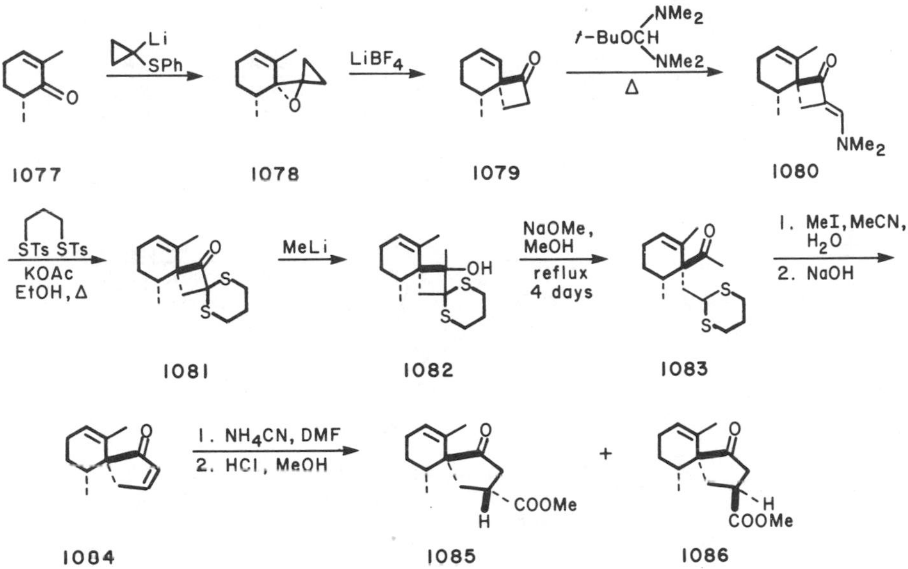

Scheme 133.   Trost's Synthesis of a Spirovetivane Intermediate

(1078) which rearranges upon treatment with lithium fluoborate to cyclobutanone 1079. The process is highly stereoselective, the reagent attack-
ing solely from the side of the cyclohexenone ring opposite the secondary methyl and the ring-expansion occurring with effective inversion of
configuration at the spiro carbon. The α-trimethylenedithio derivative
1081 is prepared via vinylogous amide 1080. Treatment of 1081 with
methyllithium gives a tertiary alcohol, which is treated with sodium
methoxide in refluxing methanol for four days to effect ring cleavage.
Ketone 1083 is obtained in 91% yield. The dithiane is hydrolyzed and
the resulting keto aldehyde is cyclized to obtain dienone 1084. Conjugate addition of cyanide gives two stereoisomers, which are hydrolyzed

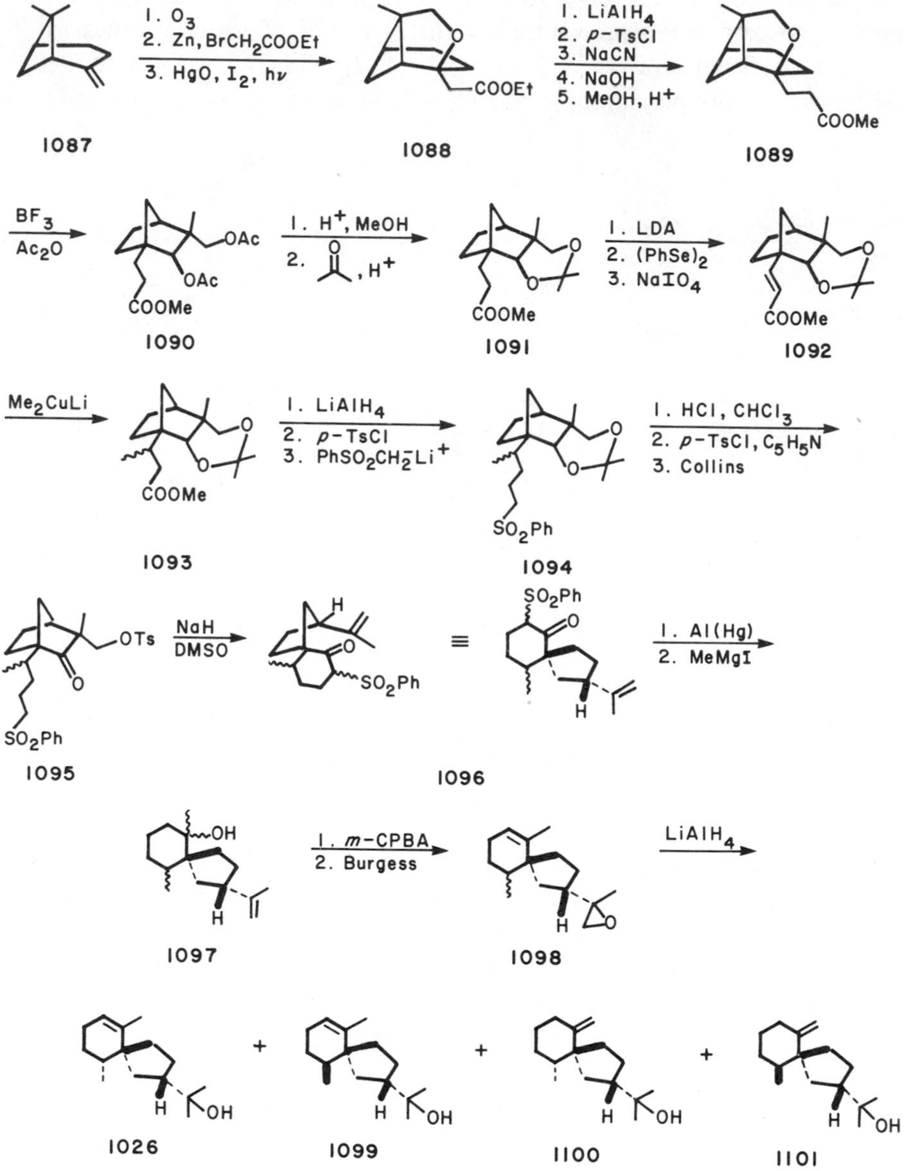

Scheme 134.   Magnus Synthesis of (+)-Hinesol

to obtain a difficult to separate 2:1 mixture of keto esters **1085** and **1086**. Although these intermediates have not been taken further, removal of the ketone carbonyl and addition of methyllithium should provide ($\pm$)-hinesol (from **1085**) and ($\pm$)-agarospirol (from **1086**).

Magnus and his co-workers have reported a synthesis (Scheme 134) which leads to a mixture of ($\pm$)-hinesol (**1026**), (+)-10-epi-hinesol (**1099**), and two exocyclic double bond isomers (**1100** and **1101**).[339] The synthesis proceeds through the bridged ether **1088**, which is homologated to ester **1089**. Wagner-Meerwein rearrangement provides the norbornane derivative **1090**, which is elaborated to α,β-unsaturated ester **1092**. Conjugate addition of lithium dimethylcuprate gives a 1:1 mixture of diastereomers **1093**. Another homologation ensues, leading to sulfone **1094**. After deprotection of the 1,3-diol system, the primary alcohol is tosylated and the secondary one is oxidized. The stage is now set for the key maneuver of the synthesis. The sulfone anion adds to the carbonyl group, generating a tertiary alkoxide which is well disposed for Grob fragmentation. The result is a spiro[4.5]decane (**1096**) in which the absolute configurations of the spiro carbon and the isopropenyl carbons are established correctly. Unfortunately, the secondary methyl carbon is not correctly fixed, and the ultimate product is a mixture of (+)-hinesol (**1026**) and (+)-10-epi-hinesol (**1099**), along with smaller amounts of the exocyclic double bond isomers **1100** and **1101**. Silica gel chromatography suffices to separate **1026** + **1099** (26% yield) from **1100** + **1101** (18% yield) and treatment of the latter fraction with acid causes quantitative rearrangement to **1026** + **1099**, thus increasing the yield of the endocyclic isomers to about 45%. However, **1026** and **1099** can be separated only by conversion to the mixture of acetates, which are separable by glpc. Hydrolysis of the separated acetate affords pure (+)-hinesol (**1026**). It should be noted that (+)-hinesol is the enantiomer of the natural product.

Dauben and Hart developed a synthesis (Scheme 135) which efficiently leads to most of the known spirovetivanes.[340] Enol ether **1031** (see also Scheme 126) is formylated to obtain **1102**. This substance gives an anion which reacts with phosphonium salt **1103** to give the spiroannelated product **1104**. It is noteworthy that the alkylation is highly

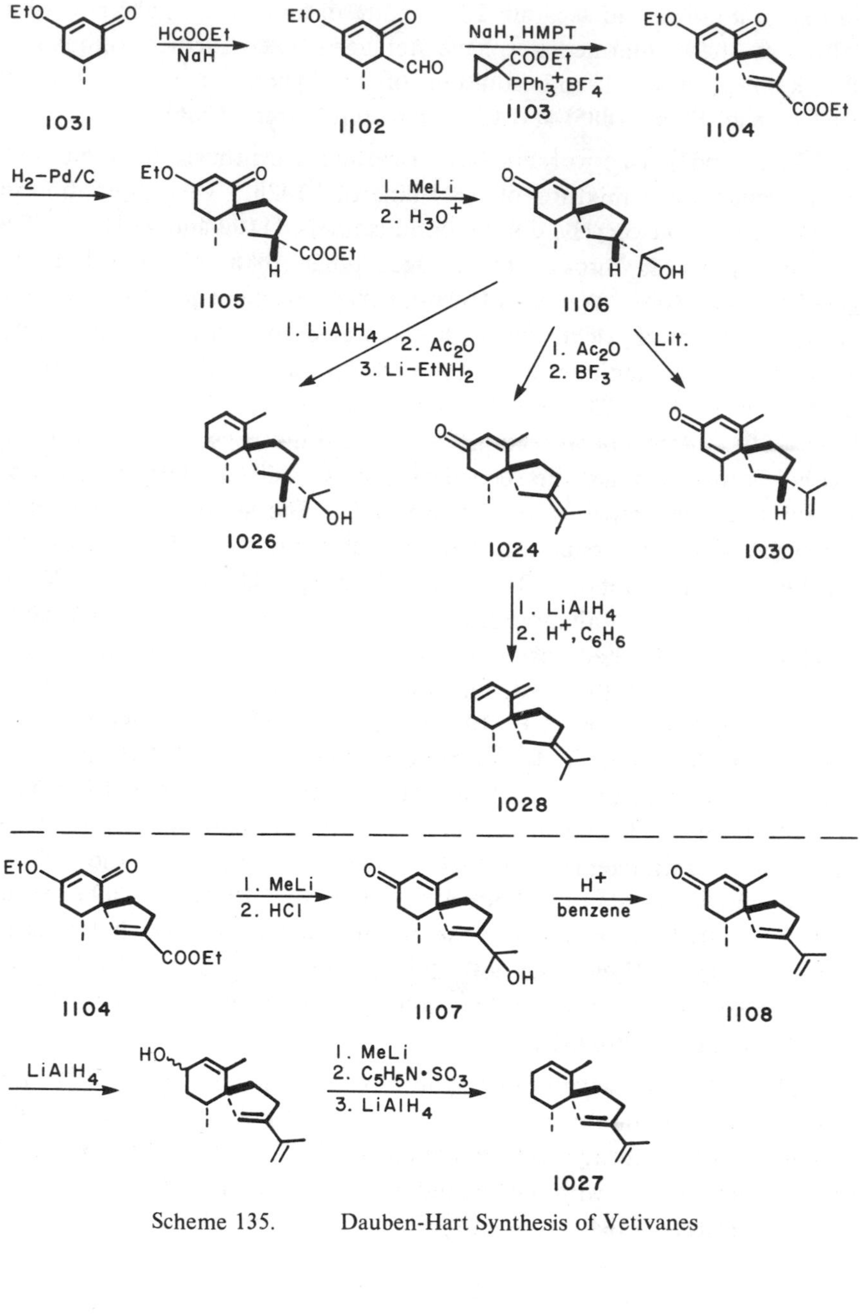

Scheme 135.   Dauben-Hart Synthesis of Vetivanes

stereoselective, occurring trans to the secondary methyl group. Thus, intermediate **1104** is obtained in stereohomogeneous form. Catalytic hydrogenation of **1104** is also stereoselective, although in this case the stereochemical outcome is perhaps more fortuitous than predictable. As in the Stork synthesis (Scheme 126), intermediate **1105** reacts with methyllithium to form the cyclohexenone system. In the present case, of course, the ethoxycarbonyl group is simultaneously changed to the isopropanol side chain. Straightforward transformations then lead to (±)-anhydro-β-rotundol (**1030**), (±)-β-vetivone (**1024**), (±)-hinesol (**1026**), and (±)-β-vetispirene (**1028**). By a slight modification, (±)-α-vetispirene (**1027**) is also readily available.

Ramage has utilized R-(+)-3-methylcyclohexanone as the starting point for a synthesis of (−)-agarospirol (**1025**), the natural enantiomer (Scheme 136).[341] The ketone is first blocked on the less hindered side and then exhaustively cyanoethylated. After hydrolysis and esterification, keto diester **1110** is obtained. Dieckmann cyclization and removal of the methoxycarbonyl group affords the spiro[5.5]-undecanedione **1111**. One of the two carbonyl groups is severely hindered and selective Wittig reaction leading to **1112** is possible. Acid-catalyzed ketalization is accompanied by double bond isomerization. Unfortunately, isomers **1113** and **1114** are obtained in a 3:2 ratio. The isomers are separated by chromatography on silver nitrate impregnated alumina. Ozonolysis of the major isomer gives a keto aldehyde which cyclizes to **1115**. Catalytic hydrogenation gives two diastereomers (**1116**), which are oxidized to acids, esterified, and deketalized. Although **1117** and **1118** (formed in a ratio of 7:3) may be separated by chromatography on alumina, this is not necessary since both react with methylenetriphenylphosphorane to give **1119**. Apparently the two isomers are in equilibrium under the conditions of the Wittig reaction and isomer **1117** reacts faster, due to the hindered environment of the ketone carbonyl in **1118**. Isomerization of the double bond gives **1120**, which is converted into (−)-agarospirol in the standard manner.

In 1976, Caine and co-workers published a synthesis of (±)-α-vetispirene by a route involving a photochemical cyclohexadienone rearrangement as a key step (Scheme 137).[342] Photolysis of

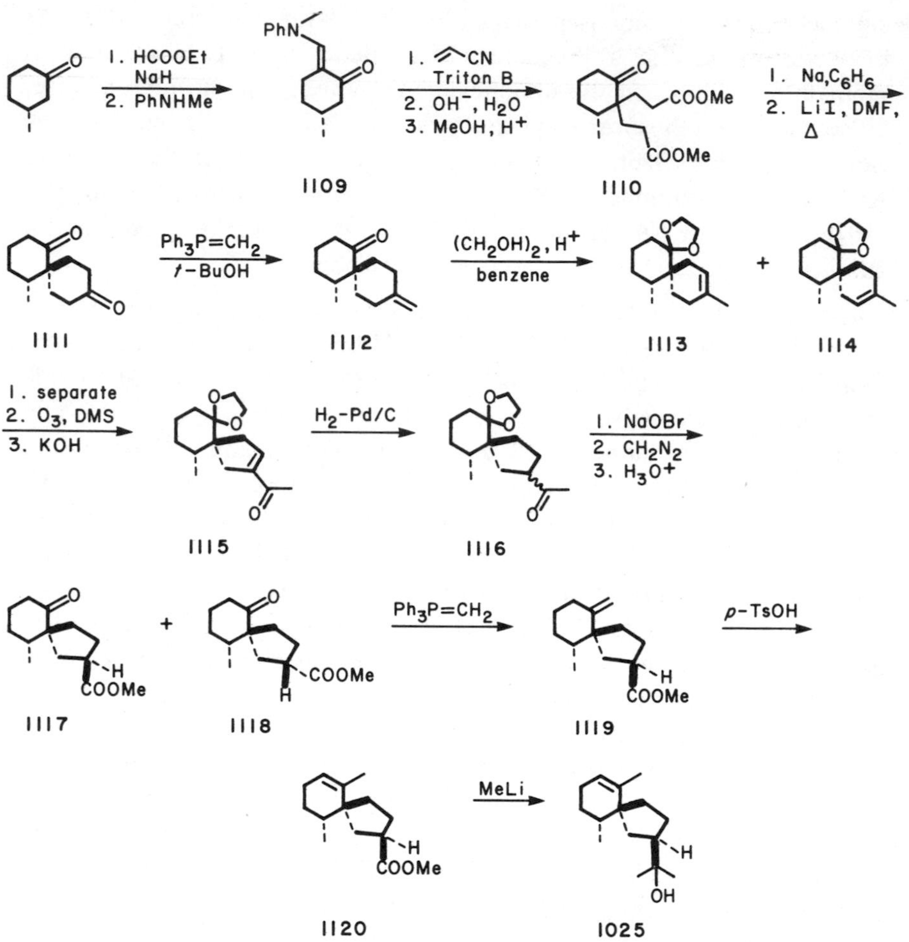

Scheme 136.   Ramage's Synthesis of (-)-Agarospirol

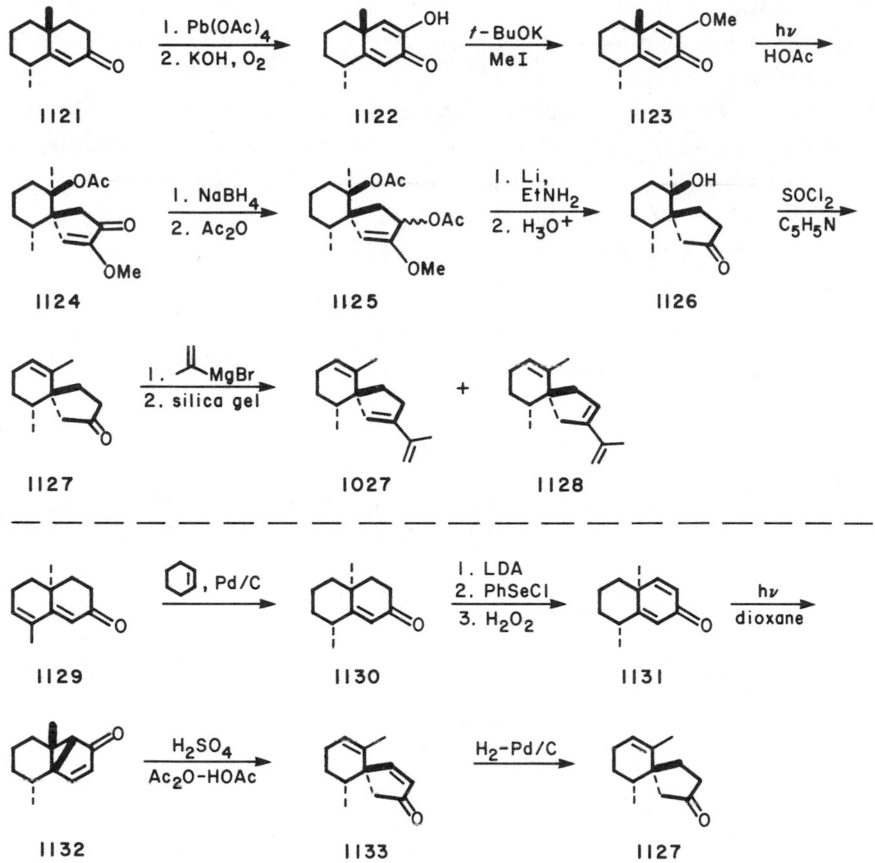

Scheme 137.   Caine's Synthesis of (±)-α-Vetispirene

methoxycyclohexadienone **1123** in acetic acid provides the spiro-
[4.5]decane intermediate **1124**, which is elaborated by a tedious but
straightforward sequence of reactions into the valuable spiro[4.5]-
decenone **1127**. Addition of isopropenylmagnesium bromide and dehydration gives (±)-α-vetispirene (**1027**) and its isomer **1128** in a ratio of
3:2. The two isomers may be separated by preparative glpc. In a later

paper, it was reported that intermediate **1127** may be prepared more conveniently from dienone **1129**, via the axial methyl isomer **1130**, which is obtained from **1129** by transfer hydrogenation.[343] Introduction of the second double bond is difficult because of the propensity of **1130** to revert to the more stable isomer **1121**. However, if **1131** is photolyzed as soon as it is prepared, the tricyclic lumiproduct **1132** may be obtained in good yield. Electrophilic cleavage of the cyclopropyl ketone gives **1133**, which is hydrogenated cleanly to **1127**.

Büchi and co-workers synthesized (±)-β-vetivone by the imaginative and stereoselective route shown in Scheme 138.[344] Cyclopentadiene condenses with keto aldehyde **1134** to give fulvene **1135** in 70% yield.

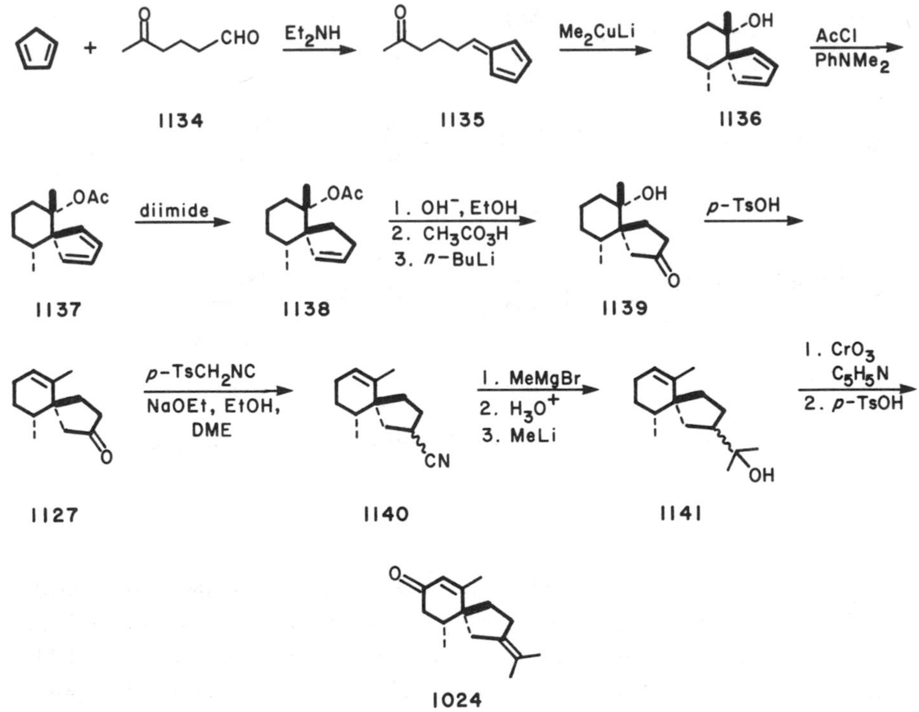

Scheme 138.   Büchi's Synthesis of (±)-β-Vetivone

Addition of lithium dimethylcuprate to **1135** regenerates a cyclopenta-dienide ion, which adds to the ketone carbonyl. The reaction is per-formed in ether solution at -20°C and gives a single carbinol, **1136**. Reduction of acetate **1137** by diimide gives the crystalline dihydro com-pound **1138** in 57% yield along with 5-10% of the tetrahydro derivative. Epoxidation of the unsaturated alcohol is highly stereoselective (presum-ably syn to the hydroxy group). Rearrangement of the epoxide with *n*-butyllithium provides the spiro[4.5]decenone **1127**. For preparation of (±)-β-vetivone, **1127** is transformed by *p*-toluenesulfonylmethyl iso-cyanide to nitrile **1140**. Conversion of this substance to the hinesol-agarospirol mixture (**1141**) to (±)-β-vetivone follows a precedented line.

Torii and co-workers have generated the spiro[4.5]decane system by base-catalyzed cyclization of a 4-(4-hydroxyphenyl)butyl tosylate, a reac-tion discovered by Winstein and Baird in 1957 (Scheme 139).[345a,b] The required substrate is built up efficiently from phenolic ester **1142** (obtained in four steps from *m*-cresol methyl ether and succinic anhy-dride). Treatment of **1144** with potassium *t*-butoxide provides **1145** as a

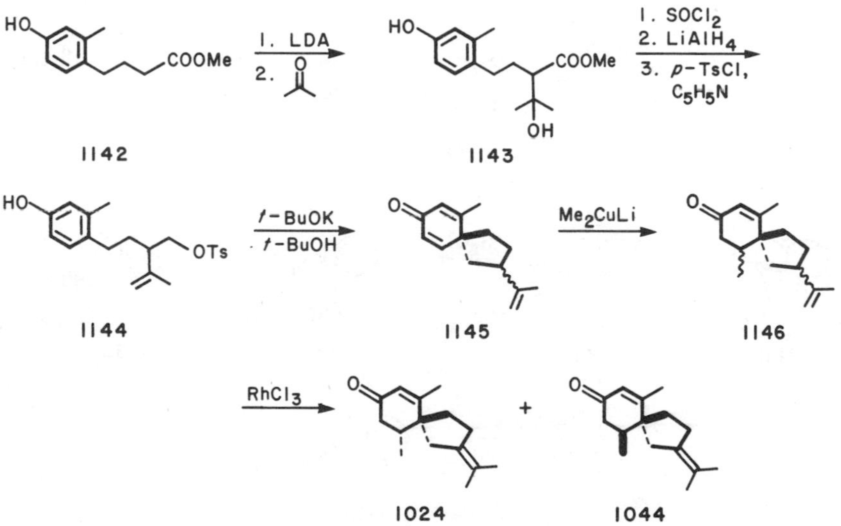

Scheme 139.   Torii's Synthesis of (±)-β-Vetivone

mixture of diastereomers. Addition of lithium dimethylcuprate occurs at the less substituted double bond of the cyclohexadienone system, giving **1146** as a mixture of several stereoisomers. Isomerization of the double bond occurs with rhodium chloride. The product is a 1:1 mixture of (±)-β-vetivone (**1024**) and its diastereomer **1044**, which may be separated by hplc.

In connection with his investigation of the reaction of enol ethers with diazomethyl ketones, Wenkert carried out the synthesis outlined in Scheme 140.[292] Compound **1147** (available by the Diels-Alder reaction of piperylene and crotonaldehyde) is converted into the enol ether **1148**, which is conjugated by treatment with strong base. Diazoacetone is decomposed in the presence of **1149** to obtain a 1:3 mixture of the desired cyclopropane **1150** and its isomer **1151**. Electrophilic cleavage of the cyclopropane rings gives a mixture of **1152** and **1153**. The mixture is separated and **1152** is aldolized to obtain **1154**, an intermediate in Marshall's synthesis of (±)-β-vetivone.[346]

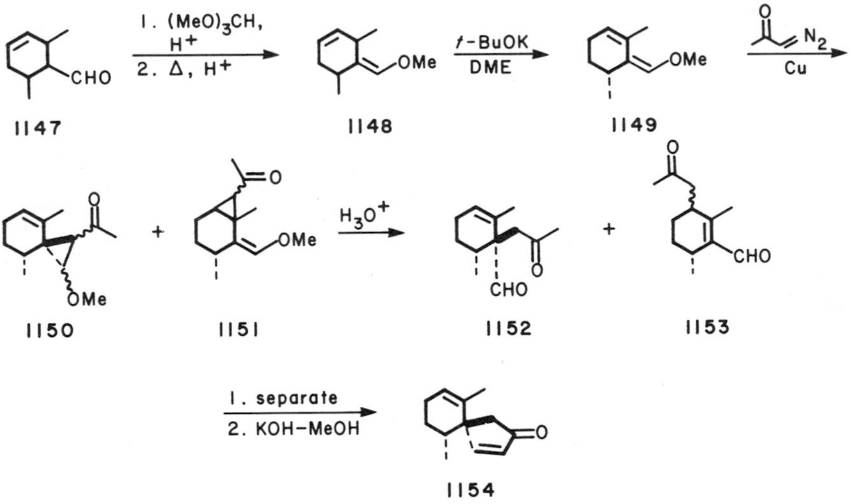

Scheme 140.   Wenkert's Synthesis of a (±)-β-Vetivone Precursor

The final spirovetivane synthesis we discuss is a 1979 synthesis leading to (±)-hinesol, (±)-agarospirol, and (±)-α-vetispirene (Scheme 141).[347a,b] Inubushi and co-workers begin with acetylacetone which is converted into diene **1155** by House's method. Diels-Alder reaction of

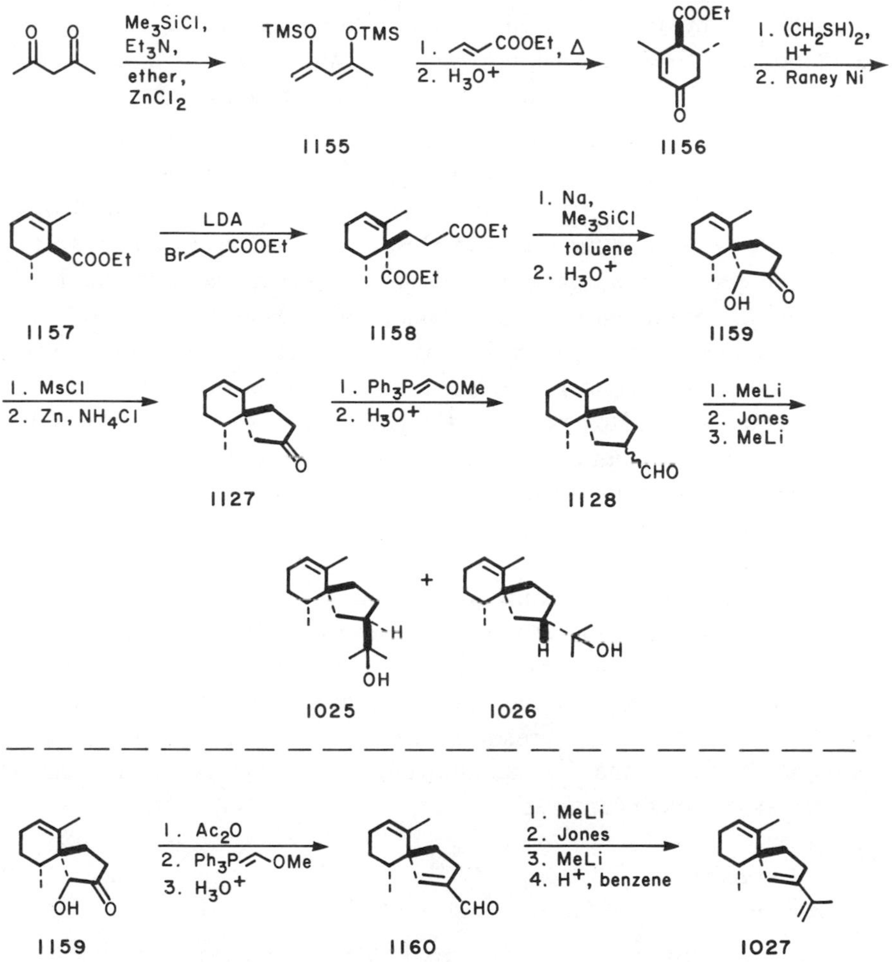

Scheme 141.   Inubushi's Synthesis of (±)-Hinesol,
(±)-Agarospirol, and (±)-α-Vetispirene

1155 with ethyl crotonate gives cyclohexenone ester 1156. Removal of the ketone carbonyl is accomplished by desulfurization of the dithioketal. Alkylation of the enolate ion of 1157 with ethyl 3-bromopropionate is stereoselective, but proceeds in only 19% yield. The resulting diester 1158 is subjected to acyloin condensation and the α-hydroxy ketone so obtained is reduced to obtain the Caine-Büchi enone 1127. This material is elaborated by the indicated sequence to obtain a 3:2 mixture of ($\pm$)-hinesol (1026) and ($\pm$)-agarospirol (1025). A minor modification allows production of ($\pm$)-α-vetispirene (1027) from acyloin 1159.

Because of the large number of spirovetivane syntheses that have been published, it is appropriate to compare them against one another. This we do in Table 2. It includes the three syntheses (Marshall's β-vetivone, Marshall's hinesol, and Deslongchamps' agarospirol) which were reviewed in Volume 2 of this work. From the standpoint of efficiency, the Stork synthesis is by far the best. However, it should be noted that this synthesis is easily applicable only to β-vetivone. The Dauben synthesis is almost as efficient in the number of steps, although not in overall yield. However, the Dauben synthesis is highly adaptable and easily leads to almost all members of the class. Many of the syntheses are clearly not useful for producing the end compound in any quantity. Such is the case with the Marshall hinesol synthesis, which was done principally to determine the stereostructure of hinesol. Other impractical syntheses serve only to illustrate some particular chemistry, such as Büchi's interesting utilization of fulvene 1135 or Trost's application of his cyclobutanone spiroannelation. At this point, it would seem that it will be hard to improve upon the Stork and Dauben syntheses. However organic chemists seem to have a particular fascination for spirocyclic systems, and we can probably expect the parade of spirovetivane syntheses to continue.

(2) Acoranes

The acoranes also contain the spiro[4.5]decane nucleus, but have a different substitution pattern from the spirovetivanes. In this group also a large number of syntheses have been recorded. In this section, we review syntheses leading to acorone (**1161**), isoacorone (**1162**), acorenone (**1163**), acorenone B (**1164**), α-acorenol (**1165**), β-acorenol (**1166**), β-acoradiene (**1167**), γ-acoradiene (also known as α-alaskene, **1168**), δ-acoradiene (also known as β-alaskene, **1169**), and the unnamed acoradiene (**1170**).

1161   1162   1163   1164   1165

1166   1167   1168   1169   1170

There are several synthetic challenges in this group. One is clearly the construction of the spiro[4.5]decane nucleus. As will be seen, this turns out to be especially difficult and many methods have been used. For acorone and isoacorone, arranging the relative stereochemistry of the methyl and isopropyl groups in the five-membered ring is no problem, since the isopropyl group is epimerizable and the natural products have the more stable trans stereochemistry at these two centers. The other methyl group is also epimerizable (the thermodynamic ratio of **1161** to **1162** is 7:3). None of the syntheses reported to date have dealt with this problem; all give a mixture of **1161** and **1162**. For the acorenones (**1163** and **1164**) the methyl and isopropyl groups must be cis, and this is usually dealt with by establishing the cis stereochemistry on a cyclopentane precursor before constructing the cyclohexenone ring. In the acorenols (**1165** and **1166**), as in the acorone isomers, the two side chains are trans. However, the major problem, one that is common to all members of the group, is establishing the correct sense of chirality of the spiro carbon relative to the centers in the cyclopentane ring. Because the six-membered ring is pseudosymmetric, this stereochemical problem some-times presents itself as a regiochemical problem. For example, the

Table 2

Principal Investigator	Year	Starting Material	Product	Steps	Yield	Isomer Separations
Marshall	1968		(±)-β-vetivone	15	4.5%	0
Marshall	1969		(±)-hinesol	31	<0.1%	1(1:4)[a]
Deslongchamps	1970		(±)-agarospirol	18	3.0%	1(1:1)
Stork	1973		(±)-β-vetivone	3[b]	27%	0
McCurry	1973		(±)-β-vetivone	11	15%	2(1:1)
			(±)-β-vetispirene	12	15%	
Yamada	1973		(±)-β-vetivone	23	10.9%	0
			(±)-hinesol	23	11.9%	0
			(±)-α-vetispirene	23	8.6%	0
			(±)-β-vetispirene	25	5.3%	0
Caine	1974		(±)-β-vetivone	6	16.0%	1(7:3)
Pesaro	1974		(±)-β-vetivone	12	3.2%	1(5:1)
Trost	1975		(±)-hinesol	13[c]	~8.0	1(2:1)

Table 2

Principal Investigator	Year	Starting Material	Product	Steps	Yield	Isomer Separations
Magnus	1975		(±)-hinesol	26	0.2%	1(1:1)
Dauben	1975	EtO	(±)-β-vetivone	7	7.4%	0
			(±)-hinesol	8	8.1%	1(9:1)
			(±)-α-vetispirene	8	9.0%	1(3:2)
			(±)-β-vetispirene	9	6.4%	0
Ramage	1975		(−)-agarospirol	18	c	1(3:2)
Caine	1976		(±)-α-vetispirene	8	11.0%	1(3:2)
Büchi	1976	CHO	(±)-β-vetivone	14	8.5%	0
Torii	1977	MeO HO +	(±)-β-vetivone	10	13.0%	1(1:1)
Wenkert	1978	CHO	(±)-β-vetivone	20[d]	<0.2%	1(1:3)[a]
Inubushi	1979	COOEt	(±)-hinesol	12	0.3%	1(3:2)
			(±)-agarospirol	12	0.3%	
			(±)-α-vetispirene	12	0.03%	0

a. Synthesis completed with minor isomer.
b. Convergent; 5 steps from unsaturated diester **1034** (Scheme 126).
c. No yields given.
d. Projected on basis of Marshall synthesis from **1154**.

acorenols (**1165** and **1166**) can be interconverted by isomerization of the double bond from one position to another. However, this process is equivalent to inversion of the spiro carbon. The same relationship exists between the acorenones (**1163** and **1164**) and the acoradienes **1168** and **1169**. As we will show, many of the reported syntheses do not address this interesting problem.

The first published report of a synthetic approach to the acorane sesquiterpenes came from Conia and co-workers, who prepared and studied the pyrolysis of ketone **1171**.[348] Intramolecular ene cyclization occurs via the enol, giving four spiro[4.5]decanes (**1172-1175**) in yields of 20%, 30%, 8%, and 22%, respectively. Conversion of **1173** into acorone or isoacorone would require ketone transposition and oxidation of the cyclopentane double bond. However, this has not been carried out.

Pinder reported a serious attempt to synthesize acorone that failed at a late stage.[349] Because the basic approach has some merit, and may be adaptable, this unsuccessful attempt is summarized in Scheme 142. *p*-Methylacetophenone (**1176**) is converted by a rather tedious nine-step sequence into acid **1178**, which is reduced by the Birch method. The side chain is further elaborated to obtain β-keto ester **1181**. This substance undergoes intramolecular Michael addition, affording a diketo ester of the gross structure **1182**. Although no attempt is made to control stereochemistry in this synthesis, Pinder notes that **1182** is probably mostly one stereoisomer since the material is "mostly crystalline".

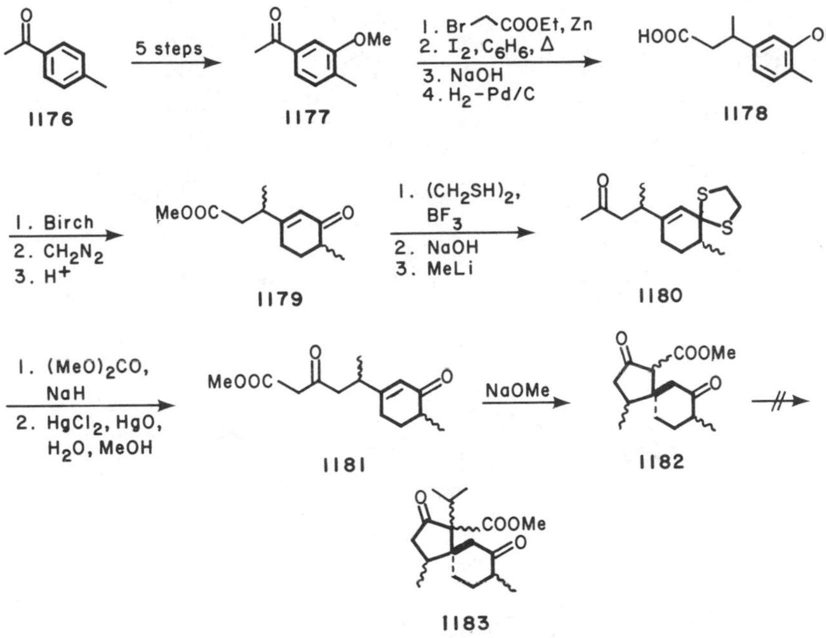

Scheme 142.   Pinder's Unsuccessful Approach to (±) Acorone

Indeed, since the Michael addition is probably reversible, all four stereo-centers in **1182** are effectively epimerizable. Thus, it might not be surprising if one or two stereoisomers predominate heavily.

The first report of a successful acorane synthesis came from Kasturi and Thomas at the Indian Institute of Science in Bangalore.[350] The reported synthesis (Scheme 143) begins with *p*-methoxyisobu-tyrophenone (**1184**) which is converted by steps similar to those used by Pinder into enone **1187**. Note that the relative stereochemistry of the methyl and isopropyl positions is established in reduction of methyl ketone **1186**—undoubtedly **1187** is a 1:1 mixture of diastereomers. Cyclization of **1188** gives **1189**. In this step, a new asymmetric carbon is created but it is removed immediately when the double bond is hydro-genated. Chirality of the spiro carbon is restored when the tertiary

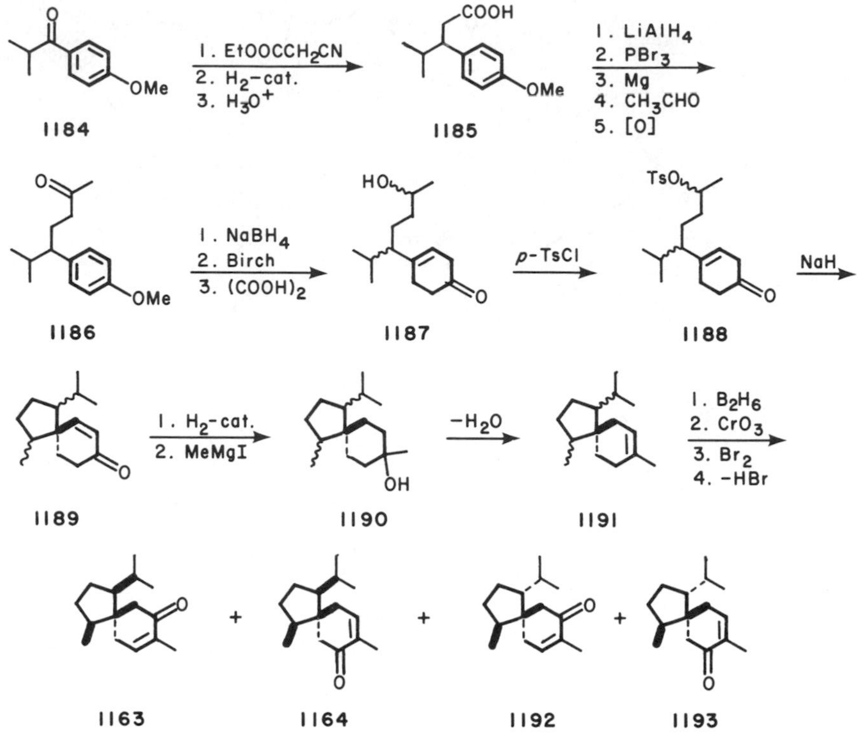

Scheme 143.   Kasturi-Thomas Synthesis of Acorenone Isomers

alcohol **1190** is dehydrated. Four final steps convert the cyclohexene ring to the correct cyclohexenone oxidation level. The final product is probably a nearly equimolar mixture of the two acorenones (**1163** and **1164**) along with the unnatural isomers **1192** and **1193**.

In 1972, Naegeli and Kaiser, of Givaudan in Switzerland, isolated acoradiene **1170** from vetiver oil. As part of their structure elucidation, they synthesized the compound as shown in Scheme 144.[351] Dehydro-linalool (**1194**) cyclizes thermally to a mixture of alcohols **1195** which is converted by Claisen rearrangement into ketone **1196**. Catalytic hydrogenation of the isopropenyl double bond is followed by addition of vinylmagnesium bromide. The resulting allylic alcohol undergoes

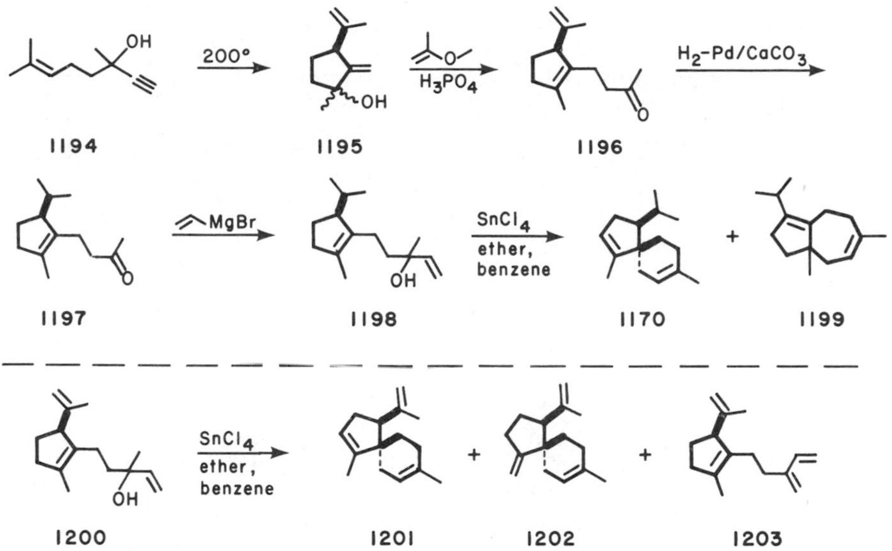

Scheme 144.   Naegeli-Kaiser Synthesis of Acoradiene **1170**

stereospecific cyclization when treated with stannic chloride. Acoradiene **1170** and the hexahydroazulene **1199** are formed in a ratio of 3:7. Similar cyclization of the dehydro derivative **1200** is much cleaner, affording a hydrocarbon mixture containing 75-80% of acoratriene **1201** in 65% yield. The principal by-products in this case are the double bond isomer **1202** and the simple dehydration product **1203**.

In 1973 Ramage and co-workers at Liverpool reported a synthesis leading to (−)-α-acorenol and (+)-β-acorenol (enantiomers of the actual natural products).[352] The syntheses (Scheme 145) begin with the diastereomeric unsaturated ketals **1113** and **1114** which have been discussed earlier in connection with the Ramage agarospirol synthesis (see Scheme 136). In the present case, the ketal grouping of **1113** is hydrolyzed and the resulting ketone is nitrosated. Treatment of the oximino ketone with chloramine gives a diazo ketone which undergoes photochemical Wolff rearrangement. The resulting acid is esterified to obtain methyl ester **1205**. Addition of methyllithium to **1205** yields

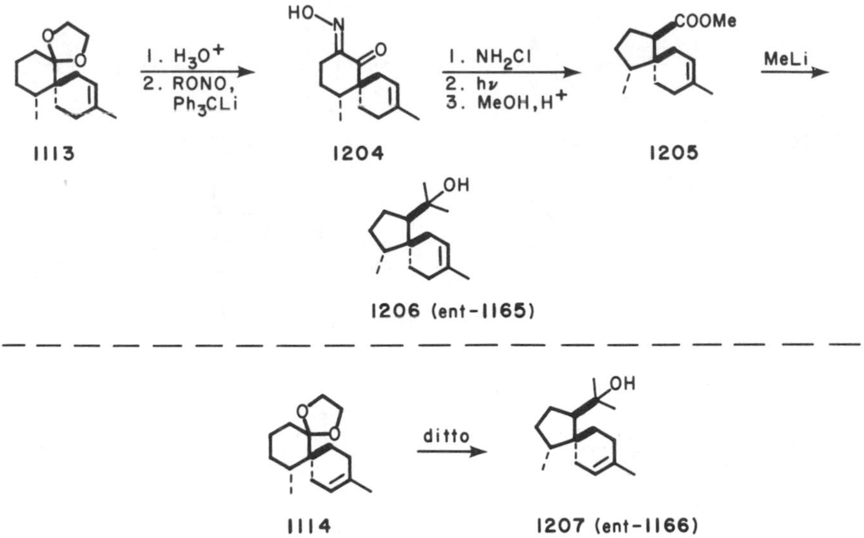

1113   1204   1205

1206 (ent-1165)

1114   1207 (ent-1166)

Scheme 145.   Ramage's Synthesis of (-)-α-Acorenol and (+)-β-Acorenol

(−)-α-acorenol, the enantiomer of natural α-acorenol (1165). Application of the same sequence to unsaturated ketal 1114 affords (+)-β-acorenol (1207). The apparent stereospecificity of the Wolff rearrangement probably results from equilibration during esterification of the initially formed acids. There is no reason to expect kinetic stereoselectivity in hydration of the ketenes (1208 and 1209) which intervene in the

1208   1209

two cases. In fact, if there is some steric factor which favors addition to one face of the ketene double bond in 1208, it should favor the other face of 1209.

In the same year, Marx and Norman also published a synthesis leading to the enantiomers of two acorane sesquiterpenes, (−)-γ-acoradiene and (+)-δ-acoradiene.[353a] They later extended the synthesis to (−)-acorone

and (+)-isoacorone, the enantiomers of **1161** and **1163**.[353b] The Marx
synthesis (Scheme 146) begins with (R)-(+)-pulegone (**1210**) which is

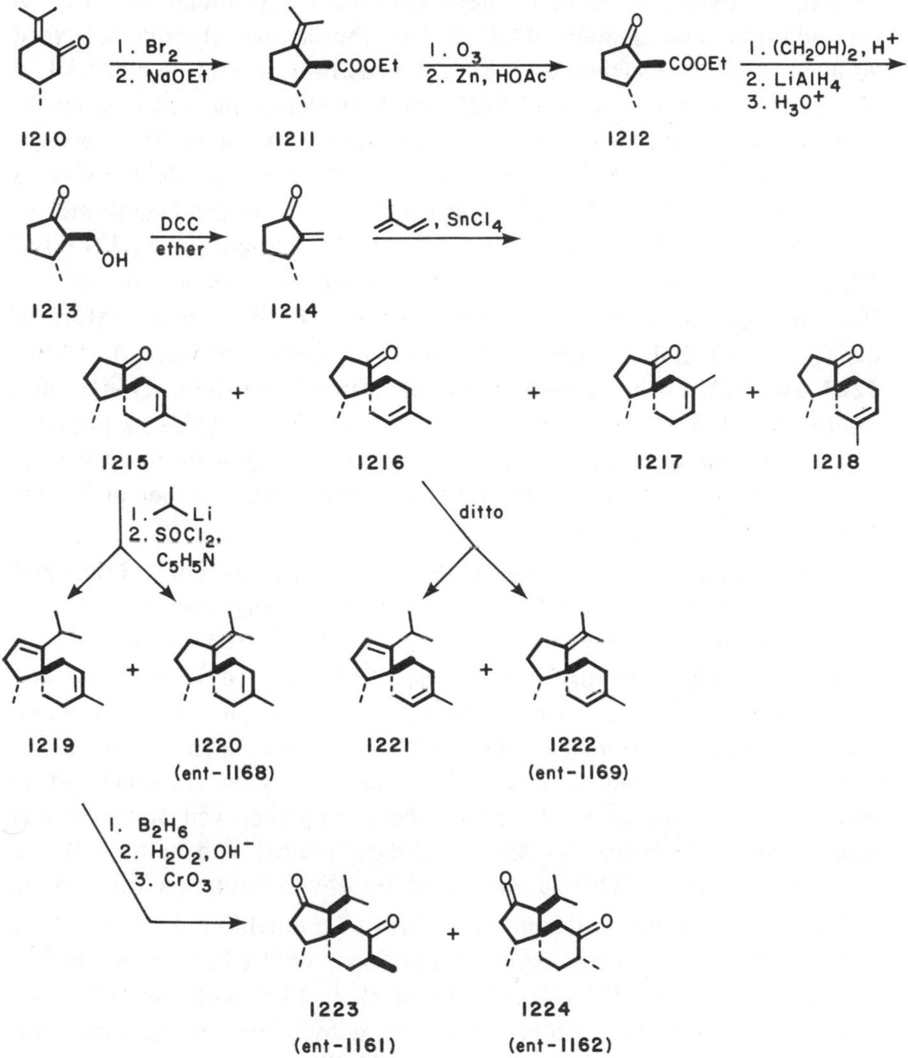

Scheme 146.   Marx's Synthesis of (-)-Acorone, (+)-Isoacorone,
(-)-γ-Acorodiene and (+)-δ-Acoradiene

transformed into β-keto ester **1212** by a literature procedure. This compound is converted by a further four-step sequence into the methylene cyclopentanone **1214** which undergoes Diels-Alder reaction with isoprene to give a mixture of stereoisomers **1215-1218**. Although the thermal cycloaddition ratio is poor (45:25:20:10), the stannic chloride-catalyzed addition is more selective, giving **1215 to 1218** in a ratio of 69:27:3:1 in 48% yield. The major isomers **1215** and **1216** were separated by preparative hplc and used to complete the synthesis. Addition of isopropyllithium to **1215**, followed by dehydration of the resulting alcohol gives a 7:3 mixture of **1219** and **1220**, which is found to be the enantiomer of γ-acoradiene (**1168**). Similar treatment of **1216** gives dienes **1221** and **1222**, the enantiomer of δ-acoradiene (**1169**), also in a ratio of 7:3. Hydroboration of both double bonds in diene **1219** yields a mixture of diastereomeric diols, which is oxidized to obtain a mixture of diones **1223** and **1224**—the enantiomers of natural acorone (**1161**) and isoacorone (**1162**), respectively. Although the Marx synthesis played a role in correlating the acoradienes with acorone and isoacorone, it is poor as a source of any of the compounds because of the number of isomer separations which are necessary.

In 1973, Oppolzer provided a synthesis of (±)-β-acorenol (**1166**) and (±)-β-acoradiene (**1167**).[354a] The basic scheme was later extended to provide syntheses of (±)-acorenone B (**1164**)[354b] and (±)-acorenone (**1163**).[354c] The Oppolzer synthesis (Scheme 147) exploits the intramolecular ene reaction of diene **1226** to create the spiro[4.5]decane skeleton. Diastereomers **1227** and **1228** are produced in a ratio of 1:1 and can be equilibrated by base to a 3:2 mixture. They are separated by silica gel chromatography. Note that the ene reaction can only give one relative stereochemistry for the secondary methyl and spiro carbons. The trans isomer (**1228**) is converted by allylic oxidation into enone **1229**, which is treated with one equivalent of methyllithium to provide a tertiary alcohol. The corresponding methoxymethyl ether is pyrolyzed to obtain cyclohexene **1231**. Treatment of ester **1231** with methyllithium gives (±)-β-acorenol (**1166**) which is dehydrated by heating with alumina to obtain (±)-β-acoradiene (**1167**). For the synthesis of (±)-acorenone B, the cis isomer **1227** is used. Conversion to the terti-

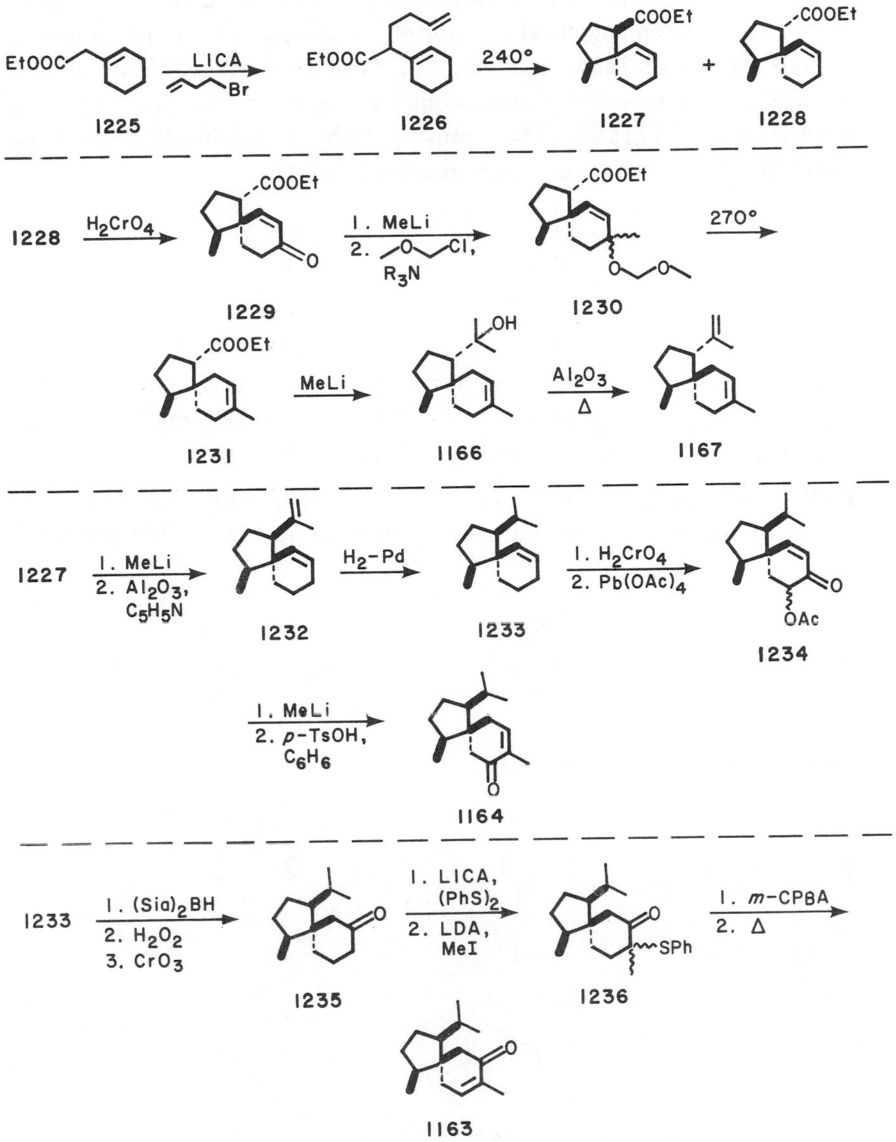

Scheme 147.   Oppolzer's Syntheses of Acoranes

ary alcohol, which is dehydrated by heating with alumina at 220°C gives diene **1232**. The more reactive isopropenyl double bond is hydrogenated and the cyclohexene allylically oxidized to obtain an enone which is acetoxylated to give **1234**. Treatment of **1234** with methyllithium gives a diol which reacts with p-toluenesulfonic acid in benzene to give (±)-acorenone B (**1164**). The unusual dehydration leading to **1164** apparently involves a pinacol-type rearrangement:

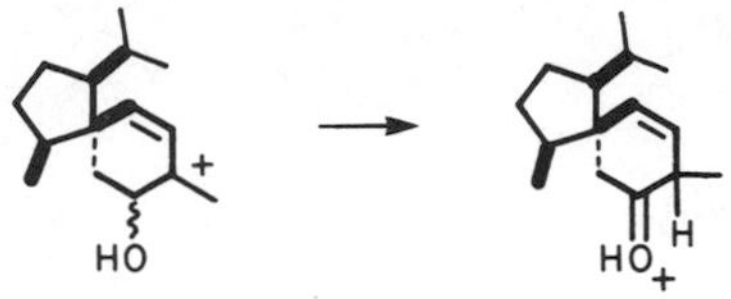

For preparation of (±)-acorenone alkene **1233** is hydroborated using disiamylborane to obtain ketone **1235**. Introduction of a methyl and the double bond provides (±)-acorenone (**1163**).

Wolf and Kolleck reported the stereorandom synthesis of (±)-acorenone B, which we summarize in Scheme 148.[355] The synthesis

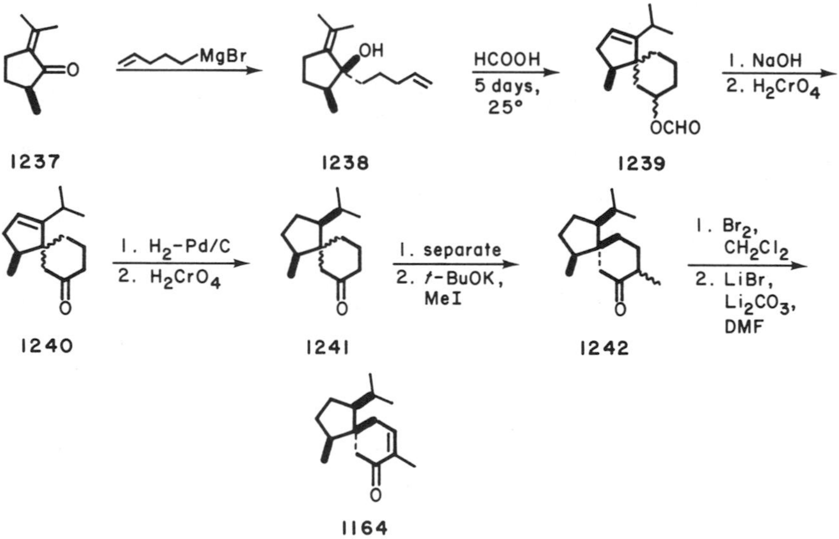

Scheme 148.   Wolf's Synthesis of (±)-Acorenone B

begins with (±)-camphorone (**1237**) which reacts with 4-pentenylmagnesium bromide to give alcohol **1238**. This substance undergoes slow solvolysis to give a mixture of diastereomeric formates (**1239**) in 32% yield. Hydrolysis of the ester and oxidation of the resulting secondary alcohol gives a 1:1 mixture of diastereomeric ketones (**1240**). After hydrogenation of the double bond, a mixture of isomers (**1241**) is obtained in which the desired isomer constitutes 46%. The mixture is separated by preparative tlc and the desired ketone methylated. After introduction of the double bond, (±)-acorenone B is obtained. It is a little perplexing that the π-cyclization leading to **1239** is not more stereoselective. One would have thought that the methyl substituent would have provided some steric hindrance. For example, see the Diels-Alder reaction of enone **1214** (Scheme 144). Wolf et al. have also reported a similar study leading to the unnatural product 4-epiacorenone B.[356]

Trost and co-workers have applied "secoalkylation" to the synthesis of (±)-acorenone B as shown in Scheme 149.[357] 2-Methylcyclopentanone is converted into 2-isopropyl-5-methylcyclopentanone (**1244**) by a three-step sequence featuring exhaustive methylation of the β-acetoxy enone **1243** by lithium dimethylcuprate. Formation and rearrangement of the oxaspiropentane gives a single cyclobutanone, **1245**. Since **1244** is a mixture of cis and trans isomers, equilibration must occur under conditions of the ylid reaction. It is reasonable that the less hindered cis isomer reacts faster than the trans and it is also reasonable that attack of the ylid occurs exclusively trans to the two alkyl groups flanking the carbonyl. After conversion of **1245** to its α-formyl derivative, acidic conditions promote reverse Claisen reaction and dehydration to the enol lactone **1246**. This material is reduced by diisobutylaluminum hydride to a lactol which is oxidized by chromic acid to lactone **1247**. Addition of methyllithium to **1247** gives a new hemiketal which reacts with ethanedithiol to give **1248**. After oxidation of the primary alcohol, regeneration of the ketone, and base-catalyzed aldolization, spiro[4.5]decenone **1250** is reached. This key intermediate is converted into (±)-acorenone B (**1164**) by a four-step sequence featuring a novel use of sulfenylation for transposing the enone chromophore by one posi-

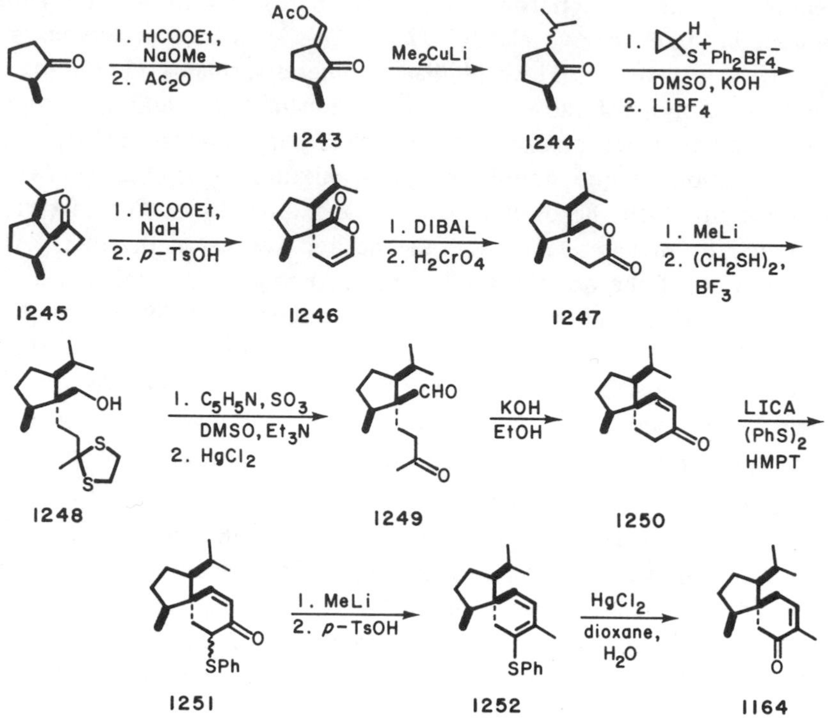

Scheme 149.   Trost's Synthesis of (±)-Acorenone B

tion. The use of the cyclopropylsulfurane to achieve spiroannelation is a novel strategy. However, the 14 steps required to reach **1250** by this procedure are rather laborious. As will be seen, intermediates of this type are more easily accessible by direct Robinson annelation of the readily available aldehyde **1255** (see Schemes 150 and 151).

White and co-workers developed the interesting stereoselective synthesis of (−)-acorenone B which is documented in Scheme 150.[358] The synthesis begins with (R)-(+)-limonene (**1253**), which is transformed

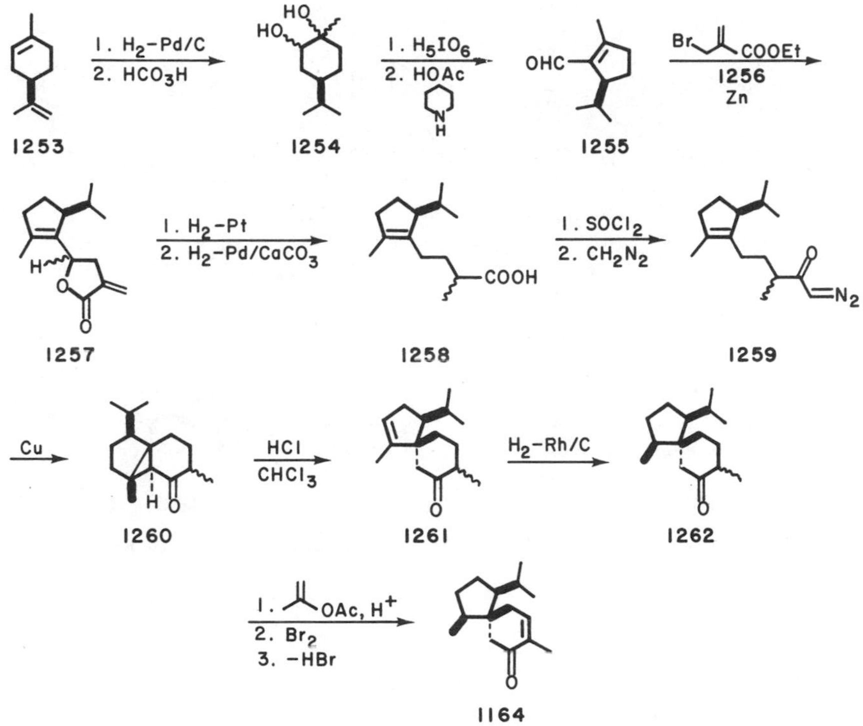

Scheme 150.   White's Synthesis of (-)-Acorenone B

by a precedented four-step sequence into unsaturated aldehyde **1255**. Reformatsky reaction of this aldehyde with unsaturated bromo ester **1256** gives a lactone (**1257**) which is subjected to successive hydrogenation and hydrogenolysis to obtain unsaturated acid **1258**. This compound is transformed into a diazoketone, which is decomposed by copper powder. The carbene attacks the double bond exclusively trans to the isopropyl group, giving rise to tricyclic ketone **1260**. Electrophilic cleavage of the cyclopropane ring gives enone **1261**, which undergoes

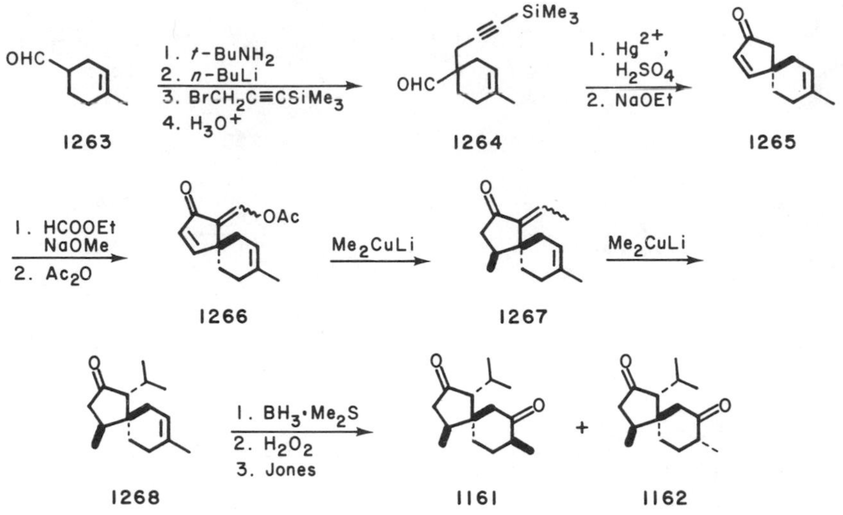

Scheme 151.   Dolby's Synthesis of (±)-Acorone and (±)-Isoacorone

hydrogenation trans to the isopropyl group. After introduction of the double bond, (−)-acorenone B (**1164**) is obtained.

In 1977, McCrae and Dolby reported the nice synthesis of (±)-acorone and (±)-isoacorone, which we outline in Scheme 151.[359] Unsaturated aldehyde **1263** (from isoprene and acrolein) is alkylated with 1-trimethylsilyl-3-bromo-1-propyne by Stork's metalloenamine method. The resulting acetylenic aldehyde **1264** is hydrated to a 1,4-keto aldehyde which undergoes subsequent aldolization to produce the spirocyclic dienone **1265**. The isopropyl and methyl groups are introduced by exhaustive methylation of **1266** with lithium dimethylcuprate. Hydroboration of enone **1268**, followed by oxidation with Jones' reagent gives a 55:45 mixture of (±)-acorone (**1161**) and (±)-isoacorone (**1162**). Pure **1161** may be obtained by direct crystallization from this mixture.

Lange has reported a very direct synthesis of (−)-acorenone (Scheme 152).[360] The synthesis begins with α,β-unsaturated aldehyde, **1255** (see

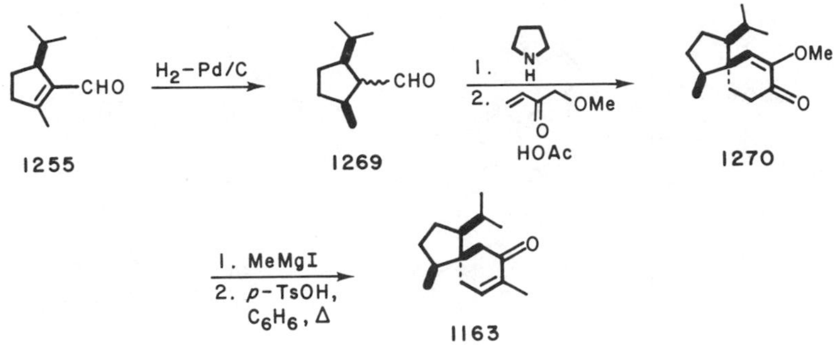

Scheme 152.   Lange's Synthesis of (-)-Acorenone

Scheme 150). Hydrogenation occurs from the side of the molecule trans to the isopropyl to give an epimeric mixture of aldehydes (1269) which is subjected to Robinson annelation with methoxymethyl vinyl ketone to obtain 1270 in 35% yield. In spite of the low yield, this represents a direct and highly stereoselective entry into the acorane skeleton. Treatment of 1270 with methylmagnesium iodide gives a tertiary alcohol which is refluxed with p-toluenesulfonic acid in benzene to obtain (±)-acorenone (1163).

Pesaro and Bachmann have reported a synthesis of (−)-acorenone and (−)-acorenone B which is very similar to the Lange synthesis (Scheme 153).[361] In their synthesis, the Robinson annelation is carried out with methyl vinyl ketone and the yield is lower (only 10%, or 20% based on reacted aldehyde). Manipulation of the cyclohexane ring features a novel hydroboration step. The initial borane adduct, being a β-hydroxy-borane, undergoes elimination to give a new cyclohexene which is hydroborated again. Upon oxidation, alcohol 1273 is produced. Introduction of the double bond leads to (−)-acorenone (1163). To prepare isomer 1164 the allylic alcohol 1272 is dehydrated to diene 1275, which is epoxidized at the more nucleophilic and less hindered double bond. Treatment of the epoxide with p-toluenesulfonic acid results in rearrangement to the β,γ-unsaturated enone 1276. The sequence from 1272

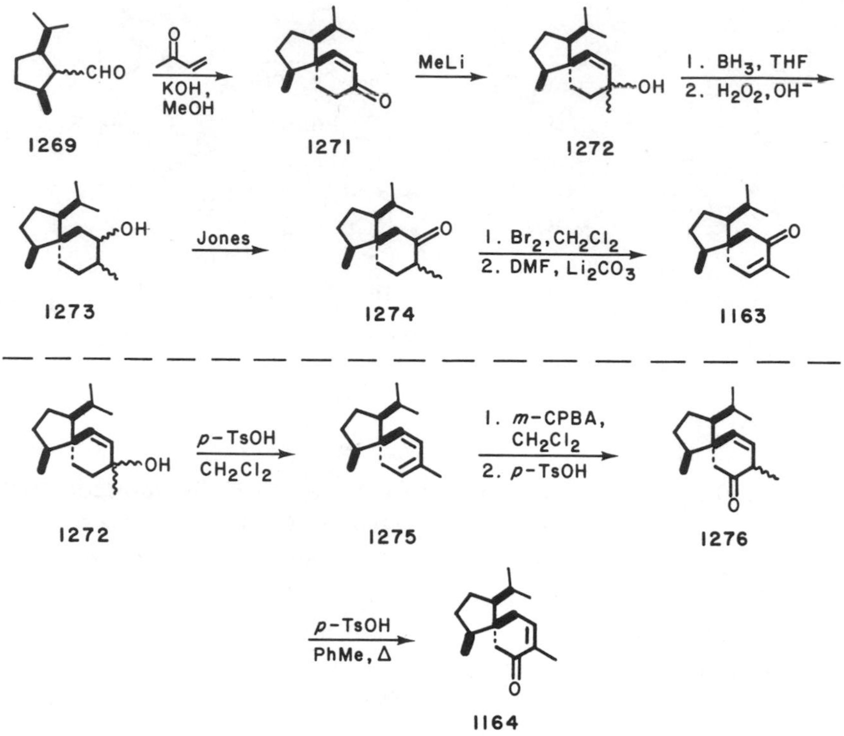

Scheme 153.   Pesaro's Synthesis of (-)-Acorenone and (-)-Acorenone B

to **1276** is efficiently carried out in one pot by treating **1272** with a mixture of *p*-toluenesulfonic acid and *m*-chloroperoxybenzoic acid in methylene chloride.   Conjugation of **1276** provides (±)-acorenone B **(1164)**.

Martin and Chou have prepared (±)-acorone and (±)-isoacorone by a route which is essentially the same as the Dolby route (Scheme 154).[362]

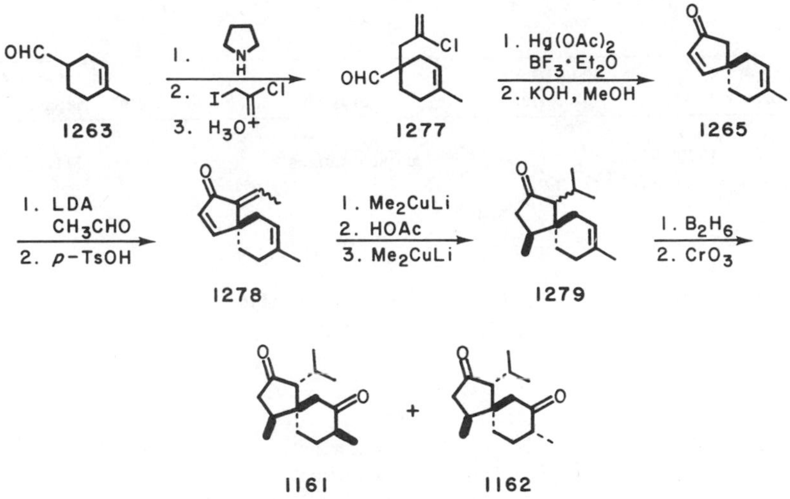

Scheme 154.   Martin's Synthesis of (±)-Acorone and (±)-Isoacorone

The principal difference is in the method used for adding the cyclopentenone ring. Rather than the propargyl group which Dolby uses, Martin employs 2-chloro-3-iodopropene and does the alkylation using the enamine of aldehyde **1263**. A similar method is also used to add the methyl and isopropyl groups to the five-membered ring. As in the Dolby synthesis, the crucial stereochemistry at the secondary methyl center results from attack of the cuprate reagent over the face of the cyclopentenone chromophore syn to the cyclohexene double bond.

Table 3 compares the successful syntheses of acorane sesquiterpenes. Note that the only syntheses which do not require isomer separations are those leading to the acorenones, which are clearly the simpler synthetic tasks.

Table 3

Principal Investigator	Year	Starting Material	Product	Steps	Yield	Isomer Separations
Kasturi	1972		(±)-acorenones	18	—	1(1:1:1:1)[a]
Naegeli	1972		(−)-acoradiene **1170**	5	~12%	1(3:7)[b]
Ramage	1973		(−)-α-acorenol (+)-β-acorenol	15	b	1(1:1)[c]
Marx	1973		(−)-γ-acoradiene (+)-δ-acoradiene	10 10	0.9% 0.9%	2(7:3, 3:7[d]) 2(3:7[d], 3:7[d])
	1975		(−)-acorone (+)-isoacorone	12 12	0.5% 0.5%	3(7:3, 7:3, 1:1) 3(7:3, 7:3, 1:1)
Oppolzer	1973		(±)-β-acorenol	7	5.5	1(1:1)
			(±)-β-acoradiene	8	5.5	1(1:1)
	1975		(±)-acorenone B	9	11.3%	1(3:2)
	1977		(±)-acorenone	11	14.5%	1(3:2)
Wolf	1975		(±)-acorenone B	9	1.2%	1(1:1)
Trost	1975		(±)-acorenone B	18	4.4%	0

Table 3

Principal Investigator	Year	Starting Material	Product	Steps	Yield	Isomer Separations
White	1976		(−)-acorenone B	15	1.5%	0
Dolby	1977	OHC	(±)-acorone	10	4.9%	1(1:1)
			(±)-isoacorone	10	4.9%	1(1:1)
Lange	1977		(−)-acorenone	8	12.2%	0
Pesaro	1978		(−)-acorenone	11	2.3%	0
			(−)-acorenone B	10	3.4%	1(3:1)
Martin	1978	OHC	(±)-acorone	10	16%	1(1:1)
			(±)-isoacorone	10	16%	1(1:1)

a. The paper does not consider stereochemistry. However, a consideration of the chemistry suggests that 4 isomers are produced in approximately equal amounts.

b. Not reported.

c. The two isomers are each employed, one for one target and one for the other.

d. The minor isomer is employed.

(3) Axisonitrile-3

(+)-Axisonitrile (**1280**) is an unusual sesquiterpene isonitrile which occurs in the marine sponge *Axinella cannabina*. Caine and Deutsch

1280

have synthesized the enantiomer of **1280** as shown in Scheme 155.[363] (+)-Dihydrocarvone is converted by a known procedure to the bicyclic ketol **1281**,[364] which is hydrogenated to obtain **1282**. Oxidation of the dienyl acetate **1283** by the Kirk-Wiles procedure gives 42% of the enone arising from simple hydrolysis of **1283** and 56% of a 6:1 mixture of alcohols **1284** and **1285**. The major isomer is obtained by chromatography on Florisil. After protection of the hydroxy group, the $\Delta^{1,2}$ double bond is introduced by the selenoxide method. Irradiation of **1287** in anhydrous dioxane gives the lumiproduct **1288**, which is hydrogenated and methylenated to obtain vinyl cyclopropane **1290**. Lithium in ethylamine reduces **1290** with inversion of configuration at the methyl center, giving the spiro[4.5]decene **1291**. The remaining steps are required to change the hydroxy group into the isonitrile function of the proper relative stereochemistry. This is a fine example of a well-planned stereorational synthesis. The sole point of uncertainty is the stereochemical outcome of the vinyl cyclopropane reduction, and this aspect was tested prior to the synthesis outlined in Scheme 155 with simple analogs of **1290**.

(4) Chamigrenes

Chamigrenes contain the spiro[5.5]undecane nucleus. In Volume 2 of this series, we reported syntheses of chamigrene (**1294**) and (±)-α-chamigrene (**1295**). We now report three more syntheses of

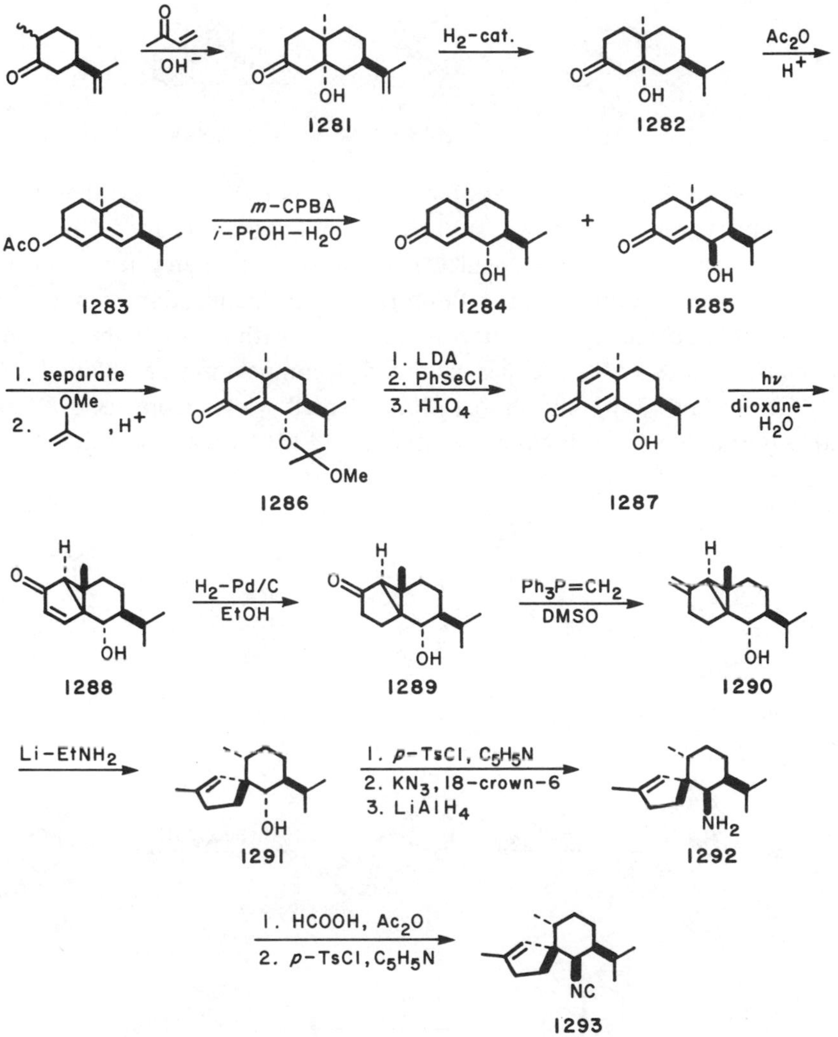

Scheme 155.   Caine-Deutsch Synthesis of (±)-Axisonitrile-3

($\pm$)-α-chamigrene as well as two syntheses of ($\pm$)-10-bromo-α-chamigrene (**1296**).

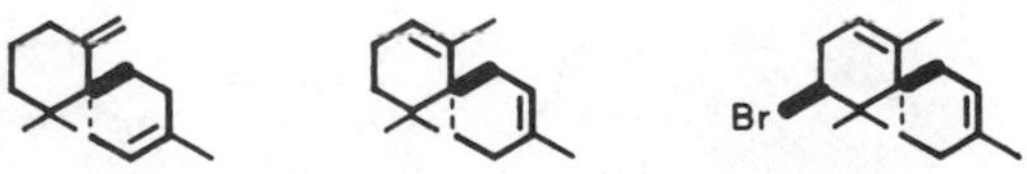

1294   1295   1296

In 1974, White, Torii, and Nogami reported the synthesis of ($\pm$)-α-chamigrene which we summarize in Scheme 156.[365] The synthesis begins with acid **1298**, which is obtained in one step from anisole and 4,4-dimethylbutyrolactone. Birch reduction, ketalization, conversion to the acid chloride, and reaction of the latter with diazoethane provide the diazo ketone **1300**. Copper-catalyzed decomposition of **1300** affords tricyclic ketone **1301** in almost 30% overall yield from acid **1298**. Lithium-ammonia reduction converts **1301** quantitatively into

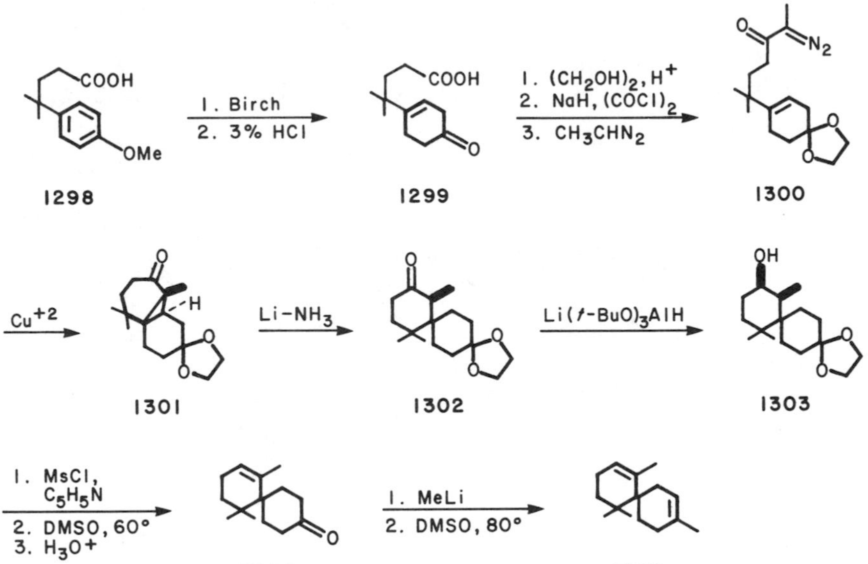

Scheme 156.   White's Synthesis of ($\pm$)-α-Chamigrene

spiro[5.5]undecanone **1302**. Reduction of the carbonyl gives the cis alcohol **1303**, which is dehydrated by heating the derived mesylate in DMSO at 60°C. Ketone **1304** is methylated to obtain a diastereomeric mixture of tertiary alcohols, which is dehydrated by heating in DMSO to obtain (±)-α-chamigrene (**1295**). The 12 step synthesis proceeds in an overall yield of about 14%.

Fráter studied the conversion of *cis-* and *trans*-β-ionol (**1305**) into (±)-α-chamigrene as shown in Scheme 157.[366] Treatment of **1305** with

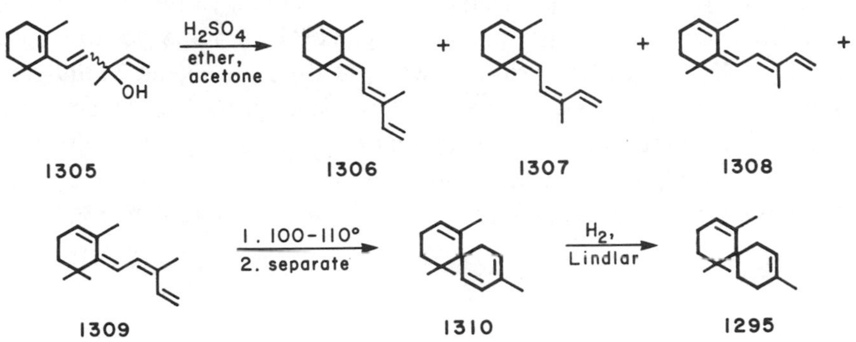

Scheme 157.   Fráter's Synthesis of (±)-α-Chamigrene

10% sulfuric acid in a mixture of ether and acetone gives a mixture of tetraenes **1306-1309**. The composition of the mixture depends upon whether *cis-* or *trans-***1305** is employed:

	*trans-***1305**	*cis-***1305**
1306	70.5	54.5
1307	22.5	36.0
1308	4.5	7.0
1309	2.5	2.5

When the mixture of tetraenes is heated at 100-110°C, electrocyclic closure of isomers **1307** and **1309** occurs to afford (±)-dehydro-α-chamigrene (**1310**), which may be obtained in a pure state by chromatography. If *trans*-**1305** is employed, the isolated yield of **1310** is 19%; if *cis*-**1305** is used, **1310** is obtained in 25%. Hydrogenation of **1310** using Lindlar's catalyst provides (±)-α-chamigrene (**1295**) and an isomer in a 3:1 ratio in quantitative yield. Pure **1295** can be obtained by chromatography.

In 1979, Iwata, Yamada, and Shinoo at Osaka University reported a synthesis of (±)-α-chamigrene which is almost a duplicate of White's 1974 synthesis (Scheme 158).[367] The only point of substantial difference in the two syntheses is that Yamada and co-workers carried out the carbene cycloaddition on the phenol ring (**1312**) leading to the spiro dienone **1313** while White and co-workers used an alkene (**1300**) and must reductively cleave the cyclopropane ring to reach the spiro[5.5]undecane system. The modification is not especially effective, since Yamada reported a yield of only 20% in the cyclization while White obtained a yield of at least 45% for the two steps from **1300** to **1302**.

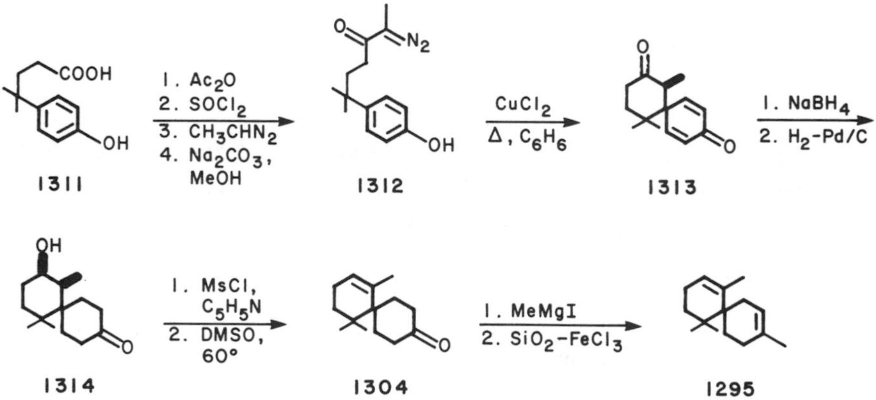

Scheme 158.   Yamada's Synthesis of (±)-α-Chamigrene

10-Bromo-α-chamigrene (1296) is the simplest of several halogenated chamigrenes of marine origin. In early 1976, Wolinsky and Faulkner reported a synthesis of the compound, although it had not actually been found in nature at that time.[368] Later, in 1976, Fenical and Howard discovered the compound in the red algae *Laurencia pacifica*.[369] The Wolinsky-Faulkner synthesis (Scheme 159) begins with geranylacetone (1315), which is treated with stoichiometric amounts of bromine and silver fluoborate to effect cyclization; bromo ether 1316 is obtained in 20% yield. Treatment of 1316 with *p*-toluenesulfonic acid in benzene

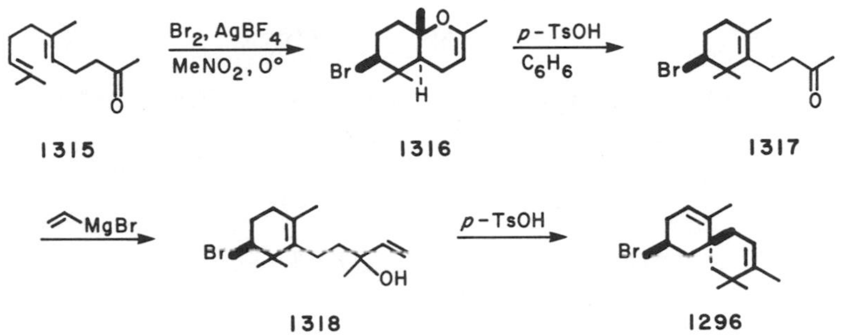

Scheme 159.   Wolinsky-Faulkner Synthesis
of ($\pm$)-10-Bromo-α- chamigrene

causes rearrangement to unsaturated ketone 1317. Treatment of the latter with vinylmagnesium bromide gives a tertiary alcohol (1318) which cyclizes upon treatment with acid to produce ($\pm$)-10-bromo-α-chamigrene in 26% yield. The cyclization is stereospecific, only one diastereomer being produced, and the synthetic product is identical with the natural product.

In a later publication, Ichinose and Kato also reported the synthesis of ($\pm$)-10-bromo-α-chamigrene (Scheme 160).[370] This synthesis also starts with geranylacetone, which is converted into a mixture of the 2E and 2Z

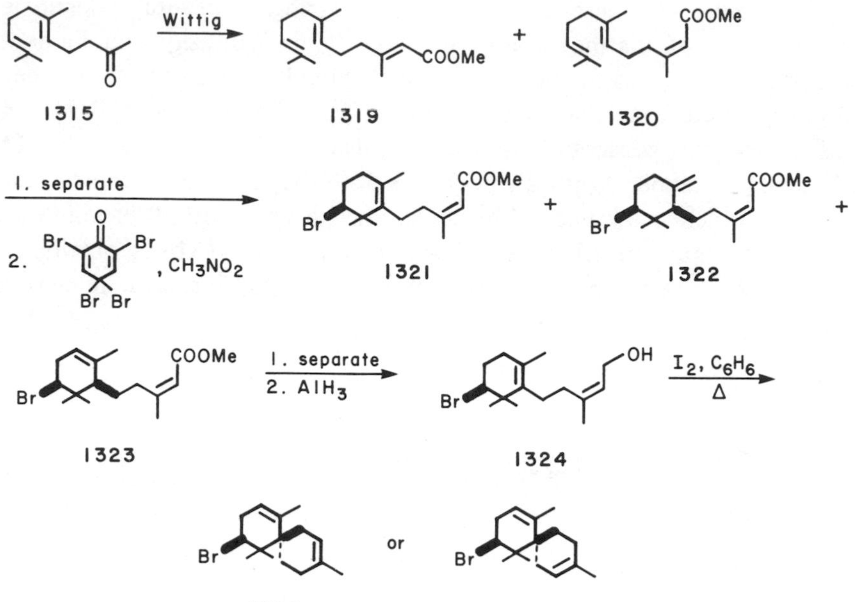

Scheme 160.   Ichinose-Kato Synthesis
of (±)-10-Bromo-α-chamigrene

diastereomers of methyl farnesate (**1319** and **1320**). Although the composition is not stated in the paper, it is known that this transformation normally gives a $2E:2Z$ ratio of about 3:1. The isomers are separated by chromatography and the minor isomer (**1320**) is treated with 2,4,4,6-tetrabromocyclohexadienone in nitromethane. A mixture of cyclization products (**1321-1323**) is produced, from which **1321** can be obtained in 28% yield by crystallization from pentane. The ester group is reduced and alcohol **1324** is cyclized by treatment with iodine in refluxing benzene (conditions employed earlier by Kitahara in a synthesis of (±)-α-chamigrene).[371] From the cyclization mixture, (±)-10-bromo-α-chamigrene may be obtained in 6% yield by "repeated high pressure liquid chromatography." It is not at all clear that this synthesis, which requires three isomer separations and proceeds in an overall yield of no more than 0.4%, is an improvement over the Wolinsky-Faulkner synthesis, which gives the sesquiterpene in an overall yield of approximately 4%.

D. Fused Ring Compounds: 3,6

(1) Bicycloelemene

Bicycloelemene (**1325**) is a constituent of Bulgarian peppermint oil and also of the cold pressed peel oil of *Citrus junos*. Although its full stereostructure has not been proven, it is probably as depicted based on

1325

analogy to simple elemanes. Vig and co-workers have synthesized **1325** as shown in Scheme 161.[372] Known keto-acid **1326** is transformed by the identical sequence used earlier in the synthesis of (±)-β-elemene[373] into dibromide **1332**, which is cyclized by the method first used by Corey for the synthesis of (±)-elemol.[374] The purity and identity of the synthetic **1325** was established by tlc and glpc. The yield of pure **1325** was not given. However, in Corey's synthesis of elemol, the desired diastereomer was produced in 24% yield.[374] It should be noted that the extra rigidity of **1332** as a result of the cyclopropane ring may provide an entropic advantage for cyclization in this case.

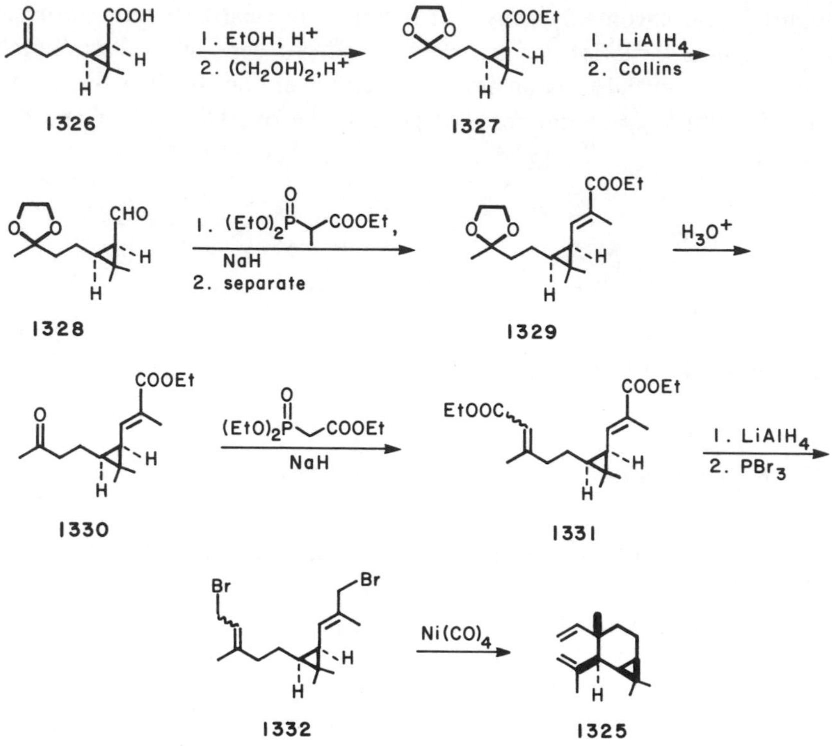

Scheme 161.   Vig's Synthesis of (±)-Bicycloelemene

(2) Sirenin and Sesquicarene

The interesting sesquiterpene sperm attractant sirenin (**1333**) and its related hydrocarbon sesquicarene (**1334**) have attracted a good deal of synthetic attention. In Volume 2 of this series, we reported five

1333   1334

syntheses of **1333** and four syntheses of **1334**. Rapoport has published
the full details of his first sirenin synthesis.[375,376] Intermediate **1335** has
now been resolved via the diastereomeric ketals formed with optically
active 2,3-butanediol. The two enantiomers corresponding to **1335** have

1335

been converted into (−)-sirenin and (+)-sirenin by the procedure
described earlier for the racemic series.[375] Correlation of natural
(−)-sirenin with natural (+)-sesquicarene shows that the two have the
same absolute configuration, as shown by **1333** and **1334**.

Garbers and co-workers prepared unsaturated ester **1343**, an intermediate in the Rapoport synthesis of (±)-sirenin,[375] from methyl (±)-peril-
late (**1336**) as shown in Scheme 162.[377] The synthesis is based on the
observation that certain alkenes undergo effective Prins addition when
treated with chloromethyl ethers and Lewis acids. Thus, **1336** reacts
with allyl chloromethyl ether to give a mixture of diastereomeric adducts
1337. Cyclization gives cyclopropanes **1338** and **1339** in a ratio of 2:3 in
30% yield, based on **1336**. After cleavage of the allyl ether and
transesterification with methanol, alcohols **1340** and **1341** are separated
by chromatography and the minor isomer is converted into tosylate **1342**.
Tosylate displacement is faster than conjugate addition when **1342** is
treated with lithium diisobutenylcuprate and ester **1343** is produced.
This material has been converted into (±)-sirenin in three steps.

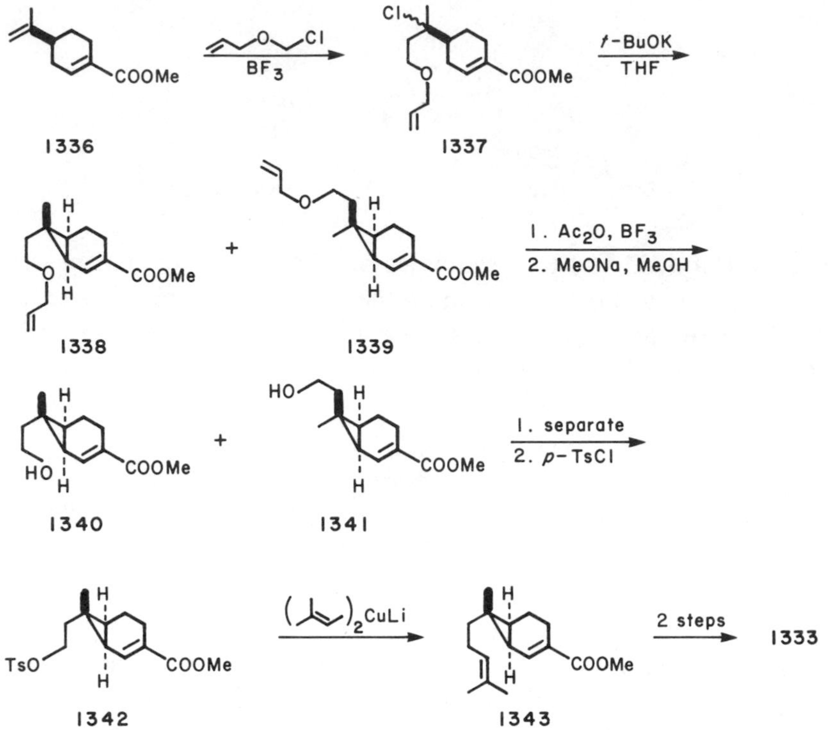

Scheme 162.   Garber's Synthesis of (±)-Sirenin

Although the yield and the isomer ratio in the key reactions (**1336→1338 + 1339**) are not especially good, this route makes sirenin available in only eight steps from a relatively available natural product. Although the Garbers synthesis was done with racemic methyl perillate, the optically active ester is available and it is known that it undergoes Prins reaction without racemization. Thus, the synthesis in Scheme 162 could probably be applied to prepare (−)-sirenin.

However, Hiyama and co-workers have discovered a route to sirenin and sesquicarene which is even more effective (Scheme 163).[378] The 7,7-dichloronorcarane **1344**, available in three simple steps from cyclohexanone, is added to an ether solution of the cuprate reagent prepared

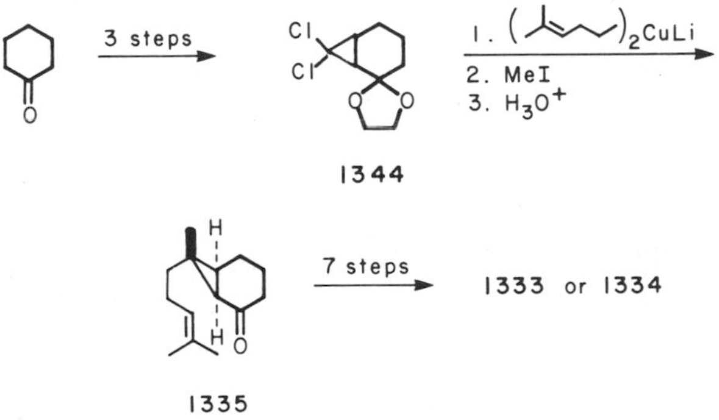

1344

1335 → 7 steps → **1333** or **1334**

Scheme 163.   Hiyama's Synthesis of (±)-Sirenin

from 5-bromo-2-methyl-2-pentene. After 1 hour at -20°C, excess methyl iodide is added and the mixture is kept at room temperature overnight. After hydrolysis of the ketal, ketone **1335** is obtained as a single stereoisomer in 44% yield! The reaction apparently involves displacement of one chlorine by the cuprate reagent to give a 7-chloro-7-alkylnorcarane, which undergoes transmetallation to the 7-lithio-7-alkyl compound. The stereoselectivity of the process must result from attack of the cuprate reagent at the more accessible exo face to give **1345**, which is stereospecifically transmetallated to give **1346** in which the lithium is endo. The latter compound must undergo displacement on methyl iodide with retention of configuration at the cyclopropyl center. Since ketone **1335** has been converted into both sirenin and sesqui-

1345   1346   1347

carene,[379,376] the Hiyama contribution makes both terpenes much more accessible.

E. Fused Ring Compounds: 5,5

(1) Pentalenolactone

Pentalenolactone (**1348**) is an antibiotic sesquiterpene containing a bicyclo[3.3.0]octane carbon framework. In addition to the construction of the intricate framework itself, there are several interesting stereochemical problems which must be reckoned with. The first is arranging

1348

for the secondary methyl group to be endo with respect to the bicyclo[3.3.0]octane unit. The second problem is at C_5—the hydroxymethyl group must be exo at this position. Finally, the most difficult problem is the stereochemistry at C_9, the spiro carbon.

The first successful synthesis was reported in 1978 by Danishefsky and co-workers (Scheme 164).[380] The readily available diene **1349** is

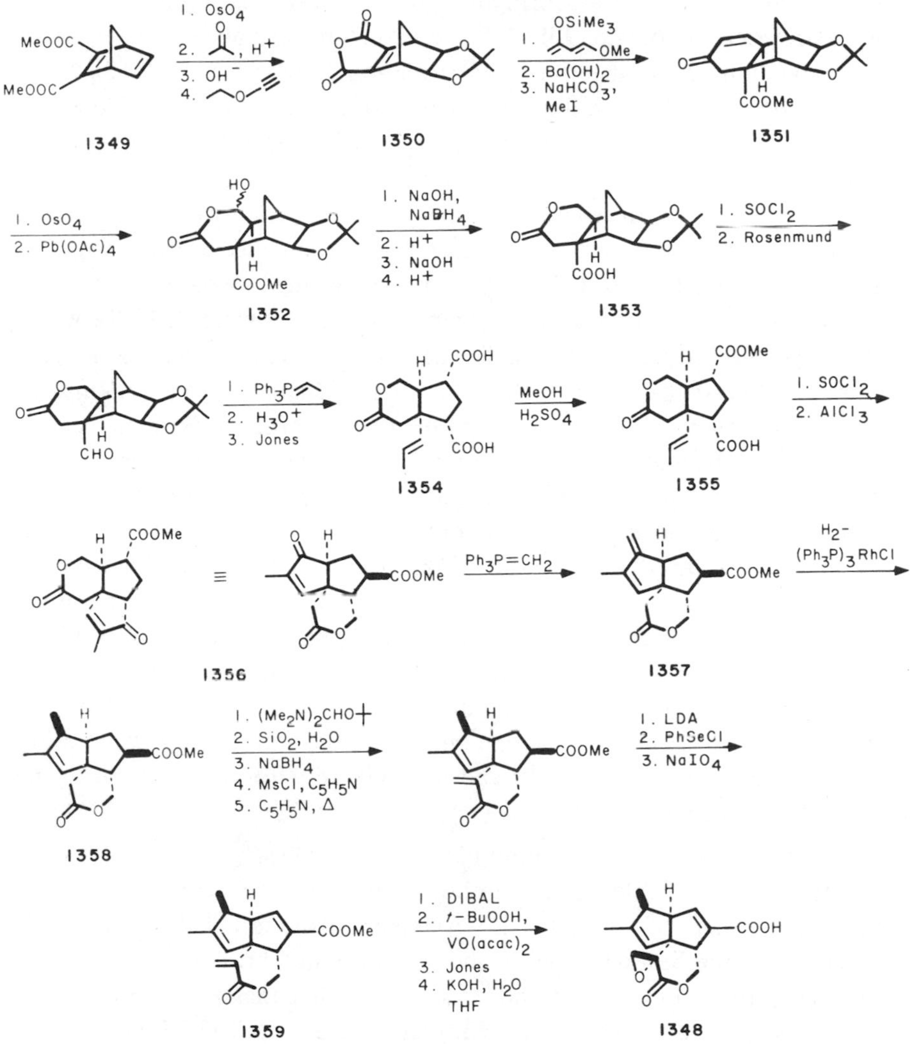

Scheme 164.   (±)-Pentalenolactone: Danishefsky's Synthesis

hydroxylated and the resulting exo diol is protected as the acetonide. The esters are saponified and the diacid dehydrated to obtain anhydride **1350**. Diels-Alder reaction with the Danishefsky diene occurs in nearly quantitative yield. When the resulting adduct is treated with barium hydroxide a series of transformations occurs, resulting in an acid, which is methylated to obtain **1351**. Scission of the cyclohexenone ring by hydroxylation and lead tetraacetate oxidation gives a lactol (**1352**) which is reduced. After saponification of the ester, acidification provides lactonic acid **1353**. The carboxy group is transformed into an aldehyde by Rosenmund reduction of the corresponding acyl chloride. Wittig reaction occurs without interference from the lactone. After hydrolysis of the acetonide, the resulting diol is oxidized to obtain diacid **1354**, which is selectively esterified to **1355**. The carboxy group is converted into an acyl chloride which undergoes Darzens acylation to yield **1356**, in which the bicyclo[3.3.0]octane nucleus is finally established. Wittig methylenation gives a diene (**1357**) which is hydrogenated using Wilkinson's catalyst to obtain solely the endo methyl diastereomer **1358**. The lactone ring is methylenated by a five-step sequence commencing with a Vilsmeier-type formylation. The second cyclopentene double bond is introduced via the selenoxide to obtain deoxypentalenolactone methyl ester (**1359**). Direct epoxidation of the methylene lactone with alkaline hydrogen peroxide fails, providing instead the 9-epi compound. However, if the lactone carbonyl is first reduced and the resulting hemiacetal is epoxidized using the Sharpless reagent, the correct stereochemistry is obtained at C_9. The synthesis is completed by oxidation of the hemiacetal back to the lactone and saponification of the methyl ester. The synthesis requires 33 steps and proceeds in 0.7% overall yield to the methyl ester of pentalenolactone.

A second synthesis was communicated by Schlessinger and co-workers in 1980.[381] The Schlessinger synthesis (Scheme 165) commences with the enol ether of 2-methyl-1,3-cyclopentanedione (**1360**), which is alkylated by the Stork-Danheiser method with allyl bromide. The kinetic enolate is then regenerated and added to diethyl maleate. The product (**1361**) is a 1:1 mixture of diastereomers. The bicyclo[3.3.0]octane system is now established by reaction of **1361** with sodium hydride in the

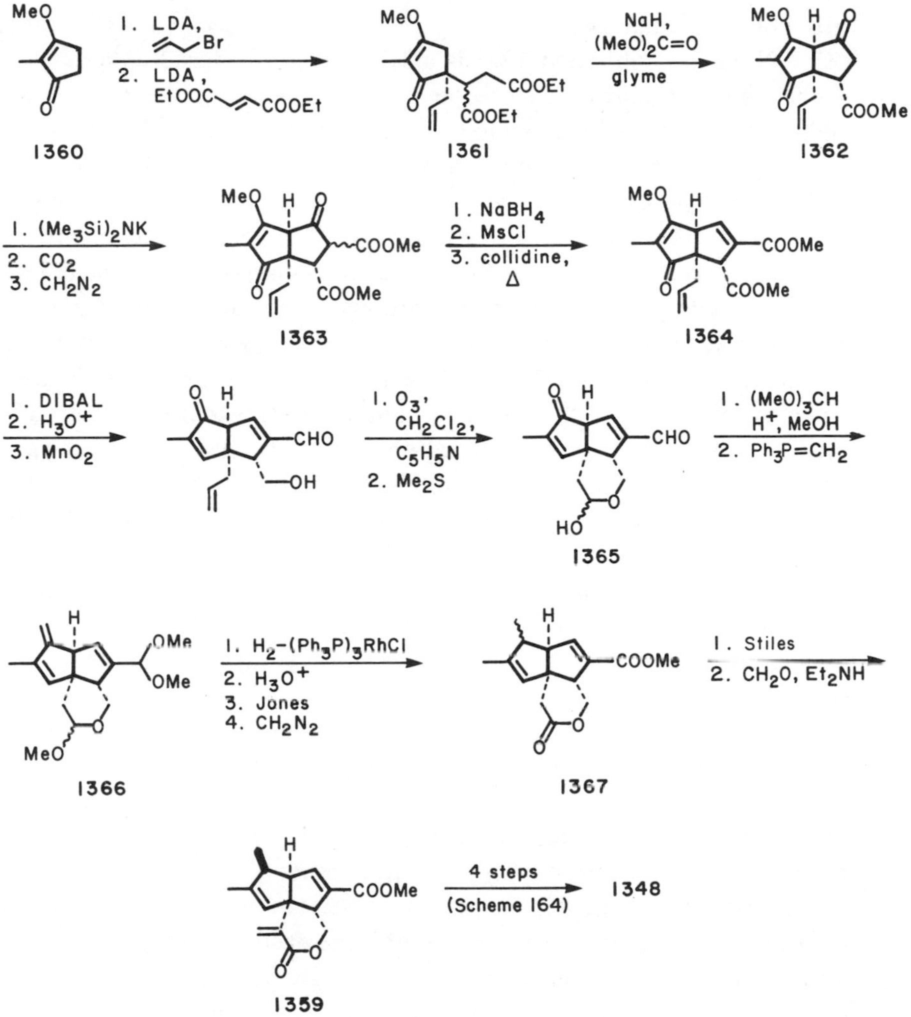

Scheme 165.   (±)-Pentalenolactone: Schlessinger's Synthesis

presence of dimethyl carbonate. This is perhaps the key reaction in the Schlessinger sequence, as any number of things might (but don't) go wrong. The resulting highly functionalized bicyclo[3.3.0]octane **1362** is thus obtained in a two-step sequence which proceeds in 68% overall yield. Note that, although **1361** is a 1:1 mixture of diastereomers, **1362** is obtained as a single diastereomer. It is clear from the yield in the Claisen step that epimerization occurs at some stage. The cyclopentanone is carbonated by reaction of its enolate (formed under conditions of thermodynamic control) with CO_2. Esterification of the resulting β-keto acid with diazomethane gives **1363**. The more reactive carbonyl is reduced and the resulting β-hydroxy ester is eliminated to obtain **1364**. At this point, all three carbonyl groups are reduced by diisobutylaluminum hydride and the enol ether is hydrolyzed, with subsequent dehydration of the β-hydroxycyclopentanone. The sequence is completed by oxidation of the allylic alcohol with manganese dioxide. This important three-step series of reactions proceeds in 65% overall yield. Ozonolytic cleavage of the allyl double bond gives a hemiacetal (**1365**) which is treated with methyl orthoformate and then with methylenetriphenylphosphorane to obtain diene **1366**. This compound is hydrogenated using Wilkinson's catalyst, as in Danishefsky's synthesis. Surprisingly, both the endo and the exo methyl diastereomers are obtained. After removal of the protecting groups, oxidation of the aldehyde and hemiacetal, and esterification, **1367** is obtained as a 2:1 mixture of isomers, with the endo isomer predominating. It is postulated that the endo methoxycarbonyl group in Danishefsky's hydrogenation substrate (**1357**) provides significant assistance in directing the catalyst to the exo face of the bicyclo[3.3.0]octane system. The methylene group is introduced by carbonating the lactone with magnesium methyl carbonate, and treating the resulting acid with aqueous formaldehyde and diethylamine. Methylene lactone **1359** can be converted into ($\pm$)-pentalenolactone by the Danishefsky method. The Schlessinger synthesis is considerably shorter than the Danishefsky synthesis (24 steps) and gives pentalenolactone methyl ester in 1.2% overall yield.

F. Fused Ring Compounds: 5,6

(1) Hypolepins and Pterosin E

The hypolepins are a group of indanones which were isolated from the fern *Hypolepis punctata* Mett. Three compounds, hypolepin A (**1368a**), hypolepin B (**1368b**), and hypolepin C (**1368c**) comprise the class. Hypolepin B has been converted into hypolepin A (POCl$_3$-C$_5$H$_5$N) and hypolepin C (*t*-BuOK-CH$_3$I), thus correlating the three compounds.

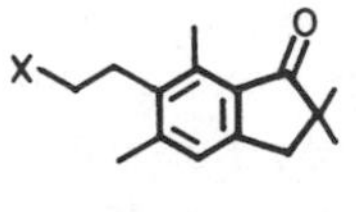

```
1368a: X = Cl
1368b: X = OH
1368c: X = OCH₃
```

Hayashi and co-workers synthesized hypolepin B as shown in Scheme 166.[382] Ester **1369** is acylated with acetyl chloride to obtain ketones **1370**

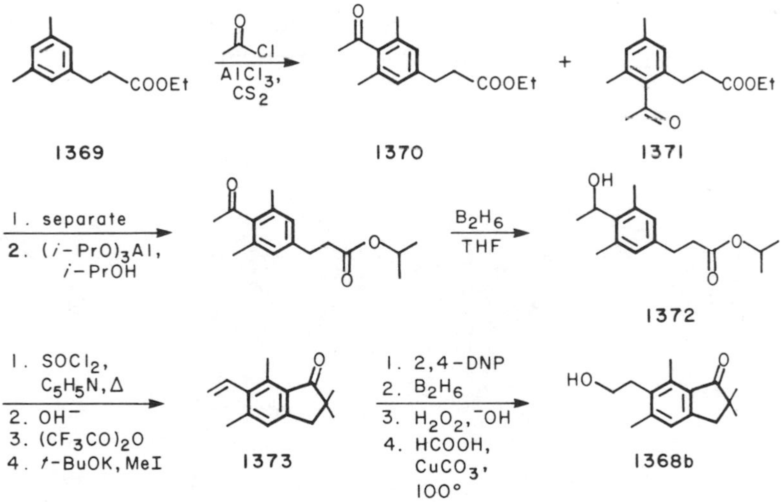

Scheme 166.   Hayashi's Synthesis of Hypolepin B

and **1371** in a ratio of 2:1. The major isomer is transesterified using aluminum isopropoxide in isopropyl alcohol. Reduction of the ketone with diborane gives an alcohol (**1372**) which is dehydrated by thionyl chloride in refluxing pyridine. Hydrolysis of the ester and cyclization of the resulting acid gives an indanone which is permethylated to obtain **1373**. The carbonyl is protected as the 2,4-dinitrophenylhydrazone and the vinyl group is hydrated by hydroboration and oxidation. Hydrazone hydrolysis is accomplished by heating with formic acid and in the presence of copper carbonate to give hypolepin B (**1368b**).

Pterosin E (**1374**) is a nor-sesquiterpene which is structurally related to the hypolepins. It may be isolated from the leaves of bracken, *Pteridium aquilinum* var. *Latiusculum*, along with a large number of other indanones of similar structure. For a time there was disagreement as to the position of the two-carbon side chain on the nucleus (C_4 or C_6). To

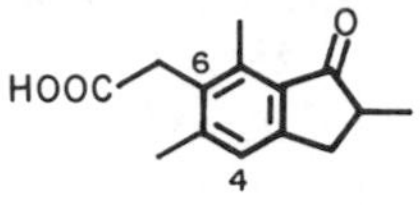

1374

settle this question Krishna Rao synthesized pterosin E as shown in Scheme 167.[383] 3,5-Dimethylbenzyl bromide is used to alkylate diethyl methylmalonate to obtain **1374**, which is chloromethylated to **1375**. The

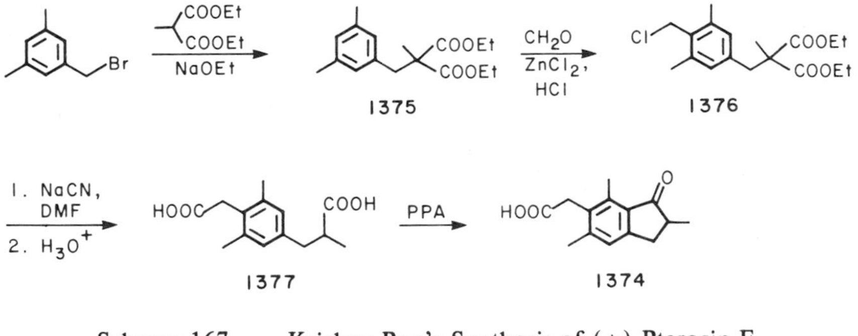

Scheme 167.   Krishna Rao's Synthesis of (±)-Pterosin E

benzylic chloride is displaced and the cyano-diester hydrolyzed and decarboxylated to diacid **1376**. Friedel-Crafts cyclization results from treatment of **1376** with polyphosphoric acid; the product is (±)-pterosin E (**1372**).

(2) Bakkenolide A

Bakkenolide A (**1378**) is an interesting spiro lactone which has been isolated from the flower stems of *Petasites japonicus* Max. It was suggested that it may arise biogenetically from the eremophilane fukinone (**434**) by some path such as the following:[384]

In fact, Hayashi, Nakamura, and Mitsuhashi succeeded in preparing bakkenolide A from fukinone in just this way (Scheme 168).[385] Alkaline epoxidation followed by treatment of the resulting epoxide with ethanolic sodium hydroxide gives a mixture of acids which is esterified with diazomethane. Compounds **1379-1381** are produced in a ratio of 42:42:16. Column chromatography gives pure **1379**, which is dehydrated to obtain **1382**. This compound is oxidized by selenium dioxide in acetic acid to obtain 33% of bakkenolide A and 57% of aldehyde **1383**. Treatment of the latter compound with sodium borohydride gives more bakkenolide A. Ester **1382** has also been prepared by Naya and co-workers by a similar route from fukinone.[386]

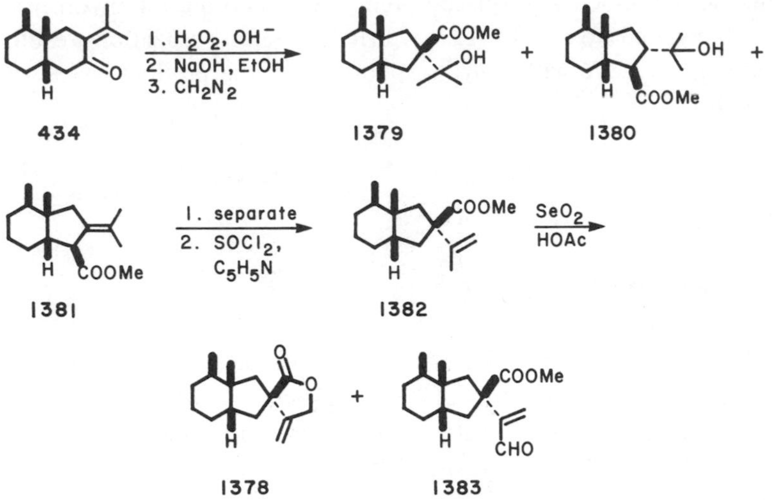

Scheme 168.   Hayashi's "Biomimetic" Synthesis of Bakkenolide A

Evans and co-workers prepared (±)-bakkenolide A by the novel method summarized in Scheme 169.[387a,b] The 4.5:1 mixture of ketone **1384**, prepared by the Piers method,[388] is subjected to Lemieux-Johnson oxidation to obtain diones **1385** and **1386**. The major isomer is separated by column chromatography and cyclized by treatment with strong base. Hydrogenation of the resulting enone gives the cis-fused hydrindanone **1387**. This compound reacts with isopropenyllithium to give a mixture of alcohols which is treated with PBr$_3$ to obtain allylic bromide **1388**, which is probably a mixture of E and Z diastereomers. Treatment of **1388** with the sodium salt of p-toluenesulfonyl-S-methylcarbazate gives the hydrazone **1389**. Treatment of the latter with sodium hydride initiates the sequence of events shown leading to dithioester **1390**. The key event is the highly stereoselective [2,3]sigmatropic rearrangement of the intermediate carbanion **1389a**. Compound **1390** is hydrolyzed and oxidized by selenious acid to obtain (±)-bakkenolide A (**1377**). In an earlier paper, Sarma and Sarkar reported the preparation of ketone **1387** by the same route.[389]

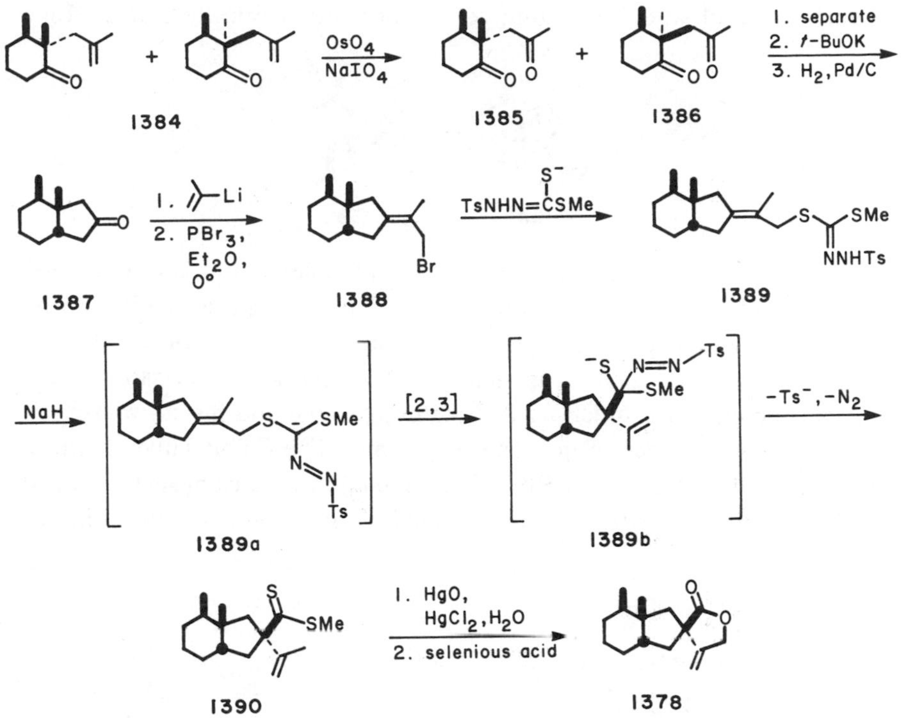

Scheme 169.   Evans' Synthesis of (±)-Bakkenolide A

(3) Oplopanone

Oplopanone (**1391**) is a rearranged cadinane which has been synthesized by Caine and Tuller at Georgia Institute of Technology[390] and by Taber and Korsmeyer at Vanderbildt University.[391] It appears at the outset that there may be a major problem in creating the trans-fused hydrindanone nucleus. However, ketone **1392** had been prepared in the course of the structure proof and was found to be thermodynamically stable. Like-

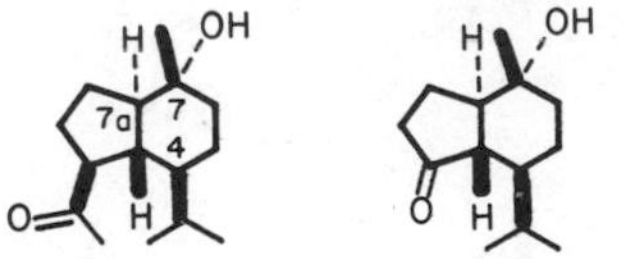

1391   1392

wise, the epimerizable acetyl side chain of oplopanone was previously shown to be in the more stable position. Thus, the stereochemical problem reduces to controlling the relative stereochemistry of C_4, C_7 and C_{7a}. Caine and Tuller applied photochemical isomerization of a cyclohexadienone, a much used reaction by Caine and his co-workers for the synthesis of other sesquiterpene systems. The Caine-Tuller synthesis (Scheme 170) begins with Robinson annelation of 2-isopropyl-5-methyl-cyclopentanone (**1393**)[392]; unfortunately, enone **1394** is produced in only

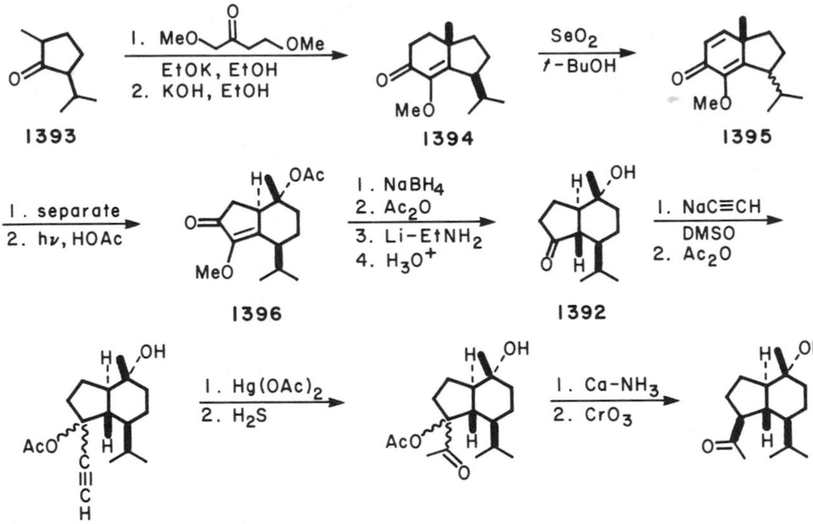

Scheme 170.   Caine-Tuller Sythesis of (±)-Oplopanone

17% yield. Although **1394** is a homogenous substance, the dienone produced by oxidation with selenium dioxide in *t*-butyl alcohol is a 5:1 mixture of diastereomers. The isomers are separated by careful silica gel chromatography and the major isomer is photolysed in acetic acid. A single acetoxy ketone, **1396**, is produced in 91% yield. Thus, the three important asymmetric centers of oplopanone are established in the correct sense. Removal of the oxygen is accomplished by a precedented sequence. Ketone **1392** is indeed found to be the thermodynamically more stable isomer, since careful hydrolysis initially yields an isomer of **1392** which rearranges to it upon extended treatment with acid or upon refluxing with sodium methoxide in methanol. The acetyl side chain is introduced into **1392** via the acetylene **1397**. After hydration of the triple bond, reductive cleavage of the acetoxy group provides (±)-oplopanone (**1391**). In all, the synthesis requires 13 steps and proceeds in 0.6% overall yield.

The Taber-Korsmeyer synthesis is summarized in Scheme 171. Reductive alkylation of the *o*-methoxybenzoic acid **1399** with β-bromophene-

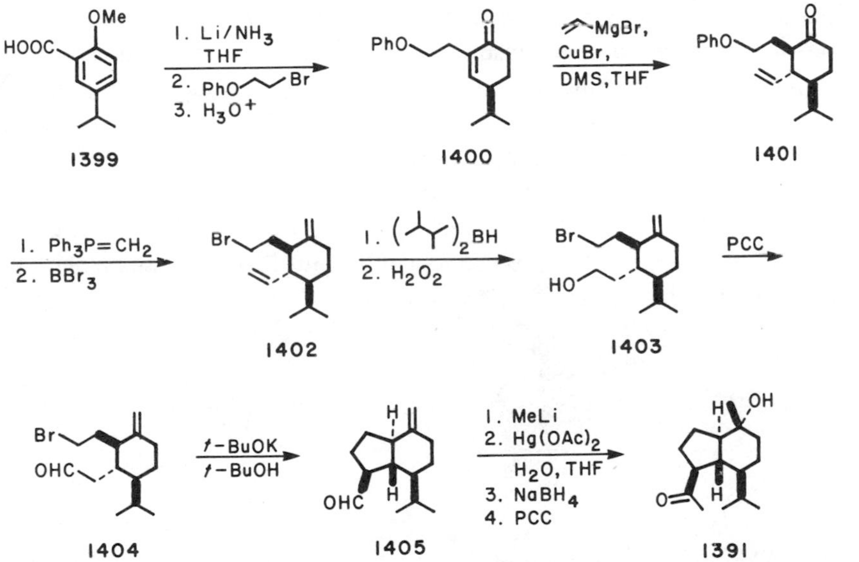

Scheme 171.   Taber-Korsmeyer Synthesis of (±)-Oplopanone

tole gives cyclohexenone **1400** which adds vinylmagnesium bromide stereoselectively trans to the isopropyl group. Equilibration of the epimerizable position affords the more stable all-trans isomer **1401** in which three of the stereocenters of oplopanone are correctly established. After Wittig methylenation and cleavage of the phenoxy group, the more accessible double bond is selectively hydroborated to obtain bromo alcohol **1403**. This compound is oxidized to aldehyde **1404** which is cyclized by potassium *t*-butoxide. In this reaction, a fourth stereocenter is fixed thermodynamically since, as discussed earlier, the natural configuration at this position is the more stable. The final stereocenter is introduced by oxymercuration of the exocyclic methylene group, which apparently occurs by attack of water on the mercurinium ion solely from the equatorial direction. The Taber-Korsmeyer synthesis requires 11 steps and produces **1391** in 2.9% overall yield.

(4) Picrotoxinin

Picrotoxinin (**1406**), a constituent of the long-known "picrotoxin", is one of the more complex sesquiterpenes to have been synthesized. Of its numerous interesting structural features, not the least intriguing and challenging from a synthetic standpoint is the "prow-stern" interaction between the methyl and isopropenyl groups on the cyclohexane boat. The first and only successful synthesis of picrotoxinin was reported by Corey and Pearce in 1979.[393] Several syntheses of the sesquiterpene alkaloid dendrobine, which presents similar but less challenging synthetic problems, will be discussed in section 7.

1406

The Corey-Pearce synthesis (Scheme 172) begins with (−)-carvone (**1407**) which is converted into the dimethylhydrazone and alkylated with

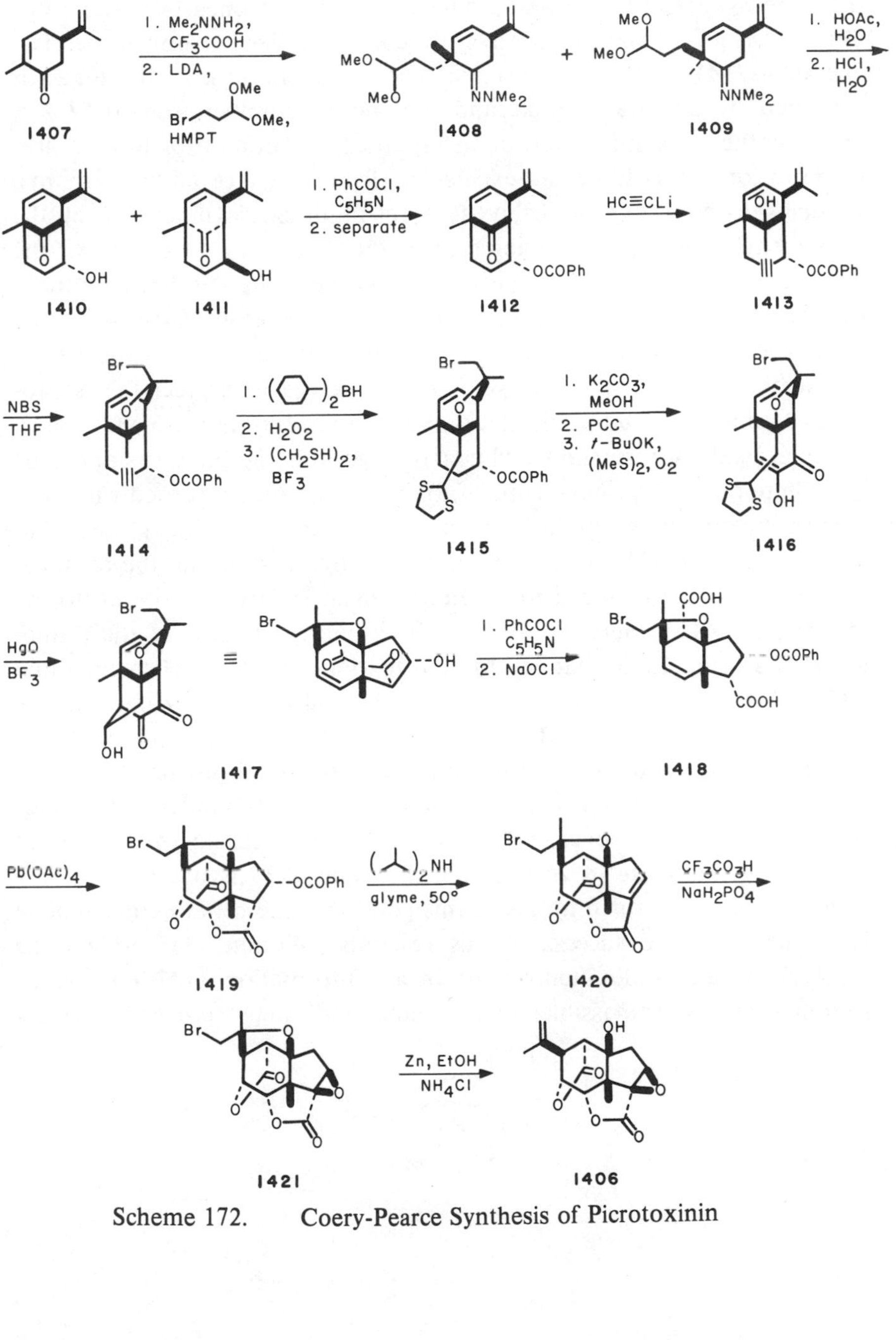

Scheme 172.   Coery-Pearce Synthesis of Picrotoxinin

the dimethyl acetal of 3-bromopropanal. The hydrazones **1408** and **1409**, obtained as a 3:2 mixture, are hydrolyzed and cyclized to obtain the bicyclic aldols **1410** and **1411** in a 3:2 ratio. The mixture of benzoates is separated by preparative hplc and the more abundant isomer **1412** is used for the remaining steps of the synthesis. Addition of lithium acetylide to ketone **1412** occurs exclusively from one face of the carbonyl, yielding alcohol **1413**, probably as a result of steric hindrance by the isopropenyl group on the other face of the carbonyl. At this point the newly created hydroxy is added to the isopropenyl double bond by treatment of **1413** with N-bromosuccinimide. Bromo ether **1414** is formed stereoselectively in quantitative yield. This transformation is the key to the synthesis. The ether bridge not only serves to protect the isopropenyl double bond, but it also locks the cyclohexene ring, which it spans, into one rigid conformation. At a later stage, this rigidity turns out to be quite important. The next steps of the synthesis are concerned with converting the ethynyl group into an acetaldehyde unit, which is protected as a dithioacetal (**1415**). The benzoate is now hydrolyzed and the resulting secondary alcohol oxidized to a ketone, which is further oxidized to the diosphenol **1416**. Mercuric oxide suffices both to hydrolyze the dithioacetal and bring about aldol cyclization to yield the tetracyclic compound **1417**. After benzoylation of the secondary alcohol the α-diketone is cleaved to obtain diacid **1418**. At this point, the synthesis almost foundered, as numerous straightforward methods for bringing about the required bis lactonization failed due to the extremely hindered nature of the double bond. However, cyclization occurs when diacid **1418** is treated with lead tetraacetate in acetonitrile; dilactone **1419** is obtained in 99% yield. Again, it is worth noting at this point that the bromo ether moiety is essential for the success of this reaction. Without this bridge, the molecule would undoubtedly exist in a conformation (**1422**) in which lactonization is impossible. The remaining steps proceed without

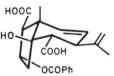

1422

incident. After elimination of benzoic acid, the double bond is epoxidized and the bromo ether cleaved to obtain picrotoxinin. The synthesis requires only 18 steps, and proceeds in a very creditable 7.8% overall yield. All in all, it is a very professional piece of work.

G. Fused Rings: 5,7

(1) Guaiazulenes; Bulnesol, α-Bulnesene, Guaiol, Dehydrokessane, and Kessanol

In Volume 2 of this series, a notably brief section was devoted to the synthesis of hydroazulenic sesquiterpenes. We noted that general methods for the synthesis of the hydroazulene skeleton were few, and problems of stereochemical control about this flexible nucleus were largely unsolved. During the 10 years that have elapsed since our original review article, the progress in this area has been truly impressive and some 30 papers have appeared detailing the synthesis of one or more natural products.

In this section we begin by discussing the synthesis of the guaiazulenes bulnesol (**1423**), α-bulnesene (**1424**), guaiol (**1425**) and the kessane derivatives **1426** and **1427**.

1423   1424   1425

1426   1427

Andersen has reported the synthesis of bulnesol (**1423**) outlined in Scheme 173.[394] Homologation of photocitral-A (**1428**) to **1429** and subsequent conversion to the diene-aldehyde **1430** were carried out in a precedented manner. When **1430** is exposed to perchloric acid and acetic

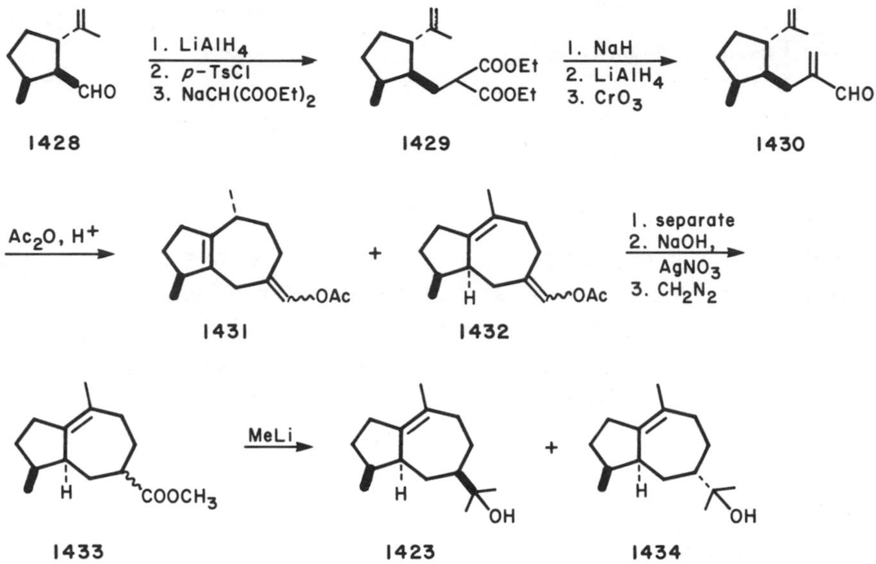

Scheme 173.   Andersen's Synthesis of (±)-Bulnesol

anhydride, two cyclization products **1431** and **1432** are obtained in 27% and 56% yields, respectively. After separation, **1432** is converted to the esters **1433**, an equilibrium mixture of C-7 stereoisomers. Treatment of this mixture with methyllithium affords bulnesol (**1423**) and 7-epibulnesol (**1434**) in a ratio of 7:3.

During the period of this review, full experimental details of a bulnesol synthesis previously abstracted[395a] by us have appeared.[395b] For completeness we include these references here.

Mehta and Singh have prepared α-bulnesene (1424) in a brief synthesis from patchouli alcohol (1435) (Scheme 174).[396] Lead tetraacetate oxidation of 1435 affords a mixture of products from which 1436 may be

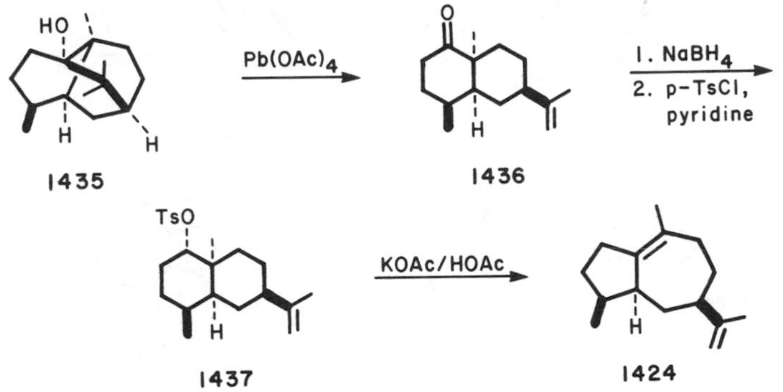

Scheme 174.   Mehta-Singh Synthesis of ($\pm$)-α-Bulnesene

isolated in approximately 20% yield. After reduction and tosylation, solvolytic rearrangement of 1437 affords α-bulnesene (1424) in unspecified yield. Solvolytic rearrangement of cis-fused and trans-fused decalyl tosylates has previously been employed for the synthesis of bulnesol and α-bulnesene.[397]

Guaiol (1425) has attracted a great deal of interest among synthetic chemists. However, the stereospecific synthesis of this molecule has been elusive. In Marshall's approach[398] (Scheme 175), a series of simple transformations is employed to convert the known octalone 1438 to keto-aldehyde 1441. After an aldol condensation and dehydration, 1442 is converted to its enolate with trityllithium. Addition of this enolate to ethanol containing sodium borohydride followed by mesylation affords 1443, in which the double bond has migrated to the six-membered ring. Note that the stereochemical relationship between the substituents at what will be C-7 and C-10 in guaiol has been established by the direction of kinetic protonation of the enolate, apparently quite efficiently.

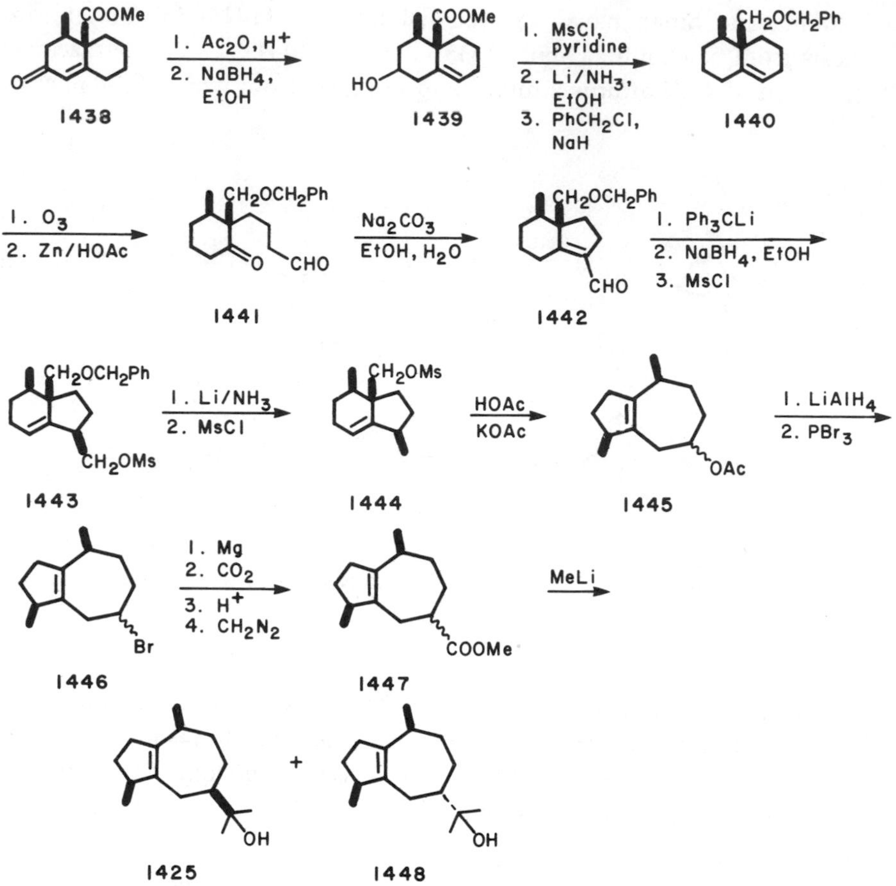

Scheme 175.   Marshall's Synthesis of (±)-Guaiol

After further transformation to **1444**, the stage is set for an ingenious solvolytic rearrangement affording the hydroazulene **1445**. The mechanism of this solvolysis presumably involves participation of the π system of the double-bond giving the cyclopropyl carbinyl cation **1449** which rearranges to **1450**, the more stable tertiary cation. The latter species suffers nucleophilic attack by solvent leading to **1445**, which is converted to ester **1447** as shown in the scheme. Unfortunately, a 1:1

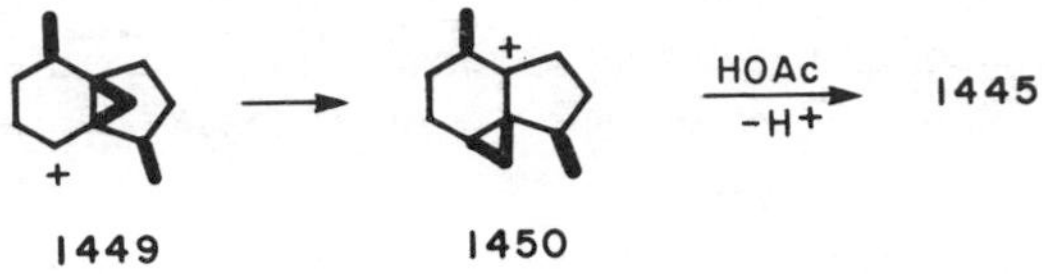

mixture of isomers at C-7 is obtained and thus addition of methyllithium gives an equimolar mixture of guaiol (**1425**) and 7-epiguaiol (**1448**).

Buchanan and Young reported the preparation of guaiol outlined in Scheme 176.[399] Condensation of Mannich base **1451** with 2-methyl-cyclopentanone under thermal conditions affords a 68% yield of **1452**, where Michael reaction has occurred at the less substituted carbon of the cyclopentanone. Upon treatment with HCl, **1452** undergoes aldol condensation and lactonization. Lactone **1453**, a mixture of stereoisomers, is converted to a single enone **1454** in 35% overall yield upon heating

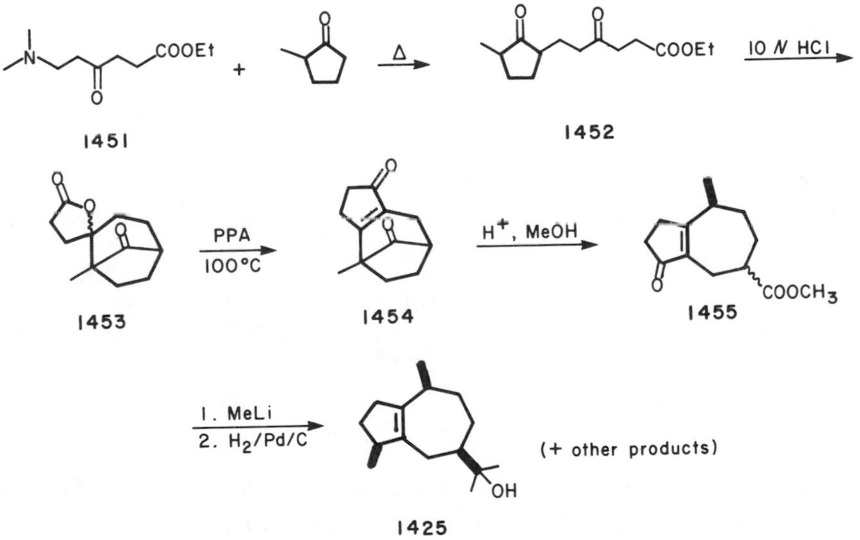

Scheme 176.   Buchanan-Young Synthesis of (±)-Guaiol

with polyphosphoric acid. After cleavage of the vinylogous β-diketone **1454** with acidic methanol, ester **1455** is produced as a 1:1 mixture of stereoisomers. After exposure of this mixture to methyllithium and hydrogenolysis of the resulting allylic alcohol, a mixture of four stereoisomers is obtained from which guaiol (**1425**) can be isolated by gas chromatography in 10% yield.

The final guaiol synthesis is by Andersen and Uh (Scheme 177).[400] This synthesis is founded on an observation made earlier in the Andersen bulnesol synthesis (Scheme 173). When diene aldehyde **1430** is treated with acetic anhydride and perchloric acid, a minor cyclization product **1431** is isolated. This material results from stereospecific hydride migration in a cyclization intermediate such as **1456**. It was therefore

reasoned that cyclization of **1457** would afford a guaiol precursor. When the cyclization reaction was carried out, **1460** was in fact a minor product. In a series of transformations identical to those employed in the

bulnesol synthesis (Scheme 173) the cyclization products were converted to a mixture of four products of which guaiol (**1425**) was a minor component.

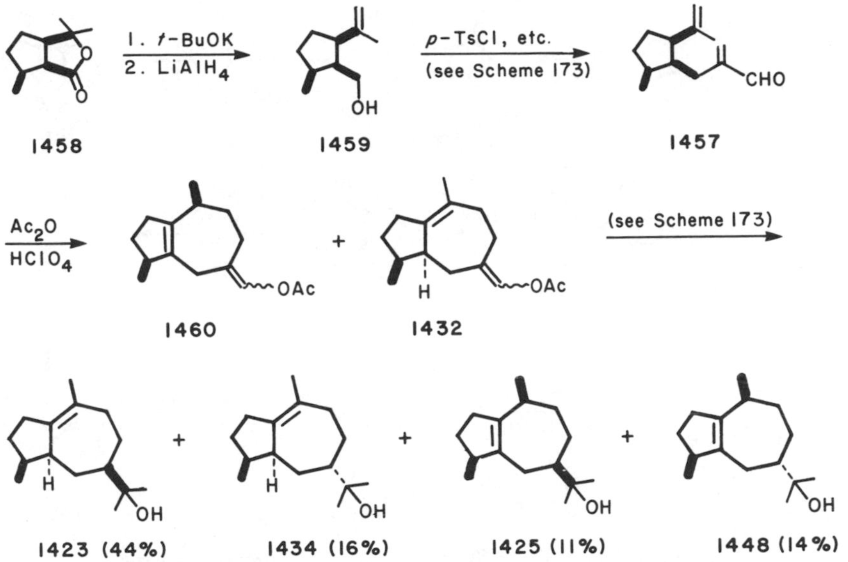

Scheme 177.   Andersen's Synthesis of (±)-Guaiol
and Related Compounds

Two syntheses of kessane derivatives have appeared in the literature. Liu and Lee prepared dehydrokessane (**1426**) by the route shown in Scheme 178.[401] Photocycloaddition of **1461** and **1462** followed by acid-catalyzed elimination of acetic acid affords tricyclic enone **1463** in 60% yield. Copper-catalyzed conjugate addition of methylmagnesium bromide occurs stereospecifically. After reductive removal of the ketone, the esters in **1464** are reduced with lithium aluminum hydride. Fragmentation occurs upon treating the intermediate diol with tosyl chloride in pyridine yielding hydroazulene **1466**. After introduction of the isopropylol group at C-7, as shown in the scheme, the ether bridge is closed by oxymercuration. It is interesting to note that stereoisomer **1467** is the principal product of the sequence shown (57% yield). After reduction of **1468**, the resulting alcohol is treated with POCl$_3$. Elimination and double bond migration occur under the reaction conditions leading to dehydrokessane (**1426**).

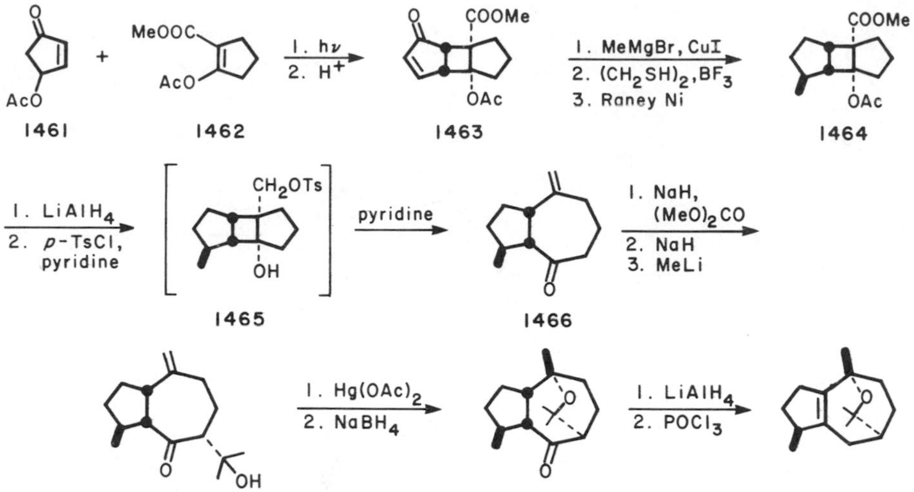

Scheme 178.   Liu's Synthesis of (±)-Dehydrokessane

Andersen has synthesized kessanol (**1427**) from photocitral-A (**1428**) as shown in Scheme 179.[402] Straightforward transformations afford unsaturated aldehyde **1471**. Prins cyclization and silylation give a mixture of products from which **1472** can be isolated from **1428** in approximately 14% overall yield. The double bond in **1472** is protected as its dibromide and the amide is converted to the acetate by nitrosation. After regeneration of the double bond by zinc in acetic acid, **1473** is isolated in 20% yield. The isopropyl side chain is fabricated as shown, and the ether linkage closed by oxymercuration. The correct stereochemistry at C-8 is established by conversion to the ketone **1476** and subsequent stereoselective reduction with an unspecified hindered borane.

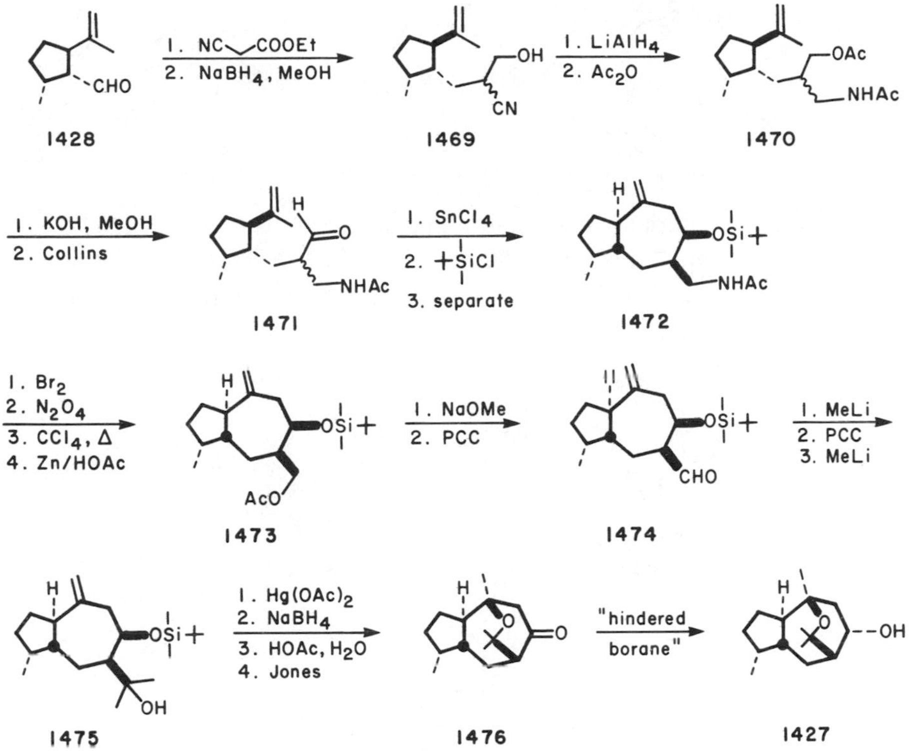

Scheme 179.   Andersen's (±)-Kessanol Synthesis

(2) Guaianolides: Dihydroarbiglovin and Estafiatin

At this juncture it is appropriate to note that in all of the syntheses thus far discussed in this section, problems of stereocontrol have remained largely unsolved. In the syntheses of the guaianolides dihydroarbiglovin (**1477**) and estafiatin (**1478**) these stereochemical problems are solved by employing the highly elaborated decalinic sesquiterpene santonin (**1479**)

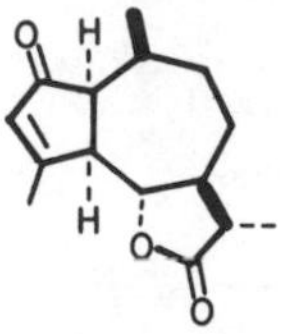

1477 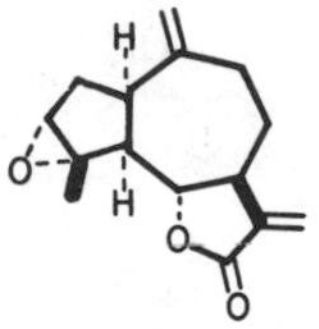   1478 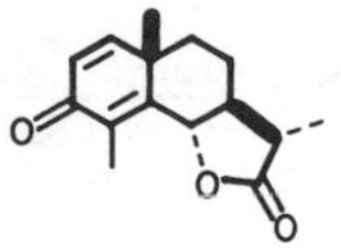   1479

as a starting material. Marx has prepared dihydroarbiglovin (**1477**) by way of santonin derivative **1480** (Scheme 180),[403] the preparation of which we described in Volume 2 of this series.[404] Methanolysis of the

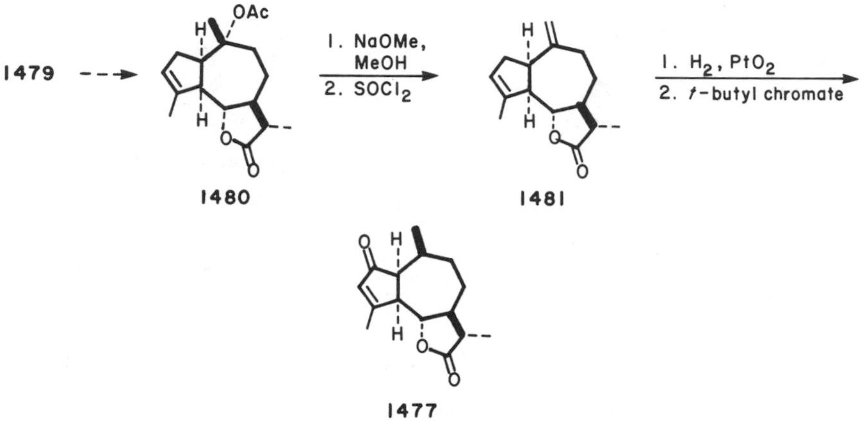

Scheme 180.   Marx's Synthesis of Dihydroarbiglovin

tertiary acetate **1480** followed by dehydration with thionyl chloride gives the diene **1481** as the major product. Selective hydrogenation of **1481** is readily accomplished, and is stereospecific. Allylic oxygenation with *t*-butyl chromate gives dihydroarbiglovin (**1477**).

In Crabbé's synthesis [405] of (−)-estafiatin (**1478**) an alternative synthesis of **1481** is employed, and this material is converted to the natural product (Scheme 181). The most interesting aspect of this synthesis is the use of a large excess of sodium borohydride in pyridine to effect

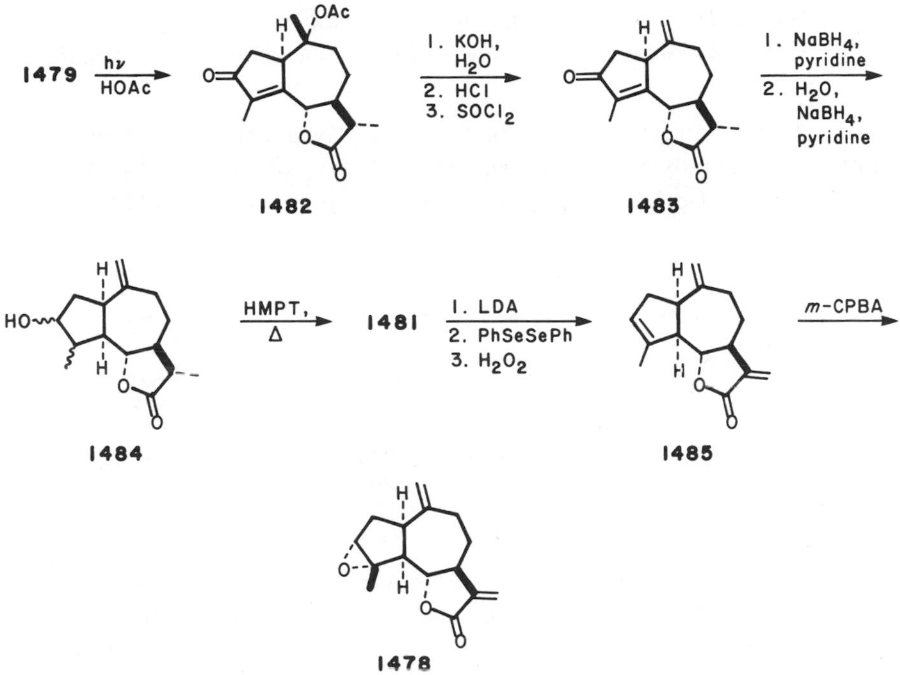

Scheme 181.   Crabbe's Synthesis of (-)-Estafiatin

reduction of enone **1483** in the presence of the exocyclic double bond. After 1,4-reduction is complete, water is added to the reaction mixture leading to **1484** as the isolated product. Dehydration of **1484** is accomplished in approximately 25% yield by the agency of hot HMPT, and the synthesis is completed as shown in the scheme. The final epoxidation is stereo- and chemoselective giving (−)-estafiatin in 51% yield along with about 10% of the β-epoxide.

(3) Guaiazulenes with an additional cyclopropane ring:
Cyclocolorenone, 4-Epiglobulol, 4-Epiaromadendrene, and Globulol.

A few guaiane derivatives containing a cyclopropane ring are known and four such materials have been synthesized: cyclocolorenone (**1486**), globulol (**1487a**), 4-epiglobulol (**1487b**) and 4-epiaromadendrene (**1488**).

1486	1487a 4α methyl	1488
	1487b 4b methyl	

Earlier approaches to cyclocolorenone (**1486**) confronted the problem of the C-1 epimer of **1486** being thermodynamically favored over the natural isomer.[406] Thus Büchi's attempted synthesis of **1486** failed when it was found that the reaction conditions required to convert **1489** to **1486** resulted in isomerization to the 1α isomer **1490**. Caine and Ingwal-

son have devised an elegant photochemical solution to this problem shown in Scheme 182.[407] Maalione (**1491**) is converted by formylation, dehydrogenation and Jones oxidation to the dienone-acid **1492**. Irradiation of this cross-conjugated cyclohexadienone causes rearrangement and concomitant decarboxylation to afford **1493** directly. Hydrogenation of this product gives **1486** in good yield.

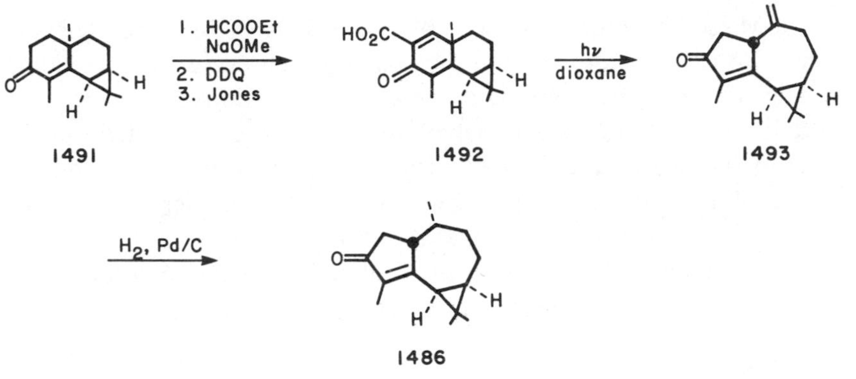

Scheme 182.   Caine-Ingwalson Synthesis of (−)-Cyclocolorenone

Marshall has prepared globulol (**1487a**) by solvolytic cyclization of a cyclodecandienol (Scheme 183).[408] The starting material has previously been prepared by Yoshikoshi in connection with his bulnesol synthesis,[409] and is available (eight steps) in good overall yield from the

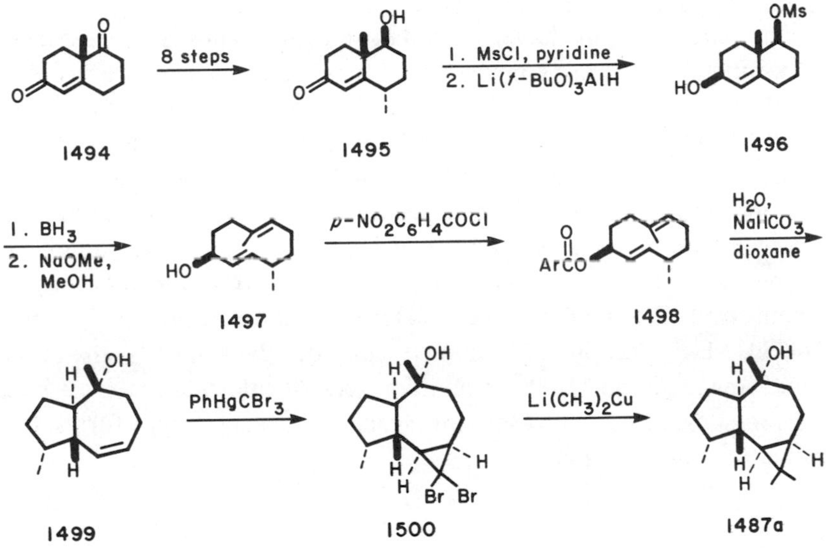

Scheme 183.   Marshall's (±)-Globulol Synthesis

Wieland-Miescher ketone (**1494**). Mesylate **1496** is hydroborated and subjected to the action of sodium methoxide which causes fragmentation via the intermediate boronate ester. After conversion to the *p*-nitrobenzoate ester, **1498** is solvolyzed in buffered aqueous dioxane leading to the hydroazulene **1499**. The trans-fused ring system and the presence of unsaturation in the seven-membered ring serve to fix the conformation of this system (**1499a**) such that reaction with dibromocarbene occurs stereoselectively from the convex face of the molecule yielding **1500**. Methylation of **1500** affords (±)-globulol (**1487a**).

1499a

Caine and Gupton have exploited the well-known photochemical transformation of cross-conjugated cyclohexadienones to prepare 4-epiglobulol (**1487b**) and 4-epiaromadendrene (**1488**, Scheme 184).[410] Compound **1501** was prepared as an intermediate in the Caine α-cyperone synthesis (see Scheme 5). Treatment of **1501** with base induces cyclization and the product is dehydrogenated with DDQ. Dienone **1502** undergoes the expected santonin-type rearrangement upon photolysis in aqueous acetic acid. Dissolving metal reduction affords material with the β-configured C-4 methyl group. Deoxygenation affords 4-epiglobulol (**1487b**). Dehydration with thionyl chloride gives mostly the desired 4-epiaromadendrene (**1488**) in addition to a 10-chloro derivative which is dehydrohalogenated with sodium acetate in acetic acid, furnishing a further quantity of **1488**.

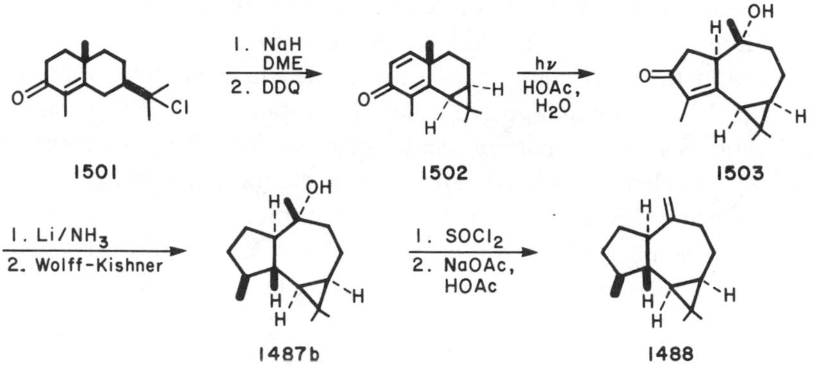

Scheme 184.   Caine-Gupton Synthesis of (-)-4-Epiglobulol

(4) Pseudoguaianolides: The Ambrosanolide Family; Deoxydamsin, Damsin, Ambrosin, Psilostachyn, Stramonin B, Neo-ambrosin, Parthenin, Hymenin, Hysterin, Damsinic Acid, and Confertin.

The tremendous success in the total synthesis of pseudoguaianolides is remarkable. Prior to 1975 only a single synthesis of a pseudoguianolide structure had been achieved, Hendrickson's preparation of the nonnaturally-occurring compound **1504**. In the ensuing five year period some 21

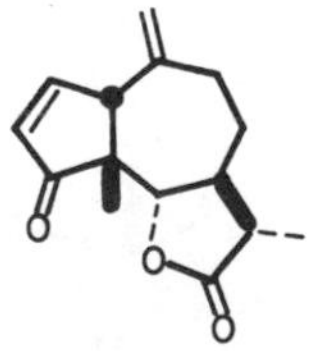

1504

papers appeared detailing the syntheses of more than a dozen of these natural products. In part, this sudden flurry of interest was prompted by the observation that many pseudoguaianolides, in common with other sesquiterpene α-methylene lactones, possess antineoplastic and cytotoxic activity.

The pseudoguaianolides have been classified into two broad subgroups depending on the configuration of the secondary (C-10) methyl group. Those with the 10-α configuration are named helenanolides while those with the 10-β configuration are called ambrosanolides. We consider the total synthesis of members of the latter class first. The ambrosanolides that have yielded to syntheses are the non-naturally occurring compound deoxydamsin (1505), damsin (1506), neoambrosin (1507), parthenin (1508), hymenin (1509), ambrosin (1510), stramonin B (1511), psilostachyin C (1512), damsinic acid (1513), confertin (1514) and hysterin (1515).

1505   1506   1507   1508

1509   1510   1511   1512

1513   1514   1515

Marshall's synthesis of deoxydamsin (Scheme 185), while not leading to a natural product, is an important contribution to pseudoguaianolide synthesis since the methodology developed here has found wide application in other syntheses.[411] The well-known hydroxy enone 1516 is con-

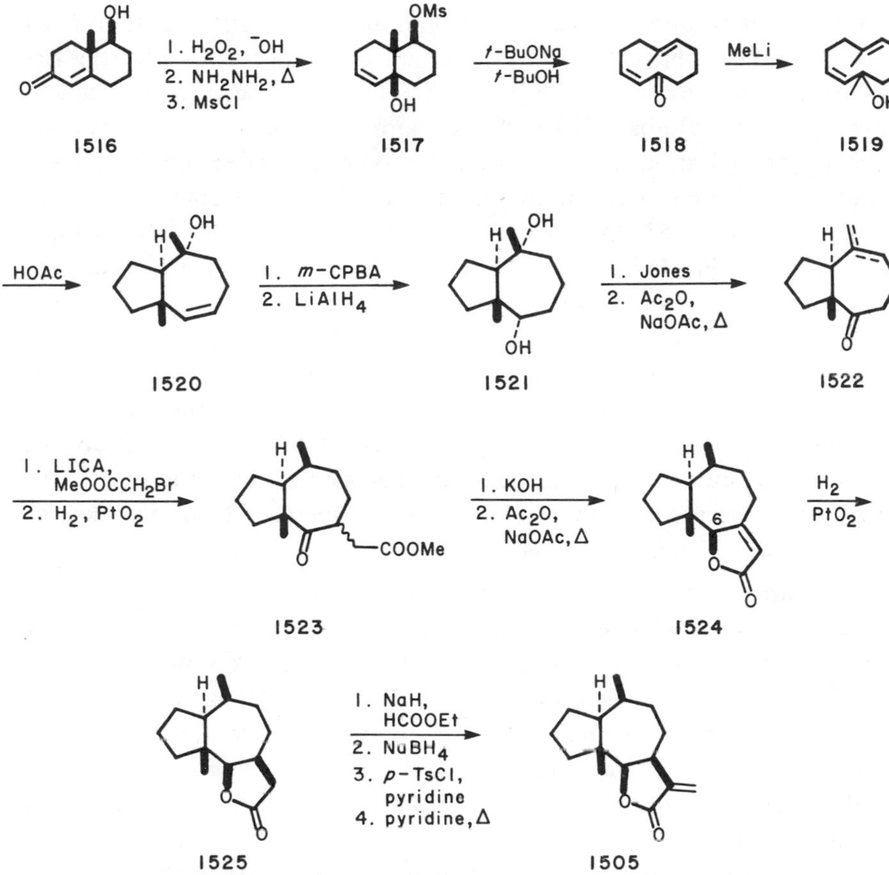

Scheme 185.   Marshall's Synthesis of (±)-4-Deoxydamsin

verted in an unexceptional fashion to the cyclodecadienol **1519**. Cyclization (to **1520**) occurs readily in acetic acid, a reaction which finds precedent in Marshall's earlier work leading to globulol (see Scheme 183).
Epoxidation of **1520** occurs stereoselectively, trans to the angular methyl
group. Reduction of the epoxide is regiospecific, the angular methyl
group directing the hydride reagent to the less hindered (C-7) position.

After oxidation of the secondary alcohol and dehydration, a mixture of alkenes **1522** is obtained. The two-carbon fragment required for the lactone is added by alkylation with methyl bromoacetate. Subsequent hydrogenation affords **1523** with excellent stereoselectivity. Again the angular methyl group appears to exert a dominant effect on the stereochemical outcome of reactions on the pseudoguiane framework. After saponification of **1523** the product acid is converted, via the enol-lactone, to the butenolide **1524**. The stereochemistry at C-6 may arise from the preferred direction of kinetic protonation of the intermediate enol-lactone or may reflect a thermodynamic preference for the β-configuration. Hydrogenation of **1524** is stereoselective, affording **1525** in 90% yield. The lactone is methylenated by the method of Minato and Horibe providing deoxydamsin (**1505**). Unfortunately the strategy employed in the Marshall deoxydamsin synthesis is not readily adapted to the preparation of naturally occurring pseudoguaianolides, all of which are oxygenated at C-4.

Kretchmer's synthesis of damsin (Scheme 186) is the first of the successful syntheses of naturally occurring pseudoguaianolides.[412] Octalone **1530**, prepared as shown in the scheme, affords triketone **1531** upon ozonolysis. This substance undergoes sequential aldolization, dehydration, and methylation upon exposure to mild base in the presence of methyl iodide. The major product of this transformation is **1532** (36% yield from **1526**) which is separated from its C-7 epimer (18% yield) by column chromatography and crystallzation. Reduction of **1532** affords a triol. Selective oxidation of the primary alcohol and saturation of the double bond provides hydroxy lactone **1534** in moderate yield. The conversion of **1534** to damsin (**1506**) is uneventful.

Vandewalle's synthesis of damsin (Scheme 187) provides an interesting construction of the hydroazulene ring system.[413] Photocycloaddition of 2-methylcyclopentenone (**1536**) with **1537** affords the tricyclic ketone **1538**. After reduction of the carbonyl and desilylation, the intermediate vicinal diol is oxidatively cleaved to obtain **1539**. Protection of the secondary alcohol and base-catalyzed epimerization affords a 6:1 mixture of the trans-fused hydroazulene **1540** and its cis diastereomer. The mixture is separated and the major isomer is carried on, in a manner remin-

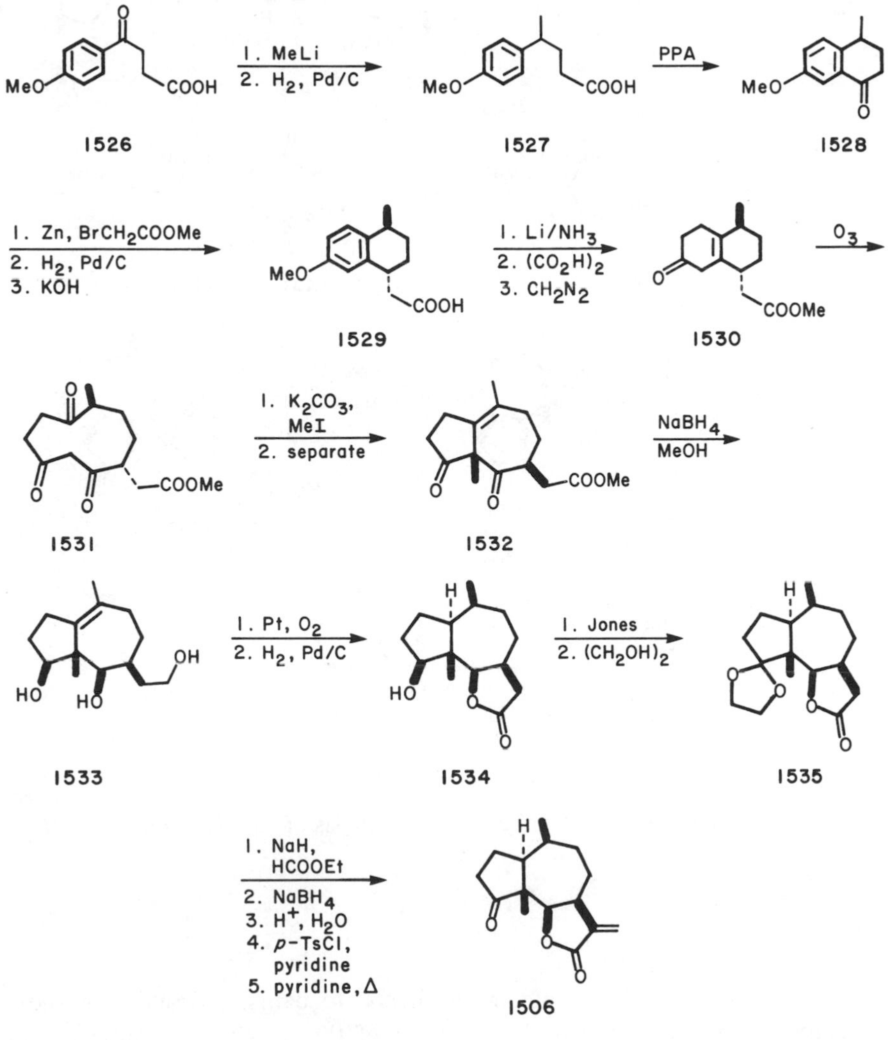

Scheme 186.   Kretchmer-Thompson Synthesis of (±)-Damsin

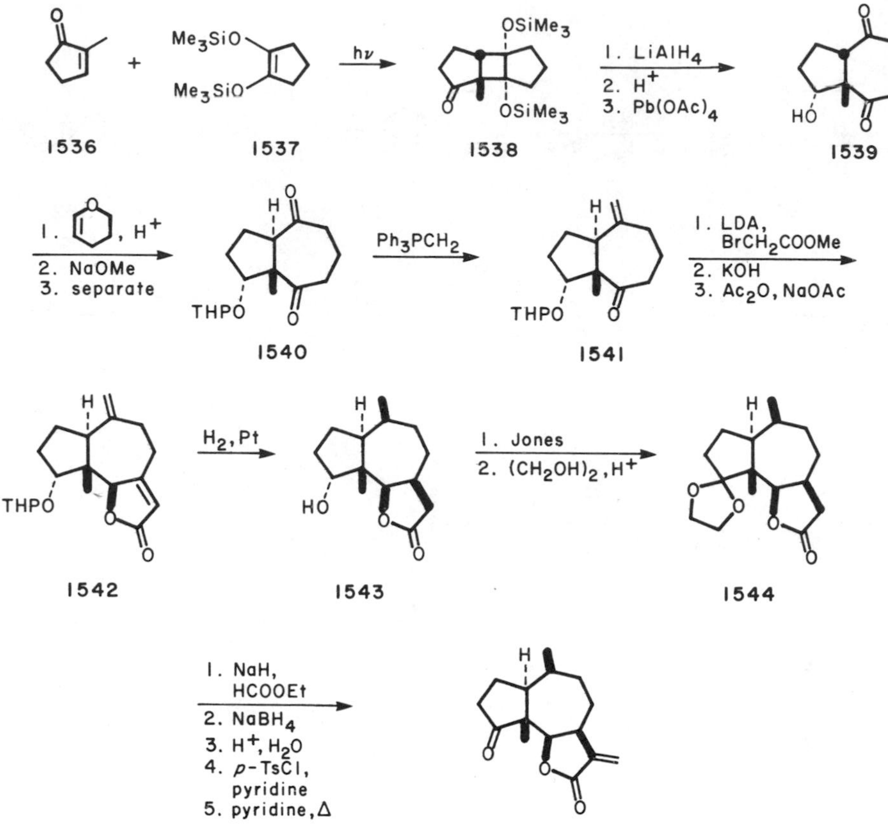

1536   1537   1538   1539

1540   1541

1542   1543   1544

1506

Scheme 187.   Vandewalle's Synthesis of (±)-Damsin

iscent of Marshall's deoxydamsin synthesis, to damsin (**1506**). Of special note in this sequence is the cleavage of the tetrahydropyranyl ether of **1542** under hydrogenation conditions.

In an extension of this work, Vandewalle has also reported the conversion of **1543** to neoambrosin (**1507**), parthenin (**1508**) and hymenin

(1509) as shown in Scheme 188.[414] Ketalization of the enone 1545 (derived as shown from 1543) occurs with concomitant migration of the double bond. After methylenation 1546 is obtained. The latter may be

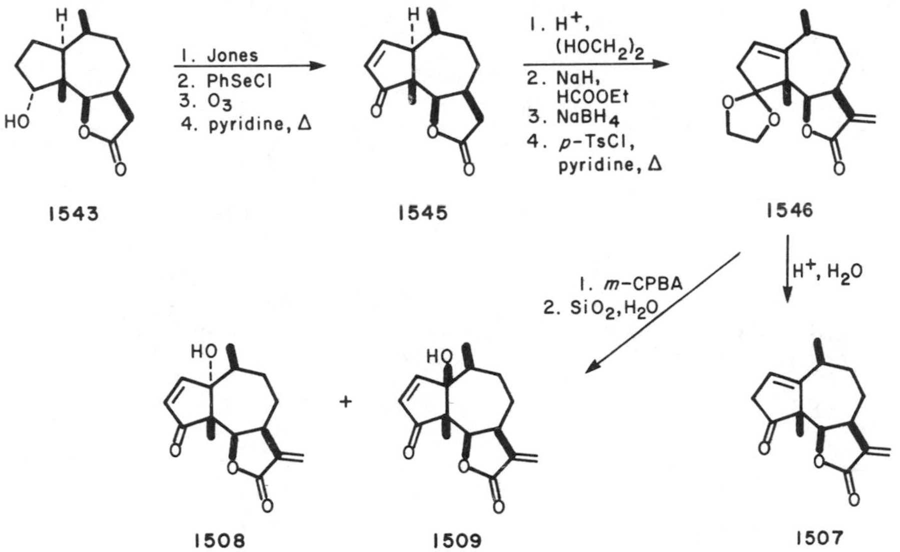

Scheme 188.   Vandewalle's Synthesis of (±)-Neoambrosin, (±)-Parthenin and (±)-Hymenin

hydrolyzed directly to yield neoambrosin (1507). Epoxidation of 1546 affords a mixture of two ketal epoxides which undergo hydrolysis and elimination to a 7:3 mixture of γ-hydroxy enones parthenin (1508), and hymenin (1509) upon attempted chromatography on silica.

Grieco has contributed impressively in the pseudoguaianolide area. His first report deals with the total synthesis of damsin (1506, Scheme 189).[415] Technology developed by Grieco and others[416] in connection with prostaglandin synthesis is used to convert norbornadiene to the functionalized bicyclo[2.2.1]heptane 1547. This material undergoes

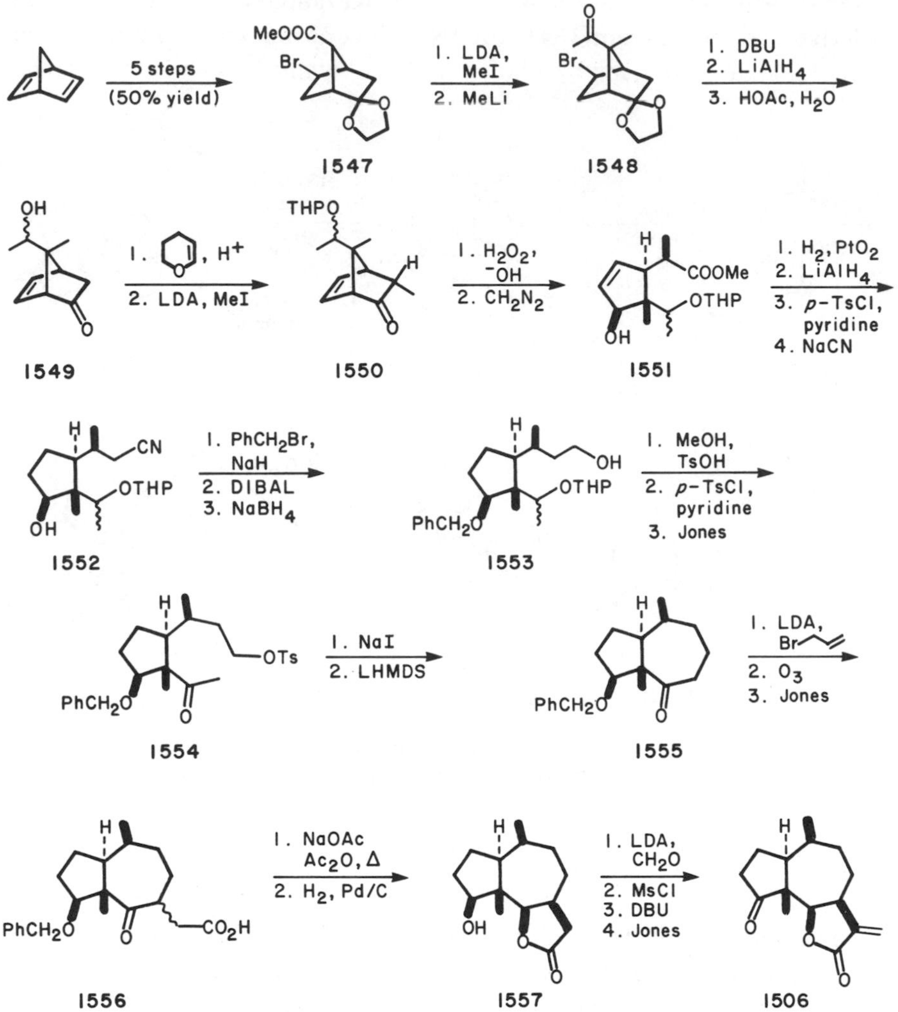

Scheme 189.   Grieco's (±)-Damsin Synthesis

remarkably stereoselective alkylation upon treatment with lithium diisopropylamide and methyl iodide. Note that the ring fusion stereochemistry of the eventual product is established in this reaction. The alkylation product is converted to the methyl ketone (**1548**) by treatment with methyllithium. Subsequent transformations lead to **1549** which is protected as the tetrahydropyranyl ether. The crucial C-10 methyl group is now introduced stereospecifically by akylation of the derived enolate from the less hindered endo face. Baeyer-Villiger oxidation of **1550** followed by esterification leads to **1551** in which three of the five asymmetric centers of damsin are correctly established. The keto-tosylate **1554**, derived from **1551** in ten straightforward steps, is converted to the iodide and the seven-membered ring is closed by alkylation. The conversion of **1555** to damsin relies on the same strategy employed by Marshall in his synthesis of deoxydamsin.

Grieco has also converted damsin into several other natural products, thus completing total syntheses of these compounds. Ambrosin (**1510**) is prepared by acid catalyzed selenylation of damsin followed by selenoxide elimination. Stramonin B (**1511**)[417] is prepared by conversion of damsin to the corresponding epoxide in two steps followed by introduction of unsaturation to the five-membered ring as outlined above. Finally, modified Baeyer-Villiger oxidation of damsin leads to psilostachyin (**1512**).[415]

Schlessinger has provided an interesting and novel solution to the problem of preparing a functionalized hydroazulene nucleus, exemplified here by his total synthesis of (±)-damsin (Scheme 190).[418] The known enone **1558** is converted by Barton's nitrone version of the Beckman rearrangement to the enamide **1559**. Treatment of **1558** with dimethyl lithiomethylphosphonate affords an intermediate aldimino-β-ketophosphonate anion, which is hydrolyzed to the aldehyde **1560**. The latter compound readily cyclizes to the enone **1561**. The stereochemistry of the ring fusion is determined in the conversion of **1559** to **1560**. The

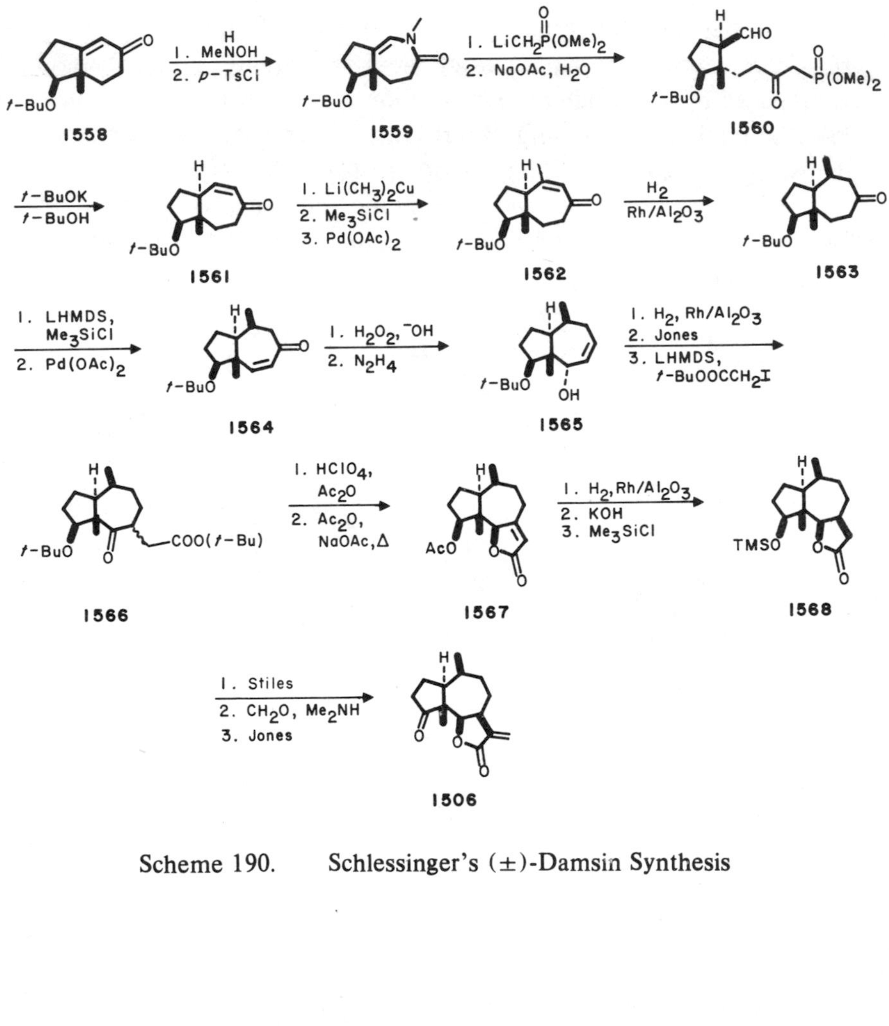

Scheme 190.   Schlessinger's (±)-Damsin Synthesis

intermediate in that conversion (1569) undergoes kinetic protonation trans to the angular methyl group. This is presumably an intramolecular process, the proton being delivered from the same side of the molecule that the β-ketophosphonate group resides on. Addition of lithium

1569

dimethylcuprate and reintroduction of a 9,10-double bond affords 1562. Hydrogenation of this enone is highly stereoselective affording the compound with the 10-β configured methyl group, 1563. Deprotonation of 1563 occurs regioselectively, allowing the preparation of 1564. This compound is converted by Wharton's method to 1565 and then, in a precedented fashion, to damsin (1506). The use of the Stiles' reagent-Mannich sequence was found by Schlessinger to be a superior method for the introduction of the α-methylene group.

Two syntheses of the simplest naturally occurring pseudoguaiane, damsinic acid (1513), have been reported. Lansbury's synthesis[419] relies on a cation-acetylene cyclization to form the seven-membered ring (Scheme 191). The required substrate for this cyclization, 1572, is prepared from 2-methylcyclopentane-1,3-dione (1570) as shown in the scheme. Formic acid induces cyclization to the hydroazulene, 1573. This cyclization is noteworthy, in that the seven-membered ring ketone is formed to the exclusion of the possible six-membered ring aldehyde 1578. In most

1578

ring forming reactions, enthalpic and entropic considerations favor formation of a cyclohexane ring over a cycloheptane. However, formation of 1579 proceeds via a primary vinyl cation and thus enthalpy strongly

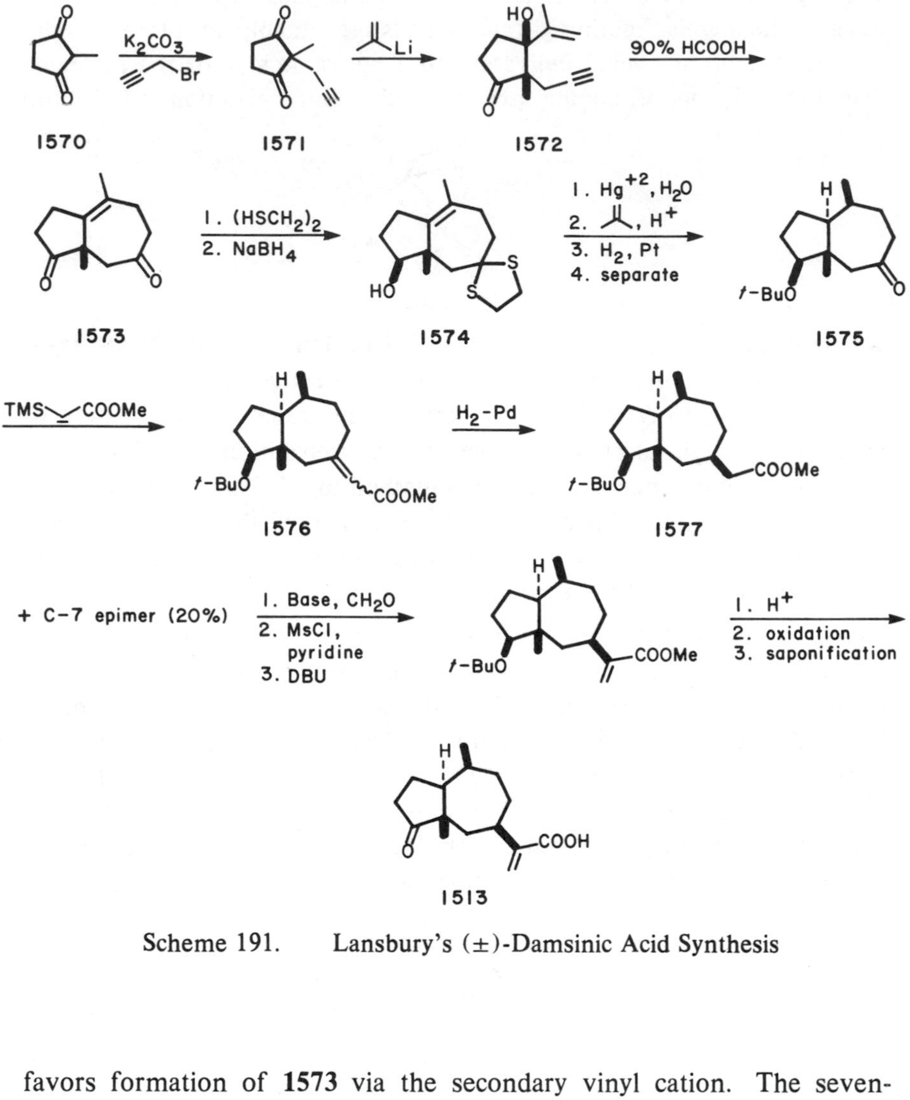

K_2CO_3

Br

90% HCOOH

1570

1571

1572

1. (HSCH$_2$)$_2$
2. NaBH$_4$

1. Hg^{+2}, H$_2$O
2. $\triangle$, H$^+$
3. H$_2$, Pt
4. separate

1573

1574

t-BuO

1575

TMS COOMe

t-BuO COOMe

H$_2$-Pd

t-BuO COOMe

1576

1577

+ C-7 epimer (20%)

1. Base, CH$_2$O
2. MsCl, pyridine
3. DBU

t-BuO COOMe

1. H$^+$
2. oxidation
3. saponification

COOH

1513

Scheme 191.   Lansbury's (±)-Damsinic Acid Synthesis

favors formation of **1573** via the secondary vinyl cation. The seven-membered ring carbonyl of **1573** is selectively protected as the dithio-ketal and the cyclopentanone is reduced.   After thioketal hydrolysis and protection of the secondary alcohol as the t-butyl ether, hydrogenation of the double bond affords a 6:1 mixture of two stereoisomers.   The major product, **1575**, was isolated by crystallization. The lithium derivative of methyl trimethylsilyl acetate converts **1575** to **1576** in nearly quantitative yield. Hydrogenation of **1576** affords **1577** in 70-80% yield along with 10-20% of the C-7 epimer. Unfortunately, the two isomers proved to be inseparable. The conversion of **1577** to damsinic acid (**1513**) is straightforward.

Wender's imaginative approach to damsinic acid (**1513**) employs a divinylcyclopropane rearrangement to fuse a seven-membered ring onto an existing five-membered ring (Scheme 192).[420] Reaction of enol ether **1579** with the lithium reagent **1580** (a mixture of stereoisomers) affords **1581** as a mixture of isomers. Upon thermolysis, isomer **1581c**, in which the vinyl groups are cis on the cyclopropane ring, readily rearranges to the desired hydroazulene **1582**. The trans isomer **1581t** does

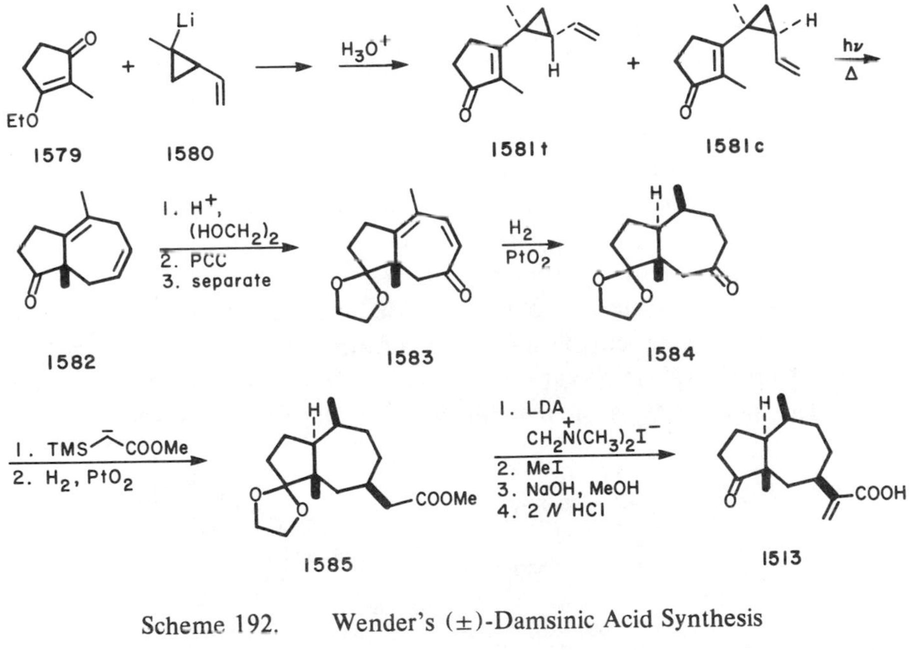

Scheme 192.   Wender's (±)-Damsinic Acid Synthesis

not give the desired rearrangement, a homo-[1,5]hydrogen migration leading to **1586** being preferred. The preparation of stereochemically

1586

homogeneous **1581c** is possible but is somewhat laborious. Wender makes use of the observation that the stereoisomers of **1581** may be equilibrated through photolysis. Since the rearrangement of **1581c** to the desired hydroazulene occurs at temperatures where **1581t** is stable to thermolysis, simultaneous irradiation and heating converts both stereoisomers cleanly to **1582**. The ketone is protected as its ethylene ketal and the diene is oxidized to a mixture of the linear dienone **1583**, and the cross-conjugated dienone oxygenated at C-9. The major isomer, **1583**, isolated in 60% overall yield from **1582**, is converted to **1584** with 90% stereoselectivity. Conversion of the latter to damsinic acid (**1513**) follows precedented lines.

Several syntheses of the pseudoguaianolide confertin (**1514**) have been reported. The first is by Marshall[421] (Scheme 193). Enone **1587** is alkylated and reduced to **1588**. The allylic alcohol moiety allows regio- and stereoselective epoxidation of the tetra-substituted olefin. Reductive elimination of the derived mesylate provides alcohol **1589**. Simmons-Smith cyclopropanation is also regio- and stereoselective, the directive effect of the allylic hydroxyl group again being exploited. The key intermediate **1590** is then prepared by oxidative cleavage of the allyl group. Solvolysis of **1590** in aqueous perchloric acid leads to the hydroazulene **1591**. The stereochemical course of this reaction is of interest. Compound **1591** may exist in either of two conformations **1591a** or **1591b**. The latter is presumably favored in this equilibrium since the

1591a   1591b

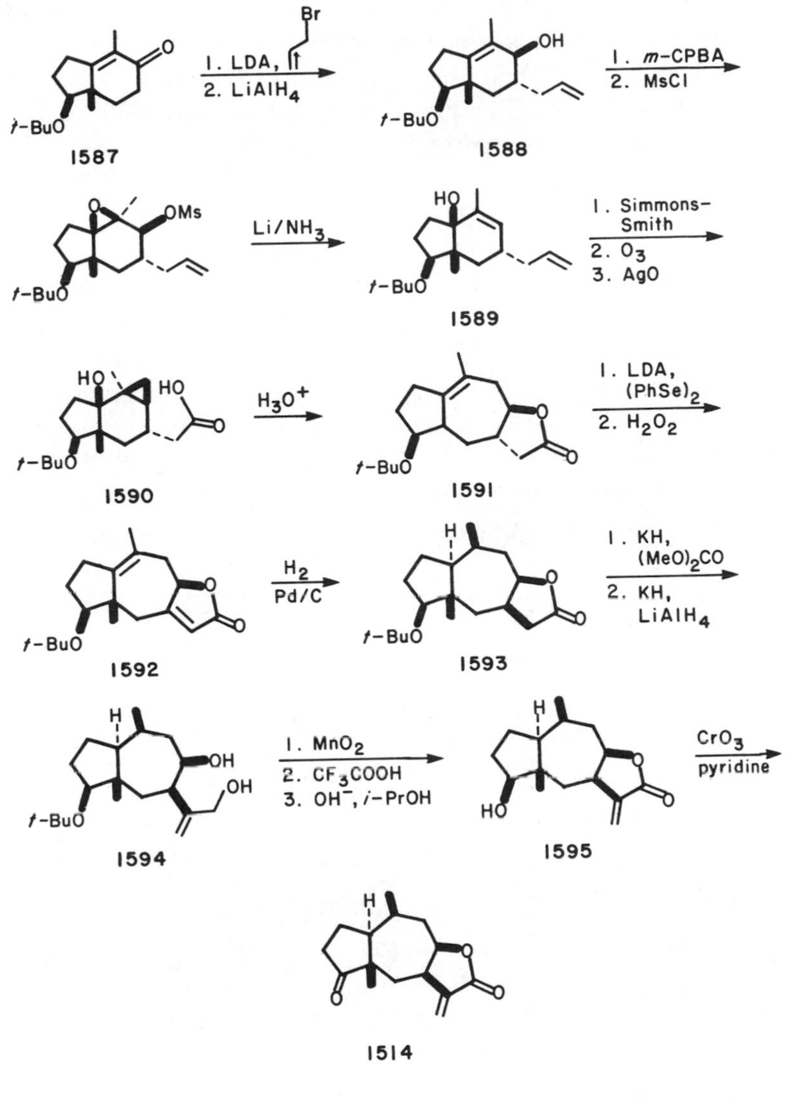

Scheme 193.   Marshall's Synthesis of (±)-Confertin

acetic acid side chain is equatorial. Moreover, only in conformation **1591b** can the carboxyl group be involved in nucleophilic participation in a concerted process in which the cyclopropane ring is cleaved. Thus rearrangement via **1591b** is energetically favored. In any event, the synchronous nature of this rearrangement ensures the stereospecific introduction of the C-8 oxygen in the β-configuration as is required for the synthesis of confertin. The correct stereochemistry at C-7 is established

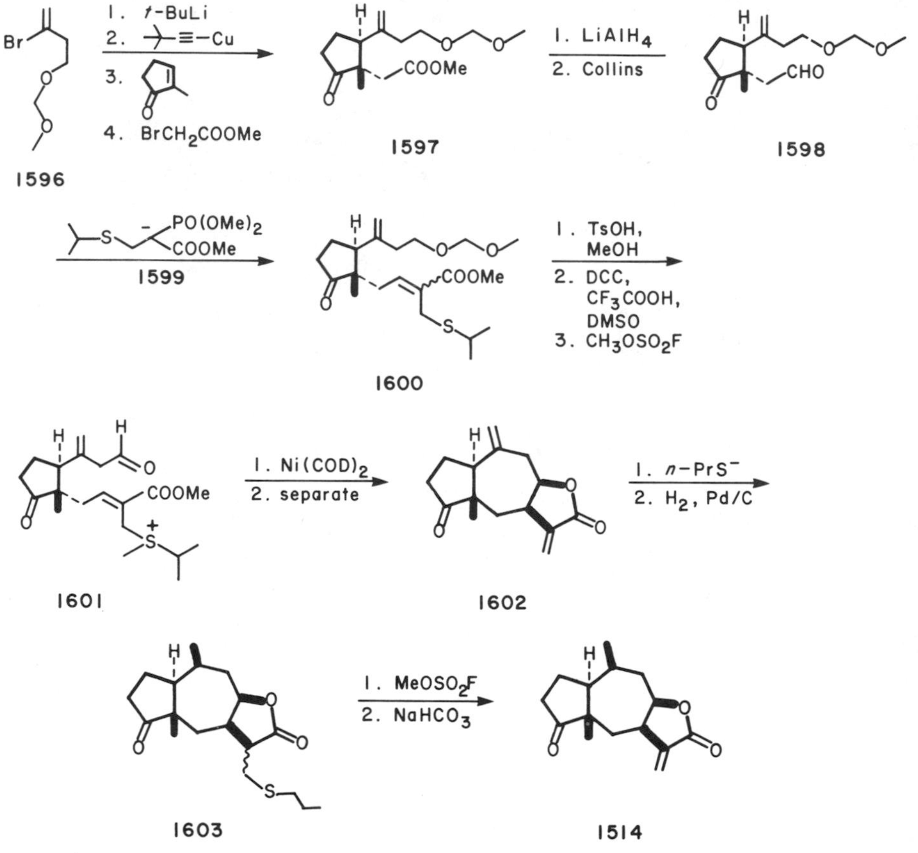

Scheme 194.   Semmelhack's (±)-Confertin Synthesis

by conversion of **1591b** to the butenolide **1592**, which is hydrogenated to obtain **1593** in 70% yield. The methylenation of **1593** is carried out by initial carbonation and subsequent reduction of the β-keto ester anion affording the unsaturated diol **1594**. The allylic alcohol is oxidized and the *t*-butyl ether cleaved with trifluoroacetic acid. Mild hydrolysis serves to cleave the intermediate trifluoroacetate ester providing **1595**. Collins oxidation provides confertin (**1514**).

Semmelhack's unique approach to hydroazulene synthesis is shown in Scheme 194.[422] The preparation of **1598** from **1596** proceeds uneventfully. Wadswoth-Emmons reaction of **1598** with **1599** (generated *in situ* from trimethyl 2-phosphonoacrylate and sodium isopropylmercaptide) provides **1600** as a 1:4 mixture of *E* and *Z* stereoisomers. The methoxymethyl group is cleaved and Moffatt oxidation provides the corresponding aldehyde. The sulfonium salt **1601** is prepared by the agency of methyl fluorosulfonate. The salt is converted by Ni° to a mixture of α-methylene lactones. The major isomer **1602** is formed in approximately 25% yield from **1600**. Unfortunately, selective hydrogenation of **1602** to confertin (**1514**) is not possible, necessitating the sequence shown. Confertin (**1514**) is produced in low yield by this method.

Schlessinger has prepared confertin by the route shown in Scheme 195.[418] Alkylation of **1563** (Scheme 190) is regiospecific, and the resulting ester is converted to the enollactone **1604**. Hydrogenation affords

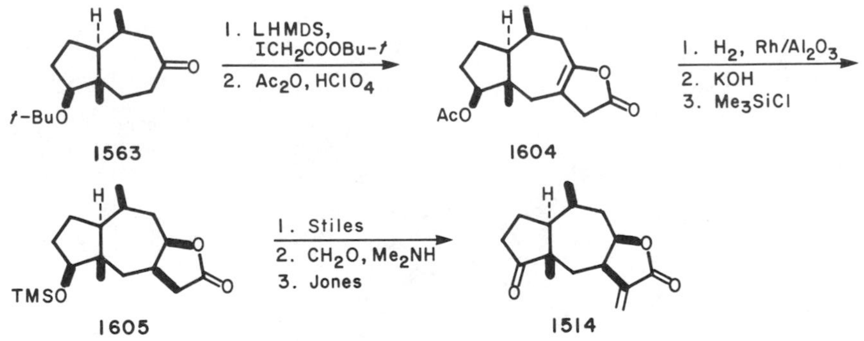

Scheme 195.   Schlessinger's Synthesis of (±)-Confertin

the desired lactone **1605** in 63% yield overall. Conversion of the latter to confertin (**1514**) is straightforward.

Heathcock and co-workers have also prepared the key intermediate **1604** and provided an alternate synthesis of confertin from it (Scheme 196).[423] The well-known enone **1606** is converted to the tosylate **1607** in a simple sequence. The latter readily rearranges to **1608** (after reacetylation of the alcohol). Direct conversion of **1608** to the key enone **1609** is accomplished by the action of trimethylaluminum and methyllithium. The mechanism for this transformation is interesting. Apparently **1612** is initially formed by β-elimination. Trimethylaluminum facilitates this

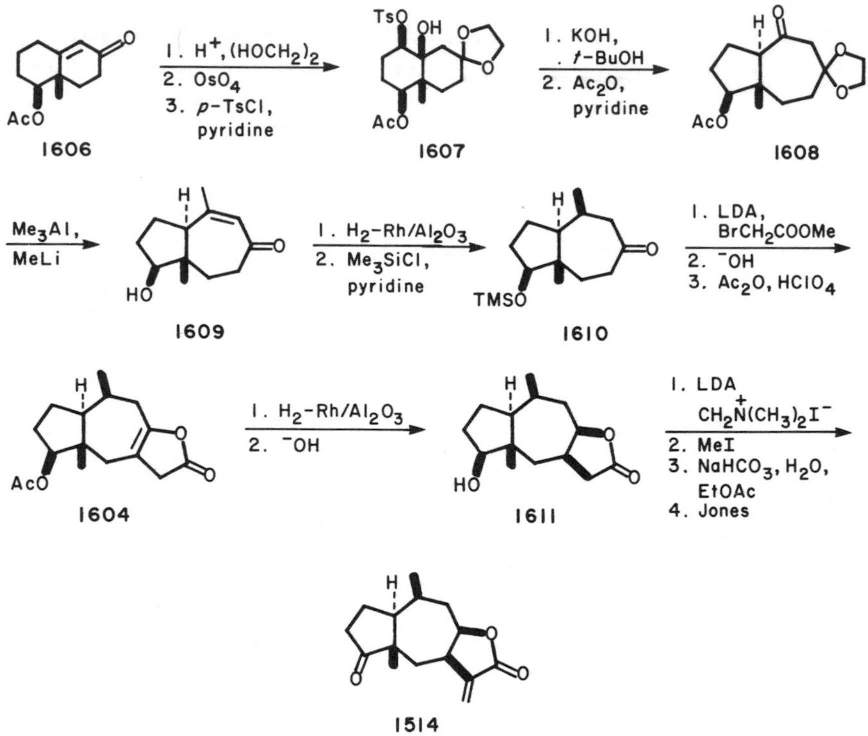

Scheme 196.   Heathcock-Graham-DelMar Synthesis of (±)-Confertin

process by coordinating with the ketal oxygens. Subsequent addition of methyllithium provides **1613** which hydrolyzes directly to **1609** at pH 9 in 63% yield. A variant of the Marshall-Schlessinger strategy affords the

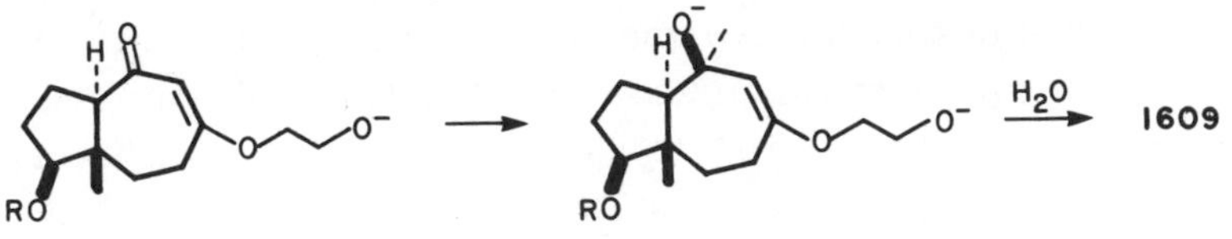

lactone **1611** which is converted to confertin (**1514**) by Poulter's methylenation sequence and Jones' oxidation.

Wender's confertin synthesis (Scheme 197) proceeds from the intermediate **1583**, the preparation of which we have discussed previously in connection with the damsinic acid synthesis (see Scheme 192). Epoxidation and modified Wittig olefination afford **1614** which is converted by

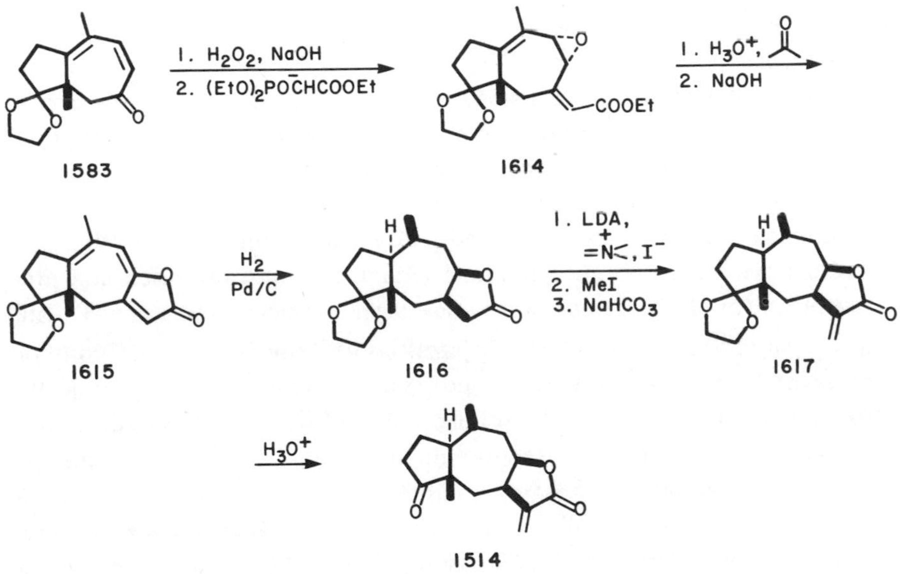

Scheme 197.   Wender's (±)-Confertin Synthesis

treatment with acid to a hydroxy lactone and thence by base treatment to the enol lactone **1615**. Hydrogenation of this intermediate proceeds with a high degree of stereoselectivity, lactone **1616** comprising greater than 80% of the product mixture. Thus the angular methyl group again proves valuable in dictating asymmetry in the generation of new stereocenters on the pseudoguaiane framework. Poulter's olefination sequence is employed to prepare the α-methylene lactone **1617** which affords confertin (**1514**) upon hydrolysis of the ketal.

The preparation of the C-10 hydroxymethyl pseudoguaianolide, hysterin (**1515**), has been reported by Vandewalle and co-workers (Scheme 198).[424] The key intermediate **1540** (see Scheme 187) is converted to enol ether **1618** by Wittig reaction. This material is elaborated by precedented means to the butenolide **1619**. Reduction of **1619** is only moderately stereoselective and, surprisingly, affords the 10α-isomer **1620** as the major component of a 7:3 mixture (compare the reduction of **1542**, Scheme 187). It is also noteworthy that the tetrahydropyranyl ether is hydrogenolytically cleaved during the course of this reduction. The mixture of **1620** and **1621** is carried on in the manner shown to a mixture of 10-epihysterin (**1625**) and hysterin (**1515**).

Heathcock and co-workers have recently developed a route to the ambrosanolides which proceeds via cis-fused hydroazulene intermediates (Scheme 199).[425] Cyclohexenone is converted into 2,4-dimethylcycloheptenone (**1628**) by a procedure developed by Ito. When enone **1626** is subjected to copper-catalyzed conjugate addition of Grignard reagent **1627**, trans adduct **1628** and its cis diastereomer are obtained in a ratio of 85:15 (both are produced as 1:1 epimeric mixtures at C-2). Ozonolysis of **1628** gives a keto aldehyde which cyclizes to an aldol upon treatment with HCl in acetic acid. Under the conditions of the cyclization the aldol is acetylated to give a keto acetate. The acetyl group is removed and the resulting aldol (**1629**) is protected as the tetrahydropyranyl ether. At this point, the product is purified chromatographically to remove the small amount of product resulting from the 3,4-cis isomer of **1628**. Alkylation of ketone **1630** occurs stereospecifically, affording ketone **1631**. Base-catalyzed epimerization of **1631** gives the diastereomeric ketone **1635** (cf. compound **1467**, Scheme

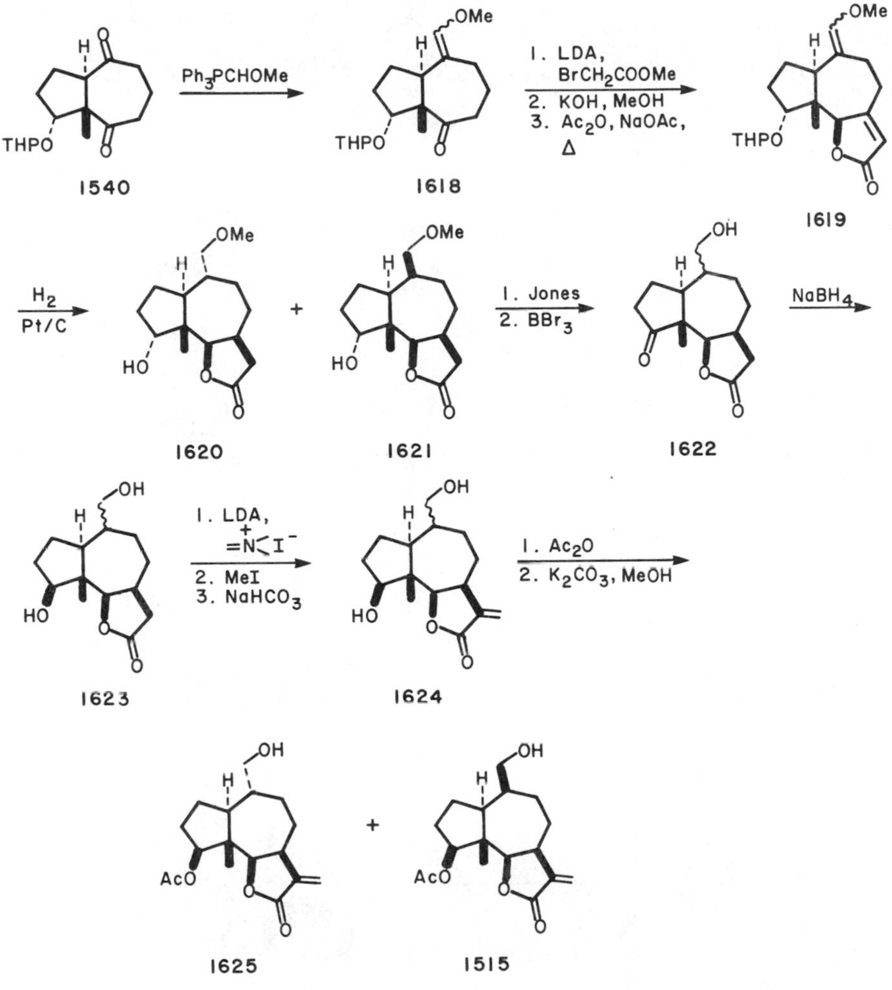

Scheme 198.   Vandewalle's Synthesis of (±)-Hysterin

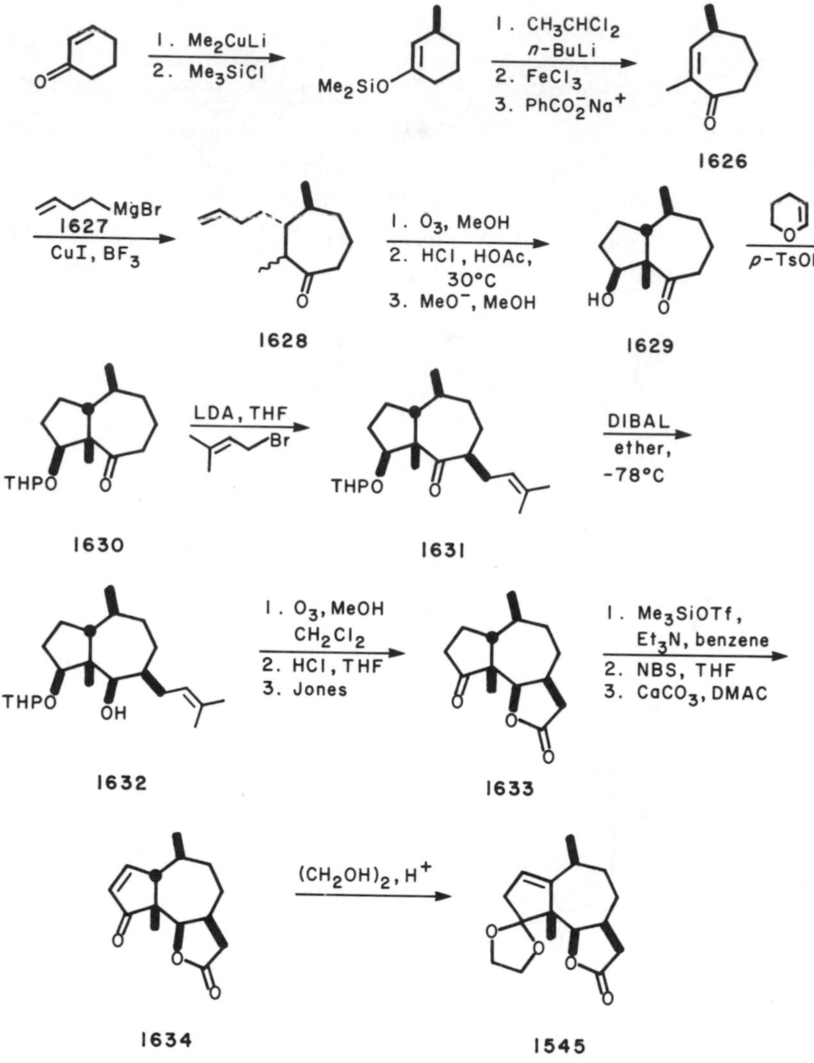

Scheme 199.   Heathcock-Germroth-Tice Synthesis of (±)-Parthenin

1631   1635

178). Reduction of ketone **1631** with diisobutylaluminum hydride in ether at -78°C gives predominantly (>95%) one alcohol, **1632**. Ozonolysis of the prenyl group in methanol gives a cyclic acetal, which is deprotected and oxidized to obtain keto lactone **1633**. A double bond is introduced into the five-membered ring and the resulting cyclopentenone is ketalized to obtain the deconjugated ketal **1545**, which had been previously converted into (±)-parthenin and (±)-hymenin (see Scheme 188).

(5) Pseudoguaianolides: The Helenanolide Family; Helenalin, Mexicanin, Linifolin, Bigelovin, Carpesiolin, Aromaticin, and Aromatin.

Notably fewer syntheses of pseudoguaianolides in which the C-10 methyl group is of the α-configuration—the so-called helenanolides—have been reported. The seven members of this family that have thus far yielded to total synthesis are helenalin (**1636**), bigelovin (**1637**), mexicanin I (**1638**), linifolin A (**1639**), carpesiolin (**1640**), aromaticin (**1641**), and aromatin (**1642**).

1636   1637   1638: R = H
1639: R = Ac

1640   1641   1642

Grieco's synthesis of helenalin (1636) is shown in Scheme 200.[426a] The key cyclopentenol 1551 (prepared in 14 steps from norbornadiene, Scheme 189) is converted to the lactone 1643 in four steps. During the lactonization reaction, the methyl group undergoes epimerization to the

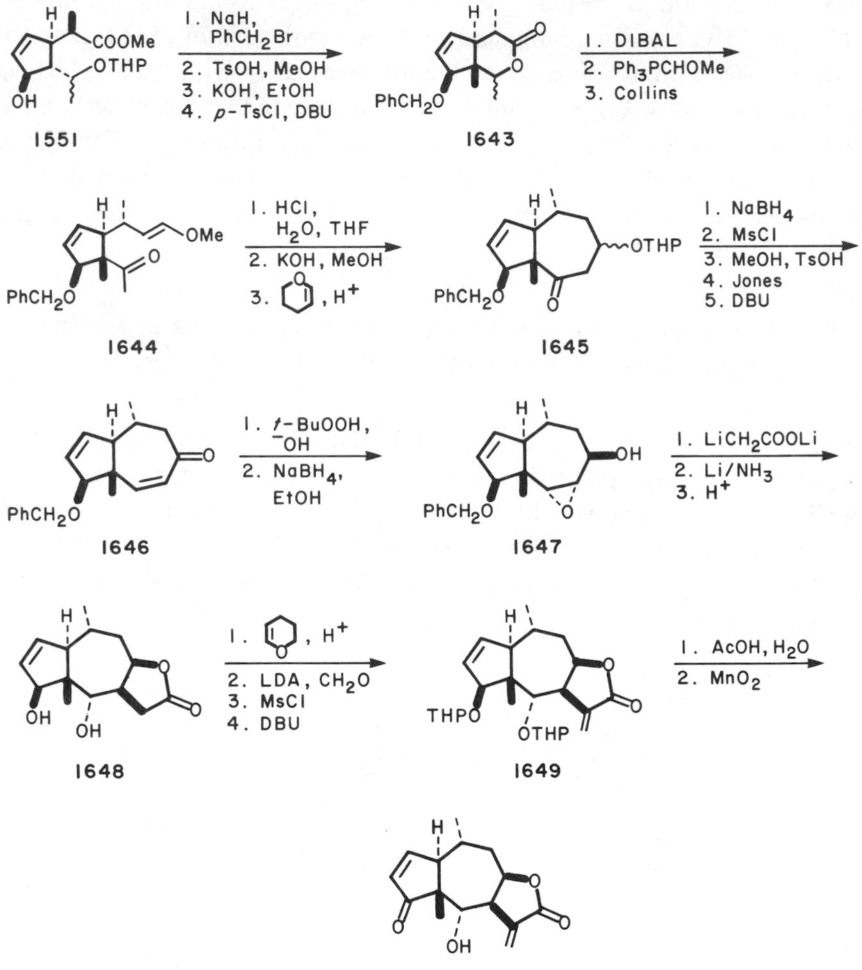

Scheme 200.   Grieco's (±)-Helenalin Synthesis

more stable equatorial position. The sequence of reduction, Wittig olefination and oxidation furnishes **1644**. The enol ether is hydrolyzed, the resulting keto aldehyde is cyclized by aldol condensation and the aldol protected to obtain **1645**. Enone **1646** is prepared from **1645** in the straightforward five-step sequence shown. Epoxidation of **1646** is stereoselective as is sodium borohydride reduction to **1647**. Epoxide **1647** is opened by treatment with dilithio acetate, the benzyl group is removed by lithium/ammonia and the resulting product acidified to obtain the cis lactone **1648**. After protection of the alcohols, the α-methylene unit is introduced in a typical sequence furnishing **1649**. Deprotection and oxidation of the allylic alcohol provide (±)-helenalin (**1636**) in 1.5% overall yield and 40 steps from norbornadiene. The remarkable achievement of an average yield of 90% per step is a tribute to the tenacity and skill of the four chemists who carried out this work.

In an extension of the foregoing methodology it was discovered that reduction of the enone **1646** prior to epoxidation allows the preparation of hydroxy epoxide **1650**, a precursor to bigelovin (Scheme 201).[426b]

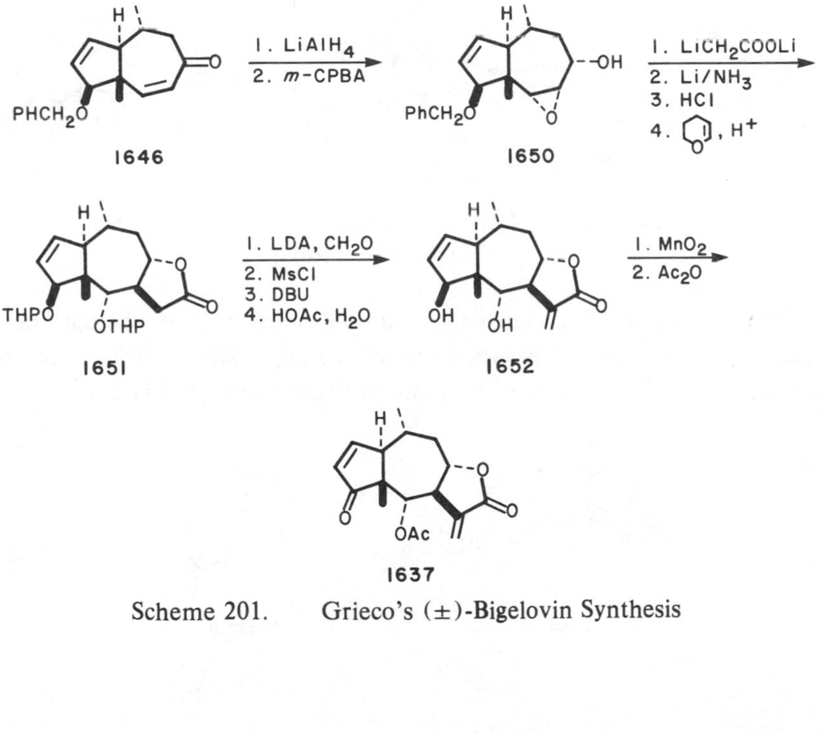

Scheme 201.   Grieco's (±)-Bigelovin Synthesis

The conversion of **1650** to bigelovin (**1637**) follows an established line. One interesting observation in that sequence is the regiospecificity of formation of the lactone ring which occurs exclusively to the C-6 hydroxy group.

In Scheme 202 the further conversion of **1650** to mexicanin I (**1638**) and subsequently to linifolin A (**1639**) is shown.[427] Reduction of **1653** is

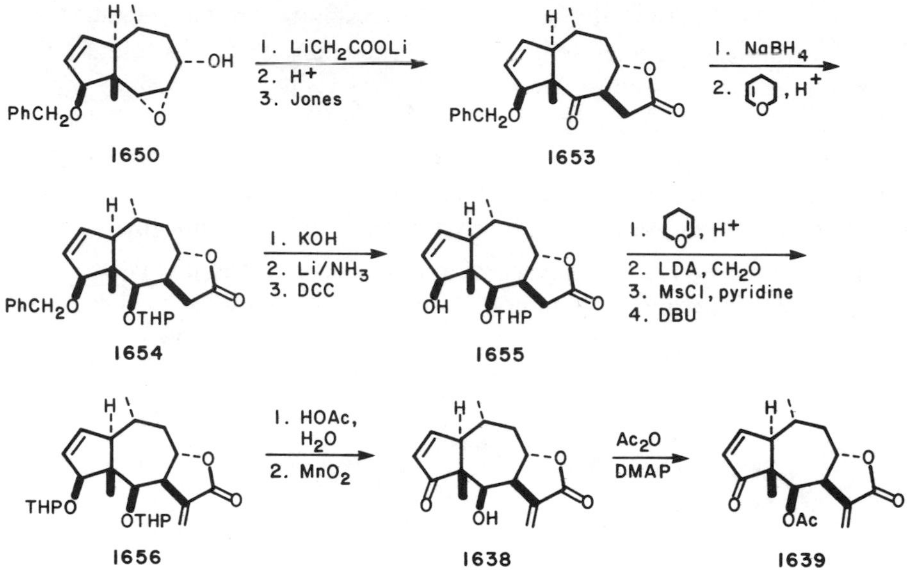

Scheme 202.   Grieco's Synthesis of ($\pm$)-Mexicanin-I and ($\pm$)-Linifolin A.

stereoselective, hydride delivery occurring trans to the angular methyl group. The C-6 alcohol is protected at this point since it was discovered that the acid **1657** cyclizes exclusively to the cis lactone **1658**:

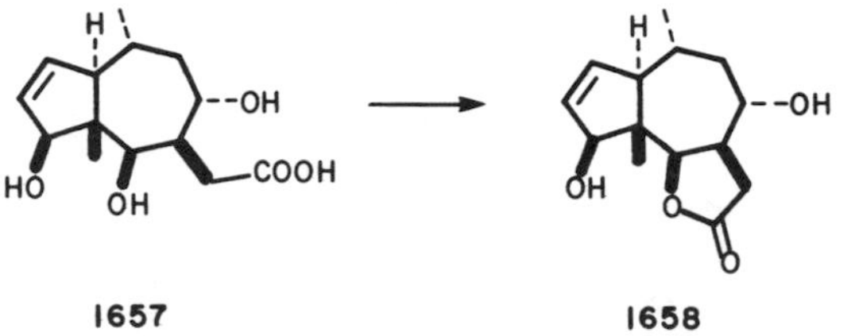

The conversion of **1654** to mexicanin I (**1630**) follows a now familiar line. Linifolin A (**1639**) is readily prepared from **1638** by acetylation.

Schlessinger's synthesis[418] of helenalin (Scheme 203) is much more direct than Grieco's approach. The enone **1561**, prepared as shown in Scheme 190, undergoes copper-catalyzed conjugate addition of methyl-magnesium bromide in a highly stereoselective fashion yielding **1659**. A straightforward sequence converts **1659** to **1660**. Deprotonation of **1660**

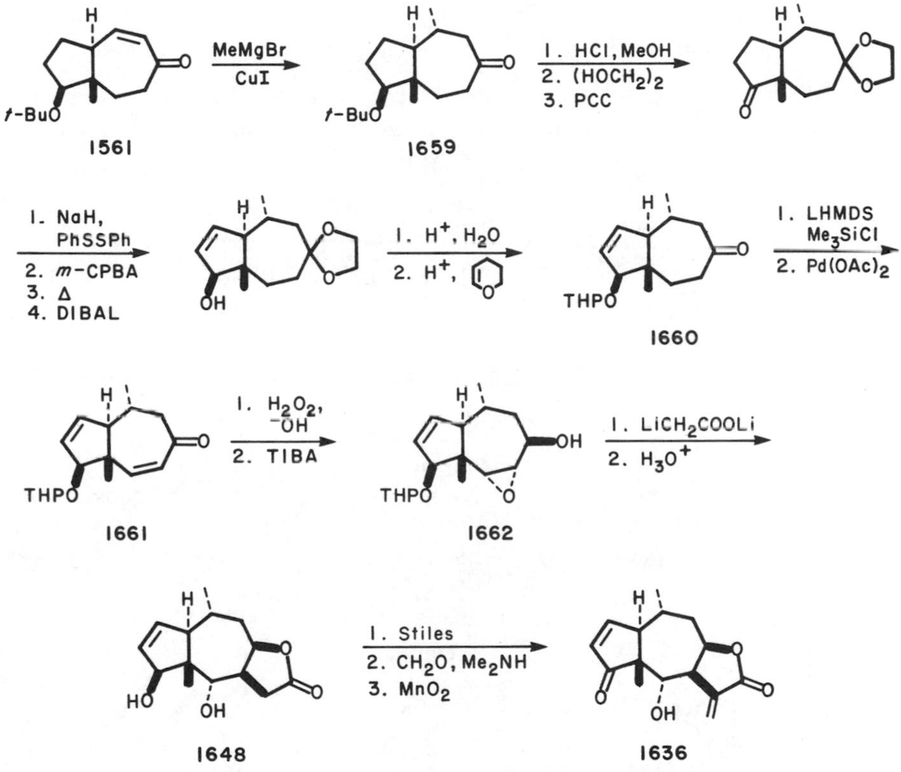

Scheme 203.   Schlessinger's Synthesis of (±)-Helenalin

with lithium hexamethyldisilazide is regiospecific, allowing formation of the enone **1661**. Conversion of the latter to helenalin (**1636**) follows a precedented line (*vide supra*). Heathcock and co-workers also prepared the Schlessinger ketone **1659** by dissolving metal reduction of the key enone **1609** (Scheme 196) used in their confertin synthesis.[423]

Vandewalle[428] has found that hydrogenation of **1663** (prepared from **1541**, Scheme 187) over palladium on charcoal takes a different stereo-chemical course than the reduction of **1542**, affording the 10-α compound **1664** in approximately 66% yield (Scheme·204). The 10-β com-

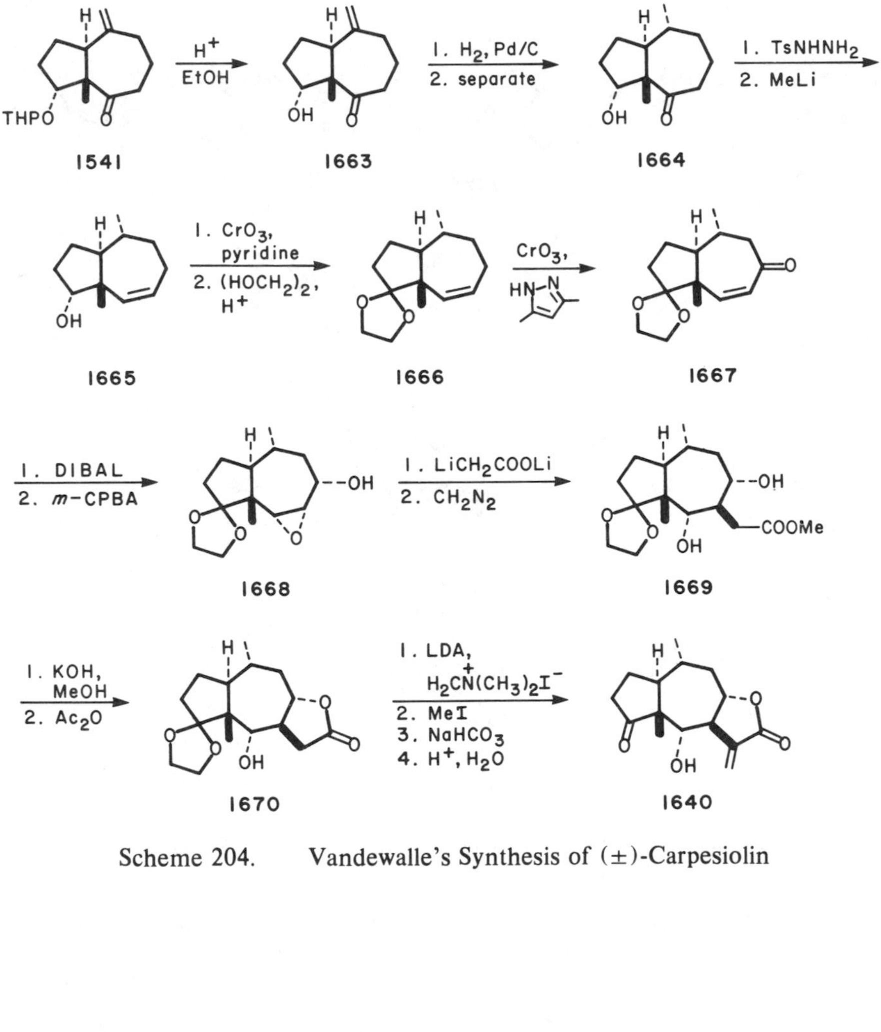

Scheme 204.   Vandewalle's Synthesis of (±)-Carpesiolin

pound is formed in 10-20% yield. No unambiguous reason for this behavior has been provided. The major isomer may be isolated by chromatography and is converted by Bamford-Stevens reaction to unsaturated alcohol **1665**. The alcohol is oxidized and the resulting ketone protected. Allylic oxygenation gives enone **1667**, which Vandewalle independently discovered may be converted to the α-epoxy-α-hydroxy system **1668** (compare Grieco's bigelovin synthesis, Scheme 201). The preparation of carpesiolin (**1640**) from **1668** finds ample precedent in other schemes of this section.

Lansbury's synthesis of aromaticin (Scheme 205)[429] proceeds from ketone **1575** which had been prepared earlier in connection with the synthesis of damsinic acid (see Scheme 191). Sulfinylation of **1575** under equilibrating conditions affords **1671**. Pummerer rearrangement gives the unsaturated sulfide **1672**. Equilibration of the methyl group is now possible using diazabicyclononane as base catalyst. (Attempts to carry out the equilibration without the presence of the thiophenyl substituent fails since the β,λ enone **1679** is thermodynamically favored in that system). Having served its purpose, the vinyl sulfide moiety is removed,

1679

affording **1674**. Reaction of **1674** with the anion derived from methyl trimethylsilylacetate provides an α,β unsaturated ester which is deconjugated by kinetic deprotonation followed by kinetic protonation of the resulting ester dienolate. The ester **1675** is the sole product of this sequence. Hydroboration-oxidation is regio- and stereoselective, but simultaneously causes reduction of the ester to the primary alcohol. Selective reoxidation provides the lactone **1676**. A novel α-methylenation sequence is then employed to prepare **1677** which is converted to aromaticin (**1641**) in four straightforward steps.

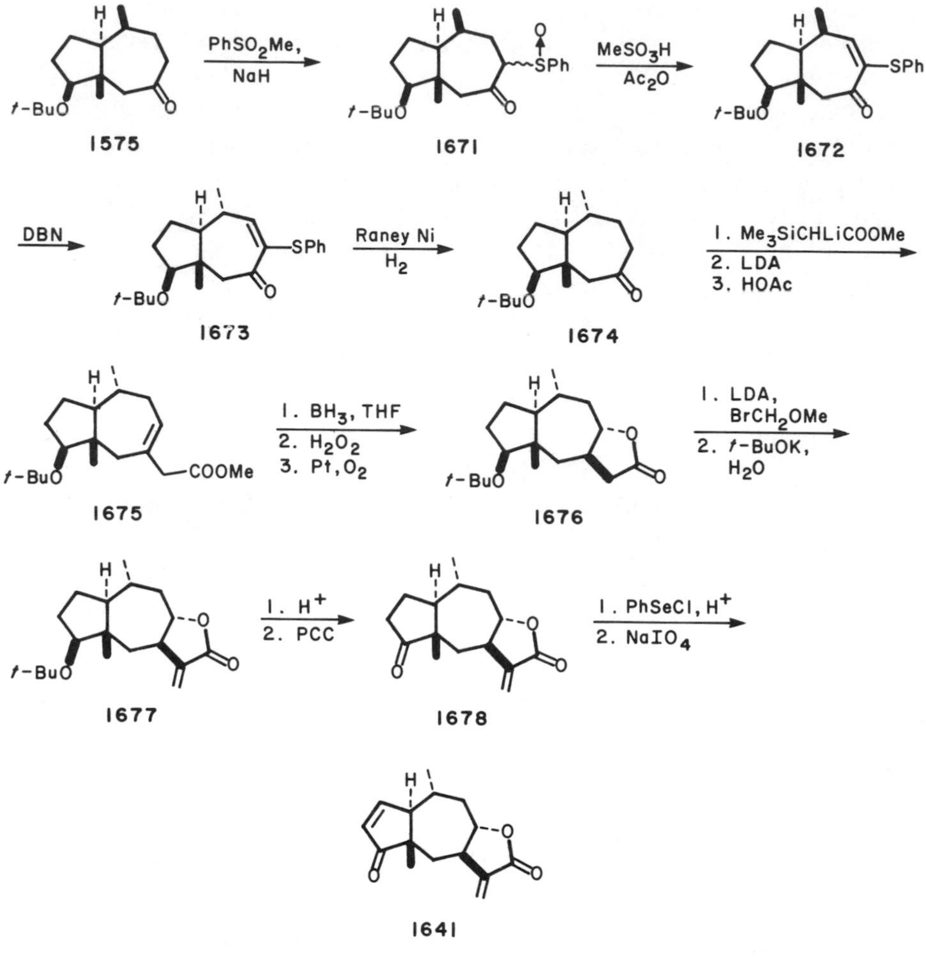

Scheme 205.   Lansbury's (±)-Aromaticin Synthesis

Lansbury has also prepared aromatin[429] (**1642**), which differs from aromaticin (**1641**) only in configuration at C-8. Lactone **1676** is subjected to the sequence shown in Scheme 206 which produces the desired cis lactone **1680**. The conversion of the latter to aromatin is carried out as shown in the scheme.

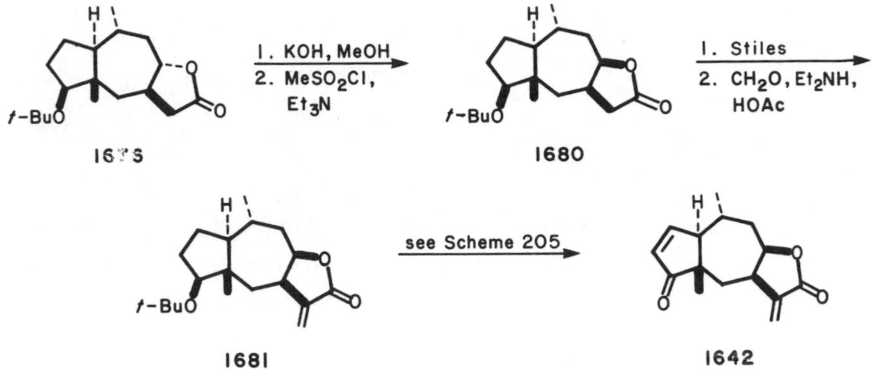

Scheme 206. Lansbury's Aromatin Synthesis

(6) Other Hydroazulenes: Daucene, Daucol, and Carotol.

The three sesquiterpenes daucene (**1682**), daucol (**1683**) and carotol (**1684**) all possess the unusual carotane skeleton, and the biosynthesis of these materials clearly is unique among the hydroazulenes. The more

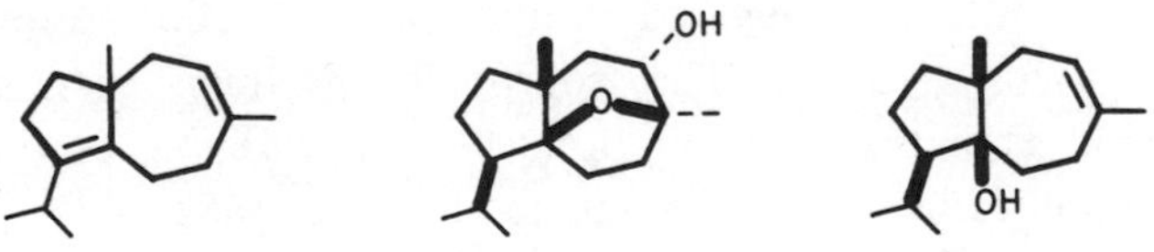

1682 1683 1684

common guaiane derivatives are formed from farnesyl pyrophosphate (**1685**) by initial cyclization to a 10-membered ring intermediate (**1686**) and subsequent formation of the hydroazulene nucleus. On the other

trans-1685   1686   guaiane
                                  (pseudoguaiane)

hand, the carotane skeleton appears to form by simultaneous closure of the five- and seven-membered rings via the hydroazulenic cation **1687**. A 1,3-hydride shift is proposed to lead to **1688** which is the precursor for the isolated materials.[430]

cis-1685   1687   1,3 H-shift   1688

Yamasaki was the first to prepare one of the carotanes, daucene (Scheme 207).[431] Monoepoxidation of (+)-limonene **1689** and mild acid hydrolysis provides the diol **1690**. Periodate cleavage of **1690** gives a keto-aldehyde which is cyclized to the unsaturated aldehyde **1691** in approximately 25% overall yield. The aldol condensation of **1691** with acetone proceeds in 37% yield to trienone **1692**. Selective reduction of the less-substituted, conjugated double bond is effected by treatment with triphenyltin hydride. The isopropenyl group is then saturated to afford **1693**. The ethynyl carbinol derived from **1693** is converted to the allylic alcohol **1198** which cyclizes to daucene **1682** (42% yield) upon exposure to formic acid. Four unidentified products (48% yield) were also reported.

At about the same time Naegeli and Kaiser reported the synthesis of racemic **1198** (see Scheme 144). They found that exposure of **1198** to stannic chloride in benzene affords essentially only two compounds, the

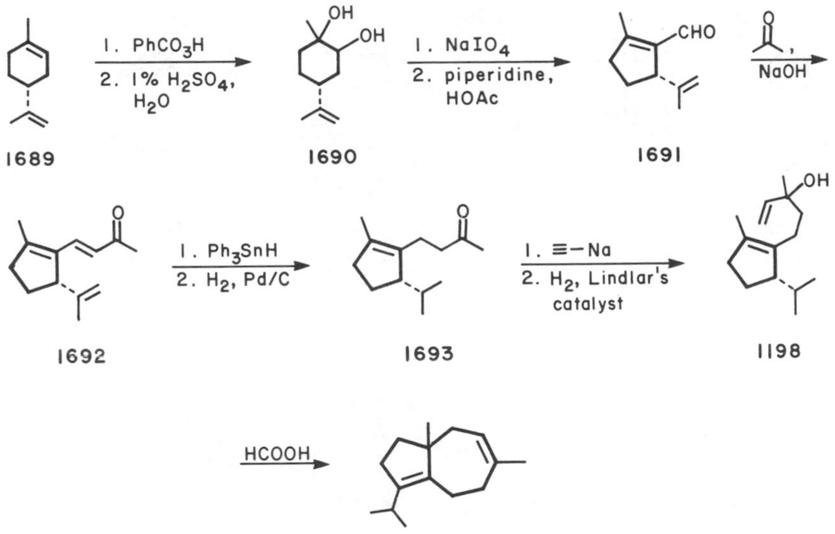

Scheme 207.   Yamasaki's Synthesis of (-)-Daucene

major component (70% yield) being identified as (±)-daucene. The minor component is (±)-acoradiene (**1170**).

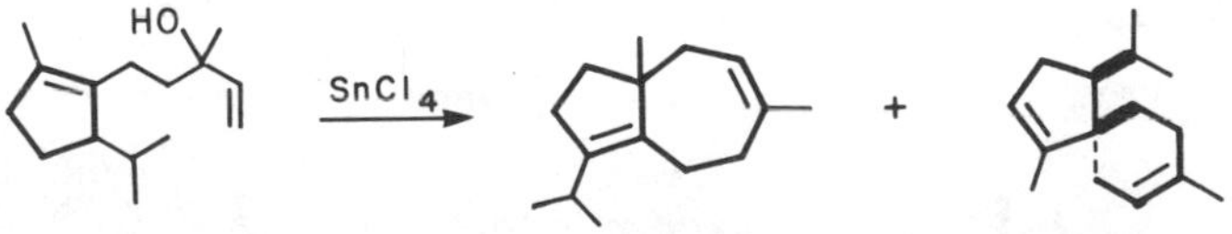

Within a few weeks after the appearance of the Naegeli and Kaiser report, Levisalles and co-workers reported the total syntheses of daucene (**1682**), daucol (**1683**) and carotol (**1684**) (Scheme 208).[432] (+)-Dihydrocarvone (**1694**) is converted in the straightforward manner shown to diol **1697**. After acetylation of the secondary alcohol, oxidation with lead tetraacetate cleaves the isopropyl group to afford, after saponification, the

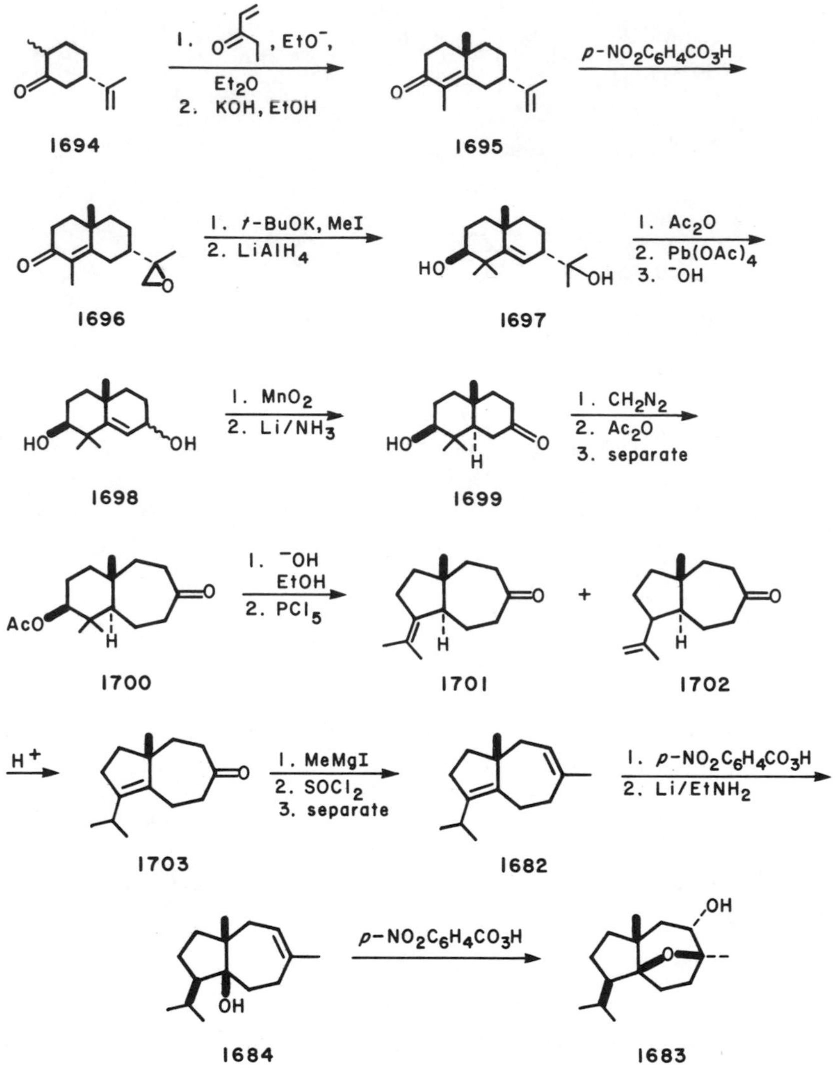

Scheme 208.   Levisalle's Synthesis of Three Carotanes

diol **1698**. The allylic alcohol is selectively oxidized and the resulting enone reduced. Diazomethane ring expansion of **1699** gives an inseparable 2:1 mixture of keto alcohols. Chromatography of the derived acetates, however, allows isolation of the desired ketone **1700** which is the minor component of the mixture. After hydrolysis of the acetate, phosphorus pentachloride induces dehydration which is accompanied by skeletal rearrangement affording a mixture of **1701** and **1702**. These are isomerized by acid to **1703**. Addition of methylmagnesium iodide to **1703** and subsequent dehydration affords a mixture of compounds from which daucene (**1682**) may be isolated in unspecified yield. Epoxidation of daucene followed by reduction affords carotol (**1684**) in approximately 5% yield, some daucene, and four unidentified alcohols. Upon further oxidation, carotol is converted to daucol (**1683**) in unspecified yield.

(7) Other Hydroazulenes: Velleral and Pyrovellerolactone

Several sesquiterpenes with the unique lactarane skeleton have been isolated. Three of these, pyrovellerolactone (**1704**), velleral (**1705**), and vellerolactone (**1706**) have been successfully synthesized by Froborg and Magnusson.[433a,b,c] In referring to the original literature, the reader is

1704   1705   1706

cautioned that the earlier papers dealing with these three substances[433a,b] contain erroneous assignments of the structures of the natural products and of many of the synthetic intermediates. The last paper of the series[433c] provides the correct structures and with that knowledge we have corrected the earlier reports in editing them for inclusion in this section. The first report details the preparation of pyrovellerolactone (**1704**, Scheme 209).[433a] Cycloalkylation of **1707** with bischloromethyl-furan (**1708**) provides **1709** in high yield. After reduction and mesyla-

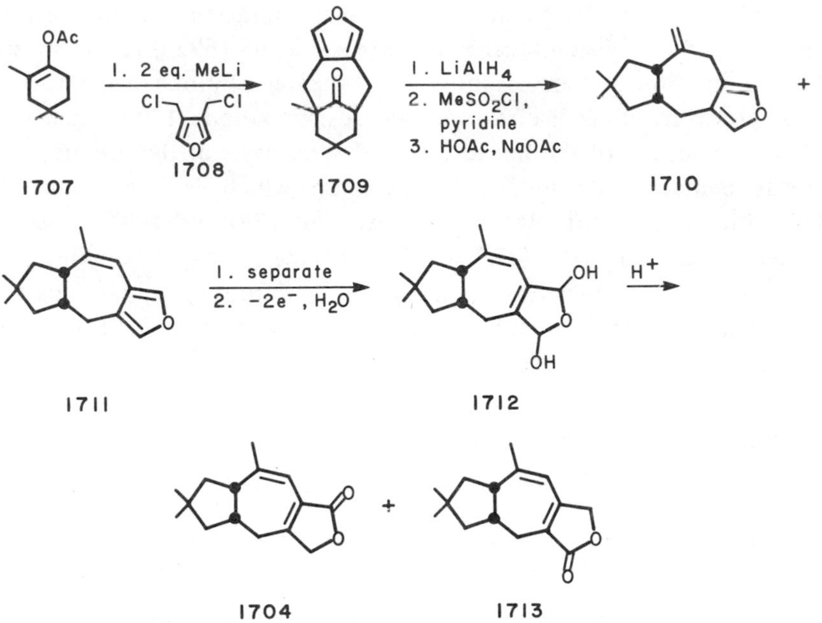

Scheme 209.   Magnusson's First Synthesis of (±)-Pyrovellerolactone

tion, solvolysis leads to a mixture of hydroazulenes **1710** (53%) and **1711** (26%). (This method for synthesizing the hydroazulene ring system was developed by Marshall in connection with his bulnesol synthesis and has been discussed in detail in Volume 2 of this series[434].) For the synthesis of pyrovellerolactone the minor isomer (**1711**) is electrochemically oxidized. Subsequent dehydration of **1712** affords pyrovellerolactone **1704** in 12% yield along with the regioisomer **1713** in approximately 1% yield.

A successful approach to the synthesis of all three natural products was reported in two papers in 1976[433b] and in 1978[433c] (Scheme 210). The known β-keto ester **1714** is condensed with 2-nitro-1-butene and subsequently decarboxylated. Nef reaction gives the diketone **1716** in 47% overall yield from **1714**. Aldol cyclization occurs in fair yield and the

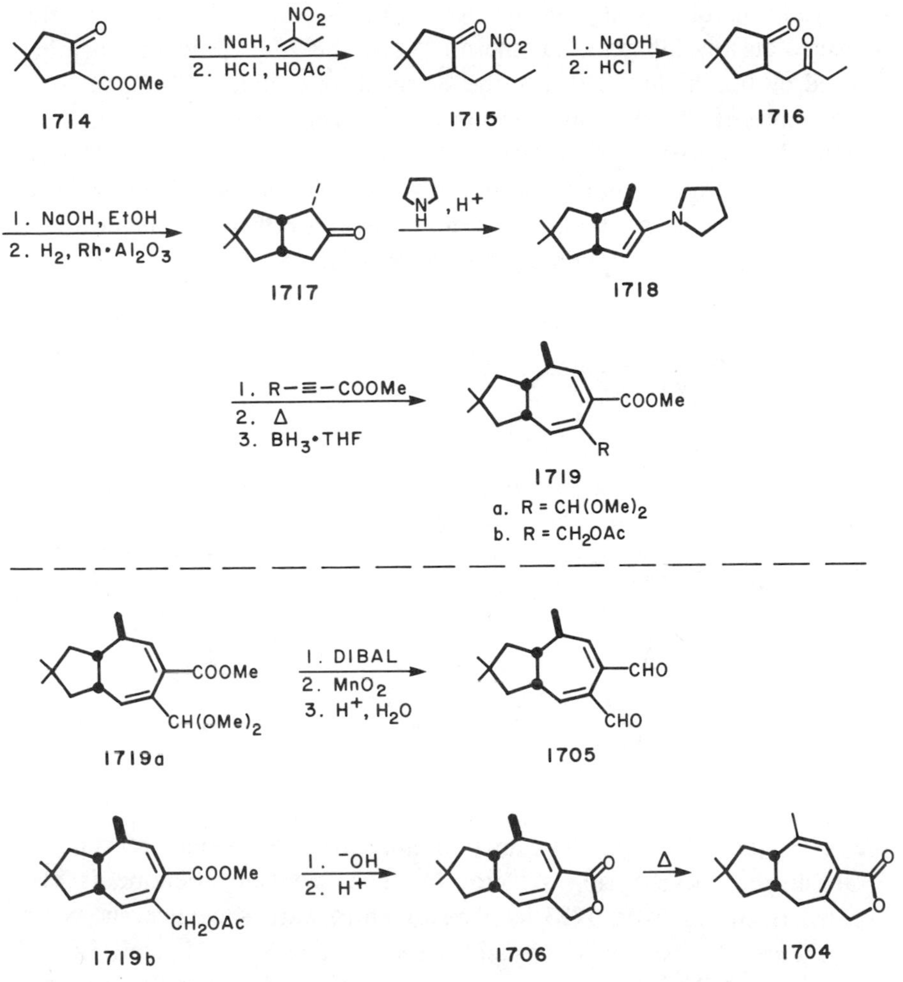

Scheme 210.   Magnusson-Froborg Synthesis of Velleral,
Vellerolactone and Pyrovellerolactone

product is hydrogenated to **1717**. Enamine formation also leads to epi-merization of the methyl group to the more stable exo configuration. Cycloaddition of the appropriate acetylenic ester occurs smoothly, the intermediate cyclobutene is thermolyzed and the product enamine cleaved by borane in THF. For the synthesis of velleral (**1705**), the oxi-dation state of the ester in **1719a** is first adjusted and the acetal subse-quently hydrolyzed. Vellerolactone (**1706**) is derived from **1719b** by sequential treatment with hydroxide, to saponify the esters, and acid to catalyze lactonization. Heating vellerolactone in refluxing toluene for 4 hours causes isomerization to pyrovellerolactone (**1704**).

H. Fused Ring Compounds: 6,7

(1) Himachalenes

Several syntheses of α- and β-himachalene (**1720** and **1721**) and ar-himachalene (**1722**) have been reported. The Wenkert-Naemura syn-

1720   1721   1722

thesis (Scheme 211) employs an intramolecular Diels-Alder reaction to establish the bicyclic ring skeleton.[435] The requisite trienone is ela-borated from aldehyde **1723** by the straightforward sequence shown in the scheme. Cyclization is brought about with Lewis acid catalysis; cis-fused enone **1728** is apparently the only isomer produced in the reaction. This compound reacts with methyllithium to give 7-isohimachalol (**1729**), which is dehydrated by phosphorus oxychloride and pyridine to a 4:1 mixture of α-himachalene (**1720**) and β-himachalene (**1721**). Pure α-himachalene may be prepared by allowing ketone **1728** to react with methylenetriphenylphosphorane.

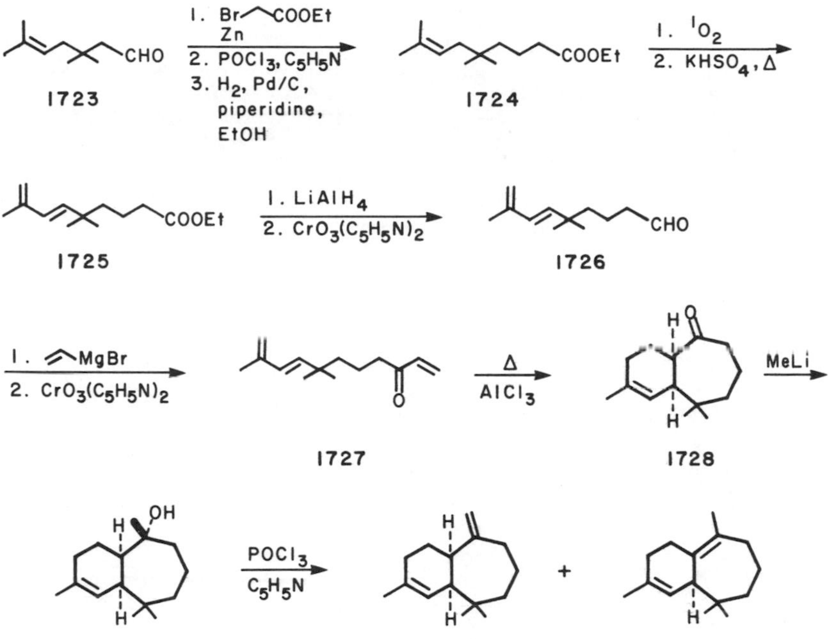

1723 1724 1725 1726 1727 1728 1729 1720 1721

Scheme 211. Wenkert-Naemura Synthesis of the Himachalenes

In 1974, Mehta and Kapoor reported the synthesis summarized in Scheme 212.[436] The synthesis begins with (*E*)-ω-bromolongifolene (**1730**), available in two steps from natural longifolene. Solvolysis of **1730** in anhydrous trifluoroacetic acid gives a mixture of isomeric hydrocarbons and bromo ester **1731** (53% yield). The mechanism of this interesting reaction involves Wagner-Meerwein rearrangement, followed by a transannular hydride shift. Hydrolysis of the ester and oxidation of the resulting secondary alcohol gives ketone **1732** (55% yield), accompanied by 27% of a diketone which is formed by over-oxidation of **1732**. The enolate formed upon treatment of bromo ketone **1732** with dimsyl sodium undergoes fragmentation to give a mixture of three isomeric

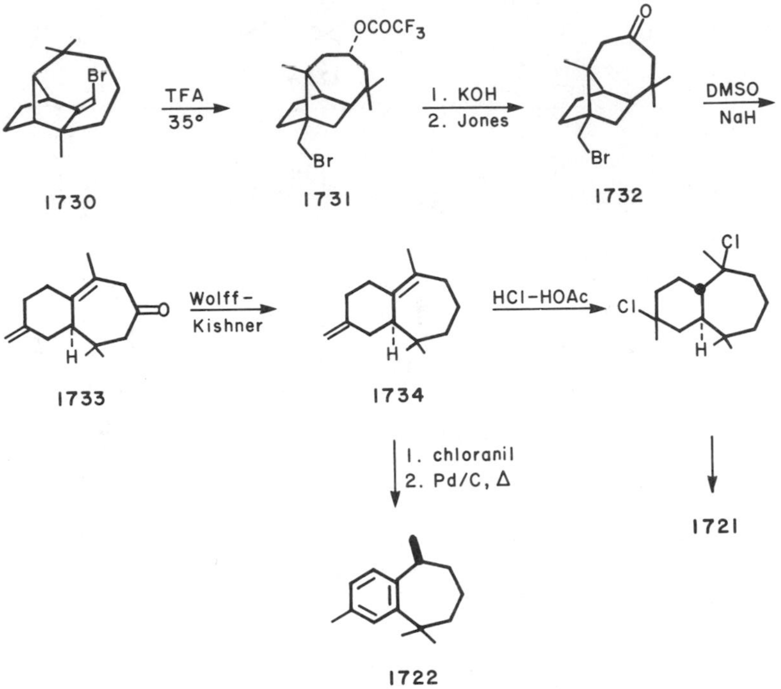

Scheme 212.   Mehta-Kapoor Synthesis of (±)-Himachalene
Dihydrochloride and (±)-ar-Himachalene

dienones, of which isomer **1733** is the major component (75%). Wolff-Kishner reduction of **1733** gives diene **1734**, a himachalene isomer. This compound gives a dihydrochloride, which has previously been converted into β-himachalene (**1721**). The most interesting aspect of this synthesis is the conversion of diene **1734** to (+)-ar-himachalene (**1722**), a transformation which apparently occurs *with complete retention of optical activity*. From this result it can be deduced that the chloranil dehydrogenation of **1734** gives only triene **1735**, since any **1736** formed would lead to loss in optical activity. Furthermore, the aromatization step must take place by suprafacial transfer of hydrogen from one end of the allylic system to the other, in a process not allowed on the basis of the Woodward-Hoffman rules.

1734   1735   1722

                1736

Piers' synthesis of racemic β-himachalene is outlined in Scheme 213.[437] Unsaturated acetal **1737**, obtained from acrolein and 2,2-dimethyl-1,3-propanediol, is allowed to react with bromoform and sodium hydroxide under phase-transfer conditions to obtain dibromocyclopropane **1738**. When a cold solution of **1738** is treated with *n*-butyllithium in the presence of methyl iodide, a 6:1 mixture of isomers **1739** and **1740** is obtained. The isomers are separated and the major isomer (isolated in 74% yield) is hydrolyzed to obtain aldehyde **1741**. After Wittig reaction, bromide **1742** is converted into a cuprate reagent with is added to 3-iodocyclohexenone. Substituted cyclohexenone **1743** is obtained in quantitative yield; thermolysis of **1743** in refluxing xylene provides enone **1744**, which is methylated and selectively hydrogenated to obtain **1746**. The ketone is removed via the derived enol phosphonate (Ireland method) to obtain (±)-β-himachalene (**1721**).

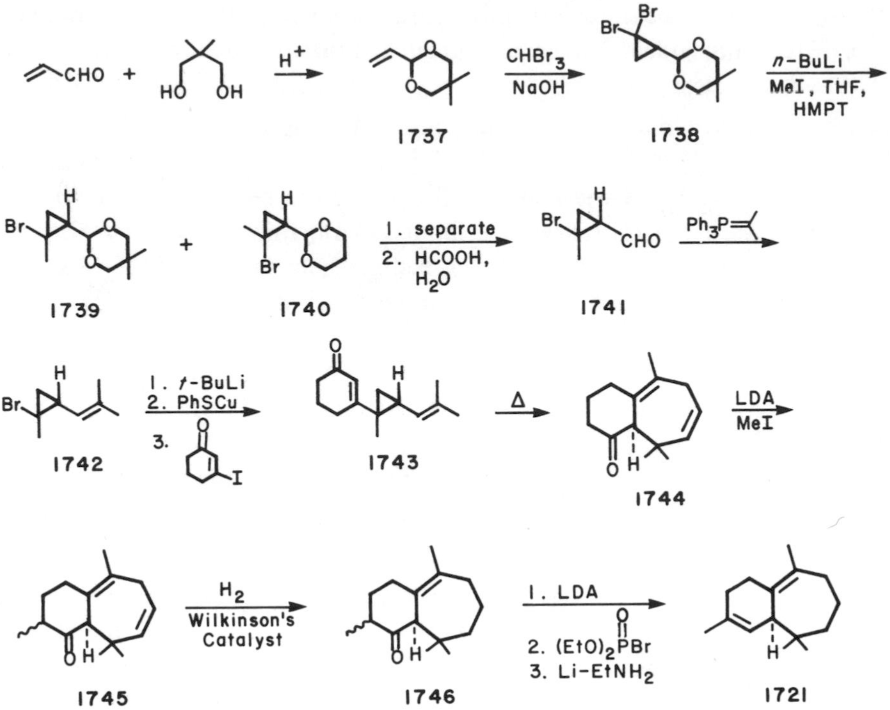

Scheme 213.   Pier's Synthesis of (±)-β-Himachalene

(2) Perforenone

Perforenone (**1752**) is a bicyclic sesquiterpene of marine origin. It has been synthesized by González and co-workers as summarized in Scheme 214.[438] Homoallylic bromide **1747**, obtained from methyl cyclopropyl ketone by Julia's method, is metalated and condensed with methyl vinyl ketone to obtain dienol **1748**. Cyclization of this substance occurs upon treatment with *N*-bromosuccinimide. When bromo ether **1749** is heated

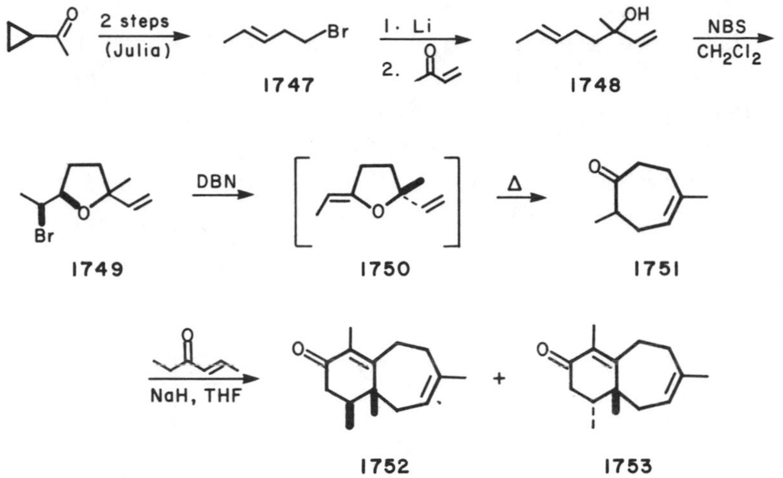

Scheme 214.   González' Synthesis of (±)-Perforenone

with DBN, elimination of HBr occurs to give enol ether **1750** which undergoes Claisen rearrangement to cycloheptenone **1751**. Robinson annelation occurs in poor yield (29%) giving a 2:1 mixture of diastereomeric enones, of which (±)-perforenone (**1752**) is the major isomer. The noteworthy feature of this synthesis is the imaginative method employed to construct the cycloheptenone ring.

(3) Widdrol

Danishefsky has synthesized widdrol (**1764**) as shown in Scheme 215.[439] The synthesis mainly serves as a vehicle to demonstrate the use of a cyclic version of Ireland's ester enolate Claisen rearrangement to generate a cycloheptane ring (i.e., **1761**→**1763**). The synthesis begins with cyclocitral (**1754**), which is converted into unsaturated ester **1757** by well-precedented transformations. Allylic bromination of this substance

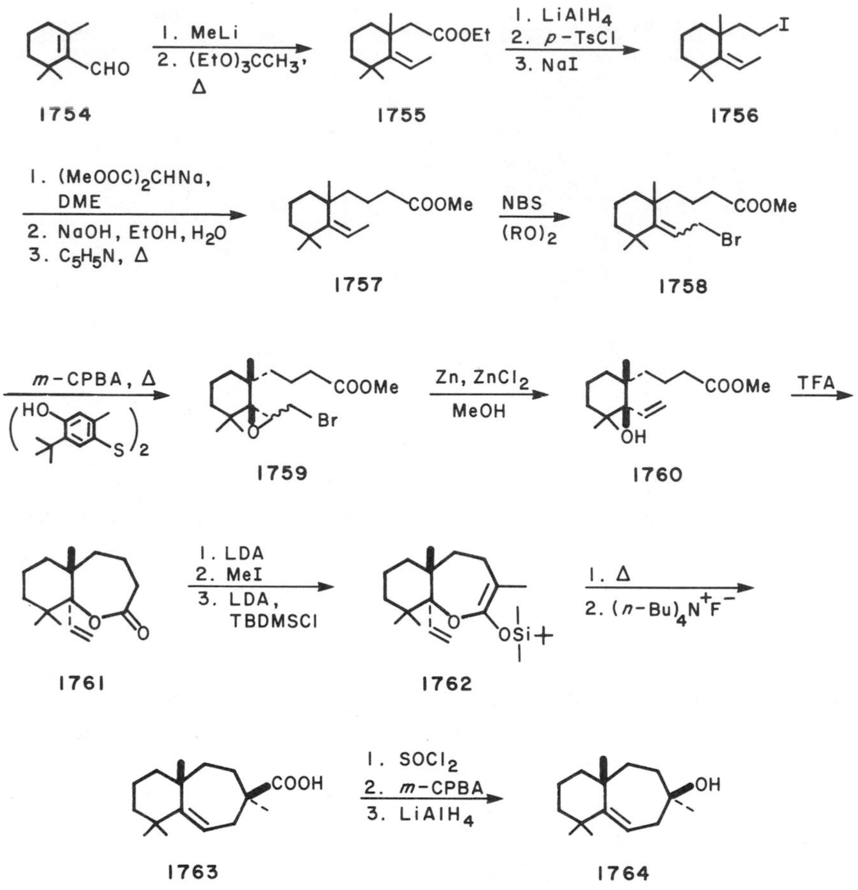

Scheme 215.   Danishefsky-Tsuzuki Synthesis of (±)-Widdrol

gives a diastereomeric mixture of allylic bromides (**1758**), which is oxidized to give epoxides **1759**. Reductive elimination of **1759** gives a single alcohol (**1760**), showing that epoxidation of **1758** occurs only from the face of the double bond cis to the methyl group. The derived lactone **1761** is then subjected to the Ireland reaction to obtain acid **1763**. The carboxy group is transformed into hydroxy by a "carboxy inversion reaction" giving racemic widdrol (**1764**).

I. Fused Ring Systems: 4,9

(1) Isocaryophyllene

In Volume 2 we reported a model study by Gras, Maurin, and Bertrand aimed at the total synthesis of caryophyllene.[440] This work has now been extended to the synthesis of (+)-isocaryophyllene shown in Scheme 216.[441] 1,5-Cyclooctadiene is converted by two successive carbene reactions into alkene **1767**, which is resolved by partial hydroboration with the borane derived from (−)-α-pinene. The dextrorotatory enantiomer

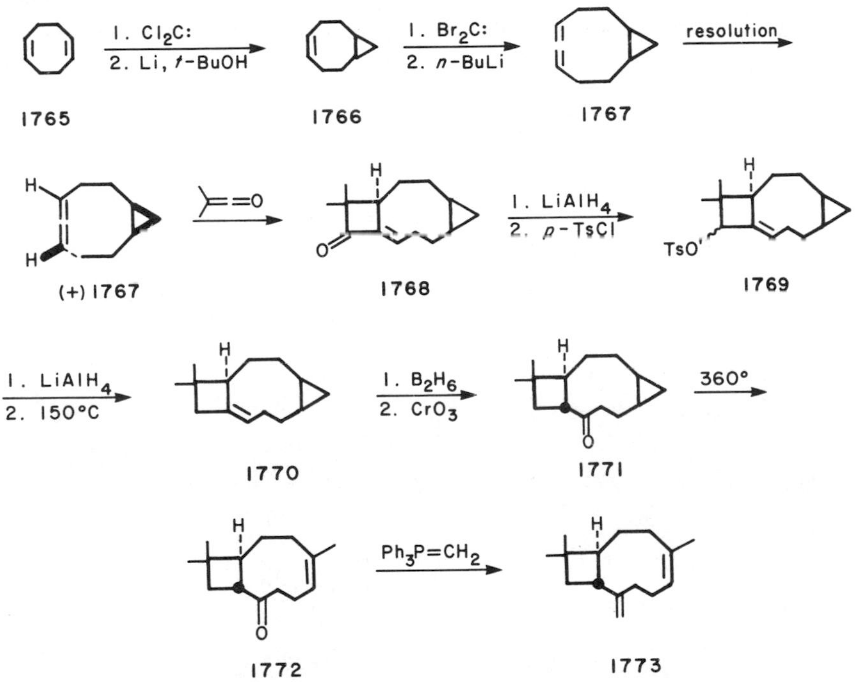

Scheme 216.   Bertrand-Gras Synthesis of (±)-Isocaryophyllene

so produced is subjected to cycloaddition with dimethylketene to obtain tricyclic enone **1768**. Although this compound is a diastereomeric mixture, each diastereomer has the R configuration at the 4,9-ring fusion position. The enone is reduced and the resulting alcohol converted to tosylate **1769**. Reduction of this allylic tosylate gives a mixture of double bond isomers. However, upon preparative glpc at 150°C, double bond isomerization occurs giving principally isomer **1770**. The derived ketone **1771** is pyrolyzed at 360°C to give enone **1772**. Wittig methylenation gives (+)-isocaryophyllene. A key reaction in this synthesis is the thermal rearrangement of cyclopropane **1771** to olefin **1772**. The isomerization actually affords three products, but **1772** is the major product and is isolated in 56% yield.

Devaprabhakara and co-workers have capitalized on the Gras-Bertrand approach by the modification outlined in Scheme 217.[442] In this approach, 1,5-cyclooctadiene is first converted to allene **1774**, which is converted into dienone **1775**. Photoannelation of this compound with

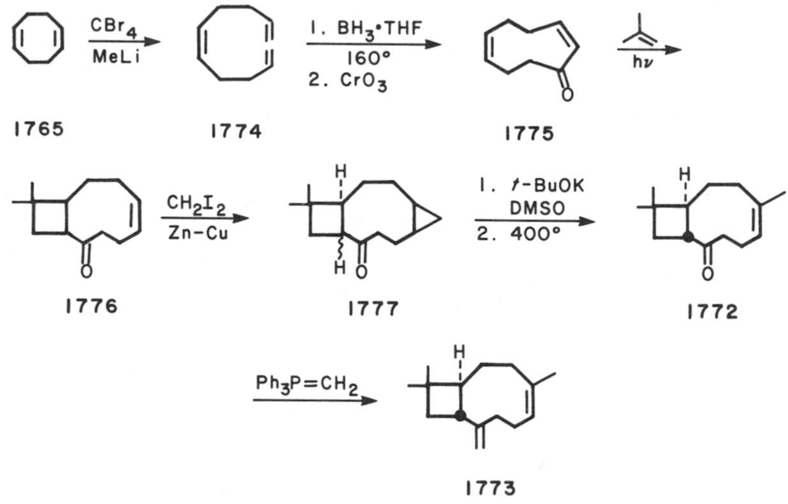

Scheme 217.   Devaprabhakara's Synthesis of (±)-Isocaryophyllene

isobutylene gives **1776** which is subjected to Simmons-Smith methylenation to obtain ketone **1777** (presumably a mixture of **1771** and its cis-fused isomer). The synthesis is completed by the Gras-Bertrand method. Devaprabhakara has reported an alternate synthesis of enone **1775**.[443]

6. TRICARBOCYCLIC AND TETRACARBOCYCLIC SESQUITERPENES

A. Fused Systems

In this section, we will discuss syntheses of tricarbocyclic systems in which the three rings are fused side-to-side. Syntheses of cubebol (**1**) and the cubebenes (**2** and **3**), which were discussed in Volume 2 of this series, have been reported in full.[444-446] New syntheses which will be discussed are those of illudol (**4**), the protoilludanols **5** and **6**, the protoilludenes **7** and **8**, marasmic acid (**9**), hirsutene (**10**), hirsutic acid C (**11**), coriolin (**12**) and isocomene (**13**).

8   9   10   11

12   13

(1) Illudol, Protoilludanols, Protoilludenes

In Volume 2 of this series we reported a synthesis of the protoilludane skeleton from the group of T. Matsumoto at Hokkaido University in Sapporo.[447] The synthesis has subsequently been extended to a synthesis of (±)-illudol (4) as shown in Scheme 1.[448] Keto ester 14[444] reacts with allylmagnesium bromide to give a diol (15) which is transformed by the three-step sequence shown into dihydroxy ester 16. Cleavage of the diol gives an intermediate diketo ester which closes to give a mixture of aldol products 17 and 18. Dehydration provides an unsaturated keto ester (19) which is reduced by sodium hydridobis(2-methoxyethoxy)aluminate to obtain diol 20 (47%) accompanied by 17% of its diastereomer. Treatment of diol 20 with acetone and acid yields a keto ketal (21) which is reduced and deketalized to obtain (±)-illudol (4). The stereochemistry of the two secondary alcohols was assigned on the basis of reduction of the corresponding ketones 19 and 21 from the less hindered direction in each case. The synthesis provided the first concrete evidence in favor of the relative stereochemistry of illudol at these centers.

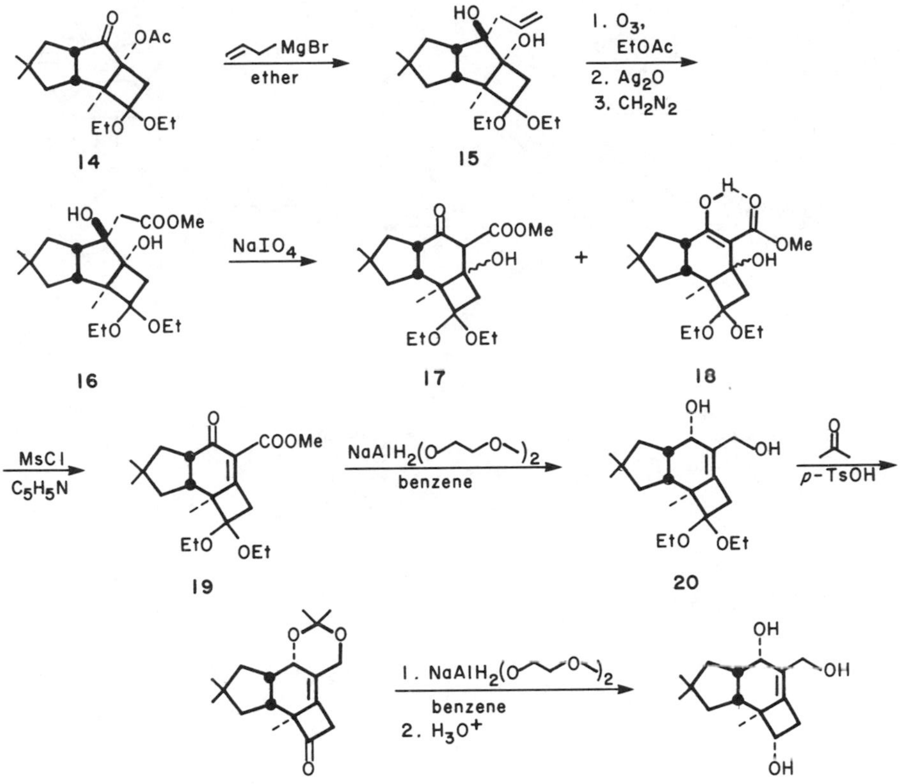

Scheme 1.   Matsumoto's Synthesis of (±)-Illudol

In 1975 the Matsumoto group reported a synthesis of the protoillu-danols **5** and **6** and the protoilludene **7** (Scheme 2).[449] Enone **22**[447] undergoes photochemical cycloaddition with ethylene to provide the cis,anti,cis tricyclic ketone **23** in 75% yield. Wittig methylenation of **23**

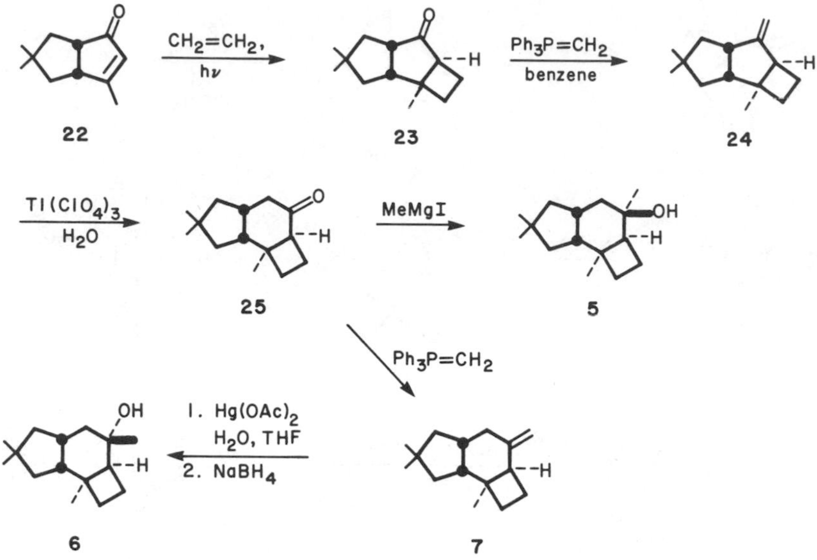

Scheme 2.   Matsumoto's Synthesis of Protoilludanes

gives alkene **24**, which is ring-expanded with thallium perchlorate to obtain ketone **25**. This substance is converted into alcohols **5** and **6** and alkene **7** by the straightforward methods illustrated in the scheme.

Takeshita and co-workers have also investigated the synthesis of illudanes by photoaddition of ethylene and 1,1-dimethoxyethylene to a hydrindenone.[450] In the Takeshita approach (Scheme 3) the hydrindenone **27** is prepared by DeMayo's method. The four-membered ring is added by photoaddition of 1,1-dimethoxyethylene. Keto ketal **28** is treated with methylmagnesium iodide and the resulting alcohol is converted into unsaturated thioketal **5** by treatment with 1,2-ethanedithiol

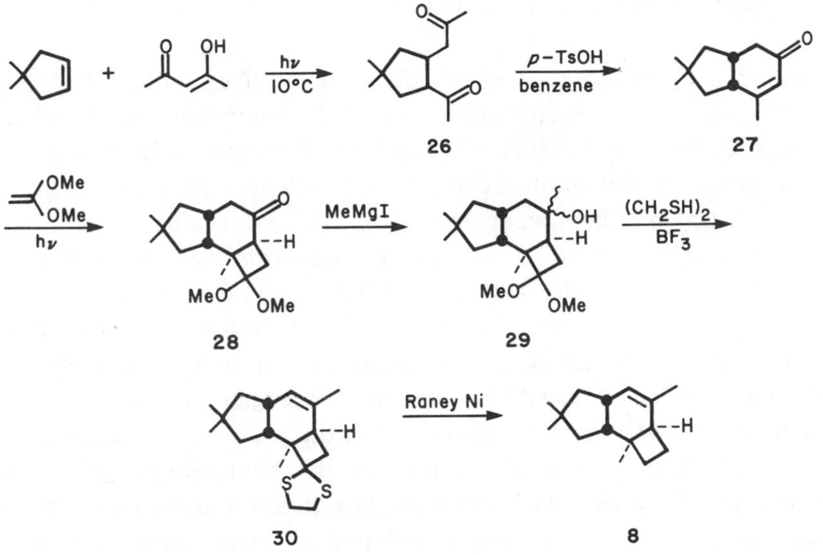

Scheme 3.   Takeshita's Synthesis of Protoillud-7-ene

and boron trifluoride. Desulfurization of **30** provides protoilludene **8**. Related studies showed that trans-fused hydrindenone **31** undergoes photoaddition to ethylene but that its cis-fused isomer **32** does not.

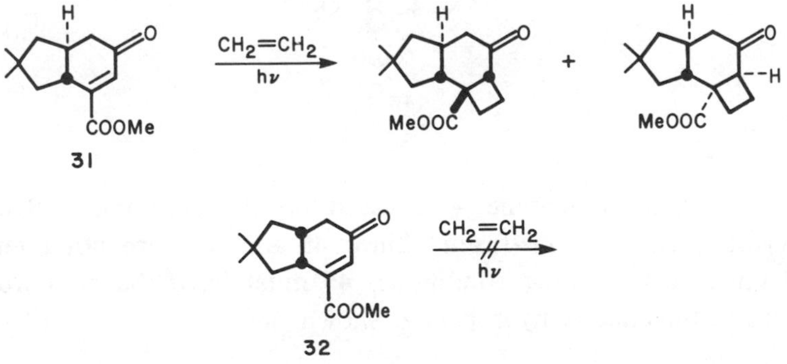

(2) Marasmic Acid and Isomarasmic Acid

Marasmic acid (9) has been the synthetic target of a number of investigations. De Mayo and co-workers took the imaginative photochemical approach outlined in Scheme 4.[451] Spiro[4.2]heptene 33 is irradiated in the presence of the enol acetate of 1,2-cyclopentanedione to obtain adduct 35, which is hydrogenolyzed to 36. After introduction of a double bond in the cyclopentanone ring, the acyloin is rearranged by treatment with 1% methanolic sodium hydroxide. The product of this rearrangement, hydroxy ketone 38, is cleaved with lead tetraacetate. After esterification, the enone system is regenerated by treatment of 39 with acid. Enone 40 is subjected to oxidation with singlet oxygen to obtain a hydroperoxide (41) which is photolyzed in the presence of vinylene carbonate. Catalytic hydrogenation reduces the hydroperoxy group and ketone carbonyl, giving 42. The secondary hydroxy is protected as the pivalate and the tertiary alcohol dehydrated to obtain unsaturated ester 43. Addition of diazomethane to 43 affords a pyrazoline which loses nitrogen upon irradiation to give 44. After saponification of the carbonate linkage, cleavage of the resulting glycol, and elimination of pivalic acid methyl isomarasmate (45) is obtained. Although the stereostructure of 45 at the allylic position was not rigorously defined, the compound is a single isomer and is definitely different from methyl marasmate (46), obtained from authentic marasmic acid. The stereo-

46

chemistry shown in Scheme 4 is based on the assumption that the hydrindane skeleton is cis-fused. Since 45 and 46 were not interconverted under equilibrating conditions, 45 must have the cyclopropane ring trans to the angular hydrogen, as shown.

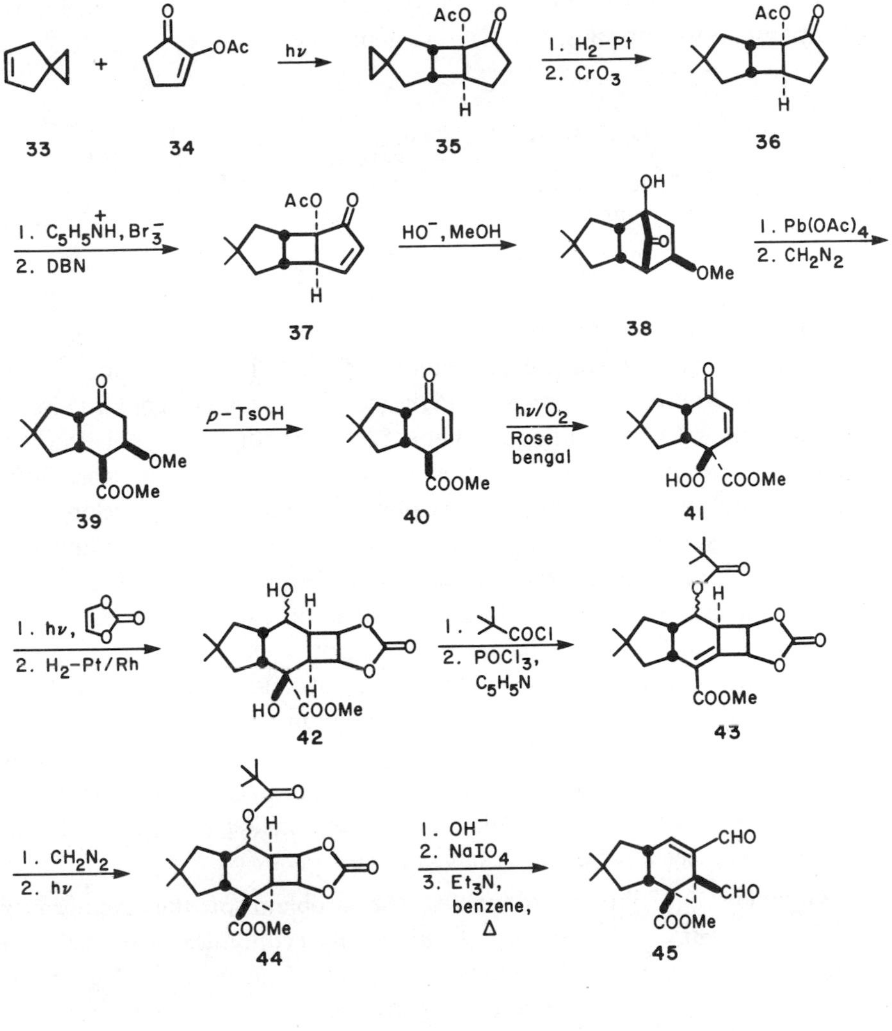

Scheme 4.   de Mayo's Synthesis of Methyl (±)-Isomarasmate

In 1973, Wilson and Turner reported an approach to marasmic acid which utilized as a key reaction addition of diazomethane to a bicyclic triester (47).[452] After photolysis of the adduct, a single cyclopropane was formed and was presumed to have structure 48.

In 1976, Greenlee and Woodward reported an investigation which casts doubt on the Wilson-Turner stereochemical interpretations.[453] The Greenlee-Woodward approach (Scheme 5) involved construction of intermediate 53 by addition of diene 52 to dimethyl acetylenedicarboxylate. As in the Wilson-Turner work, Greenlee and Woodward found that photolysis of the initial pyrazoline affords a single cyclopropane, which was subjected to straightforward manipulations to obtain anhydride 55. Reduction with disodium tetracarbonylferrate gives a 1:1 mixture of lactols, which were separated and converted into their acetates. Hydrolysis of one of these isomers gave isomarasmic acid (58), which was converted by esterification into methyl isomarasmate, identical with the de Mayo sample. Thus, it is clear that addition of diazomethane to the α,β-unsaturated ester shows a strong preference for addition anti to the angular hydrogen, both in de Mayo's intermediate 43 and in the Greenlee-Woodward intermediate 53. A convincing explanation for this highly stereoselective addition has not been advanced.

Greenlee and Woodward solved the problem of the cyclopropane stereochemistry with the elegant and highly economical synthesis summarized in Scheme 6.[453] Reduction of dienal 51 gives dienol 59, which is treated with bromomethylmaleic anhydride at room temperature to obtain a 1:1 mixture of lactone acids 60 and 61. The derived t-butyl esters are treated with potassium t-butoxide to obtain a single cyclopropane (62), which is produced in 44% overall yield from dienol 59. This material is converted into (±)-marasmic acid (9) by the sequence of

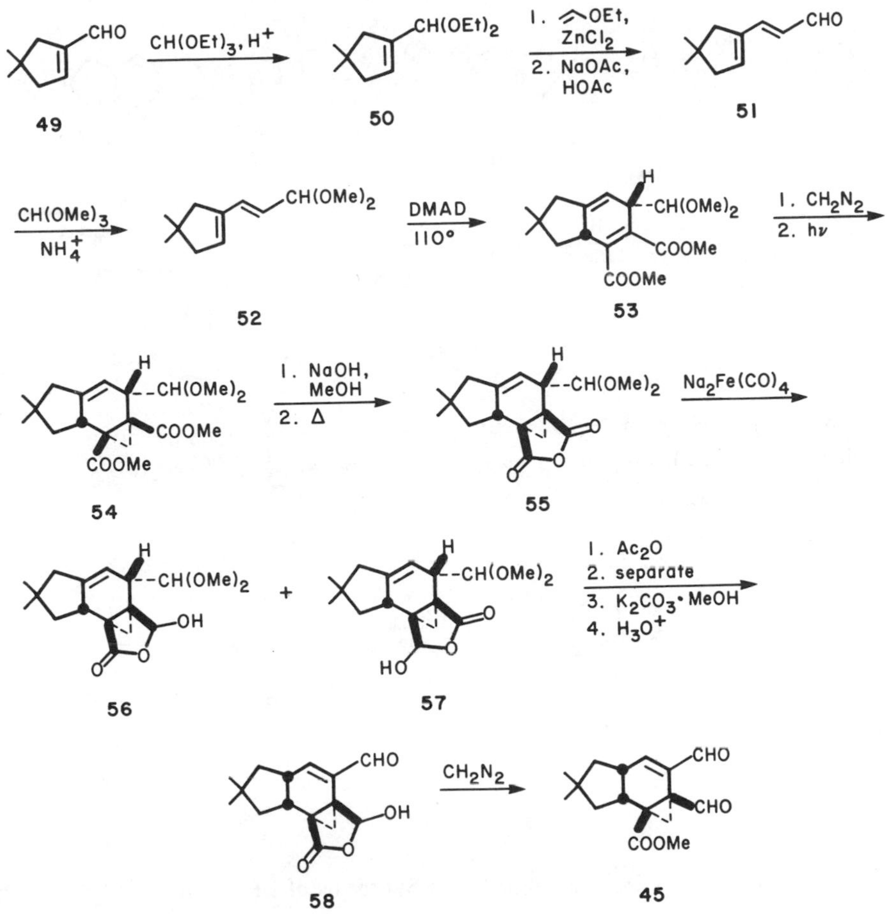

Scheme 5.   Greenlee-Woodward Synthesis of (±)-Isomarasmic Acid

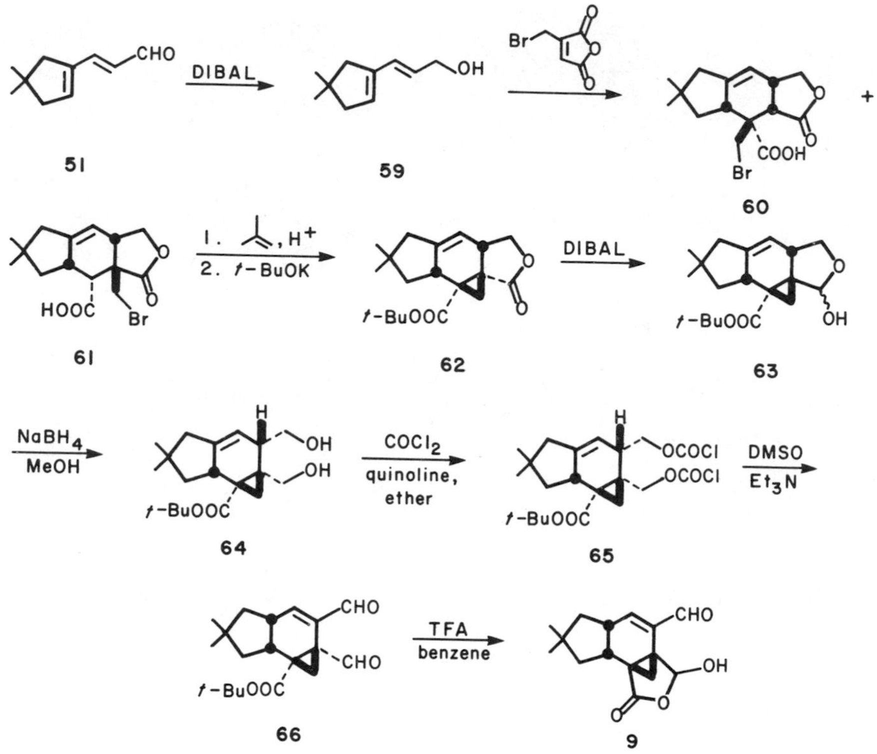

Scheme 6.   Greenlee-Woodward Synthesis of (±)-Marasmic Acid

steps shown in the scheme. The key step in the successful Greenlee-Woodward synthesis presumably involves intramolecular Diels-Alder reaction of esters **67** and **68**, which are formed by reaction of alcohol **59** with the anhydride. The fact that both adducts eventually afford marasmic acid shows that both cycloadditions proceed by way of an endo transition state:

HOOC

Br   H

67 $\longrightarrow$ **60**

HOOC

H   Br

68 $\longrightarrow$ **61**

In all, the Greenlee-Woodward synthesis requires only 12 steps from the known aldehyde **49** and is completely regio- and stereoselective.

In 1980, Boeckman and Ko reported the marasmic acid synthesis which is summarized in Scheme 7.[454] Unsaturated aldehyde **69** is transformed into the monoprotected dimethyladipaldehyde **70**, which is transformed by Wadsworth-Emmons reaction into unsaturated ester **71**. Further elaboration of this substance provides aldehyde **74**, which is condensed with phosphonate **75** to obtain the crucial intermediate **76**. The intramolecular Diels-Alder reaction is achieved by heating **76** in toluene in a sealed tube. Under these conditions isomers **77** and **78** are produced in a 1:1 ratio. This unfortunate situation is a result of poor stereoselectivity in the Diels-Alder reaction, which appears to proceed by way of endo and exo transition states with equal facility. The isomers are separated after conjugation of the double bond of **77** (the isomer resulting from exo addition conjugates under the conditions of cycloaddition). After conversion of the primary acetoxy group into a better leaving group, the

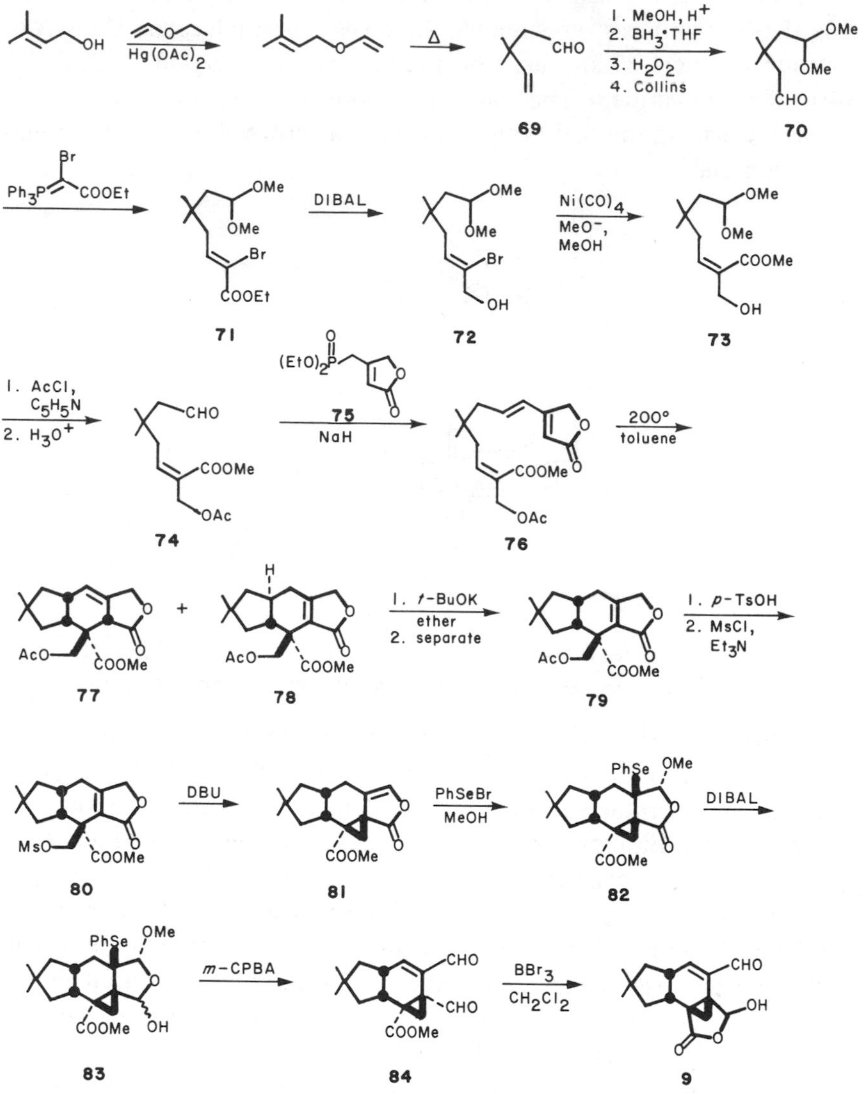

Scheme 7.   Boeckmann-Ko Synthesis of (±)-Marasmic Acid

cyclopropane ring is formed in the same manner as was employed by Greenlee and Woodward. The remainder of the synthesis involves some rather nice functional group manipulations, resulting in the eventual synthesis of (±)-marasmic acid (9). The Boeckman-Ko synthesis requires a total of 20 steps and requires separation of diastereomers after the Diels-Alder step. Although this isomer separation is necessary, Boeckman and Ko were able to convert isomer **78** by an alternate sequence of reactions also into (±)-marasmic acid.

(3) Hirsutic Acid, Isohirsutic Acid, Hirsutene, and Coriolin

The first published work directed toward the synthesis of hirsutane sesquiterpenes came from Lansbury in 1971 (Scheme 8).[455] Alkylation of the known 3-methyl-3-methoxycarbonylcyclopentanone with 3-bromo-2-butanone provides diketo esters **86** and **87** in 50% and 10% yields, respectively. Base-catalyzed aldolization of the major isomer is accompanied by rapid double bond isomerization, yielding keto acid **88** as a diastereomeric mixture. After esterification and hydrogenation of the double bond, ketone **89** is subjected to Claisen alkylation using (Z)-2-chloro-2-butenol. Isomers **91** and **92** are produced in 55% and 36% isolated yields. The major isomer is treated with 90% sulfuric acid to hydrolyze the vinyl chloride and then with potassium t-butoxide to effect cyclization. Acidic dehydration of aldol **93** affords enone ester **94**. Although compound **94** is a potential precursor to hirsutic acid C, the substance has apparently not been converted further.

Subsequently, Lansbury and co-workers reported a modification of this basic approach which provides isohirsutic acid (**99**), a substance formed by rearrangement of hirsutic acid C.[456] The synthesis begins with the keto ester **89**, which is alkylated by the Claisen route with allylic alcohol **95** to obtain a 3:2 mixture of keto esters **96** and **97** (Scheme 9). After separation, the major isomer (**96**) is treated with 90% sulfuric acid. Hydrolysis of the vinyl chloride is followed by elimination of propanethiol, leading to enedione **98**. This material is treated with mild base to effect ring closure to (±)-isohirsutic acid (**99**). There is some

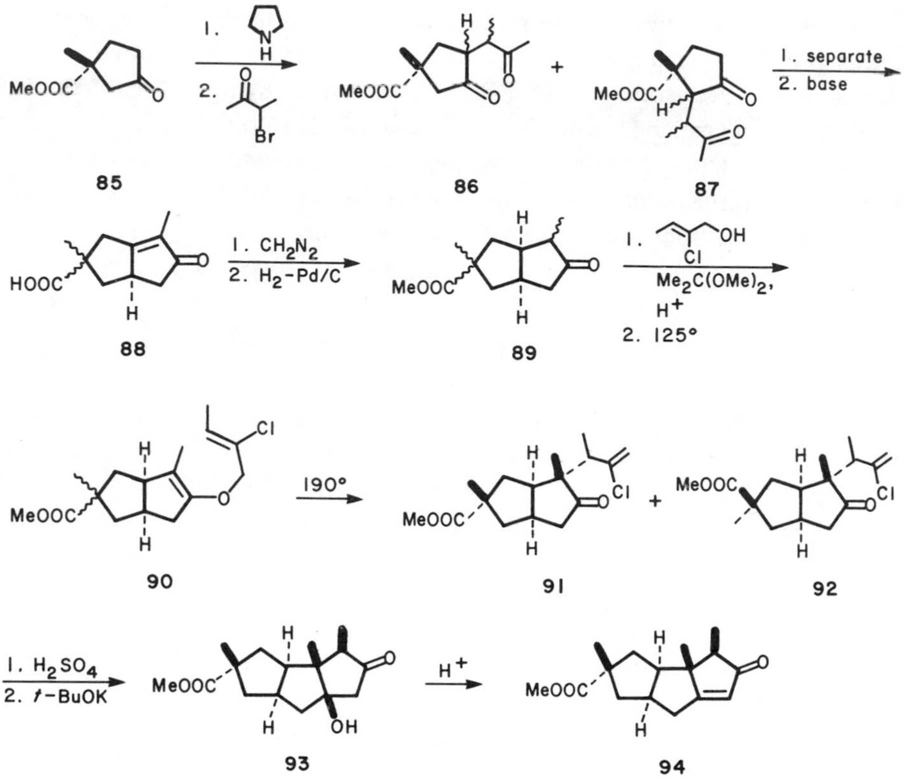

Scheme 8.   Lansbury's Approach to Hirsutic Acid

uncertainty regarding the identity of this material with isohirsutic acid derived from natural sources. Lansbury and co-workers report that the spectral data obtained from synthetic **99** "...correspond only in part with those reported". However, the naturally derived isohirsutic acid may have been impure.

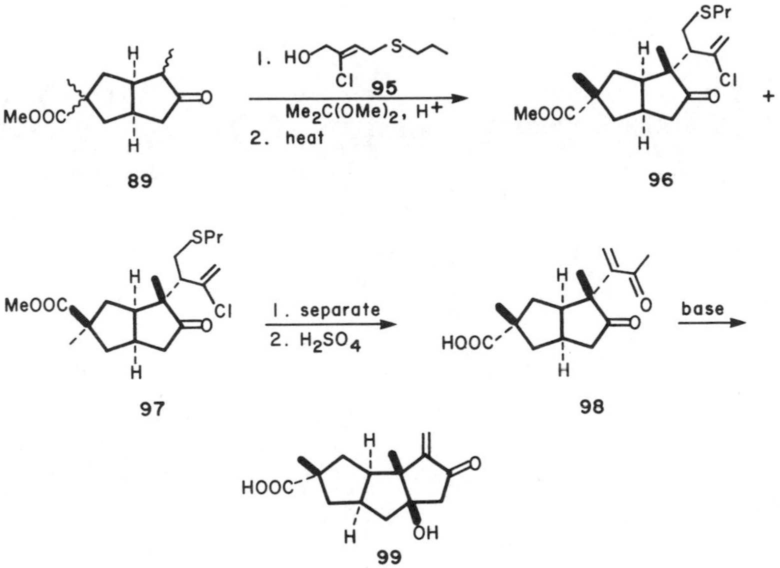

1. HO— ... S— ... Cl **95**

Me₂C(OMe)₂, H⁺

2. heat

MeOOC

89

SPr

MeOOC ... Cl

+

96

SPr

MeOOC ... Cl

97

1. separate

2. H₂SO₄

HOOC

98

base

HOOC ... OH

99

Scheme 9.   Lansbury's Synthesis of (±)-Isohirsutic Acid

The first successful synthesis of hirsutic acid C came from Matsumoto and co-workers in 1974.[457,458] As shown in Scheme 10, the dione monoketal **100** is methylenated and the resulting alkene allowed to react with the carbenoid obtained from methyllithium and α,α-dichloromethyl methyl ether. Cyclopropanes **101** and **102** are obtained in a ratio of 2:1. The isomers are separated and the major isomer (**101**) is treated with acid to effect the ring opening of the methoxycyclopropane unit. The resulting aldehyde is oxidized and esterified to obtain keto ester **103**. The cyclopentanone is indirectly methylated and the product alkylated with methallyl chloride to obtain, after ozonolytic cleavage, dione **105**. Aldolization is brought about with potassium *t*-butoxide, leading to

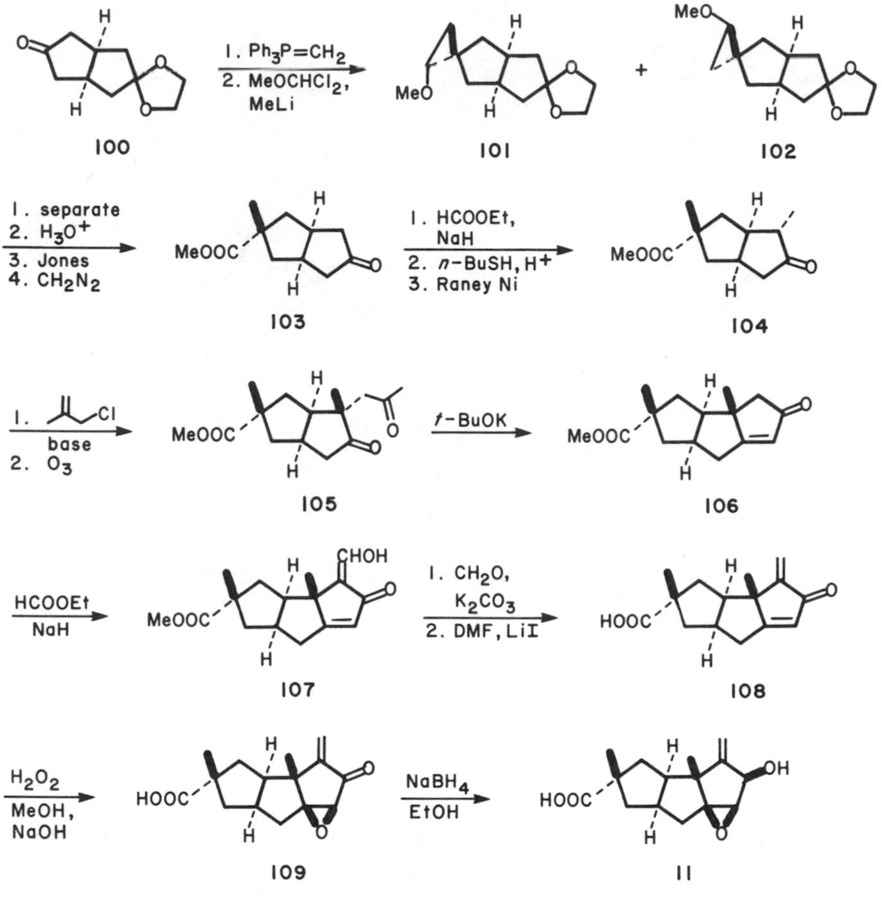

Scheme 10.   Matsumoto's Synthesis of (±)-Hirsutic Acid C

enone ester **106**. Installation of the α-methylene function is accomplished by formylation, followed by condensation with formaldehyde. After nucleophilic cleavage of the ester function, dienone **108** is treated with hydrogen peroxide in alkaline methanol. The major product of this epoxidation, complicatic acid (**109**, 60%) is accompanied by 20% of a β-methoxy ketone, which can be converted back into **108** by treatment with sodium hydroxide. Reduction of **109** with sodium borohydride provides ($\pm$)-hirsutic acid (**11**). The Matsumoto synthesis is reasonably efficient (16 steps from ketone **100**) and is marred by the necessity of only one isomer separation. However, the non-stereocontrolled step occurs early in the synthesis.

Trost and co-workers have also synthesized ($\pm$)-hirsutic acid C, by the route outlined in Scheme 11.[459] In the Trost synthesis, the desired relative stereochemistry is achieved by employing a tetracyclic intermediate (**118**) which is ozonized to generate the tricyclic system. Compound **118** is prepared from the known cyano ketal **110** by the multi-step sequence illustrated in the scheme. The most noteworthy feature of this series of steps is the use of Stettler's method to close the cyclopentanone ring (**115**→**117**). Ozonolysis of **118** provides tricyclic dialdehyde **119**, which is necessarily of the correct stereostructure, since only one stereoisomer of **118** is capable of existence. After reduction of the aldehyde groups to methyls, the lactone ring of **120** is manipulated so as to obtain dione **105**, an intermediate in the Matsumoto synthesis. As has already been stated, the Trost synthesis is of necessity stereospecific. However, a high price is paid, as the synthesis requires 33 steps, more than twice the number necessary in the Matsumoto synthesis.

Tatsuta and co-workers have developed a route to hirsutanes which has been applied in the synthesis of ($\pm$)-hirsutene (**10**).[460] As shown in Scheme 12, 2,2-dimethylcyclohexanone is converted into unsaturated ketal **121** by the Garbisch procedure. After introduction of an allylic acetoxy group, **122** is irradiated in the presence of the enol acetate of 2-methyl-1,3-cyclopentanedione to obtain tricyclic adduct **123** in 35% yield. After reduction of the carbonyl and protection of the resulting secondary alcohol, the two ester groups are hydrolyzed. The key reaction of the Tatsuta approach is solvolytic rearrangement of the mono-

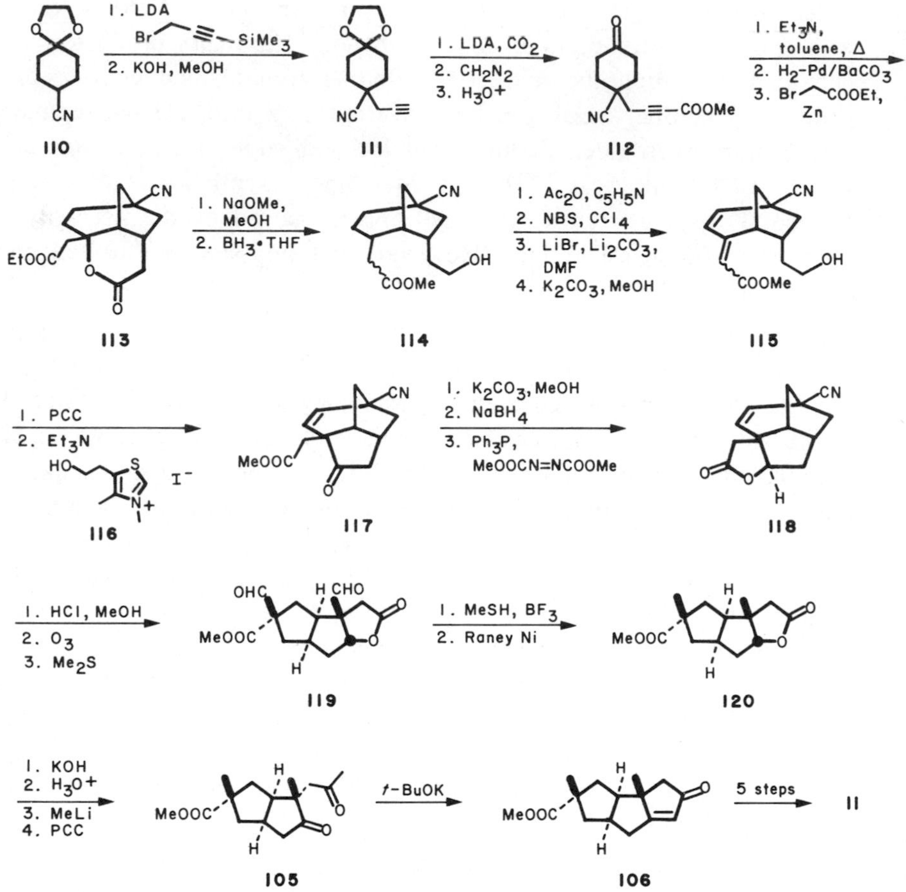

Scheme 11.   Trost's Synthesis of (±)-Hirsutic Acid C

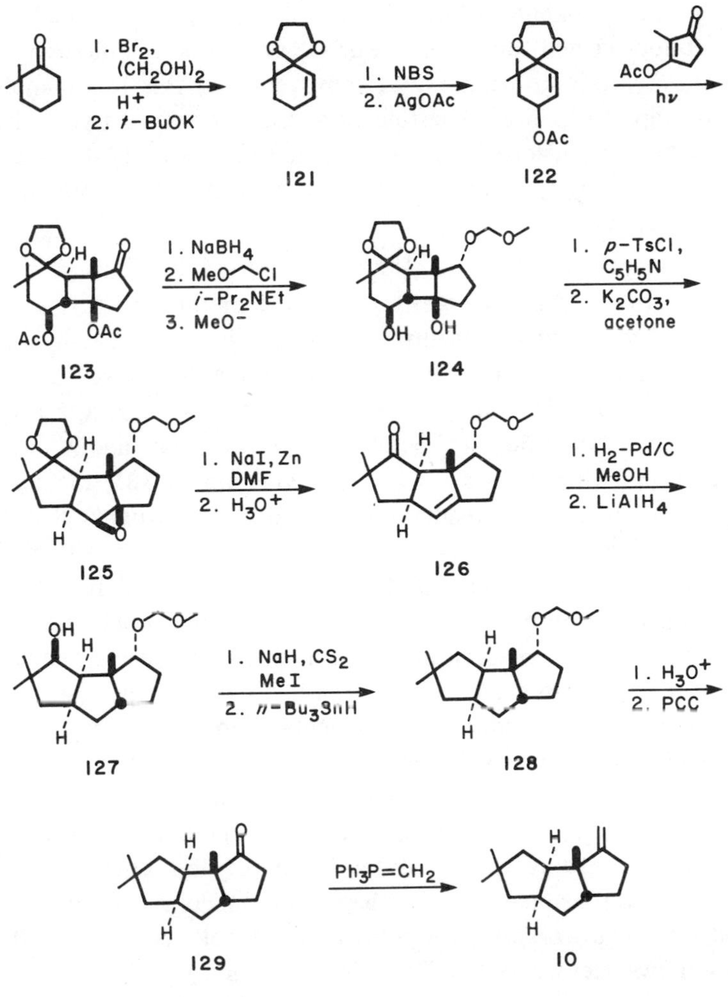

Scheme 12.   Tatsuta's Synthesis of (±)-Hirsutene

tosylate of diol **124**, which provides epoxide **125**. Deoxygenation of this material gives an alkene (**126**), which is hydrogenated and reduced to obtain monoprotected diol **127**. For a synthesis of hirsutene, the superfluous hydroxy group must be removed, and this is accomplished by the Barton procedure. Deprotection and oxidation give a ketone (**129**) which is methylenated to obtain (±)-hirsutene (**10**). Although this synthesis is long (19 steps), it is easily adapted to the more challenging target, coriolin.

Such an adaptation (Scheme 13) was reported by Tatsuta in early 1980.[461] Epoxide **125** is deoxygenated and the methoxymethyl and ketal groups removed to obtain unsaturated keto alcohol **130**. Catalytic osmylation followed by ketalization of the resulting diol gives **131**. The stereoselectivity of this oxidation results from the fact that attack on the other face of the double bond would give a highly strained *trans*-bicyclo[3.3.0]octane system. After oxidation of **131**, the diketone **132** is *bis*-sulfenylated by reaction with methyl *o*-nitrophenyl disulfide and the resulting dithioketal is transetherified by reaction with thallium trinitrate in methanol. Compound **133** is thereby obtained in 54% overall yield for the two-step process. After reaction of the more reactive carbonyl group with methyllithium the remaining ketone is reduced by lithium in ammonia to establish the desired exo stereochemistry at C-3. Hydrolysis of the acetonide, selective acetylation of the two secondary alcohols and dehydration of the tertiary alcohols give dienone **137**. After removal of the protecting groups, the epoxide functions are installed by reaction with alkaline hydrogen peroxide. Tatsuta and co-workers make no comment regarding the stereoselectivity of the final epoxidation. However, it is presumably poor, based on analogy with the results obtained by other workers (see below). However, as the quoted yield for the last two steps is only 60%, it is reasonable to assume that other stereoisomers may also have been produced. In all, the Tatsuta synthesis requires 24 steps from 2,2-dimethylcyclohexanone.

Danishefsky and co-workers have reported the synthesis of (±)-coriolin which is summarized in Scheme 14.[462] Conjugate addition of β-keto ester **138** to 5,5-dimethylcyclopentenone followed by acid-catalyzed aldolization gives enedione **139** which undergoes Diels-Alder reaction

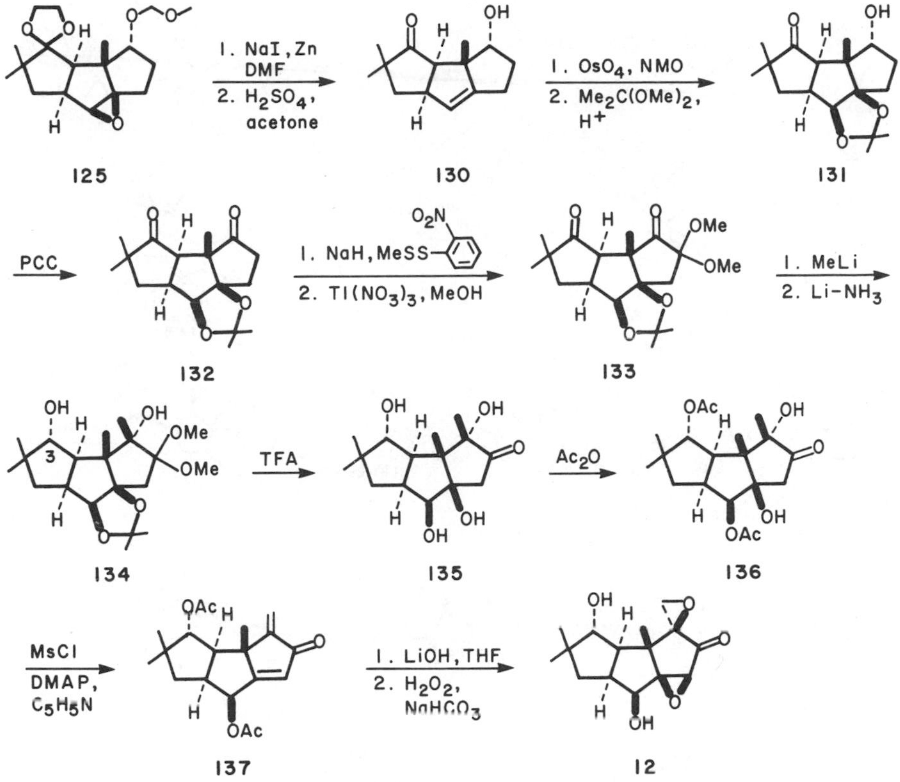

Scheme 13.   Tatsuta's Synthesis of (±)-Coriolin

with 3-trimethylsilyloxy-1,3-pentadiene. The adduct (**140**) is degraded by the interesting sequence shown to obtain triketone **144**. The overall result of this eight-step sequence is introduction of the 2-butanone-3-yl group to the eventual C-1 position. Aldolization is accomplished by treatment of **144** with potassium *t*-butoxide. The α,β-unsaturated ketone is deconjugated and the resulting enedione is selectively reduced

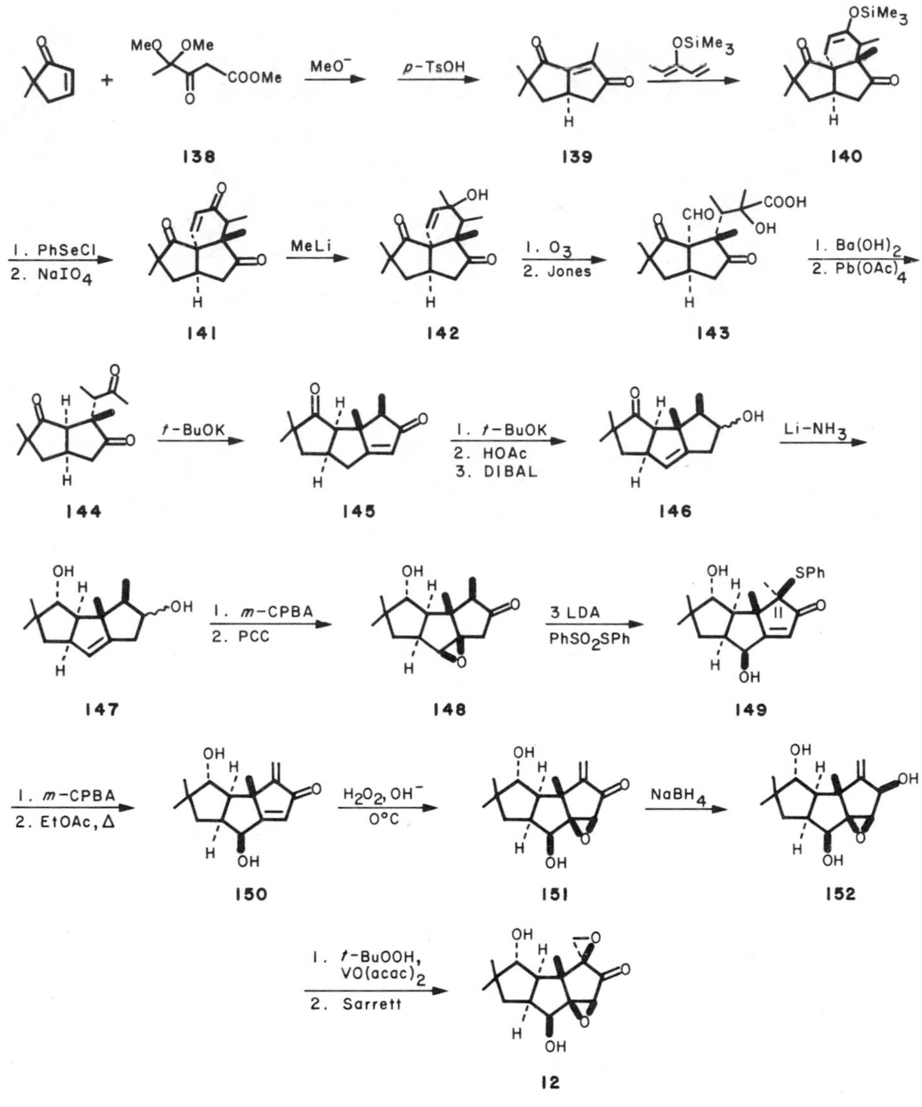
MeO OMe
COOMe
MeO⁻
p-TsOH
OSiMe3
138
139
140
1. PhSeCl
2. NaIO4
MeLi
1. O3
2. Jones
CHO
COOH
OH
1. Ba(OH)2
2. Pb(OAc)4
141
142
143
t-BuOK
1. t-BuOK
2. HOAc
3. DIBAL
Li-NH3
144
145
146
1. m-CPBA
2. PCC
3 LDA
PhSO2SPh
SPh
147
148
149
1. m-CPBA
2. EtOAc, Δ
H2O2, OH⁻
0°C
NaBH4
150
151
152
1. t-BuOOH,
VO(acac)2
2. Sarrett
12

Scheme 14. Danishefsky's Synthesis of (±)-Coriolin

by reaction with diisobutylaluminumum hydride. Lithium-ammonia reduction of the remaining carbonyl group gives a diol (147) which is epoxidized. Selective oxidation with the Corey-Suggs reagent gives keto epoxide 148. Treatment of this material with three equivalents of lithium diisopropylamide and phenyl thiophenylsulfonate results in sulfenylation at C-11 and elimination of the β,γ-epoxy ketone, yielding intermediate 149. After elimination, enedione 150 is partially epoxidized to obtain epoxy enone 151. Further epoxidation of this compound was found to give a 7:5 mixture of ($\pm$)-coriolin (12) and its diastereomer (153). However, reduction of 151 gives a triol (152) which is oxidized

153

by the Sharpless method to give only the spiroepoxide of the correct relative configuration. Selective oxidation of the resulting diepoxy triol using Sarrett's reagent has been reported by Umezawa. The remarkable aspect of Danishefsky's synthesis is that the entire 24 steps are carried out without the necessity of using a single protecting group. Considering the number of different functional groups, this is a signal achievement.

Ikegami and co-workers have also reported a synthesis of ($\pm$)-coriolin (Scheme 15).[463] 1,3-Cyclooctadiene is converted into enone 154 by an eight-step sequence in 35% overall yield. Addition of lithium dimethylcopper occurs to give 155, which is subjected to Conia alkylation to obtain 156. Lithium-ammonia reduction places the new secondary alcohol predominantly in the more stable exo position. However, both diastereomers (157 and 158) are obtained in a 5:2 ratio. After exchange of protecting groups, the secondary alcohol is oxidized and the final ring is added by the three-step sequence 159→160 . The noteworthy feature of this series of steps is use of a Wacker oxidation to convert the allyl group into an acetonyl group. Enone 160 is converted into dienone 162, which is deconjugated to obtain 163. Peracid epoxidation of this material

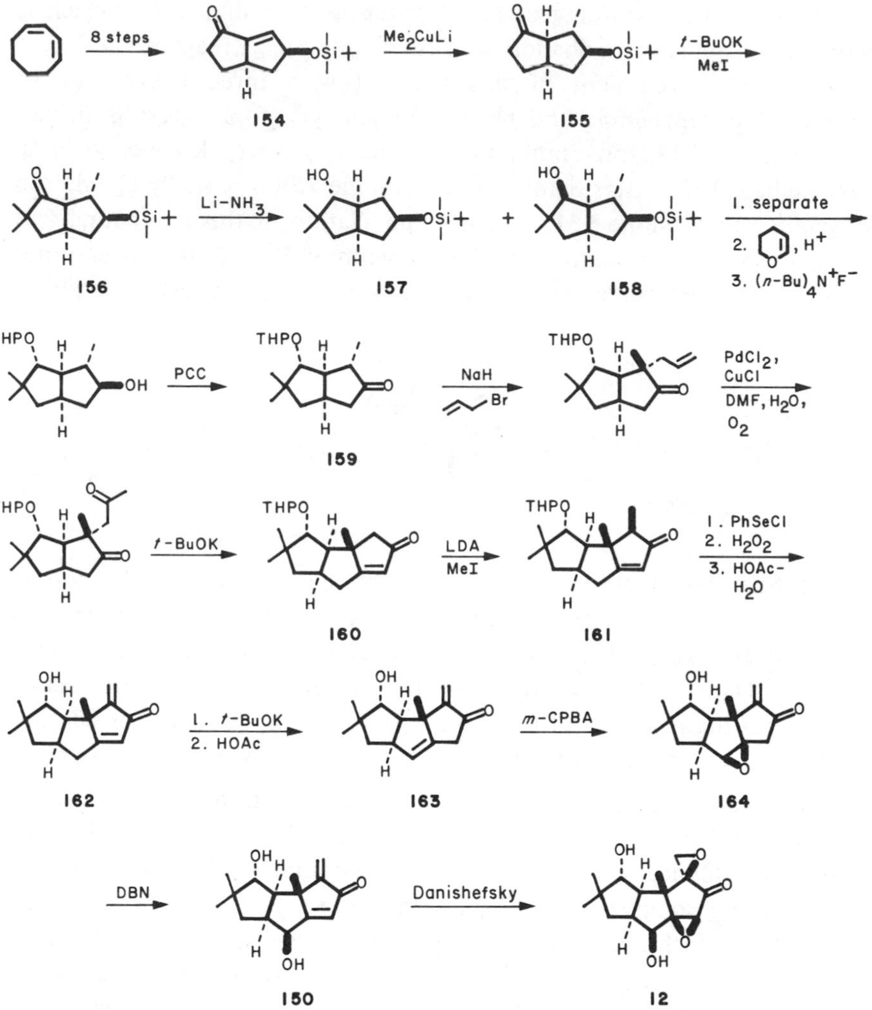

Scheme 15.   Ikegami's Synthesis of (±)-Coriolin

provides an epoxide (**164**) which undergoes elimination upon treatment with DBN. At this juncture, the Ikegami synthesis converges with Danishefsky's route. In all, the synthesis requires 28 steps and involves one isomer separation.

Greene has recently reported the remarkably efficient synthesis of hirsutene (**10**) which is outlined in Scheme 16.[464] The basic reaction employed in the Greene approach is a three-carbon annelation, which is

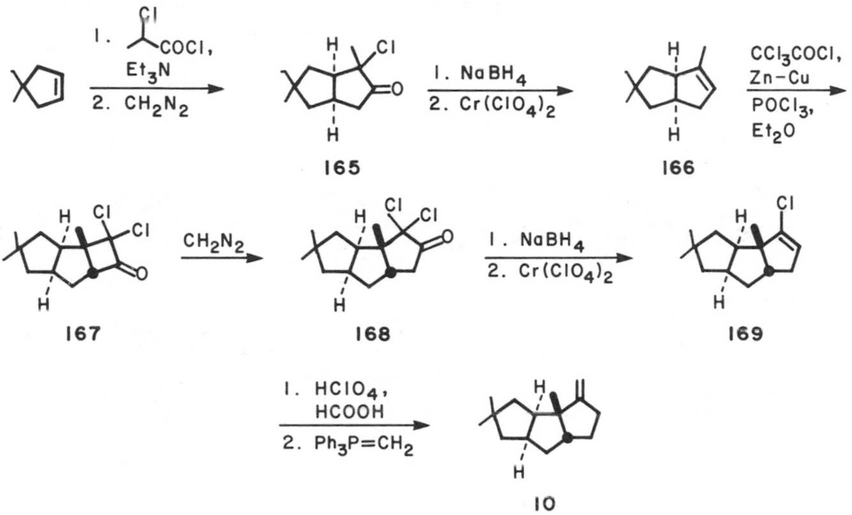

Scheme 16.   Greene's Synthesis of (±)-Hirsutene

accomplished by cycloaddition of a chloro ketene to an alkene, followed by diazomethane ring expansion. The resulting chloro ketone is transformed into a new alkene by a two-stage reduction (**165**→**166**). Repetition of the process gives a vinyl chloride (**169**) which is hydrolyzed and methylenated to obtain (±)-hirsutene (**10**).

At this point, we discuss a final synthetic endeavor in the hirsutane area. Although it has not yet led to the synthesis of a natural product, the novelty of the chemistry justifies its inclusion. Little and Miller converted 3,3-dimethylglutaric anhydride, via the known hemiacetal **170**, into unsaturated ester aldehyde **171** (Scheme 17).[465] Condensation of **171** with cyclopentadiene gives fulvene **172**, which reacts with di(2,2,2-trichloroethyl) azodicarboxylate to give an adduct, which is selectively hydrogenated to **173**. The trichloroethyl groups are removed by controlled potential electrolysis and the resulting hydrazine is oxidized to azo compound **174**. When **174** is refluxed in acetonitrile, nitrogen is

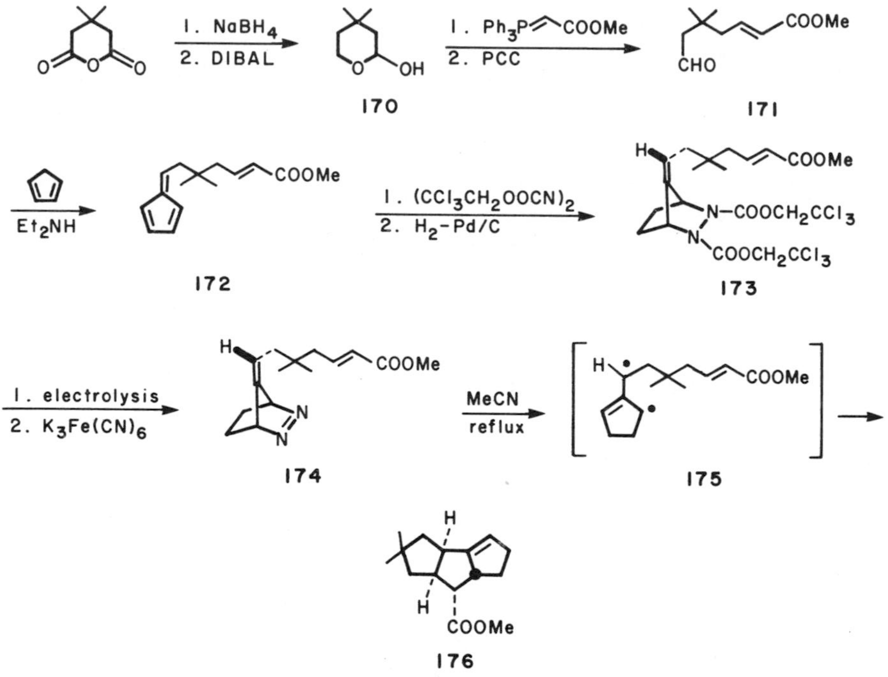

Scheme 17.   Little-Miller Hirsutane Synthesis

extruded to give a 1,3-diyl (**175**) which is trapped by the alkene linkage. The tricyclic product (**176**) is produced in 50% yield based on biscarbamate **173**.

(4) Isocomene

The tricyclic hydrocarbon isocomene (**13**) has attracted several synthetic efforts. The first published synthesis was that of Oppolzer and co-workers (Scheme 18).[466] Annelation of 2-methylcyclopentanone with dienone **177** produces a hydrindenone (**178**) which is methylated to

13

obtain **179** and its C-4 epimer **180** in a ratio of 5:1. When **179** is heated to 280°C, intramolecular ene reaction occurs to give tricyclic ketone **181**. This reaction neatly establishes the two remaining stereocenters of isocomene. Unfortunately, the yield in this process is only 17%. The synthesis is completed by ring-contraction, which is effected by photochemical Wolff rearrangement of diazo ketone **182**, reduction of the resulting methoxycarbonyl group, introduction of unsaturation, and isomerization of the double bond of **185** into the ring. In all, Oppolzer's synthesis requires 11 steps and affords (±)-isocomene in 0.7% overall yield.

Pirrung's synthesis is summarized in Scheme 19.[467] Stork-Danheiser methylation of the well-known enol ether **186** is followed by addition of the Grignard reagent **187**. The resulting dienone (**188**) undergoes smooth photoaddition to give tricyclic ketone **189** in 77% yield. Although ketone **189** is highly hindered, and undergoes predominant enolization when treated with methyllithium, it reacts with methylenetriphenylphosphorane under forcing conditions to give hydrocarbon **190**. Treatment of this compound with p-toluenesulfonic acid in benzene

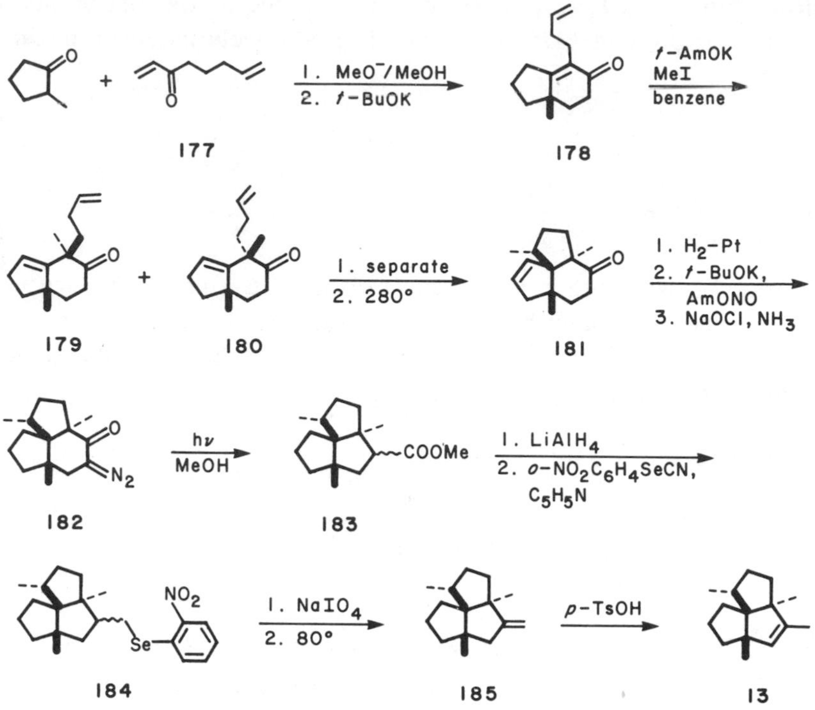

Scheme 18.   Oppolzer's Synthesis of (±)-Isocomene

brings about Wagner-Meerwein rearrangement and affords (±)-iso-comene (**13**). The synthesis requires seven steps and proceeds in 34% overall yield.

Paquette's isocomene synthesis (Scheme 20)[468] begins with a Yoshikoshi-type annelation of enol ether **191** with 2-nitro-1-butene. Conjugate addition of Grignard reagent **193** to the resulting enone **192** gives a keto acetal (**194**) which is converted into unsaturated aldehyde **195** by the efficient method illustrated in the scheme. The key step of

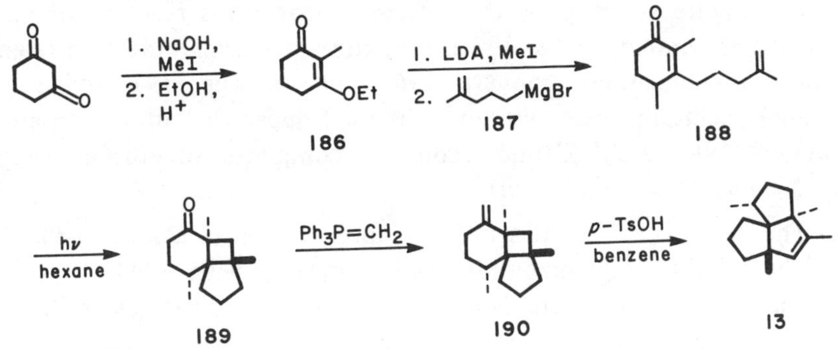

Scheme 19.   Pirrung's Synthesis of (±)-Isocomene

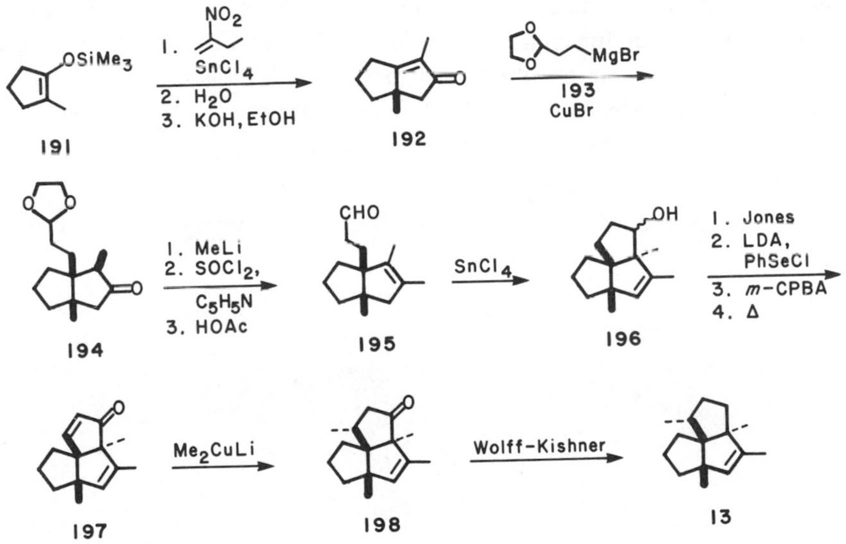

Scheme 20.   Paquette's Synthesis of (±)-Isocomene

the Paquette synthesis is the intramolecular Prins reaction which provides unsaturated alcohol **196** as a mixture of diastereomers. Straightforward manipulations convert **196** into dienone **197** which reacts stereospecifically with lithium dimethylcopper to afford unsaturated ketone **198**. Wolff-Kishner reduction completes an efficient synthesis (12 steps, 22% overall yield).

The most recent synthesis of isocomene is that of Dauben and Walker (Scheme 21).[469] Compound **199** is alkylated with ethyl β-bromopropionate and the thioacetal is hydrolyzed to obtain diketo ester **201**. Condensation of this material with dimethyl acetonedicarboxylate gives a diketo pentaester (**202**) which is hydrolyzed, decarboxylated, and reesterified to obtain diketo ester **203**. Wolff-Kishner reduction of monoketal **204** followed by deprotection gives keto acid **205**. Compound **205** undergoes smooth cyclization under acidic conditions to provide β-diketone **206** which is methylated to give the sensitive nonenolizable β-diketone **207**. Thioketalization of the more reactive carbonyl gives **208**, which is converted into alkene **209** by reaction with methylenetriphenylphosphorane under forcing conditions. After isomerization of the double bond into the ring, the dithiolane group is cleaved to give unsaturated ketone **210**, an intermediate in the Paquette synthesis. The Dauben-Walker synthesis requires 19 steps and produces isocomene in 1.6% overall yield.

An earlier report claiming the synthesis of isocomene[470] is probably erroneous.[471]

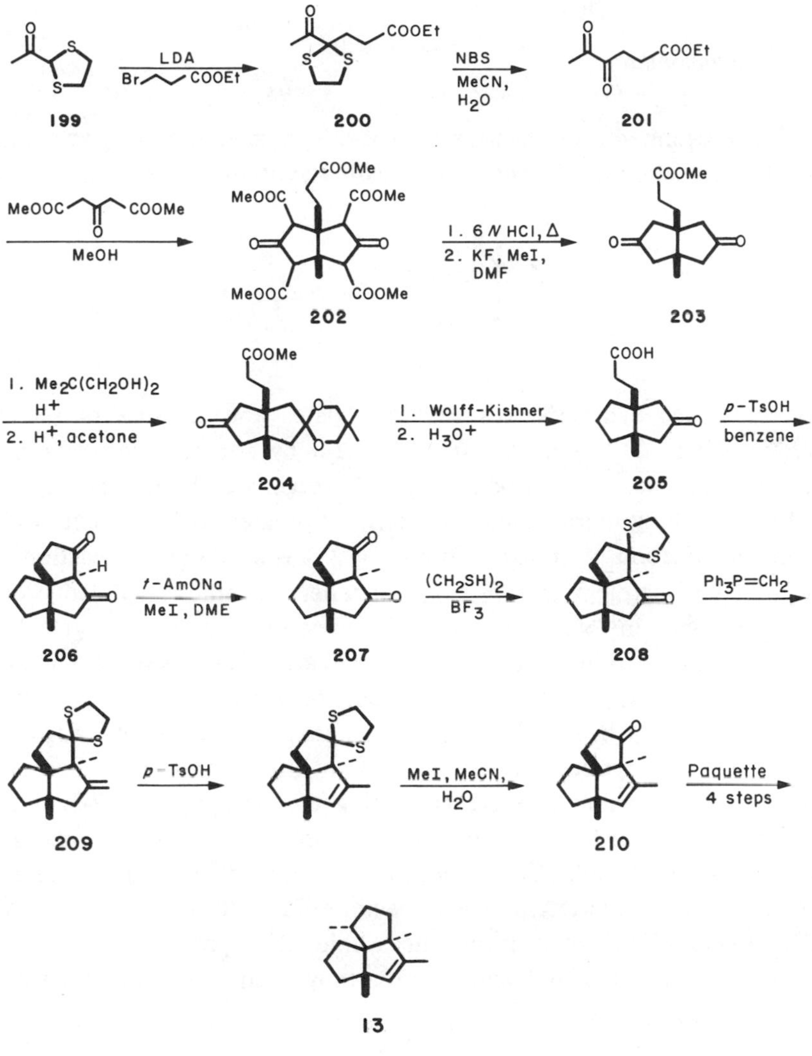

Scheme 21.   Dauben-Walker Synthesis of (±)-Isocomene

B. Bridged Systems

(1) Gymnomitrol

The interesting 4,8-methanoperhydroazulene skeleton of gymnomitrol (**211**) has elicited a number of interesting synthetic investigations. The

211

first published synthesis was that of Coates and co-workers (Scheme 22).[472] Dione **212**, available in 52% yield by condensation of biacetyl with dimethyl acetonedicarboxylate, is deoxygenated by Ireland's method and the resulting unsaturated ketone hydrogenated to obtain ketone **213**. After introduction of an α-cyano group, **214** is alkylated by reaction with the acetal of acrolein in refluxing benzene; the ethoxyallylated product **215** is obtained in 76% yield. After formation of the ethylene glycol acetal, the cyano group is removed reductively. The resulting enolate is trapped with trimethylsilyl chloride, then regenerated and methylated to obtain **217**, which is hydrolyzed to keto aldehyde **218**. Somewhat surprisingly, compound **218** is reluctant to undergo aldol condensation, giving an equilibrium mixture consisting of approximately 2-3 parts of **218** to one part of bridged ketol upon treatment with sodium carbonate in aqueous methanol. Consequently the aldehyde is oxidized and the resulting acid converted into enol lactone **219**. Reduction of **219** with diisobutylaluminum hydride produces aldol **220**, perhaps as a result of alkoxide promoted [1,3]sigmatropic rearrangement of the initially formed enol hemiacetal:

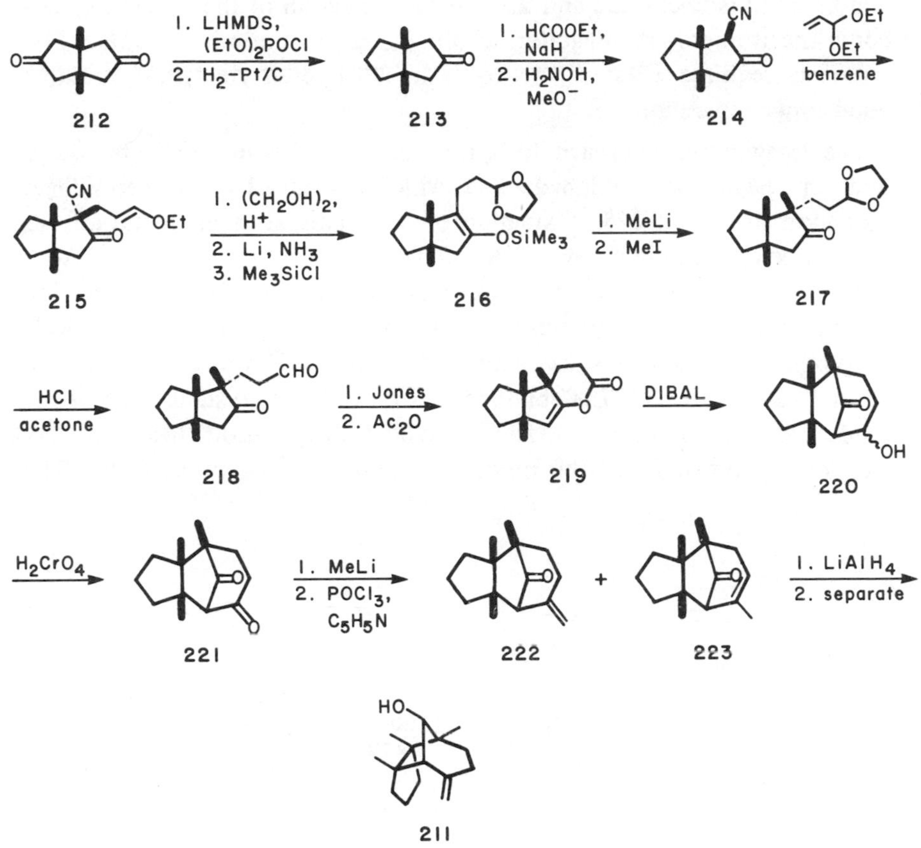

Scheme 22. Coates' Synthesis of (±)-Gymnomitrol

After oxidation of **220** to dione **221**, the final carbon is added by reaction of the more reactive carbonyl with methyllithium. Dehydration of the resulting tertiary alcohol gives a 1:1 mixture of exocyclic and endocyclic double bond isomers **222** and **223**. After reduction of the remaining carbonyl the isomers are separated to obtain (±)-gymnomitrol (**211**). The synthesis requires 16 steps, proceeds in 2% overall yield, and requires one isomer separation.

The imaginative approach to gymnomitrol (Scheme 23)[473] by Büchi and Chu begins with aldehyde **224**, which is cleaved by Baeyer-Villiger oxidation to phenol **225**. Oxidation of this material in methanol affords trimethoxycyclohexadienone **226**. This material is condensed with 1,2-dimethylcyclopentene in the presence of stannic chloride. The initially formed, highly fragile diketone is reduced with sodium borohydride to provide keto alcohols **227** and **228**, along with a larger amount of phenol **225**. Isomer **227** is obtained in 10% yield by crystallization of the crude product from ether. After catalytic hydrogenation, the secondary alcohol is protected and the methoxy group removed reductively. The

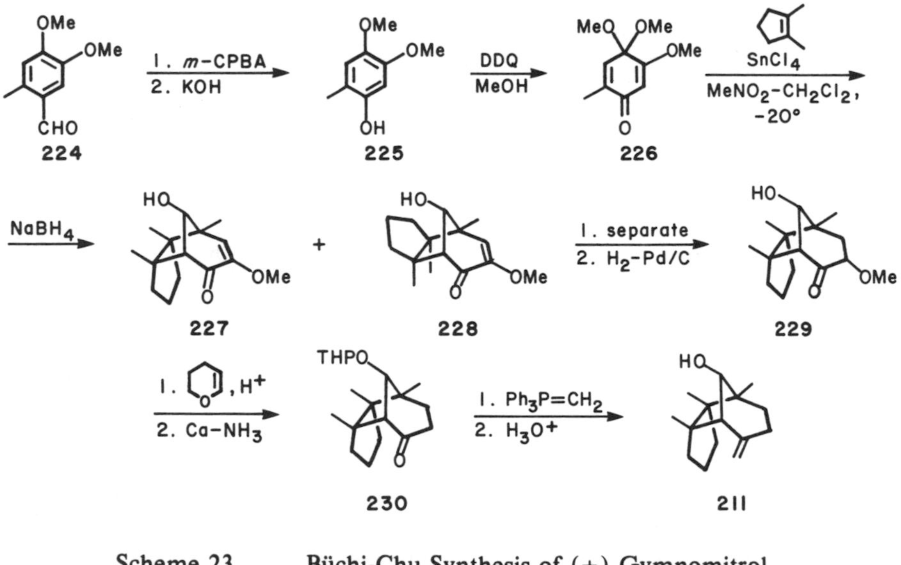

Scheme 23.   Büchi-Chu Synthesis of (±)-Gymnomitrol

synthesis is completed by Wittig methylenation and deprotection. Overall, the Büchi-Chu synthesis requires nine steps and proceeds in 2.4% overall yield.

Welch and Chayabunjonglerd synthesized gymnomitrol as shown in Scheme 24.[474] Enone **231**, obtained by straightforward annelation of 2-methylcyclopentanone, is allowed to react with lithium dimethylcopper and the resulting enolate is alkylated to obtain ketone **232**. Methylation of this material occurs from the convex face to give ketone **233**, which is transformed into keto ester **234**. Dieckman cyclization is accomplished by treatment of **234** with lithium hexamethyldisilazide. The resulting enolate is trapped by reaction with *t*-butyldimethylsilyl chloride to give keto ether **235**. After reduction of the carbonyl group and protection of the resulting secondary alcohol, the ketone is regenerated to obtain **236**. The final carbon is introduced by Wittig methylenation to give, after deprotection, (±)-gymnomitrol (**211**). The Welch synthesis is the

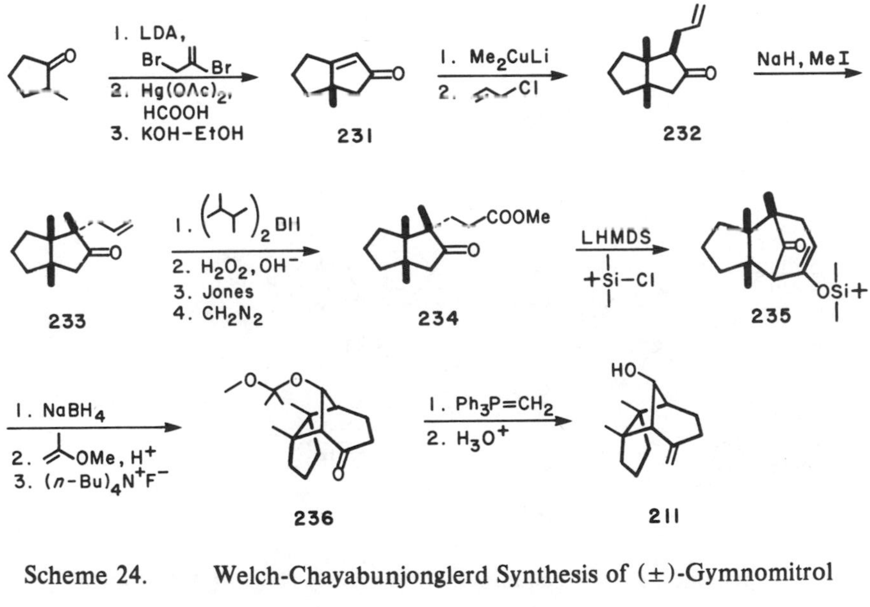

Scheme 24.   Welch-Chayabunjonglerd Synthesis of (±)-Gymnomitrol

highest-yield route thus far developed, providing **211** in 7% overall yield for the 14 steps from 2-methylcyclopentanone.

Paquette's synthesis of gymnomitrol (Scheme 25)[475] is in many respects similar to the Coates synthesis (Scheme 22). As in the Coates synthesis, the starting point is the Weiss diketone (**212**), which is deoxygenated by Wolff-Kishner reduction of the monoketal. The third ring is added by conjugate addition of 2-trimethylsilylvinylmagnesium bromide to enone **237**, followed by methylation of the resulting enolate. Peracid oxidation of **238** provides keto aldehyde **218**. As was also noted by Coates and co-workers, Paquette and Han found **218** to be resistant to aldolization. However, conditions were found whereby **218** can be converted to the extent of 43% into aldol **220**. The remainder of the Paquette-Han synthesis is identical to the Coates synthesis. Overall, the synthesis involves 11 steps and gives **211** in 2% yield.

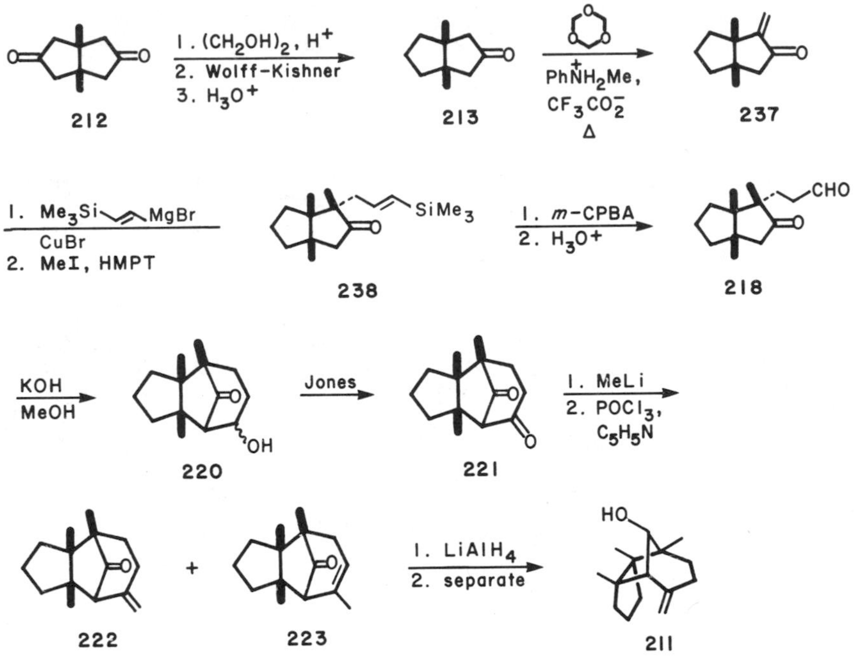

Scheme 25.   Paquette-Han Synthesis of (±)-Gymnomitrol

(2) Copacamphor, Copaborneol, Copaisoborneol,
   Copacamphene, Cyclocopacamphene, Ylangocamphor,
   Ylangoborneol, Ylangoisoborneol, Sativene, Cyclosativene,
   cis-Sativenediol, Helminthosporal, and Sinularene

This group of tricyclic compounds is characterized by the presence of a bicyclo[2.2.1]heptane system with a further three carbon bridge. The "copa" series includes copacamphor (**239**), copaborneol (**240**), copaisoborneol (**241**), copacamphene (**242**), and cyclocopacamphene (**243**).* The "ylango" series includes ylangocamphor (**244**), ylangoborneol (**245**), ylangoisoborneol (**246**), ylangocomphene (sativene, **247**), and cycloylangocamphene (cyclosativene, **248**). In addition to these 10 compounds, all of which have been synthesized, we shall discuss syntheses of *cis*-sativenediol (**249**), and its conversion product helminthosporal (**250**). Finally, although sinularene (**251**) is not biogenetically related to the foregoing group of compounds, its similar structure justifies its consideration in this section.

239

240

241

242

---

* The prefixes "copa" and "ylango" are derived from the tricyclic hydrocarbons copaene and ylangene (*vide infra*) and specify the configuration of the isopropyl-bearing carbon relative to the remainder of the molecule. Many of the compounds to be discussed in this section occur in nature in both enantiomeric forms. However, some syntheses that will be discussed employ optically active starting materials while others use racemic reactants. In cases where a specific enantiomeric series was used, we have illustrated with the correct formulas. In syntheses which used racemic reactants, we have utilized the enantiomeric series which is most convenient for drawing structures.

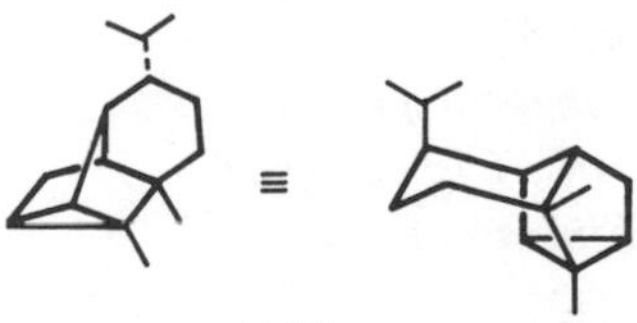

243

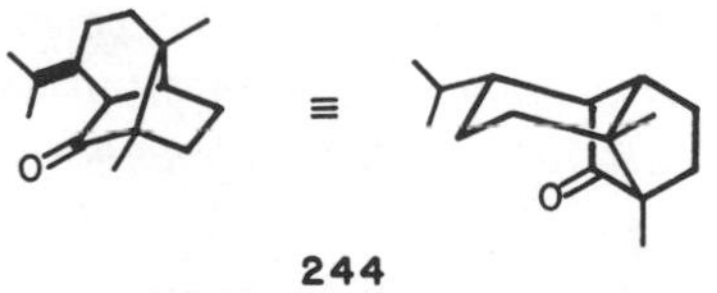

244

245

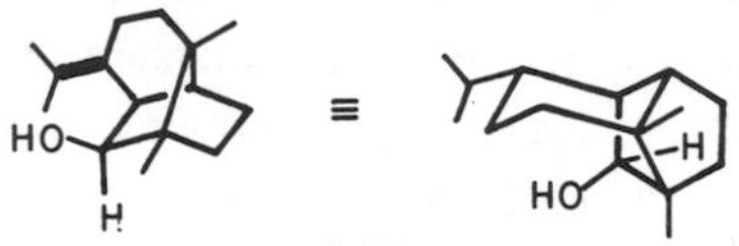

246

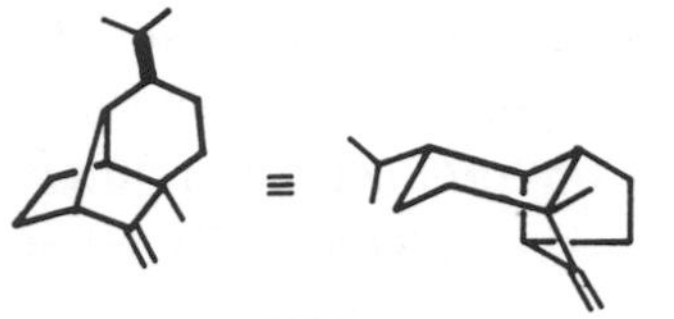

247

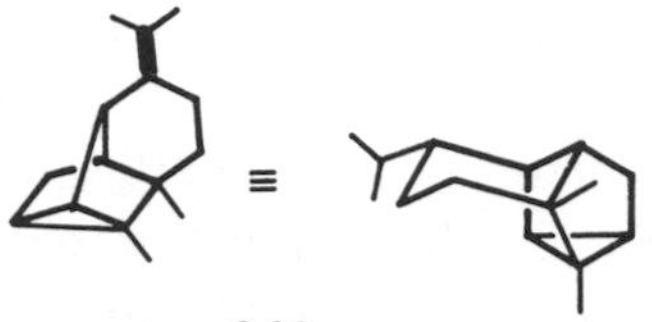

248

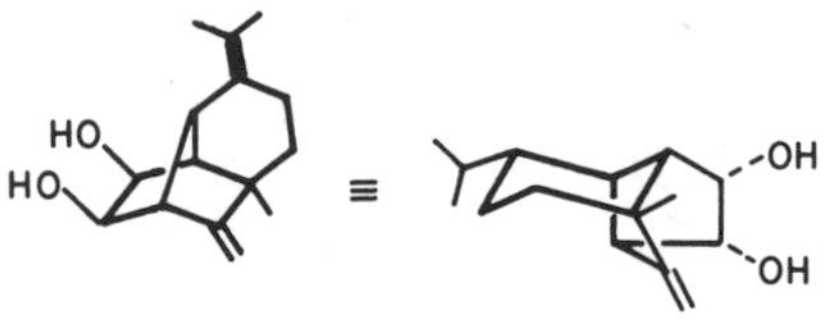

249

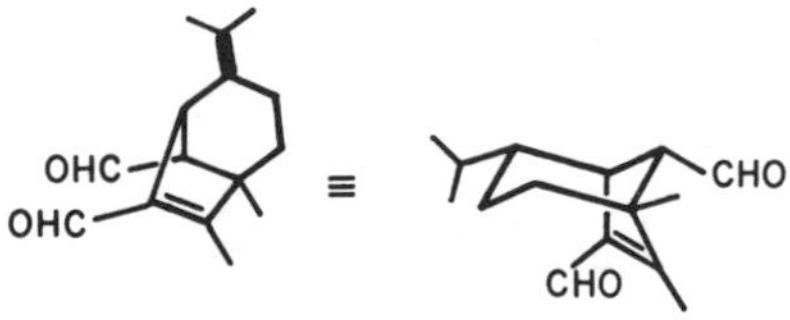

250

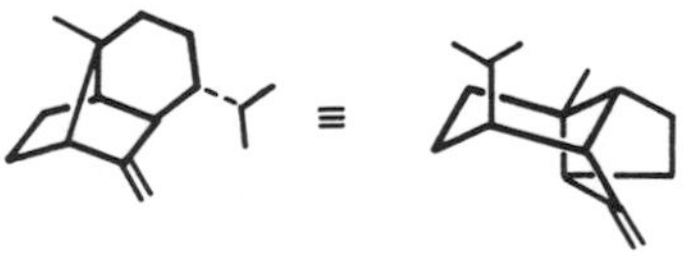

251

In Volume 2, we outlined McMurry's synthesis of (±)-sativene and (±)-cyclosativene.[476] In 1971, he reported an adaptation of his basic approach to the synthesis of (±)-copacamphene (Scheme 26).[477] Unsaturated ketone 252, an intermediate in the sativene synthesis,[476] is epoxidized stereoselectively from the convex face to give epoxy ketone 253. When this material is treated with dimsylsodium in dimethyl sulfoxide,

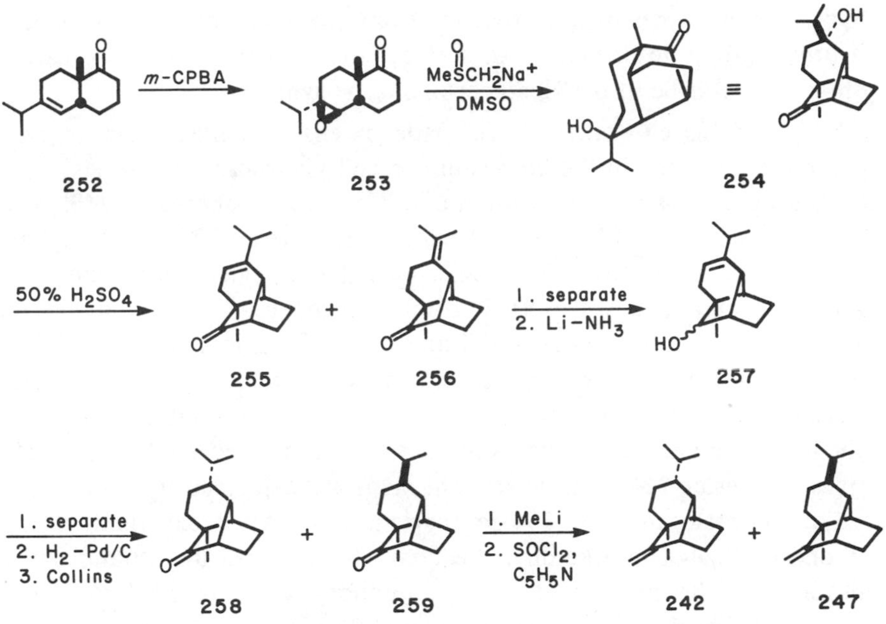

252   253   254

255   256   257

258   259   242   247

Scheme 26.　　McMurry's Synthesis of (±)-Copacamphene

intramolecular alkylation occurs to afford keto alcohol 254. Dehydration
of 254 gives a 7:3 mixture of 255 and 256, which is separated chromatographically. The crux of the synthesis is the establishment of the correct
stereochemistry of the isopropyl group. It is clear from molecular
models that catalytic hydrogenation of either 255 or 256 should occur to
produce the "ylango" rather than the "copa" configuration at the
isopropyl-bearing carbon. Indeed, experiments bore out this prediction.
In order to circumvent this problem, ketone 255 was reduced by lithium
in ammonia. A 3:2 mixture of epimeric alcohols 257 is obtained, with
the exo diastereomer predominating. Hydrogenation of this substance in
hexane solution gives an 85:15 mixture of ketones 258 and 259. Introduction of the methylene group gives a similar mixture of (±)-copacam-
phene (242) and (±)-sativene (247). The stereoselectivity of the

hydrogenation presumably results from bonding of the secondary hydroxy group to the catalyst surface and is precedented by earlier observations made in the Heathcock copaene synthesis.[478]

Piers and his co-workers have made extensive contributions to synthetic methodology in the copacamphor and ylangocamphor series. The Piers syntheses of (+)-copacamphor (239), (+)-copaborneol (240), and (+)-copaisoborneol (241) is summarized in Scheme 27.[479a,e] (+)-Carvomenthone is converted into the n-butylthiomethylene derivative 260, which is alkylated with ethyl α-iodopropionate. Alkylation occurs predominantly from the equatorial direction, affording keto acid 261 as a mixture of diastereomers at the side-chain asymmetric center. After reesterification, the keto ester is subjected to Dieckmann cyclization to obtain β-diketone 262. Hydrogenation of the derived enol acetate 263 gives keto ester 264. The next eight steps are expended in a somewhat tedious homologation of 264 to tosylate 267. After oxidation of the secondary alcohol to a ketone, the final ring is closed by intramolecular alkylation. The product, (+)-copacamphor (239), is converted into (+)-copaborneol (240) by reduction with sodium in ethanol and into (+)-copaisoborneol (241) by reduction with lithium aluminum hydride.

For the synthesis of (−)-copacamphene and (−)-cyclocopacamphor (Scheme 28),[479b,f] dione 262 is partially reduced to hydroxy ketone 269. The remaining carbonyl is then subjected to Bamford-Stevens conditions. Reoxidation provides unsaturated ketone 270. The standard homologation sequence provides unsaturated alcohol 272. The derived tosylate is solvolyzed by treatment with silica gel in pentane to obtain (−)-copacamphene (242). Tosylhydrazone 273, prepared from alcohol 272, is converted by methyllithium into a diazo compound which undergoes cycloaddition with the double bond to provide pyrazoline 274. Photochemical extrusion of nitrogen gives (−)-cyclocopacamphene (243).

For synthesis of the ylango series, Piers and co-workers condense (+)-dihydrocarvone with ethyl vinyl ketone (Scheme 29). It is well established that such annelations proceed by axial alkylation. Therefore, the methyl and isopropenyl groups in 275 are trans rather than cis, as they are in intermediate 261 (see Scheme 27). Compound 275 is transformed into dienone aldehyde 277 by standard methodology. After

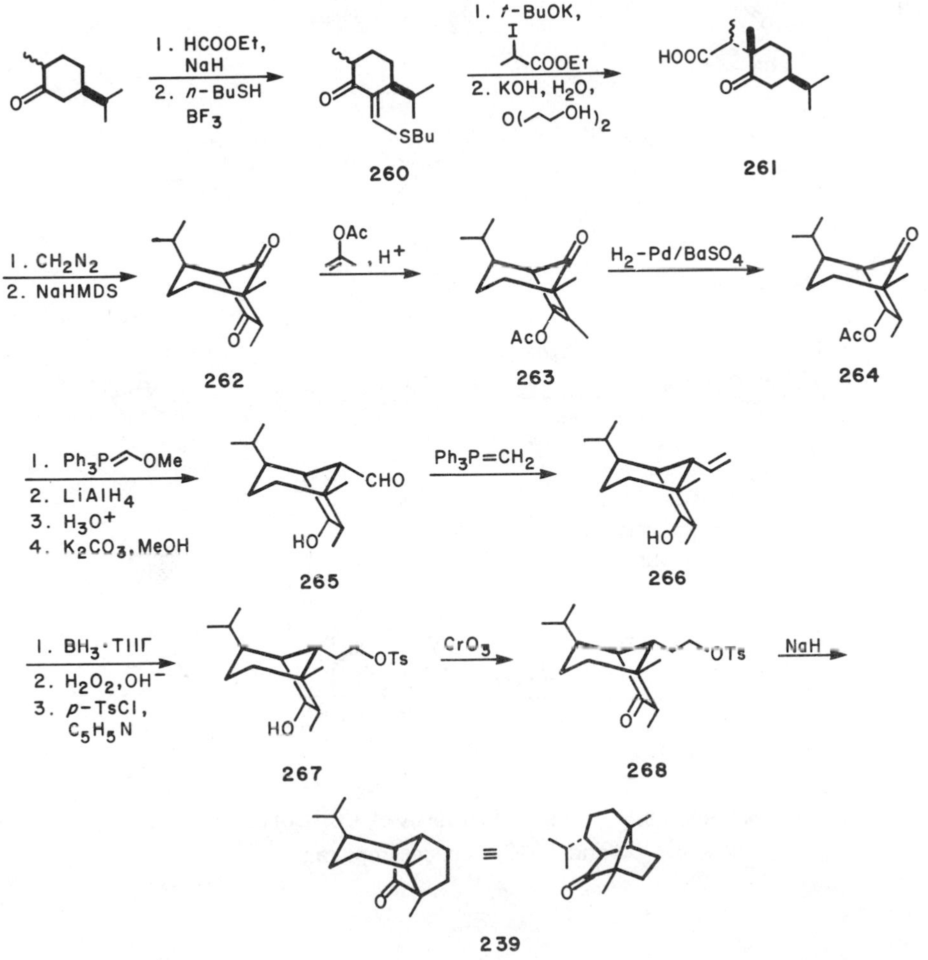

260

261

262

263

264

265

266

267

268

239

Scheme 27.   Piers' Synthesis of (±)-Copacamphor,
(±)-Copaborneol, and (±)-Copaisoborneol

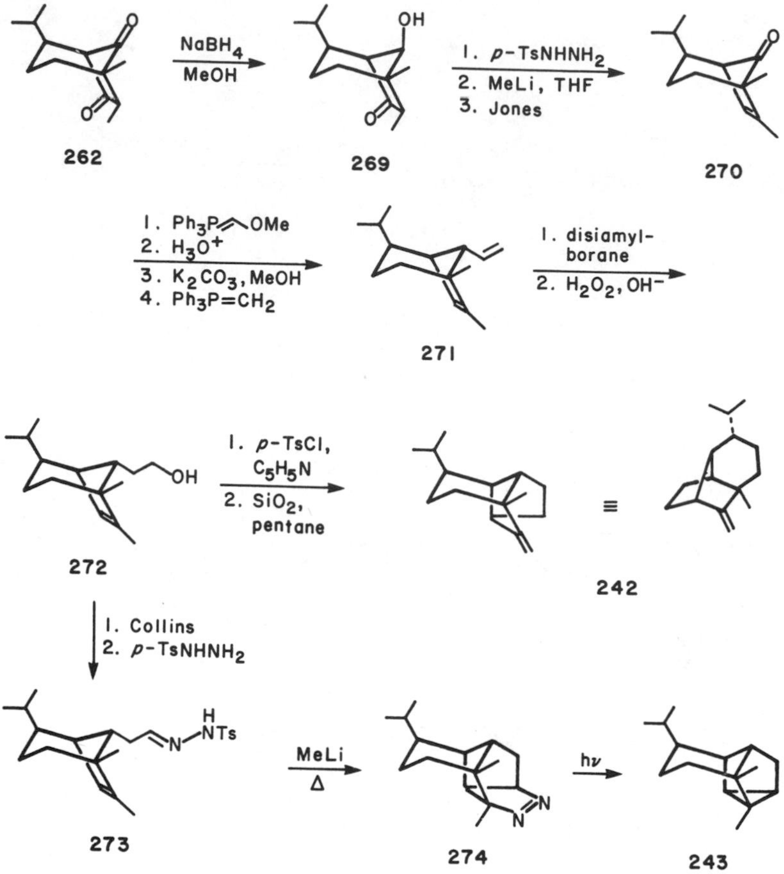

Scheme 28.   Piers' Synthesis of (±)-Copacamphene and (-)-Cyclocopacamphene

introduction of an axial methyl group with lithium dimethylcopper, the cyclohexenone ring is degraded by ozonolysis to obtain keto acid **279**, a diastereomer of **261**. The remainder of the synthesis follows lines similar to those used for synthesis of the copa series.

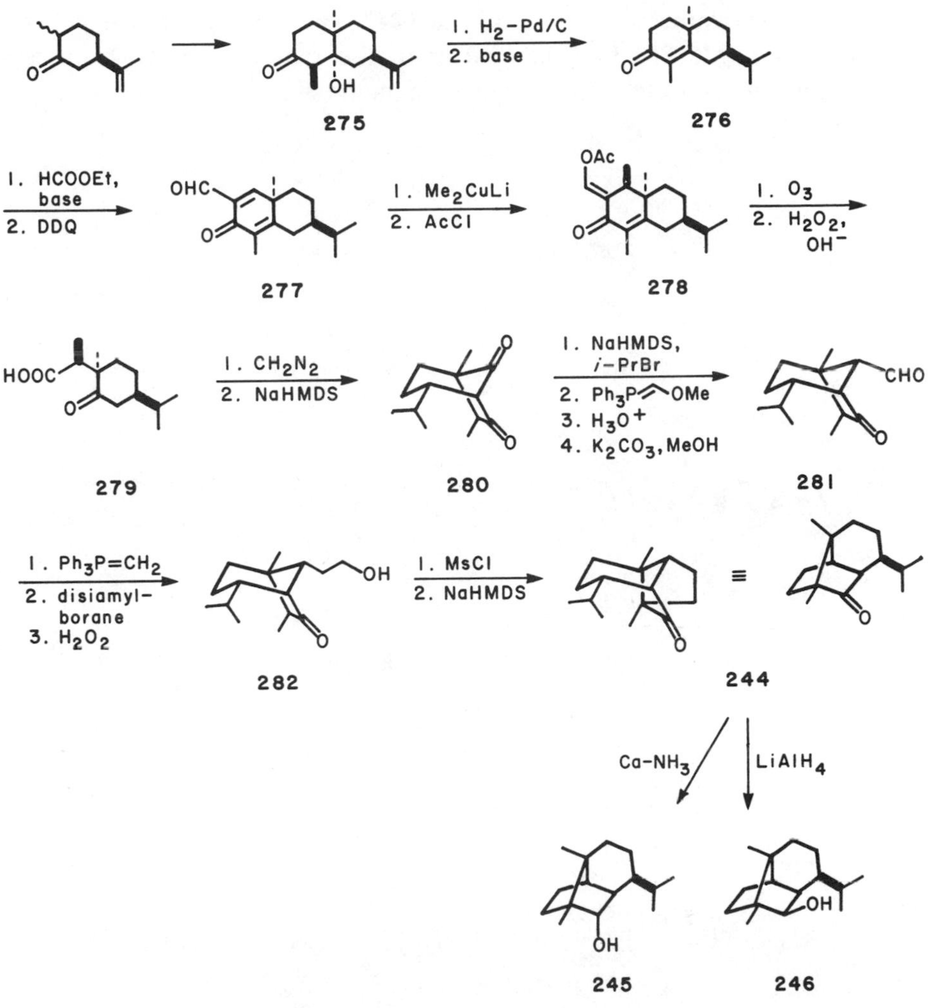

Scheme 29.   Piers' Synthesis of (-)-Ylangocamphor,
(-)-Ylangoborneol, and (-)-Ylangoisoborneol

For the synthesis of (+)-sativene (**247**) and (+)-cyclosativene (**248**), keto aldehyde **281** is elaborated as in the syntheses of (−)-copacamphene and (−)-cyclocopacamphene (Scheme 30).[479d,g]

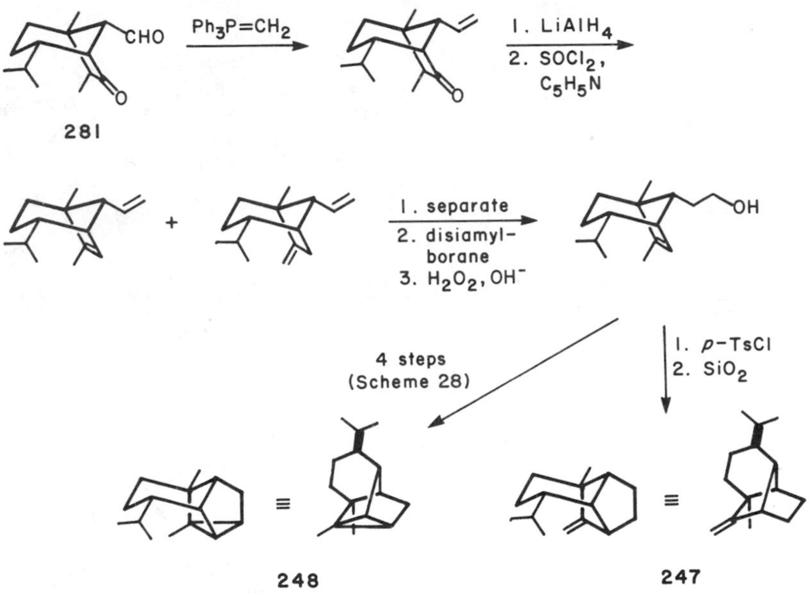

Scheme 30.   Piers' Synthesis of (+)-Sativene and (+)-Cyclosativene

Money and co-workers have prepared (±)-copacamphor (**239**) and (±)-ylangocamphor (**244**) beginning with (±)-camphorenone (**283**) as shown in Scheme 31.[480] Epoxidation of the double bond gives a diastereomeric mixture of epoxides (**284**), which undergoes cyclization

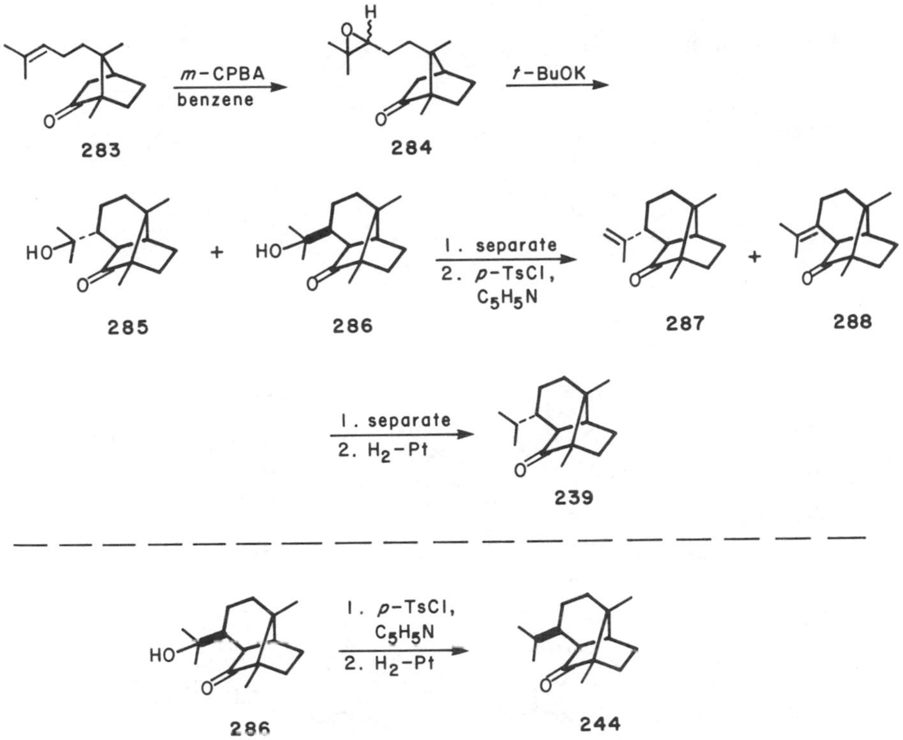

Scheme 31.   Money's Synthesis of (±)-Copacamphor
and (+)-Ylangocamphor

when treated with potassium *t*-butoxide. The resulting keto alcohols **285** and **286**, formed in a 1:1 ratio, are separated by preparative glpc. Dehydration of isomer **285** affords unsaturated ketones **287** and **288** in a 7:3 ratio. The isomers are once again separated by preparative glpc and the major isomer is hydrogenated to obtain (±)-copacamphor (**239**). Similar treatment of isomer **286** yields (±)-ylangocamphor (**244**).

Baldwin and Tomesch have synthesized (±)-cyclosativene as shown in Scheme 32.[481] Diels-Alder addition of 2,3-dimethyl-1,3-cyclopentadiene (generated *in situ* from alcohol **289**) and propynal affords an adduct

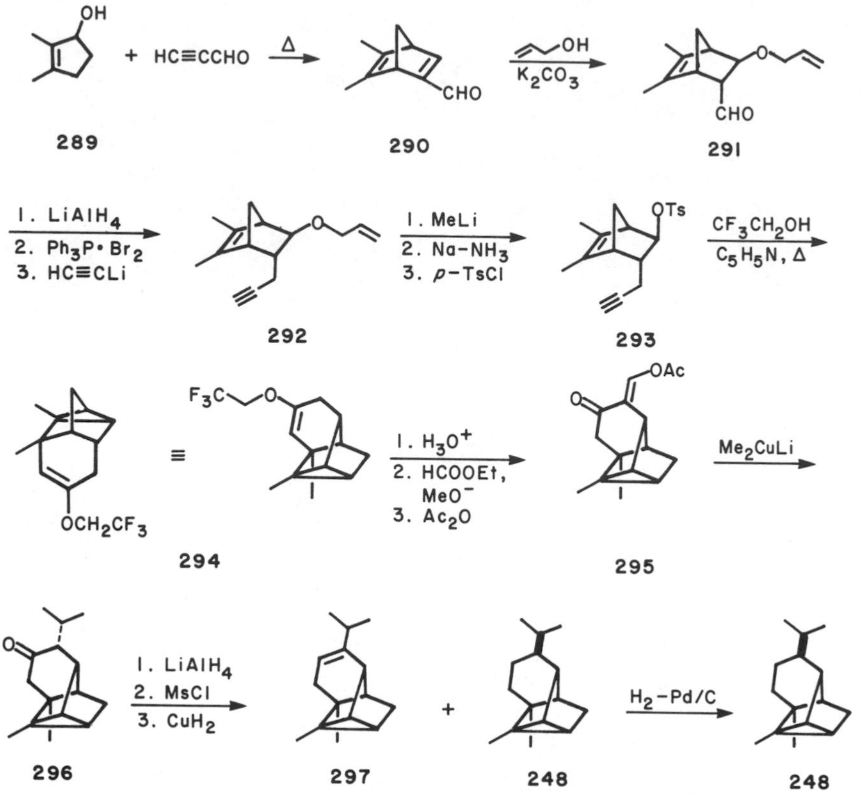

289   **290**   **291**

292   **293**

294   **295**

296   **297**   **248**   **248**

Scheme 32.   Baldwin-Tomesch Synthesis of (±)-Cyclosativene

(290) which undergoes base-catalyzed addition of allyl alcohol to give aldehyde 291. After conversion of the formyl group to propargyl, the secondary alcohol is deprotected and tosylated. The cation obtained by solvolysis of tosylate 293 is trapped by the proximal alkyne π-system to form the cyclosativene skeleton. The remaining steps are concerned with introducing the necessary isopropyl group with the proper stereochemistry. After hydrolysis of enol ether 294, the resulting ketone is formylated and converted into enol acetate 295. Reaction of 295 with excess lithium dimethylcopper gives ketone 296, which has the "copa"

stereochemistry at the isopropyl position. The ketone is reduced and the resulting alcohol converted into a mesylate which is reduced by Masamune's method. The product of this reduction is a 3:2 mixture of alkene **297** and (±)-cyclosativene. Catalytic hydrogenation of the mixture gives pure (±)-cyclosativene.

Bakuzis and co-workers have reported the synthesis of (±)-sativene and (±)-copacamphene which is summarized in Scheme 33.[482] Methylation of the known bromo ketone **298** gives ketones **299** as a 3:1 mixture of exo and endo isomers. The magnesium enolate is alkylated with allyl bromide to give a 3:4:1 mixture of ether **300** and ketones **301** and **302**.

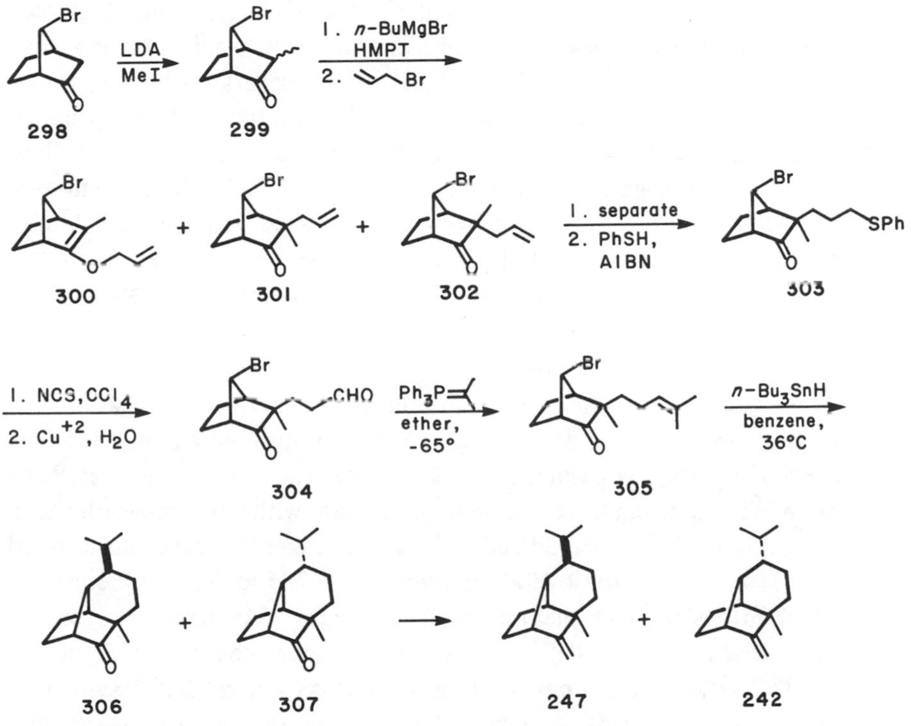

Scheme 33.   Bakuzis' Synthesis of (±)-Sativene and (±)-Copacamphene

After separation, the major product is converted into aldehyde **304** by the interesting method illustrated. Bromide **305** is treated with tri-*n*-butylstannane to give a 3:2 mixture of diastereomeric ketones **306** and **307**, which are converted by McMurry's method[476] into (±)-sativene (**247**) and (±)-copacamphene (**242**). The noteworthy feature of this synthesis is use of free-radical technology in two different transformations (**301**→**303** and **305** → **306** + **307**).

The synthesis by Matsumoto and co-workers of (±)-sativene and (±)-copacamphene is shown in Scheme 34.[483] Racemic piperitone is irradiated with 1,1-dimethoxyethylene to obtain a mixture of diastereomeric adducts **308**. When these adducts are passed through a glpc column containing the weakly acidic solid phase Shimalite at 200°C both lose methanol and rearrange to the same 55:45 mixture of **309** and **310**. When this mixture is subjected to Wittig reaction with methoxymethylenetriphenylphosphorane only isomer **310** reacts, leading to a mixture of **309** and diether **311**, which can be separated by preparative glpc. Alternatively, the mixture can be hydrolyzed and the products (keto aldehyde **312** and a diketone derived from hydrolysis of **309**) separated by silica gel chromatography. The remainder of the synthesis is very similar to the route employed by Piers (see Scheme 30). Under more vigorous conditions, the less reactive diastereomer **309** also undergoes Wittig reaction. The resulting keto ether may be converted by the same sequence of steps into (±)-copacamphene (**242**).

Piers, McMurry, and Matsumoto have all adapted their syntheses in order to prepare the interesting antibiotic *cis*-sativenediol (**249**). The Piers synthesis (Scheme 35)[484] begins with unsaturated alcohol **285** (see Scheme 30), which is oxidized to a keto aldehyde. Intramolecular Prins reaction occurs when this material is treated with trifluoroacetic acid. The resulting alcohol is oxidized by Collins reagent to give unsaturated ketone **314**. In fact, the overall conversion of **285** to **314** can be carried out without workup at the intermediate stages. Hydroxylation of the enolate of **314** with Vedejs' reagent occurs from the exo face of the bicyclo[2.2.1]heptane system, leading to hydroxy ketone **315**. Reduction of this material affords an equimolar mixture of (±)-*cis*-sativenediol (**249**) and its trans diastereomer **316**.

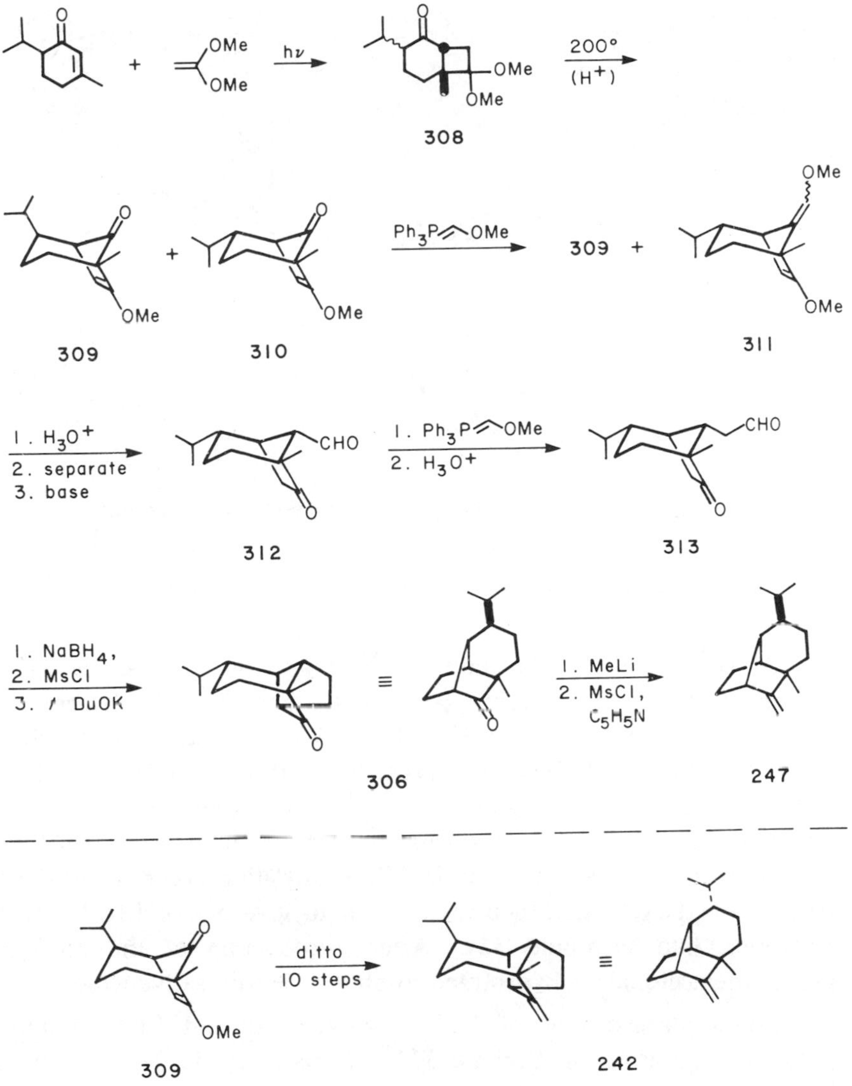

Scheme 34.   Matsumoto's Synthesis of (±)-Sativene and (±)-Copacamphene

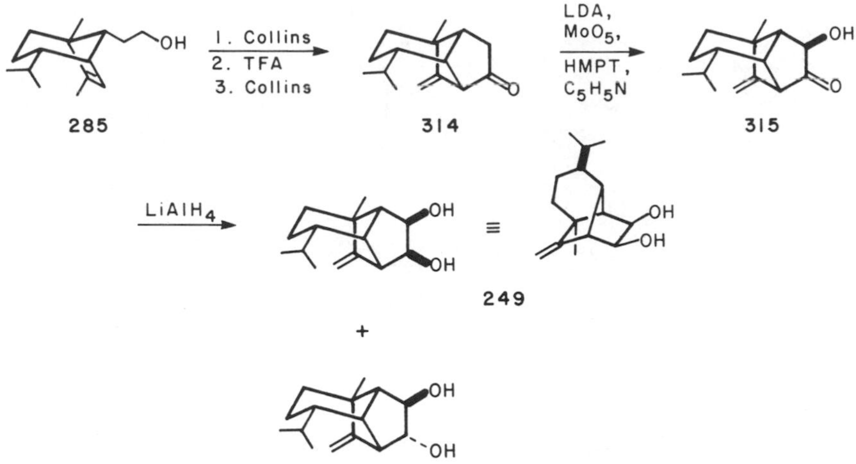

Scheme 35.   Piers' Synthesis of (±)-cis-Sativenediol

McMurry's synthesis (Scheme 36)[485] begins with hydrazone **317**, an intermediate in the McMurry sativene synthesis.[486] The hydrazone is cleaved using titanous ion in 97% yield. After introduction of a double bond by the Reich-Sharpless procedure, the secondary alcohol is tosylated.   Internal alkylation of **320** gives the cyclopropyl enone **321**, which undergoes vinyl cyclopropane rearrangement to enone **322** when passed through a quartz tube at 450°C. Osmylation occurs from the exo face of the bicyclo[2.2.1]heptane system to give a diol (**323**) which is converted into acetonide **324**. After introduction of the methylene group, the acetonide is hydrolyzed to obtain (±)-cis-sativenediol.

Matsumoto's synthesis of (±)-cis-sativenediol and (±)-helminthosporal is summarized in Scheme 37.[483] Keto aldehyde **312** is converted into an unsaturated ketone by selective Wittig methylenation of the more reactive aldehyde carbonyl. Dibromide **325** (presumably a 1:1 mixture of diastereomers) undergoes internal alkylation when treated with potassium t-butoxide to give bromo ketones **326**. Further treatment of

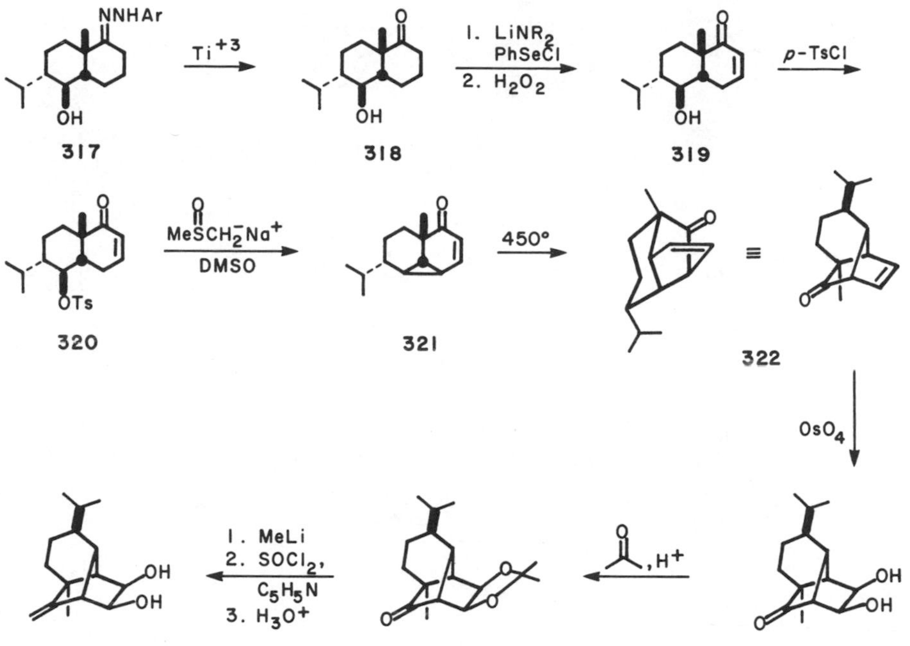

Scheme 36.   McMurry's Synthesis of (±)-cis-Sativenediol

326 with base affords unsaturated ketone 322, an intermediate in the McMurry synthesis. This material is converted into (±)-cis-sativenediol and (±)-helminthosporal (250) by the straightforward methods shown in the scheme.

The final synthesis we discuss in this section is the Collins-Wege synthesis of (±)-sinularene (Scheme 38),[486] which is patterned on Money's synthesis of copacamphor (see Scheme 31). Diels-Alder adducts 328 and 329 are treated with concentrated sulfuric acid to obtain lactone 330, which is converted into unsaturated ketal 333 by precedented procedures. The ketal is hydrolyzed, the double bond is epoxidized, and ring closure effected to obtain a 1:1 mixture of diastereomeric hydroxy ketones (see

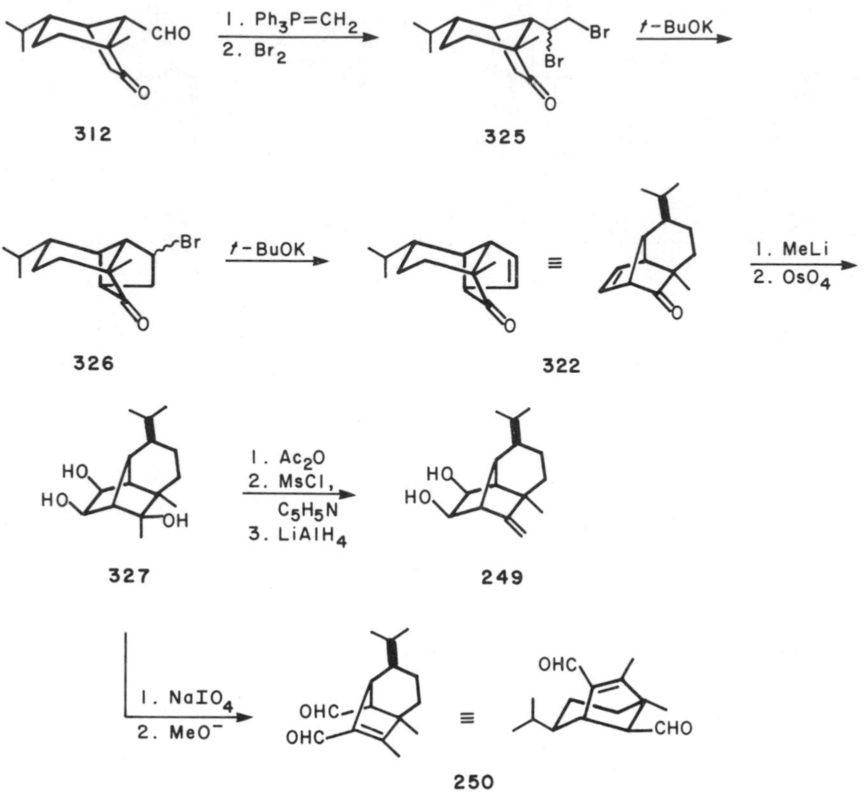

Scheme 37.   Matsumoto's Synthesis of (±)-*cis*-Sativenediol
and (±)-Helminthosporal

Scheme 31). These are separated and the exo isomer is dehydrated by pyrolysis of its acetate at 450°C. Unsaturated ketones **335** and **336**, obtained in a ratio of 95:5, are separated by preparative glpc. Catalytic hydrogenation of the major isomer gives nor-ketosinularene (**337**), which is converted into (±)-sinularene by the method indicated.

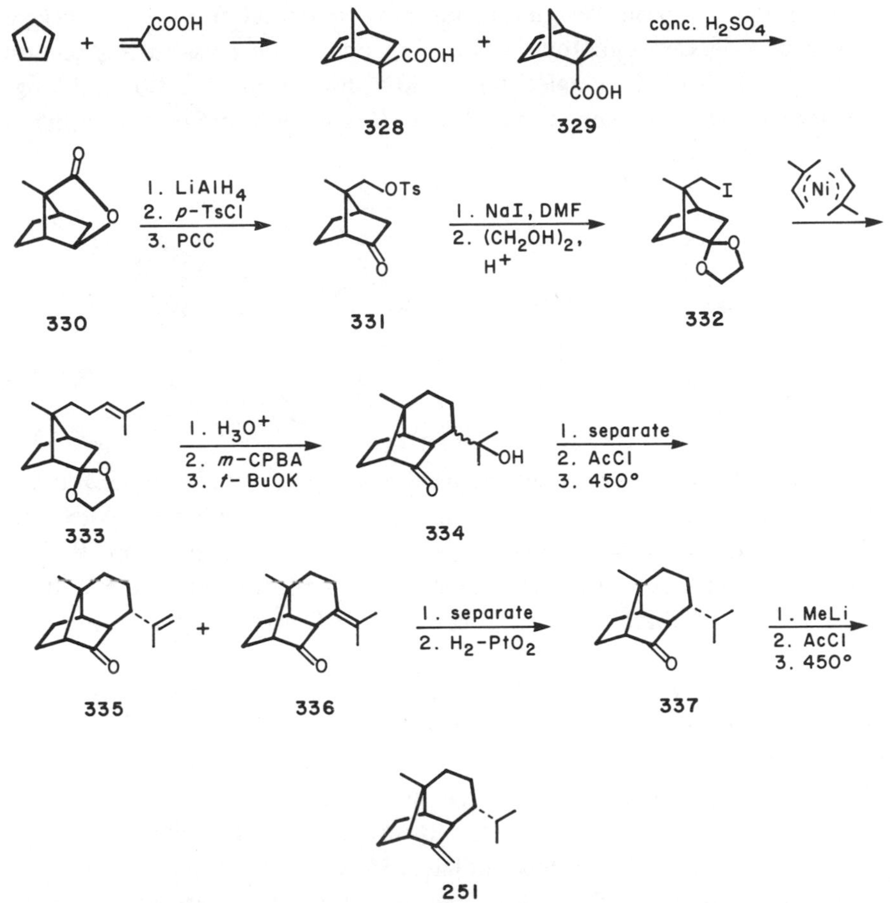

Scheme 38.   Collins-Wege Synthesis of (±)-Sinularene

(3) Longifolene, Longicyclene, Longicamphor, and Longiborneol

The "longi" tricyclic sesquiterpenes have a typical four-carbon bridge which bears a gem-dimethyl group. The ones we discuss in this section are longifolene (**338**), longicamphor (**339**), longiborneol (**340**), and longicyclene (**341**). The pioneering effort in this series was Corey's synthesis of longifolene.[487]

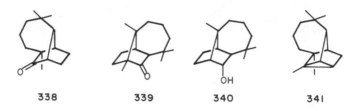

338   339   340   341

In 1972, McMurry and Isser reported a second synthesis of (±)-longifolene (Scheme 39).[488] The first four steps, beginning with the Wieland-Miescher ketone (**342**), parallel McMurry's sativene,[476] and copacamphene syntheses (see Scheme 26). Unfortunately, dehydration of the methyl carbinol derived from ketone **343** gives only a 7:3 mixture of unsaturated ketones **344** and **345**. The major isomer is epoxidized and the resulting keto epoxide cyclized to obtain hydroxy ketone **347**, which undergoes acid-catalyzed dehydration to **348**. The cyclohexene ring is expanded via dibromocyclopropane **349** and bromo alcohol **350**. After dehalogenation and oxidation of **351** to an enedione, the *gem*-dimethyl unit is completed by conjugate addition of lithium dimethylcopper. The initially formed enolate, **351a**, undergoes intramolecular aldolization to give **352**. This unexpected event is actually used to

351a

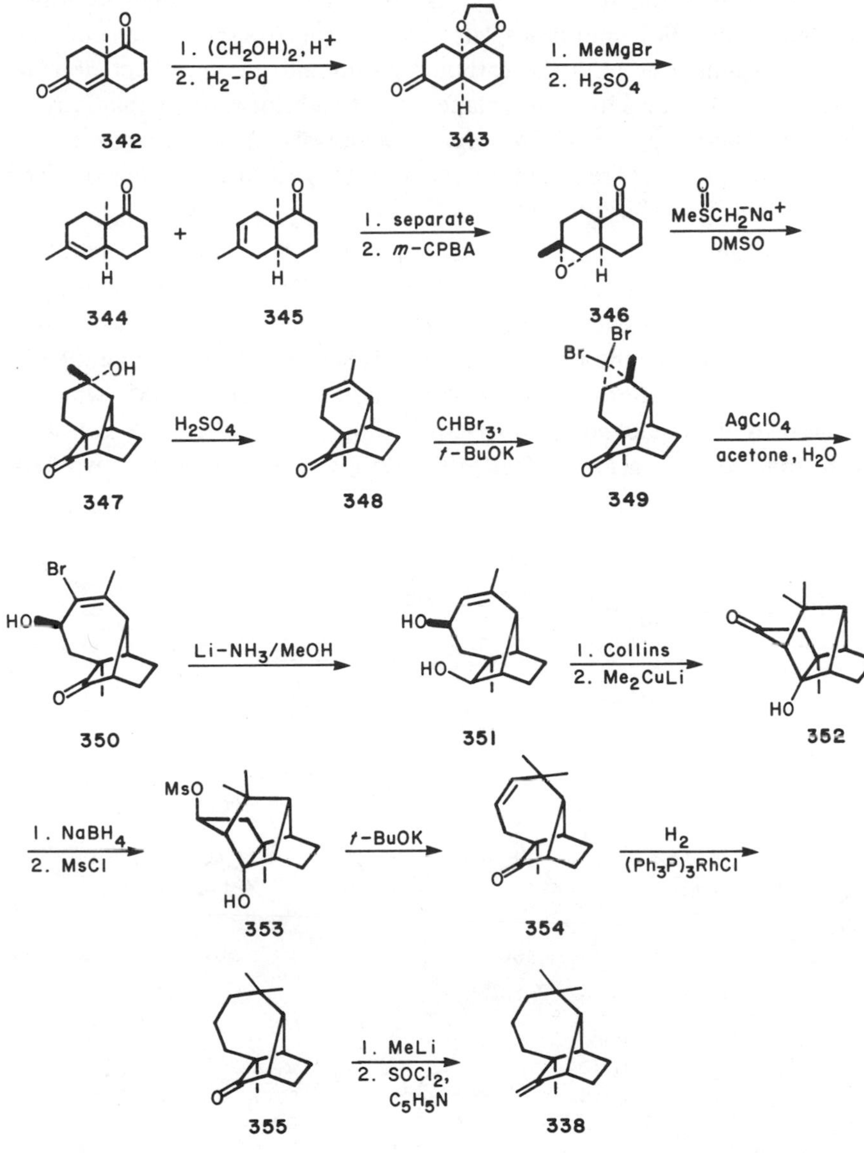

Scheme 39.   McMurry-Isser Synthesis of (±)-Longifolene

advantage, since the free carbonyl can readily be reduced and esterified to obtain the diol monomesylate **353**. Grob fragmentation of this material regenerates both the carbonyl group and the cycloheptane ring. The remaining steps to longifolene are straightforward adaptations of known methodology. In all, McMurry's synthesis entails 18 steps, somewhat more than are required in the Corey synthesis (14 steps). The overall yield is on the order of 5-6%.

Johnson and co-workers have reported the very unusual longifolene synthesis which is summarized in Scheme 40.[489] The cuprate reagent prepared from acetylenic iodide **356** is added to 2-isopropylidine-cyclopentanone and the resulting enolate is acetylated to obtain enol acetate **357**. Bromination—dehydrobromination gives enone **358**, which is reduced to allylic alcohol **359**. π-Cyclization occurs when **359** is treated with trifluoroacetic acid in ether at 0°C to give, after hydrolytic workup,

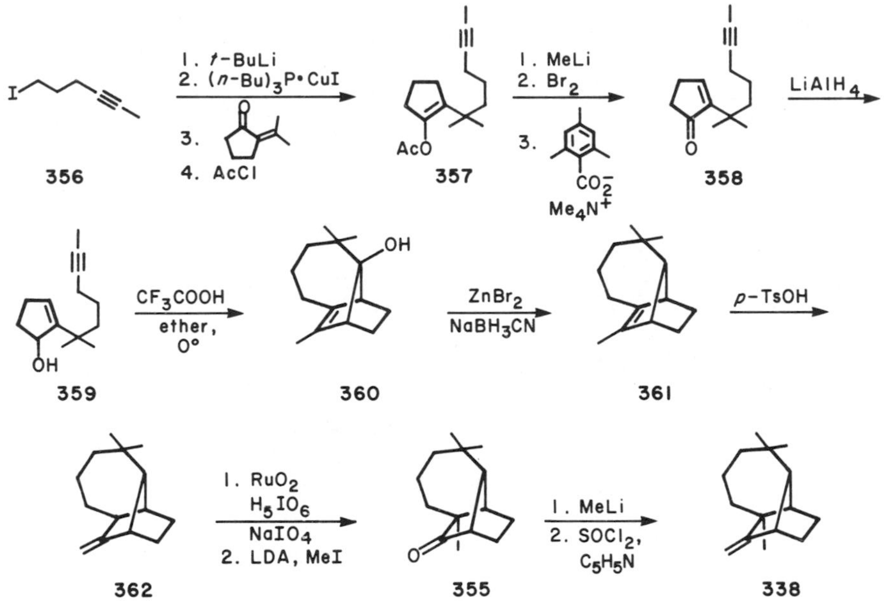

Scheme 40.   Johnson's Synthesis of (±)-Longifolene

tricyclic alcohol **360**. This remarkable transformation presumably occurs by the following mechanism:

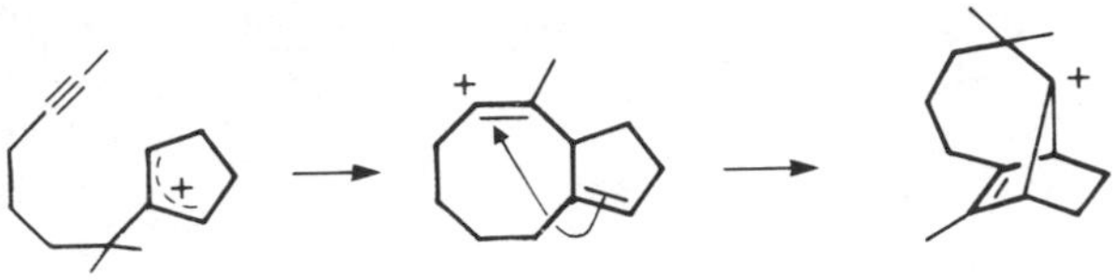

The unwanted bridgehead hydroxy is removed by taking advantage of the ease of solvolysis of 7-norbornenyl systems. Thus, when **360** is treated with sodium cyanoborohydride in the presence of zinc bromide, alkene **361** is formed in 94% yield. Treatment of **361** with *p*-toluenesulfonic acid causes isomerization of the double bond to the exocyclic position. Oxidation of the double bond gives a ketone, which is methylated to obtain **355**, an intermediate in both the Corey and McMurry syntheses. The Johnson synthesis requires 11 steps and proceeds in almost 20% overall yield.

An even more efficient longifolene synthesis was reported in 1978 by Oppolzer and Godel (Scheme 41).[490] Acylation of the morpholine enamine of cyclopentanone with 3-cyclopentenecarbonyl chloride gives an unsaturated β-diketone (**363**). Acylation of this compound with benzyl chloroformate gives an enol ether (**364**) which is photolyzed to obtain the tetracyclic keto ester **365**. Hydrogenolysis of the benzyl group gives a hydroxy ketone which undergoes *in situ* reverse aldolization to afford diketone **366**. The remaining six steps are given over to introducing the other four carbons of longifolene. The *gem*-dimethyl groups are installed by Wittig methylenation of the more reactive carbonyl, Simmons-Smith cyclopropanation, and catalytic hydrogenolysis. The synthesis is then completed as in the Johnson synthesis. The overall yield for the 10 step synthesis is an impressive 26%.

Welch and Waters have synthesized (±)-longicyclene (**341**), (±)-longicamphor (**339**) and (±)-longiborneol (**340**) as shown in Scheme 42.[491] The synthesis begins with (±)-tetrahydroeucarvone (**370**), which is readily prepared from (−)-carvone as shown. Alkylation with 4-chloro-2-pentene gives unsaturated ketone **371** in 80% yield. Cleavage of the

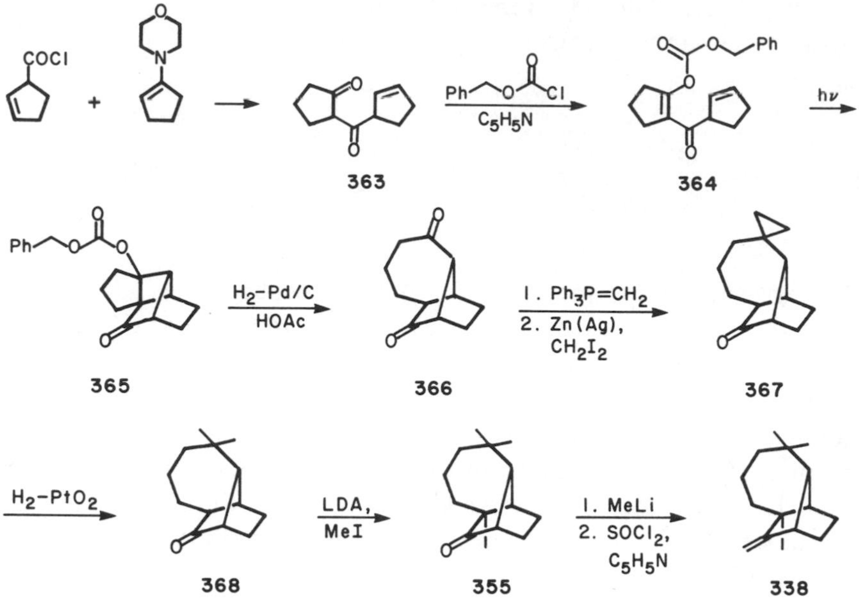

Scheme 41.   Oppolzer-Godel Synthesis of (±)-Longifolene

double bond gives keto acid **372** (mixture of diastereomers) which is cyclized to enol acetate **373**. Reduction with diisobutylaluminum hydride gives bridged ketol **374** (see also Scheme 22 and accompanying discussion). After dehydration of **374**, the resulting unsaturated ketone is manipulated more or less as in the Piers approach to copacamphene and cyclocopacamphene (see Scheme 28). The overall yield for the 19 steps from (−)-carvone to (±)-longicyclene is 8-10%. For the synthesis of (±)-longiborneol, 23 steps are required, resulting in an overall yield of about 9%.

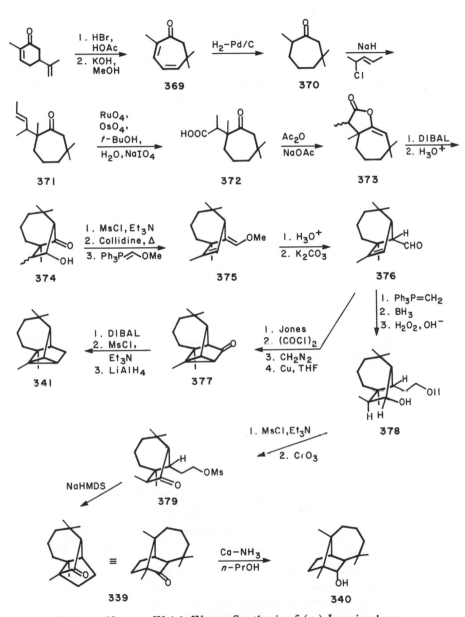

Scheme 42.   Welch-Waters Synthesis of (±)-Longicyclene,
(±)-Longicamphor, and (±)-Longiborneol

(4) Copaene, Ylangene, and Longipinene

α-Copaene (380), β-copaene (381), α-ylangene (382), β-ylangene (383), α-longipinene (384), and β-longipinene (385) are sesquiterpene analogs of the common monoterpenes α- and β-pinene. In Volume 2 of this

series, we discussed a synthesis of α-copaene and α-ylangene in which the tricyclo[4.4.0.0^{2,7}]decane skeleton is formed by base-catalyzed ring closure of a decalinic keto tosylate:[492]

In 1973, Corey and Watt reported a similar synthesis of these sesquiterpenes (Scheme 43).[493] The novel feature of the Corey-Watt synthesis is the method used to synthesize the decalinic intermediate 387. Inverse electron demand Diels-Alder reaction of pyrone 386 and 4-methyl-3-cyclohexenone gives adduct 387 in yields of up to 40%. The reaction is believed to involve the intermediate enol form of the ketone, since several other 1-methylcyclohexene derivatives do not react with pyrone 386. Compound 387 is transformed by a straightforward sequence of steps into dienone 389, which is reduced with lithium in ammonia.

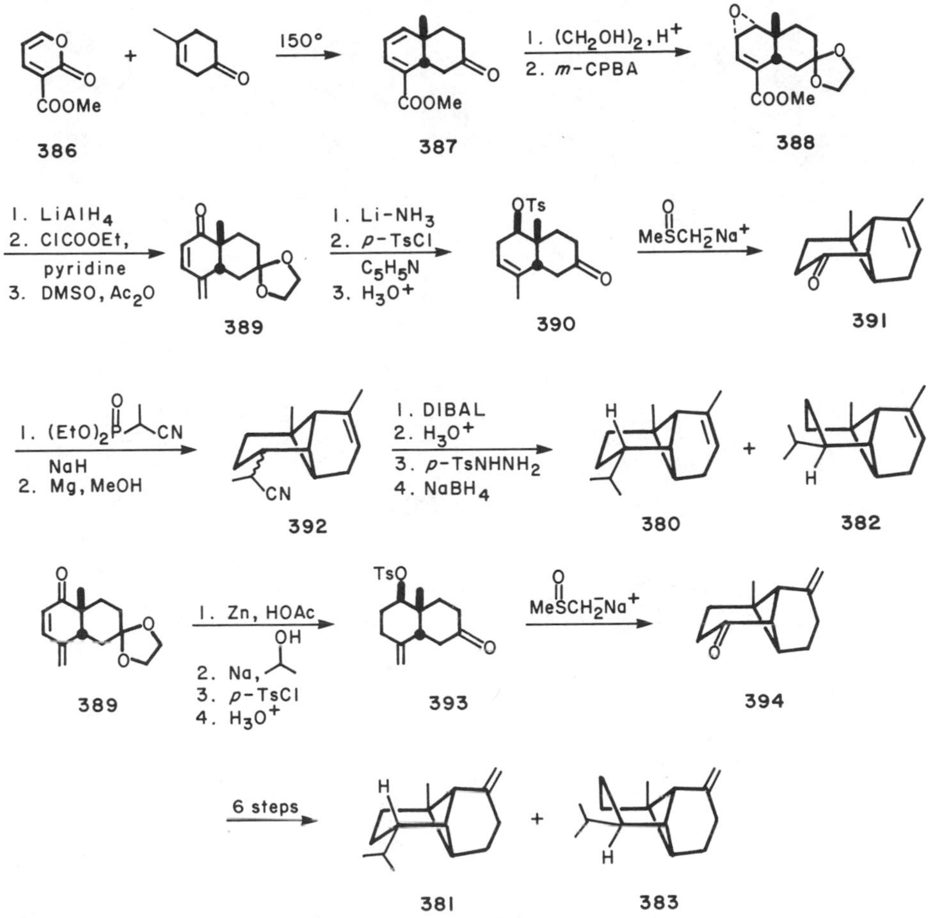

Scheme 43.   Corey-Watt Synthesis of (±)-Copaene and (±)-Ylangene

After esterification with *p*-toluenesulfonyl chloride and deprotection, keto tosylate **390** is obtained. Cyclization of **390** is accomplished in the same manner as was previously used in the Heathcock synthesis.[492] Unfortunately, tricyclic ketone **391** is obtained in only 10-38% yield. The isopropyl group is introduced by Wadsworth-Emmons reaction of **391** with α-diethylphosphonopropiononitrile, followed by reduction of the

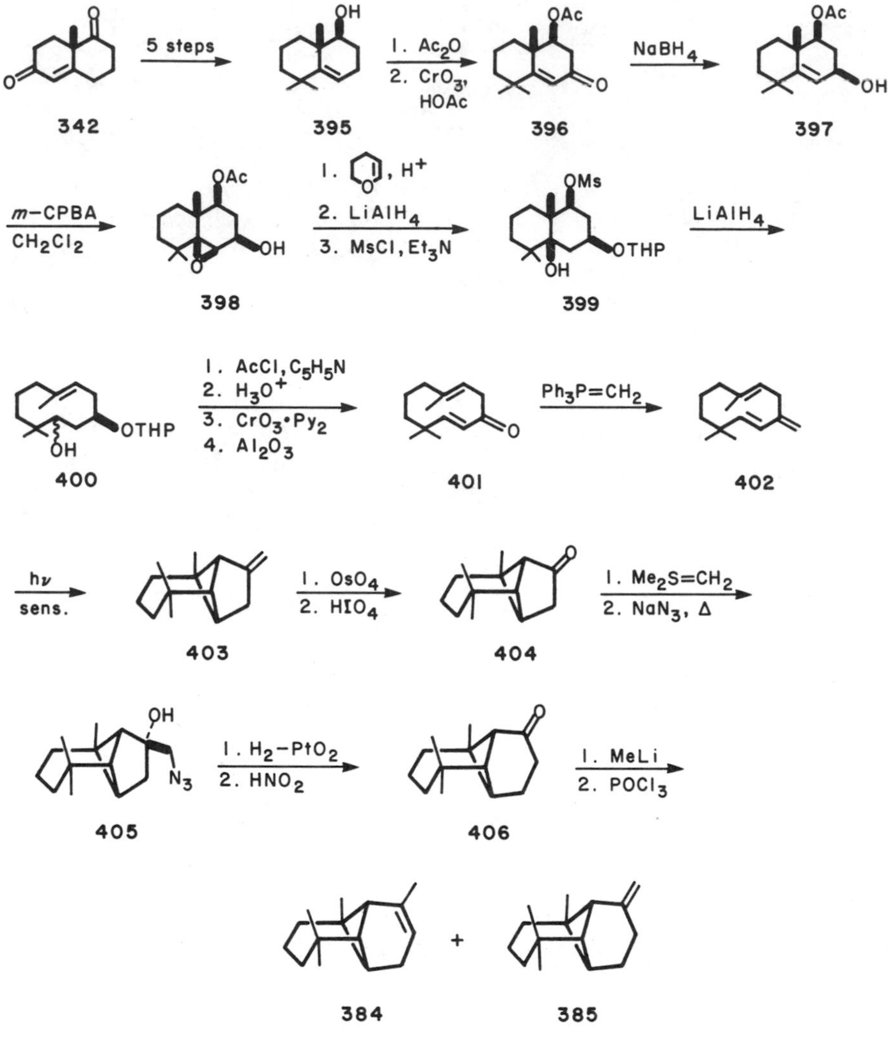

Scheme 44.   Yoshikoshi's Synthesis of (±)-Longipinene

conjugated double bond with magnesium in methanol. The nitrile group is then reduced to methyl. The final product is a 1:1 mixture of (±)-α-copaene (380) and (±)-α-ylangene (381). Zinc-acetic acid reduction of 389 gives an exocyclic double bond isomer of 390 (393) which is cyclized to unsaturated ketone 394. The latter substance is transformed by an identical sequence of steps into a 1:1 mixture of (±)-β-copaene (381) and (±)-β-ylangene (383). A disadvantage of the Corey-Watt synthesis is that it allows no control over the stereochemistry of the isopropyl group, in contrast to the Heathcock synthesis, which can be carried out so as to select either copaene or ylangene.

Yoshikoshi's synthesis of the longipinenes (Scheme 44)[494] starts out with the Wieland-Miescher ketone (342), which is transformed into keto acetate 396 by well-established procedures. Allylic alcohol 397 is epoxidized and the resulting epoxy alcohol transformed into hydroxy mesylate 399. Treatment of this material with lithium-aluminum hydride results in Grob fragmentation and reduction of the resulting ketone. Compound 400 is transformed into dienone 401 and thence into triene 402. Sensitized photoisomerization of 402 gives tricyclic alkene 403 in 75% yield.[495] The remaining steps are concerned with expanding the five-membered ring and introducing the final carbon. The final product is a 5:3 mixture of (±)-α-longipinene and (±)-β-longipinene. The overall yield of the two terpenes is 6% for the 20 steps from the known keto ester 396.

(5) Isocyanopupukeananes

9-Isocyanopupukeanane (407) and 2-isocyanopupukeanane (408) are substances which are produced by a sponge and apparently serve certain nudibranchs as defensive substances. Compounds 407 and 408 present several interesting synthetic challenges. First, there is the problem of generating the interesting tricyclo[4.3.1.0^{3,7}]decane nucleus. There is also the problem of the isopropyl group, which has the thermodynamically unfavorable endo stereochemistry. Finally, there is a problem in

407   408

establishing the correct stereochemistry of the isocyano groups in both compounds.

Two synthetic approaches have been published. Corey's synthesis of 9&isocyanopupakeanane is shown in Scheme 45.[496] Unsaturated ester **409** undergoes conjugate addition of isopropylmagnesium chloride to give compound **410**, which is cyclized to indanone **411** by treatment with polyphosporic acid. Ketone **411** is converted into ester **412** by the p-toluenesulfonylmethyl isocyanide method. Methylation of **412** gives a 6:1 mixture of diastereomers **413** and **414**. This is an important step as it is here that the eventual endo configuration of the isopropyl group is established. After cleavage of the methoxy groups the resulting phenolic acid is hydrogenated over Nishimura's catalyst to give lactone **415** in 37% yield. The stereoselectivity of the reduction presumably results from absorption of the face of the aromatic ring opposite the isopropyl group. Keto tosylate **416**, prepared from **415** by a straightforward method, is cyclized by treatment with base to give ketone **417**. Hydrogenation of the derived oxime occurs from the face opposite the angular methyl group to provide amine **418** in 80% yield. Overall, the Corey synthesis provides isonitrile **407** in 3.6% yield for the 16 steps from unsaturated ester **409**.

Yamamoto's synthesis of 9-isocyanopupukeanane is shown in Scheme 46.[497] Allylic alcohol **420** is converted by Claisen rearrangement into unsaturated aldehyde **421**. This material is further transformed into triene **424** which undergoes Diels-Alder reaction when heated to 160°C. After deprotection, hydroxy ketone **425** is obtained in quantitative yield. The ketone is protected and the secondary alcohol oxidized to obtain ketone **426**. In the Yamamoto approach, the isopropyl stereochemistry is

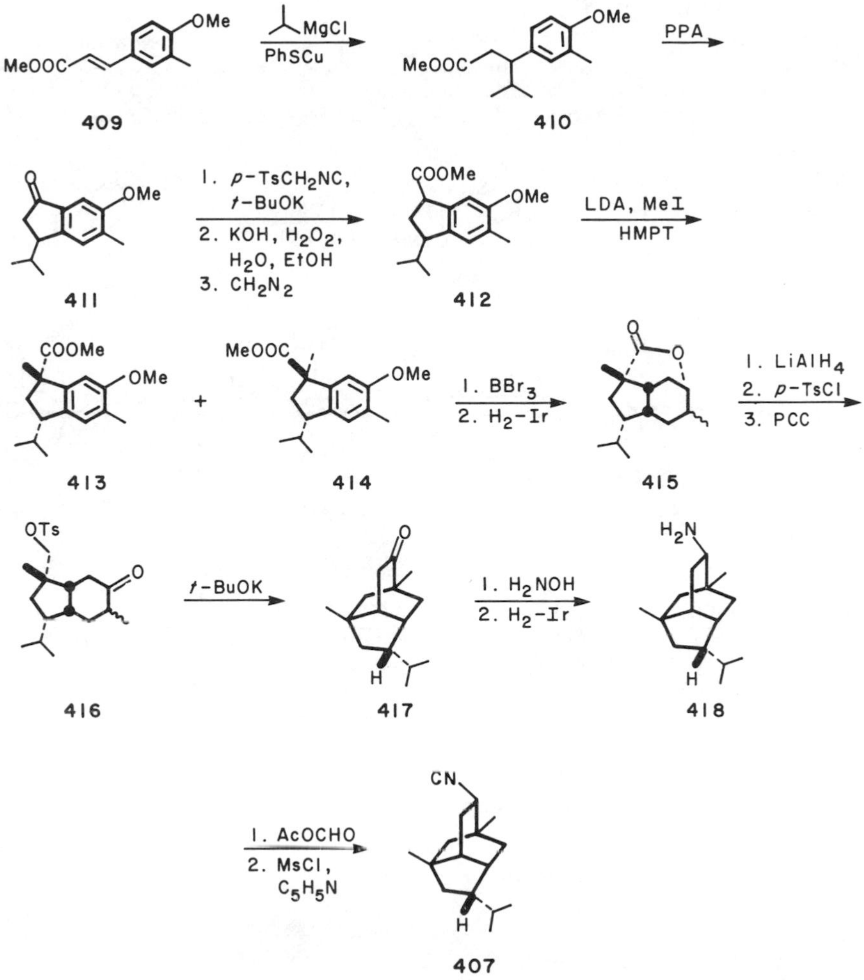

409

410

411

1. *p*-TsCH$_2$NC, *t*-BuOK
2. KOH, H$_2$O$_2$, H$_2$O, EtOH
3. CH$_2$N$_2$

412

LDA, MeI / HMPT

413 + **414**

1. BBr$_3$
2. H$_2$–Ir

415

1. LiAlH$_4$
2. *p*-TsCl
3. PCC

416

t-BuOK

417

1. H$_2$NOH
2. H$_2$–Ir

418

1. AcOCHO
2. MsCl, C$_5$H$_5$N

407

Scheme 45.   Corey's Synthesis of (±)-9-Isocyanopupukeanane

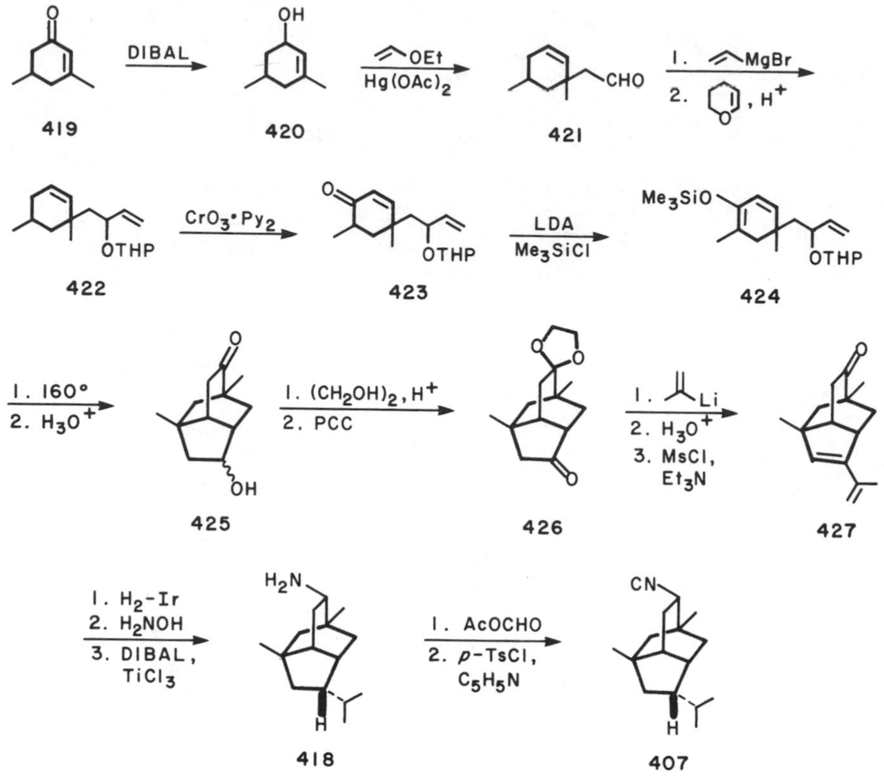

Scheme 46.   Yamamoto's Synthesis of (±)-9-Isocyanopupukeanane

established by catalytic hydrogenation of diene **427**, which occurs exclusively from the more accessible exo face. The isonitrile stereochemistry is established by the same method as is used in the Corey synthesis. The Yamamoto synthesis of **407** requires 18 steps and proceeds in 6% overall yield.

For the preparation of 2-isocyanopupukeanane, Corey and Ishiguro transformed lactone **415** into keto aldehyde **428** (Scheme 47).[498] Intramolecular aldolization of these materials affords a mixture of diastereomeric aldols **429** in 98% yield. The keto group is removed by Raney nickel desulfurization of the derived dithioketal and the hydroxy group oxidized to obtain ketone **430**. The amino group is introduced as in the earlier Corey synthesis. However, in this case hydrogenation of the oxime gives a 1:1 mixture of two diastereomers. The isomers are separated after formation of the formamides. Isomer **431** is dehydrated to (±) 2 isocyanopupukeanane (**408**).

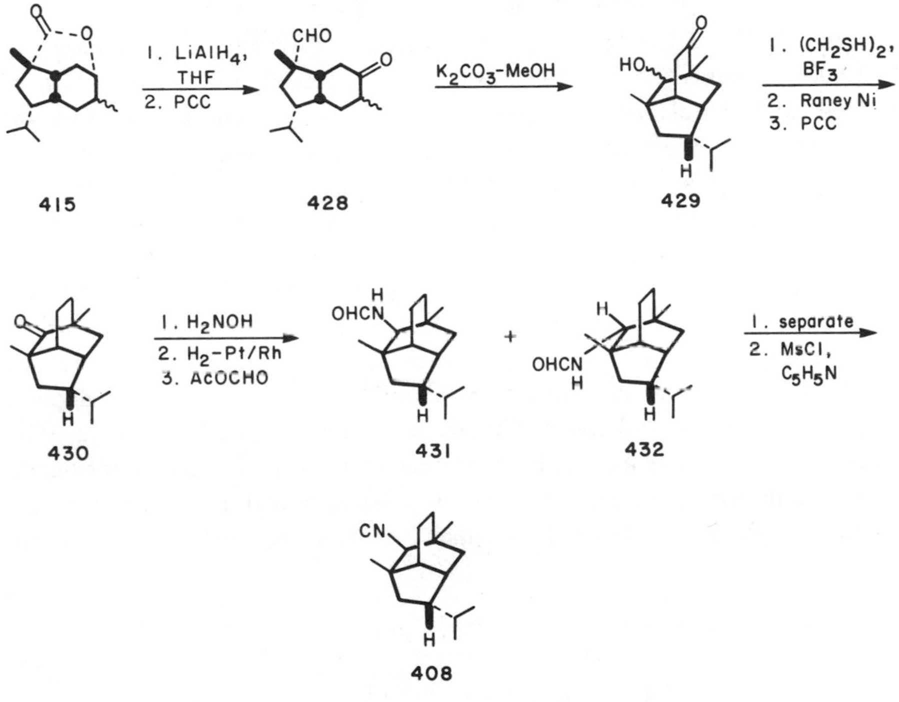

Scheme 47.   Corey's Synthesis of (±)-2-Isocyanopupukeanane

(6) Patchouli Alchohol and Seychellene

Patchouli alcohol (433) the major component of patchouli oil, is one of
the longest known terpenes. Its congeners, seychellene (434), and norpatchoulenol (435) have the same tricyclic skeleton. Previous syntheses
of 433 by Büchi and Danishefsky and of 434 by Piers were discussed in

433   434   435

Volume 2 of this series.[499] Full details of the Piers seychellene synthesis
have subsequently appeared in print.[500]

In 1970 Schmalzl and Mirrington reported a synthesis of
(±)-seychellene,[501a,c] which was later adapted for a synthesis of
(±)-patchouli alcohol.[501b] The patchouli alcohol synthesis (Scheme 48)
is similar in many respects to the Danishefsky synthesis.[502] The synthesis begins with Diels-Alder reaction of 1,3-dimethyl-
1,3-cyclohexadiene and methyl vinyl ketone, which gives unsaturated
ketone 436 as the major isomer of an 8:1 mixture. Reformatsky reaction followed by acetylation of the initially formed complex affords a
mixture of β-acetoxy esters (437) which is dehydrated by treatment with
sodium ethoxide. The resulting diastereomeric mixture (438) is reduced
with lithium in ammonia to give a 4:1 mixture of diastereomeric alcohols
439 and 440. Schmalzl and Mirrington explain the good stereoselectivity
observed in this reduction by assuming that the preferred conformation
of the intermediate radical anion is 438a and that protonation occurs
from the face away from the angular methyl group. After separation of

438a   439

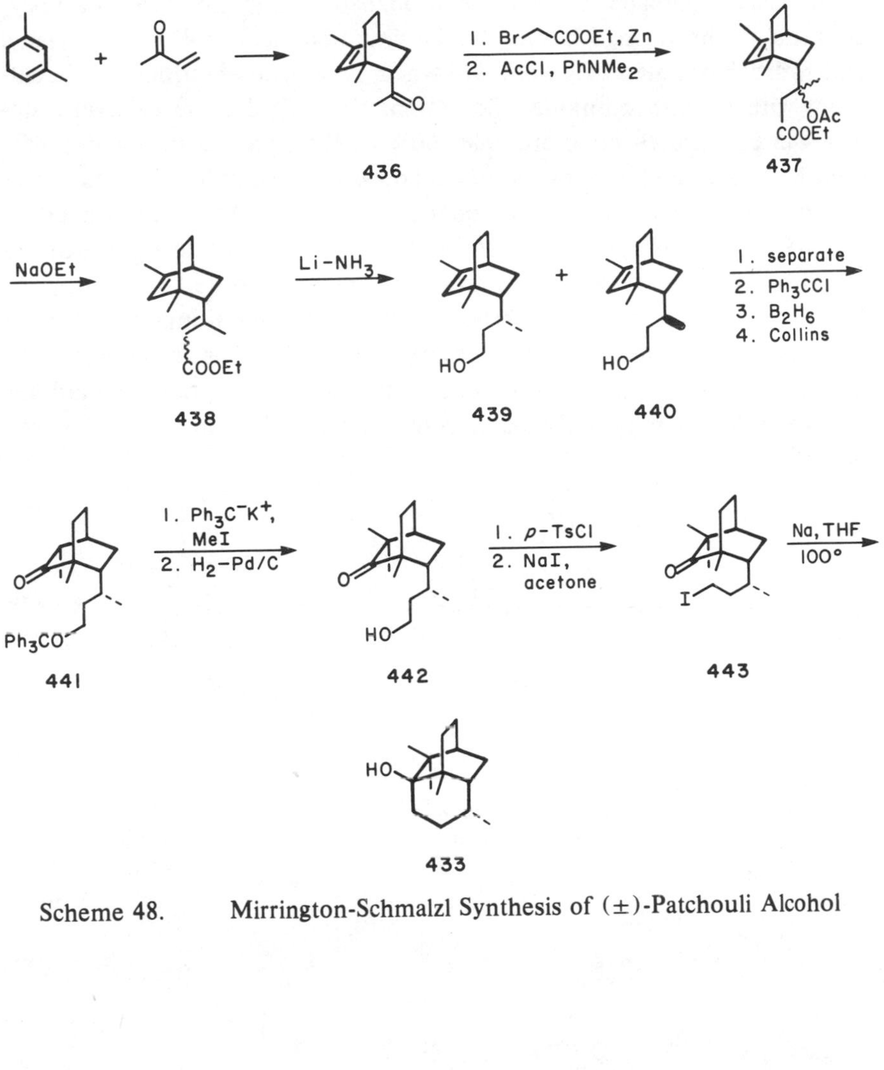

436

437

438

439   **440**

441

442

443

433

Scheme 48.   Mirrington-Schmalzl Synthesis of (±)-Patchouli Alcohol

439 and **440**, the synthesis of patchouli alcohol is completed by conversion of **439** into a ketone (**441**), which is then methylated to obtain **442**. The derived iodide **443** is cyclized by heating with sodium in THF at 100°C; racemic patchouli alcohol (**433**) is obtained in about 40% yield.

This basic approach has also been applied to a synthesis of (±)-sey-chellene as shown in Scheme 49. Unsaturated alcohol **439** is acetylated and hydrochlorinated to obtain **444** which is dehydrochlorinated by treatment with sodium ethoxide. Somewhat surprisingly, the exocyclic alkene **445** constitutes no more than 60% of the mixture of alkenes produced. More hindered bases (e.g., potassium triethylmethoxide) were found to give even more of the endocyclic isomer **439**. After separation of **445** from **439** by chromatography on silver nitrate—impregnated alumina, the primary alcohol is protected and the alkene oxidized by the Lemieux-Johnson method. After deprotection, the primary alcohol is esterified with *p*-toluenesulfonyl chloride. Cyclization is accomplished by treating the resulting keto tosylate with potassium triphenylmethide. Ketone **448** has previously been converted by Piers into (±)-seychellene.[503]

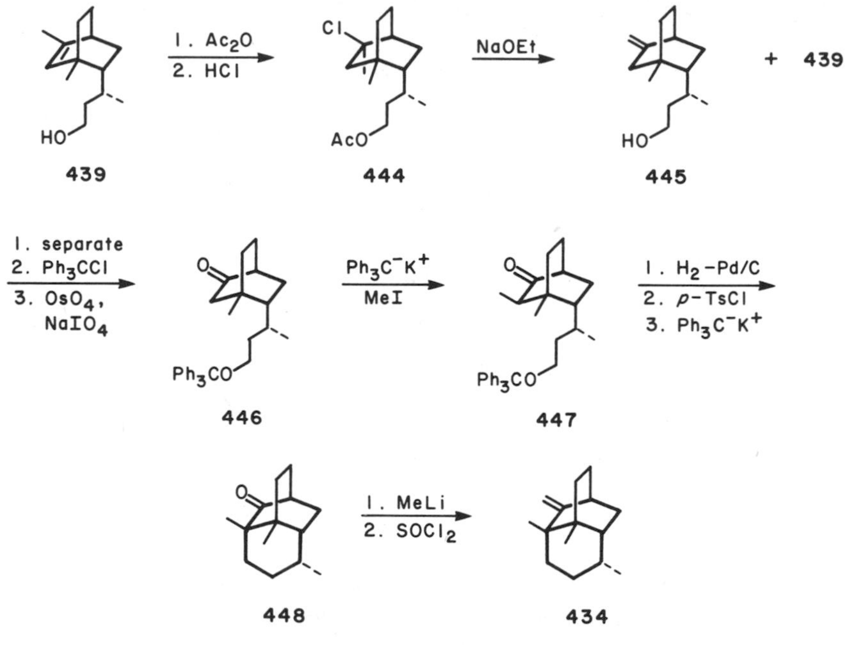

Scheme 49.   Mirrington-Schmalzl Synthesis of (±)-Seychellene

Yoshikoshi and co-workers have prepared (±)-seychellene by a route incorporating an intramolecular Diels-Alder reaction as the key step in constructing the tricyclic skeleton (Scheme 50).[504] Alkylation of 2,3-dimethylcyclohexenone with 3-methyl-1,5-diiodopentane and sodium amide in ammonia gives unsaturated ketone 449 as a 1:1 mixture of diastereomers in 35% yield. Treatment of this material with two equivalents of N-bromo-succinimide in carbon tetrachloride provides bromodienone 450 which is reduced with chromous chloride. Elimination of bromine is accompanied by deconjugation. The diene system is conjugated by treatment of 451 with acid. The third double bond is introduced by Cope elimination of the dimethylamine oxide at 430°C in a glpc apparatus. The intermediate trienone (453) undergoes *in situ* Diels-Alder cyclization affording unsaturated ketone 454. Hydrogenation of 454 gives ketone 448, which is transformed by the Piers procedure into (±)-seychellene.

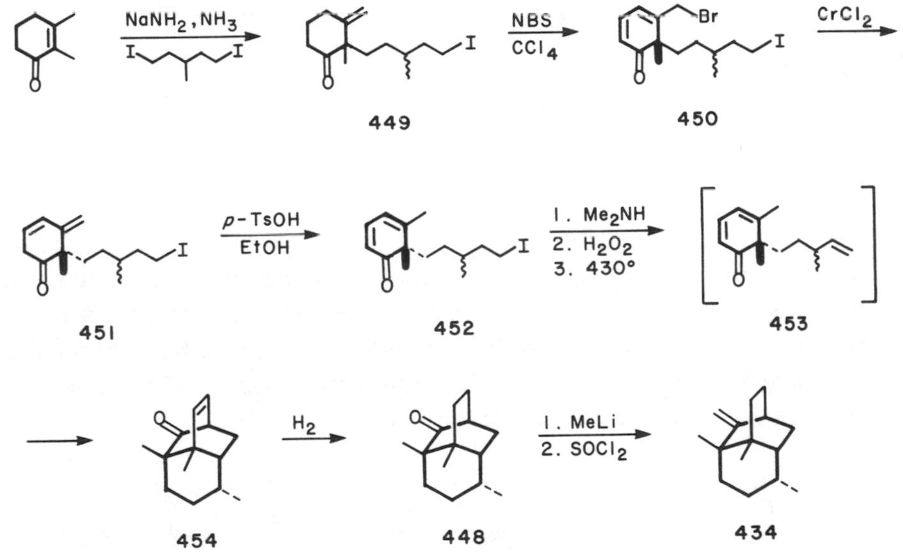

Scheme 50.   Yoshikoshi's Synthesis of (±)-Seychellene

Näf and Ohloff have reported the interesting and economical synthesis of (±)-patchouli alcohol shown in Scheme 51.[505] Cyclohexadienone 455, the starting material for Danishefsky's synthesis[502] is alkylated with 3-methyl-4-pentenyllithium to obtain a 1:1 mixture of diastereomeric alcohols 456 and 457 in 59% yield. When a decalin solution of this mixture containing 5% potassium *t*-butoxide is heated in a sealed tube for 24 hours, alcohol 458 is obtained in 60% yield based on isomer 457. The intramolecular Diels-Alder reaction appears to be remarkably stereospecific, since no adduct derived from diastereomer 456 is

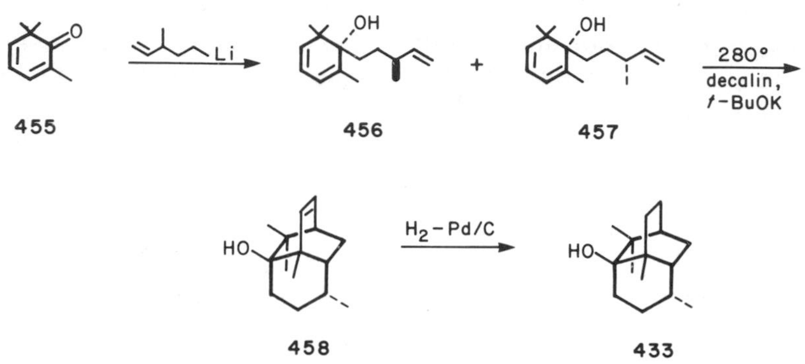

Scheme 51.   Näf-Ohloff Synthesis of (±)-Pathcouli Alcohol

obtained. The role of the base in the Diels-Alder reaction is not understood, although it appears to be necessary. It has been suggested that this may be an example of "alkoxide-accelerated cycloaddition," analogous to the well-known alkoxide-accelerated sigmatropic rearrangements. The base may also serve to catalyze interconversion of 456 and 457 by alkoxide-accelerated electrocyclic reaction:

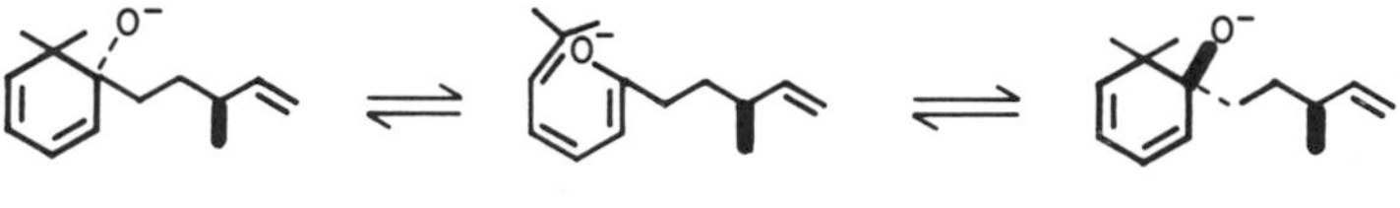

Fráter has also employed an intramoleculer Diels-Alder reaction in a synthesis of (±)-seychellene (Scheme 52).[506] In Fráter's synthesis, phenoxide **459** is alkylated with bromide **460** to obtain a 1:1 mixture of diastereomers **461** in 80% yield. When this mixture is heated at 80°C, cycloaddition occurs to give a 3:1 mixture of tricyclic ketones **462** and **463** in 60% yield. Catalytic hydrogenation of the major isomer affords a 2:1 mixture of saturated ketones **464** and **465**. When methyllithium is

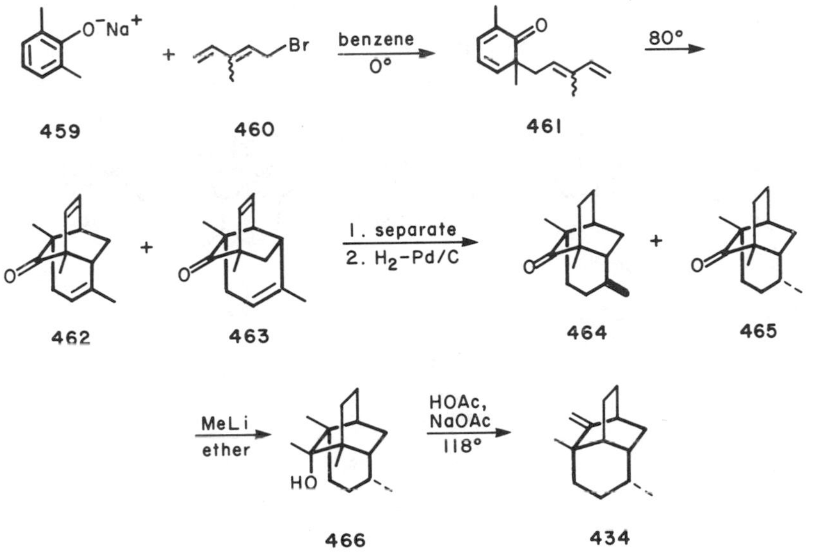

Scheme 52.   Fráter's Synthesis of (±)-Seychellene

added to this mixture only **465** reacts, giving alcohol **466** in 97% yield (based on isomer **465**). Solvolytic rearrangement of this material gives (±)-seychellene in 58% yield. Even though isomers are produced in two key steps, this short synthesis still provides seychellene in 7% overall yield.

Jung and McCombs have synthesized (±)-seychellene as shown in Scheme 53.[507] Enone **467** is transformed into the trimethylsilyloxycyclo-hexadiene **468** which undergoes regio- and stereoselective Diels-Alder

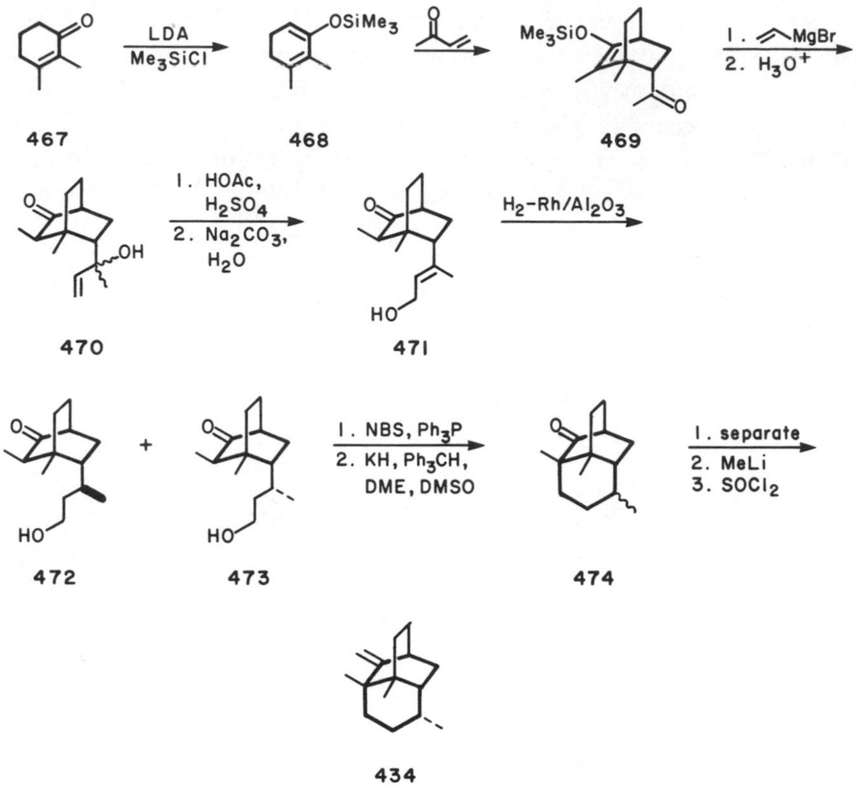

Scheme 53.   Jung-McCombs Synthesis of (±)-Seychellene

reaction to give adduct **469** in 75% yield. Homologation of this material gives allylic alcohol **471** which is hydrogenated to a 1:1 mixture of hydroxy ketones **472** and **473**. These compounds are converted into bromo ketones, which undergo cyclization to a mixture of diastereomeric ketones. After purification by silica gel chromatography the correct iso- mer is converted into (±)-seychellene by Piers' method. The Jung syn- thesis, even though stereorandom, requires only 10 steps and gives **434**

in about 10% overall yield. A similar approach has been applied by
Spitzner in a synthesis of 7-desmethyl-11-norseychellanone (**475**).[508]

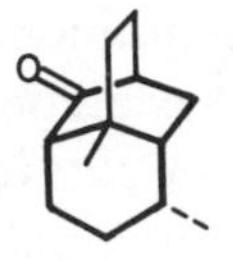

475

Yamada and co-workers have synthesized (±)-seychellene and
(±)-patchouli alcohol by a route involving a common intermediate.[509]
For the synthesis of seychellene (Scheme 54) the known keto acid **476** is

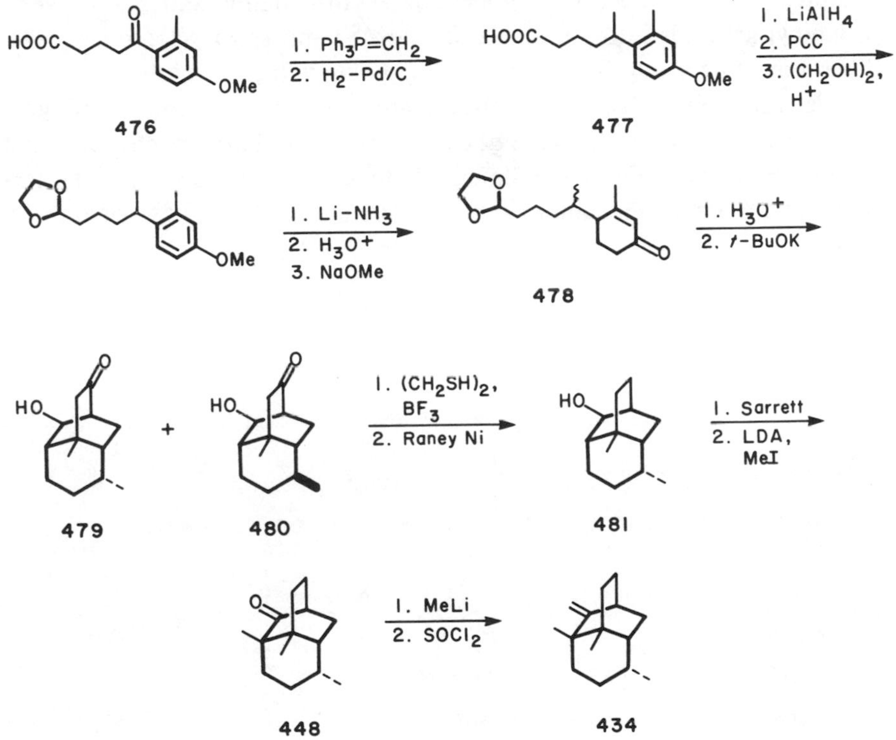

Scheme 54.   Yamada's Synthesis of (±)-Seychellene

transformed into **477** by Wittig methylenation and hydrogenation. After conversion of the carboxy group to a protected aldehyde, the aromatic ring is reduced by the Birch procedure to obtain enone **478** as a mixture of diastereomers. After deacetalization, the enone aldehyde is treated with potassium *t*-butoxide. Diastereomeric hydroxy ketones **479** and **480** are obtained in 23% and 3% yield, respectively. The major isomer is deoxygenated by Raney nickel desulfurization of the dithioketal. Alcohol **481** is oxidized with chromium trioxide in pyridine and the resulting ketone is methylated to obtain **448**, an intermediate in the Piers seychellene synthesis.

For the synthesis of (±)-patchouli alcohol (Scheme 55), Yamada and co-workers oxidize the ketone derived from **481** by the Vedejs procedure. After protecting the alcohol as the methyl ether, the *gem*-dimethyl group is introduced via alkene **483** and spirocyclopropane **484**. The protecting group is removed by oxidation with chromium trioxide in acetic acid. The Yamada syntheses are considerably longer and give lower overall yields than previous syntheses of both seychellene and patchouli alcohol. For seychellene, 12 steps are required to give the terpene in 5% yield. Patchouli alcohol is obtained in only 0.5% yield in 16 steps.

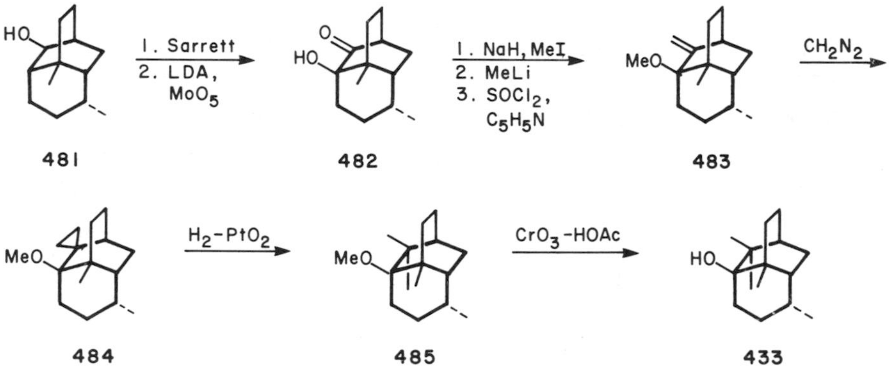

Scheme 55.   Yamada's Synthesis of (±)-Patchouli Alcohol

The interesting norsesquiterpene norpatchoulenol (**435**) has been found to be the compound actually responsible for the characteristic scent of patchouli alcohol. It was first synthesized by Teisseire by the lengthy route summarized in Scheme 56.[510] Addition of methylmag-

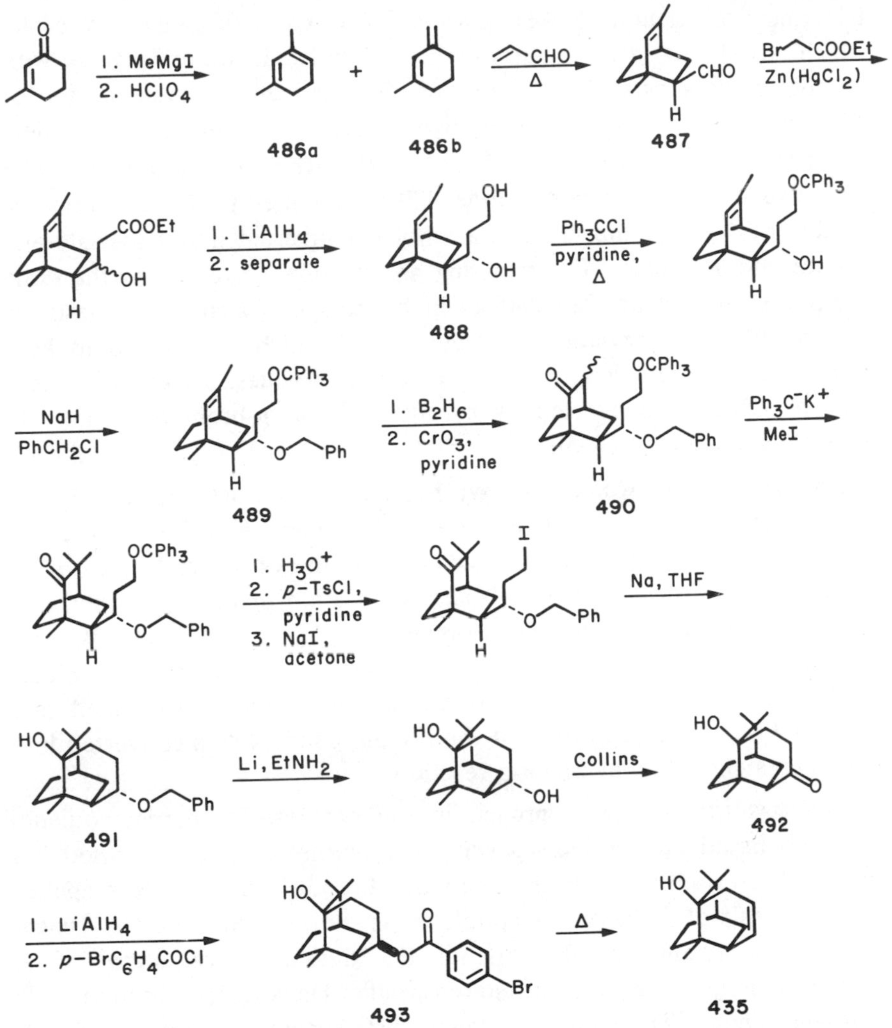

Scheme 56.   Teisseire's Synthesis of (±) Norpatchoulenol

nesium iodide to 3-methylcyclohexenone followed by acid-catalyzed dehydration of the resulting alcohol gives a 55:45 equilibrium mixture of dienes **486a** and **486b**. Treatment of this mixture with refluxing acrolien gives Diels-Alder adduct **487** and unreacted diene **486b**, which is subjected to acidic equilibration to regenerate a mixture of **486a** and **486b**. Reformatsky reaction of **487** gives a 3:2 mixture of β-hydroxy esters which is reduced to a similar mixture of diols. In order to work with homogeneous substances, the mixture is separated by crystallization of **488**, the major isomer. The two hydroxy groups are separately protected, to obtain unsaturated diether **489**, which is hydroborated with oxidative workup to obtain ketone **490**. After methylation and removal of the trityl group, the primary alcohol is converted into a tosylate and thence into an iodide. The resulting halo ketone is cyclized by the same method employed by Danishefsky in his patchouli alcohol synthesis to obtain **491**. Debenzylation provides a diol which is oxidized to keto alcohol **492**. Reduction of this compound provides the epimeric keto alcohol, which is converted by pyrolysis of the *p*-bromobenzoate into (±)-norpatchoulenol (**435**).

Oppolzer and Snowden have synthesized (±)-norpatchoulenol by the interesting route shown in Scheme 57.[511] Dienone **455** is allowed to react with 3-triethylsilyloxypentadienyllithium (**494**) and the initial adduct silylated to obtain **495**. After desilylation of the enol ether with KF in methanol, trienone **496** is obtained in 47% overall yield. When a benzene solution of **496** is heated in a sealed tube at 230°C a highly regioselective Diels-Alder cyclization occurs to produce unsaturated keto alcohol **497**. After saturation of the double bond, **498** is converted into (±)-norpatchoulenol by Shapiro reaction.

Gras has reported an approach to the synthesis of norpatchoulenol which is based on the Piers seychellene synthesis.[512] The method has not yet been used to synthesize the natural product, but has been applied in a synthesis of desoxynorpatchoulenol as shown in Scheme 58. Oxidation of the readily-available phenol **499** gives a benzoquinone (**500**) which undergoes regio- and stereospecific Diels-Alder reaction with 2,4-pentadienol. The resulting adduct (**501**) is converted into enone **504**

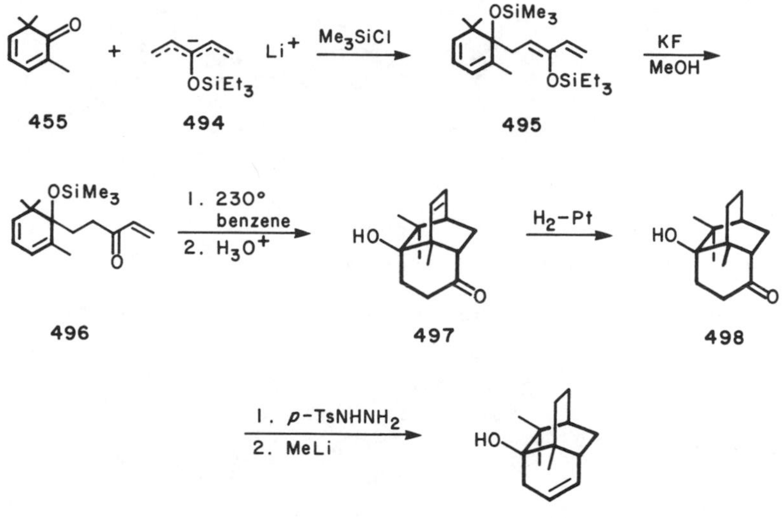

Scheme 57.   Oppolzer-Snowden Synthesis of (±)-Norpatchoulenol

by methodology originally introduced by Woodward in his pioneering steroid synthesis. The *gem*-dimethyl unit is completed by conjugate addition of lithium dimethylcopper to **504**. After replacing the tetrahydropyranyl group with tosyl, the third ring is closed using dimsylsodium as base. Wolff-Kishner reduction of tricyclic ketone **507** provides racemic desoxynorpatchoulenol (**508**).

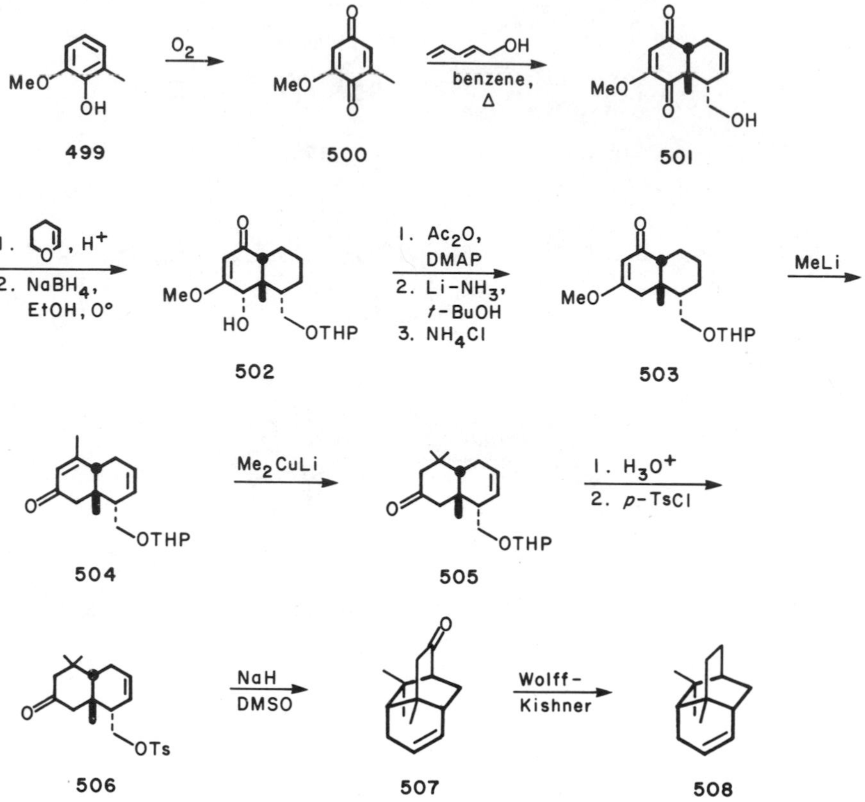

Scheme 58.   Gras Synthesis
of (±)-Desoxynorpatchouli Alcohol

(7) Zizane (tricyclovetivene), Zizanoic Acid, Epizizanoic Acid, and Khusimone

Vetiver oil contains a group of methanohydroazulenes which includes zizaene (tricyclovetivene, **509**), zizanoic acid (**510**), epizizanoic acid (**512**), and khusimone (**514**). These four natural products and the unnatural isomers of epizizaene (**511** and **513**) have yielded to synthesis. Yoshikoshi's synthesis of epizizanoic acid, which was discussed in Volume 2 of this series, has now been reported in full.[509]

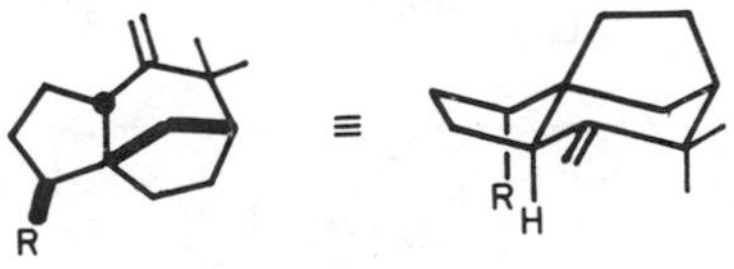

509: R = Me
510: R = COOH

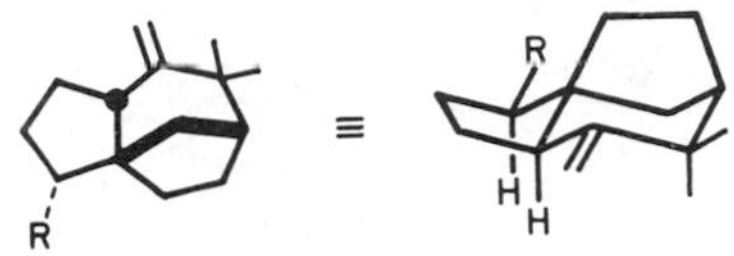

511: R = Me
512: R = COOH

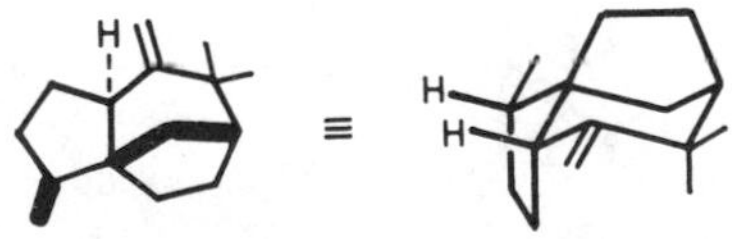

513

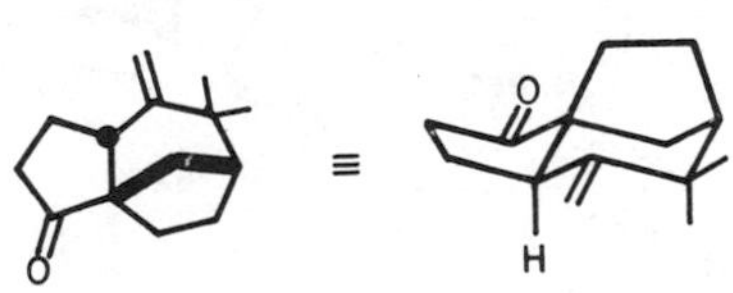

514

MacSweeney and Ramage have synthesized methyl zizanoate as shown in Scheme 59.[514] The approach is similar in many respects to

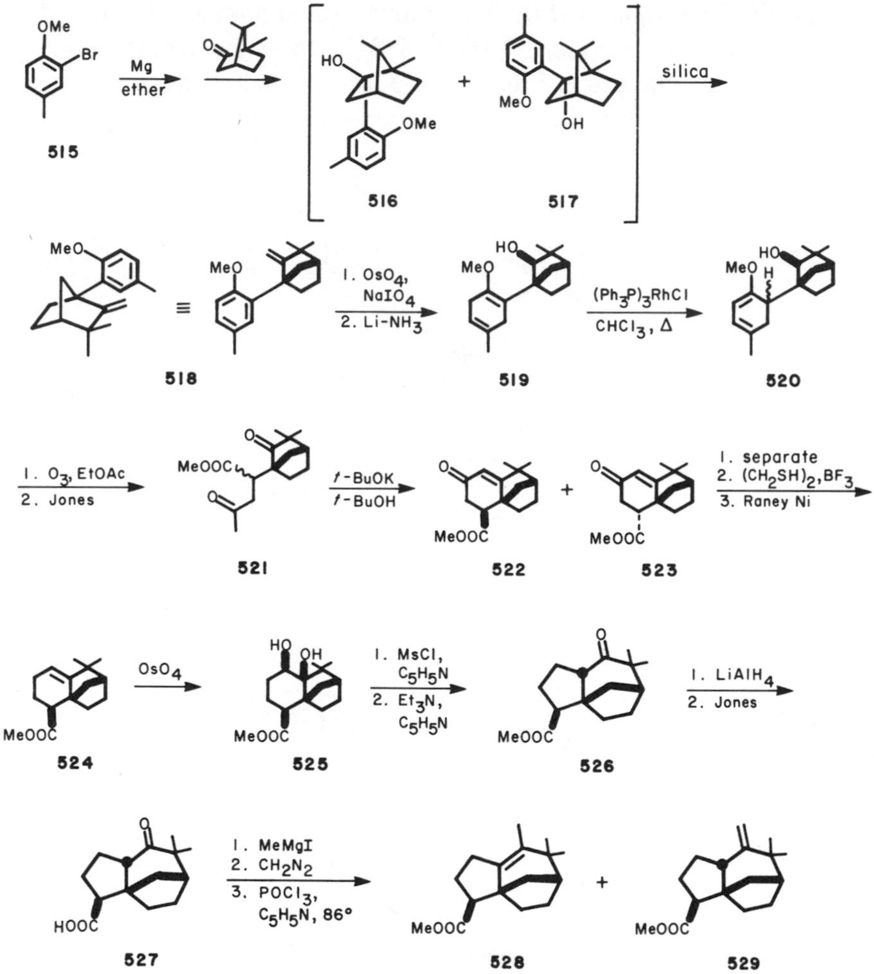

Scheme 59.   MacSweeney-Ramage Synthesis of Zizanoic Acid

Yoshikoshi's synthesis of epizizanoic acid, which begins with methyl 1-camphenecarboxylate.[515] In the MacSweeney-Ramage synthesis the Grignard reagent derived from bromide **515** is condensed with (+)-camphor to give a mixture of alcohols **516** and **517**. The mixture is not separated but is smoothly rearranged by silica gel to provide the arylcamphene **518** in 30% overall yield. The low yield is due to competing enolization in the Grignard addition step. After Lemieux- Johnson cleavage of the methylene group, the resulting ketone is subjected to Birch reduction to obtain **519**. Isomerization to the conjugated diene system is brought about by heating with Wilkinson's catalyst. Ozonolysis and Jones oxidation provide a diastereomeric mixture of diketo esters which is cyclized by base to obtain a 2:3 mixture of enones **522** and **523**. The isomers are separated chromatographically and the minor isomer deoxygenated by desulfurization of its dithioketal. The resulting unsaturated ester (**524**) is hydroxylated and the derived diol **525** rearranged by Büchi's method[516] to obtain keto ester **526**. This is a key step in the synthesis. A similar rearrangement was employed by Yoshikoshi and co-workers in their epizizanoic acid synthesis.[515] However, in that case potassium *t*-butoxide was employed as the base. The stronger base leads to equilibration of both epimerizable centers. By using more weakly basic conditions, MacSweeney and Ramage are able to realize complete stereospecificity in the rearrangement. In order to bring about saponification of **526** without epimerization, it is reduced with lithium aluminum hydride and the resulting diol is reoxidized with Jones' reagent. The methylene group is introduced by addition of methyllithium and dehydration of the resulting tertiary alcohol. The product is a 2:3 mixture of methyl isozizanoate (**528**) and methyl zizanoate (**529**) which are separated by chromatography. A similar sequence of steps beginning with epimer **523** provides methyl epizizanoate. Alternatively, this epimer may be isomerized to a 2:3 mixture of **522** and **523**, thus providing more of the former isomer for conversion into ester **529**. In all, the MacSweeney-Ramage synthesis requires 18 steps, also requires two isomer separations (in each case the minor isomer is used), and affords the final product in an overall yield of less than 1%.

Wiesner and co-workers have synthesized zizaene (509) and its two diastereomers 511 and 513 as shown in Scheme 60.[517] The synthesis begins with indanol 530, which is alkylated via Claisen rearrangement of the allyl ether. The resulting phenol is methylated to obtain 531, which

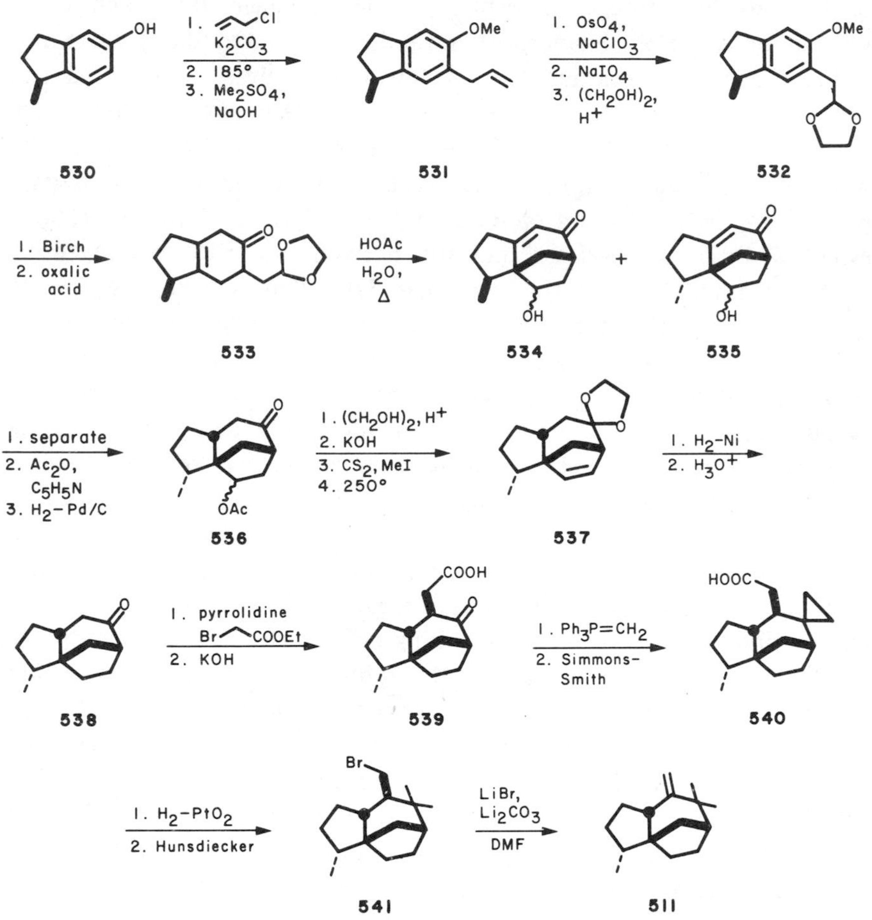

Scheme 60.   Wiesner's Synthesis of (±)-Zizaene and its Stereoisomers

Scheme 60, Continued.
Wiesner's Synthesis of (±)-Zizaene and its Stereoisomers

is converted into acetal **532**. Birch reduction followed by gentle hydrolysis yields the unconjugated enone **533**. Upon treatment with 80% acetic acid cyclization occurs to give a 2:3 mixture of hydroxy ketones **534** and **535**. The hydroxy configuration was not ascertained in either case. After separation of the isomers, compound **535** is acetylated and hydrogenated to give **536**. As indicated, hydrogenation occurs exclusively on the face of the cyclopentene ring anti to the methyl group. The ketone is masked and the unwanted ester function removed by hydrolysis, conversion of the alcohol to a xanthate, pyrolytic elimination, and catalytic hydrogenation. After deprotection of the carbonyl, ketone **538** is obtained. The three remaining carbons are introduced by Stork alkylation with ethyl bromoacetate, Wittig methylenation of the resulting keto acid, and Simmons-Smith cyclopropanation. After catalytic hydrogenolysis, the acid is subjected to modified Hunsdiecker decarboxylation and the resulting bromide eliminated to obtain (±)-epizizaene (**511**).

Isomer **534** is subjected to a similar sequence to give keto acetate **542**. Note that, as in the case of **535**, hydrogenation occurs anti to the methyl group. The remaining steps are identical to those employed for the synthesis of **511**; the final product is (±)-epizizaene **513**. For the preparation of (±)-zizaene itself, isomer **513** is cleaved by the Lemieux-Johnson method to obtain ketone **543**, which is epimerized by treatment with base to obtain a 1:1 mixture of **543** and **544**. The isomers are separated and the latter isomer converted into (±)-zizaene (**509**) as shown. The Wiesner synthesis of (±)-zizaene requires 30 steps, requires two isomer separations, and proceeds in an overall yield of 0.009%.

Coates and Sowerby have also synthesized (±)-zizaene (Scheme 61).[518] As in the MacSweeney-Ramage approach, Coates and Sowerby employ a bicyclo[2.2.1]heptane starting material. Addition of 3-phenylpropylmagnesium bromide to norbornanone gives an alcohol (**546**) which is rearranged in acetic acid to acetate **547**. The secondary methyl group is introduced by a six-step sequence which gives a 2:1 mixture of diastereomers **550** and **551**. After Wolff-Kishner deoxygenation, the isomers are separated and the major acetate (**552**, obtained in 25% overall yield from **547**) is degraded to acetoxy ester **553** in 30-35% yield. By a routine series of steps **553** is converted into keto amide **555**. This compound is further transformed into a diazo ketone which undergoes smooth cyclization to a single tricyclic ketone, **556**. The high degree of stereoselectivity observed in the cyclization step is considered to be due to exo-face attack on the carbonyl group, leading to diazonium alkoxide **557**. The gem-dimethyl group is introduced by reduction of the n-butyl-thiomethylene ketone **558** with lithium in ammonia, followed by methylation of the resulting enolate. The product of this procedure is found to

557   **556**

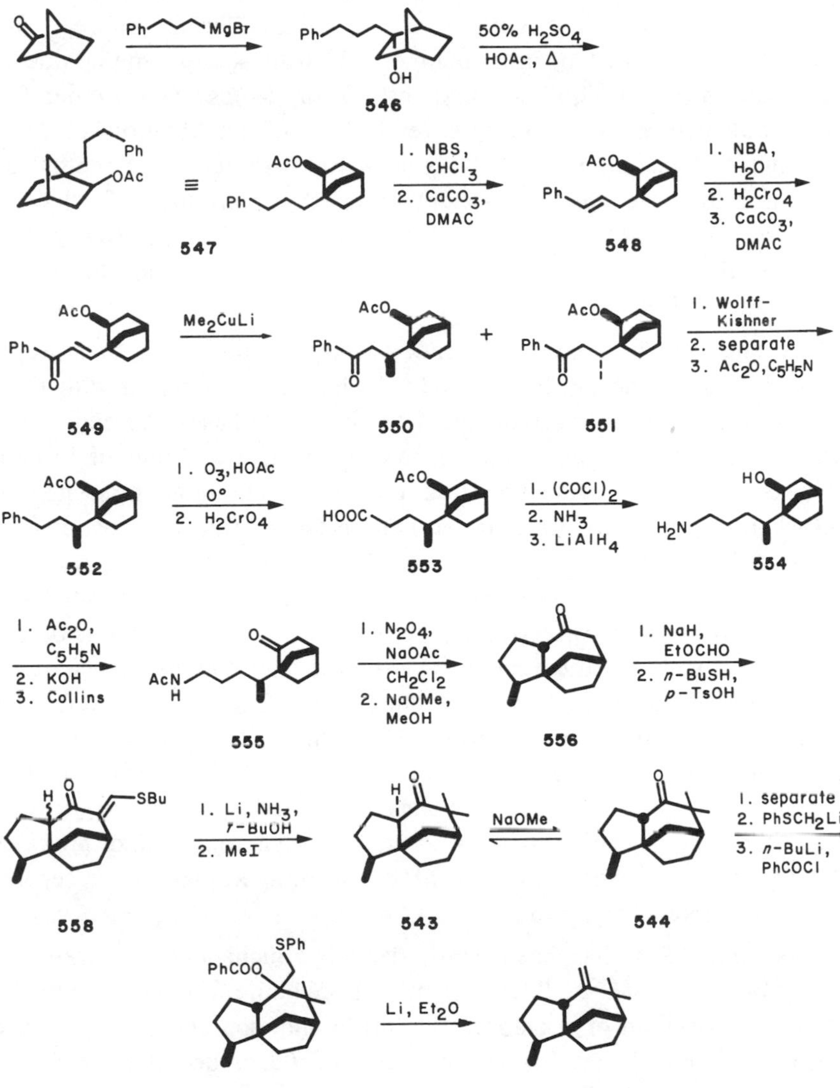

Scheme 61.   Coates-Sowerby Synthesis of (±)-Zizaene

be isomer **543**, presumably as a result of epimerization in the course of either the formylation or methylation steps. Equilibrium between **543** and **544** is established by treatment of **543** with sodium methoxide in methanol. Although equilibrium slightly favors isomer **543**, isomer **544** can be obtained in 64% yield by several cycles of equilibration, chromatography, and reequilibration of the unwanted isomer. The methylene group is introduced via the *S*-phenylthiomethyl benzoate **559** which is reductively eliminated to give (±)-zizaene. The Coates-Sowerby synthesis requires 25 steps, requires 2 isomer separations, and affords the end product in 0.2% overall yield.

Piers and Banville have prepared zizaene starting with cyclopropyl enone **560** as outlined in Scheme 62.[519] The synthesis begins with thermal vinylcyclopropane rearrangement to give, after base-catalyzed conjugation of the double bond, enone **561**. Conjugate addition of lithium divinylcopper occurs from the face of the double bond opposite the methyl group, providing unsaturated ketone **562**. After introduction of a second double bond, thiophenol is added to the enone and the resulting ketone is converted into ketal **564**. The vinyl group is hydroborated and the resulting alcohol converted into a tosylate. After oxidation of the sulfide to a sulfone, the final ring is closed and the benzenesulfonyl group removed by reduction with aluminum amalgam. The ketone is unmasked to obtain tricyclic ketone **567**, which is treated with sodium methoxide in methanol to obtain a 2:1 mixture of **567** and **556**, which is an intermediate in the Coates synthesis. The Piers synthesis requires 22 steps from enone **560**, involves 2 isomer separations, and provides (±)-zizaene in an overall yield of 0.8% (assuming a quantitative separation of isomers **567** and **556**).

Khusimone (**514**) has been synthesized by Büchi and co-workers[520] and by Liu and Chan.[521] Büchi's synthesis (Scheme 63) begins with the Diels-Alder reaction of α-chloroacrylonitrile and isoprene. The mixture of isomers **568** and **569**, formed in a ratio of 7:3, is dehydrochlorinated and the pure diene **570** isolated by fractional distillation. Addition of 5-lithio-2-methyl-2-pentene gives trienone **571**. Since this material does not undergo intramolecular Diels-Alder reaction, it is ketalized with 1,2-propanediol and the resulting diastereomeric ketals are heated at

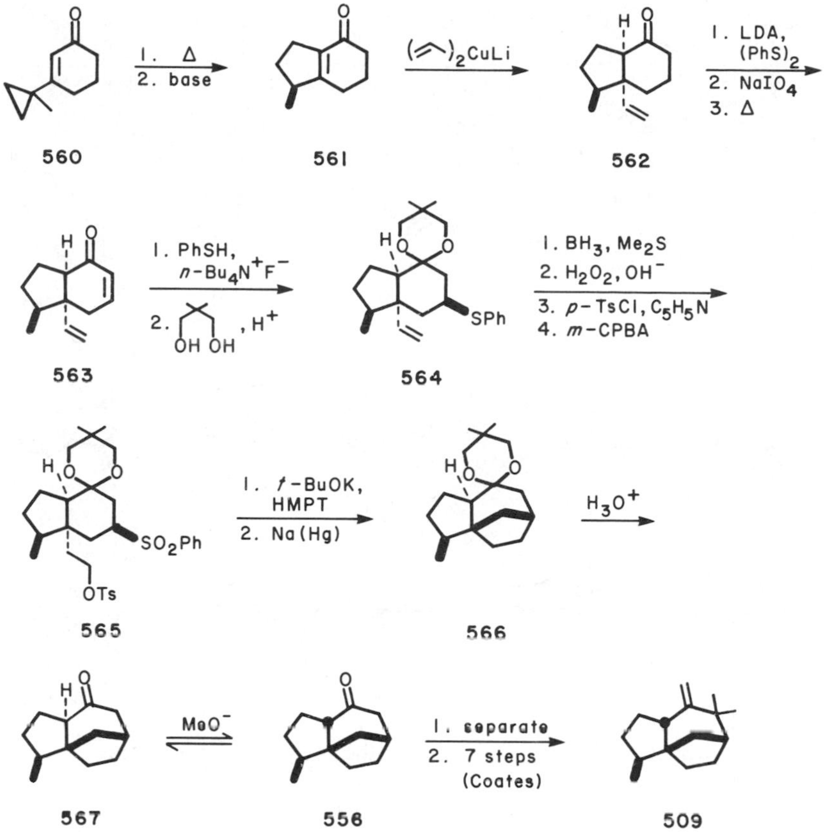

Scheme 62.   Piers-Banville Synthesis of (±)-Zizaene

250°C. The resulting adducts are separated and hydrolyzed to give tricyclic ketones **573** and **574** in a ratio of 1:3. The minor isomer is rearranged by treatment with *p*-toluenesulfonic acid in refluxing benzene; (±)-isokhusimone is obtained in 80% yield. Double bond isomerization is achieved by oxidation with singlet oxygen to give a mixture of allylic

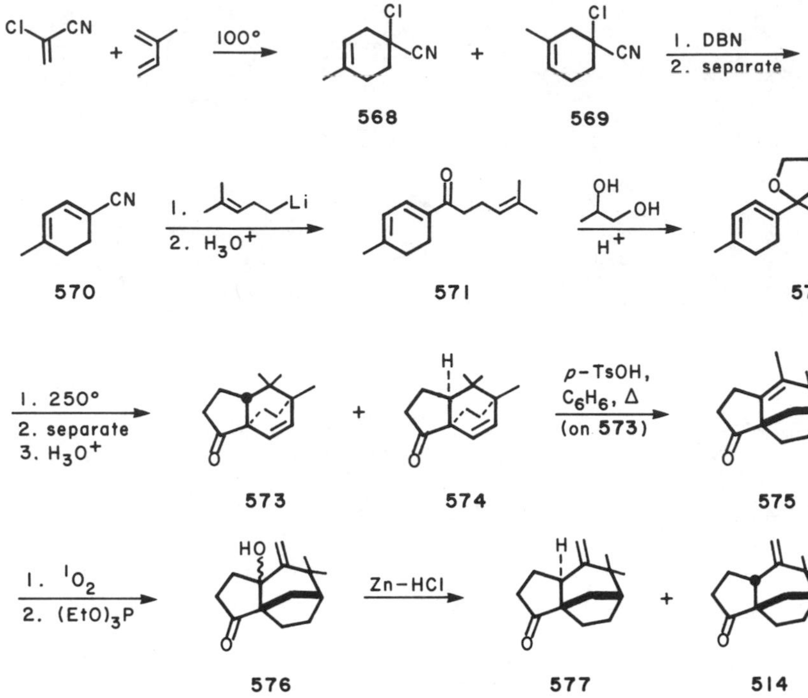

Scheme 63.   Büchi's Synthesis of (±)-Khusimone

alcohols **576**. Reduction with zinc and hydrochloric acid gives a 3:7 mixture of (±)-epikhusimone (**577**) and (±)-khusimone (**514**). The synthesis requires 10 steps, involves 2 isomer separations, and gives **514** in 1.7% overall yield.

The Liu-Chan synthesis of ketone **514** is summarized in Scheme 64.[521] The starting material is (−)-camphor-10-sulfonic acid, which is fused with sodium hydroxide to obtain (−)-α-campholenic acid (**579**) in 50% yield. After esterification, unsaturated ester **580** is ozonized and the resulting keto-aldehyde cyclized to cyclohexenone **581**. Photoaddition of

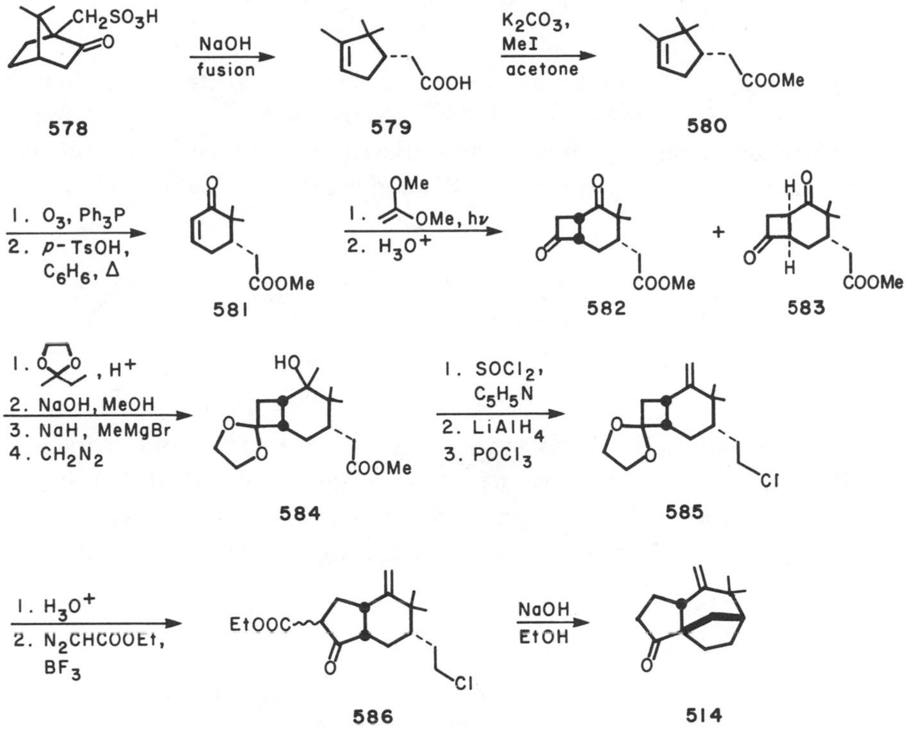

Scheme 64.   Liu-Chan Synthesis of (-)-Khusimone

ketene dimethyl acetal, followed by hydrolysis, gives a 5:8 mixture of di-keto esters **582** and **583**. After separation, the cyclobutanone carbonyl is selectively ketalized. The ester is hydrolyzed, methyllithium is added to the ketone, and the acid is reesterified. The tertiary alcohol is dehydrated, the ester reduced, and the resulting primary alcohol converted into chloride **585**. After deprotection the ketone is ring expanded by treatment with ethyl diazoacetate and boron trifluoride. The resulting keto ester (**586**) is treated with ethanolic sodium hydroxide to obtain (−)-khusimone. The synthesis requires 16 steps, involves one isomer separation, and provides the optically active natural product in 2.7% overall yield.

(8) α -Cedrene, Cedrol, Δ²-Cedrene and Cedradiene

In Volume 2 of this series, we recounted four syntheses leading to α-cedrene (**587**) and cedrol (**588**). In this section, we now discuss several additional syntheses of these tricyclic sesquiterpenes, as well as a synthesis leading to Δ²-cedrene (**589**) and cedradiene (**590**).

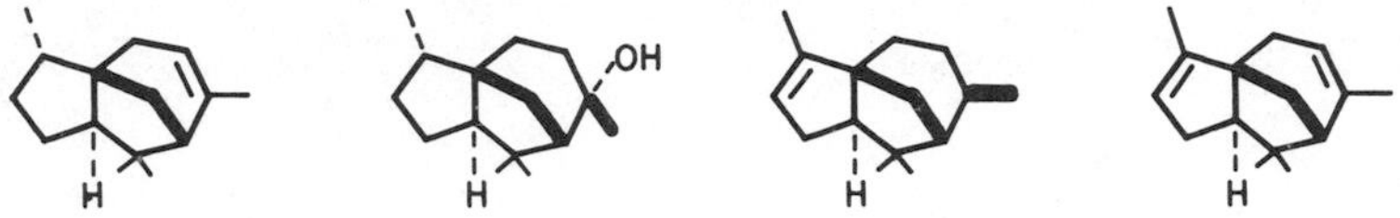

Naegeli and Kaiser have prepared Δ²-cedrene (**589**) and cedradiene (**590**) as shown in Scheme 65.[522] Thermolysis of dehydrolinalool at 200°C gives a mixture of dienols **591** and **592** which is converted into dienone **593** by treatment with methyl isopropenyl ether and acid.

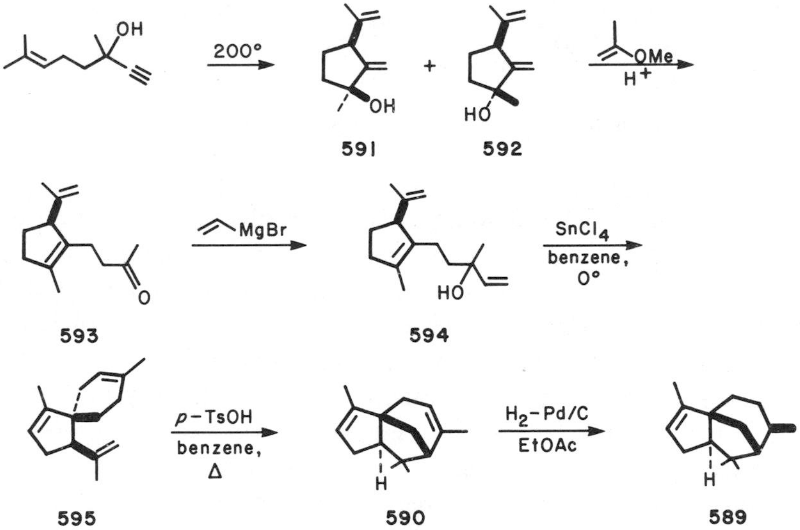

Scheme 65.   Naegeli-Kaiser Synthesis of Δ²-Cedrene

Allylic alcohol **594** is cyclized to acoratriene (**595**) by treatment with stannic chloride in benzene. Further treatment of **595** with *p*-toluenesulfonic acid in benzene gives, in 60% yield, a hydrocarbon mixture of which the major component (50%) is cedradiene (**590**). Partial hydrogenation of **590** provides **589**.

Corey and Balanson have applied a π-cyclization method developed by Stork and co-workers to the synthesis of racemic cedrone, the precursor to α-cedrene and cedrol (Scheme 66).[523] Addition of 6-lithio-2-methyl-2-heptene to enone **596** gives, after hydrolytic workup, enone **597**. Diisobutylaluminum hydride reduction gives allylic alcohol **598**, presumably as a 1:1 diastereomeric mixture. Simmons-Smith cyclopropanation, followed by Collins oxidation, gives cyclopropyl ketone **599**. Cyclization is brought about by acetyl methanesulfonate. Diastereomers **600** and **601** are obtained as a 1:1 mixture in 85% yield. Compound **600** has previously been converted into ($\pm$)-α-cedrene and ($\pm$)-cedrol.[524]

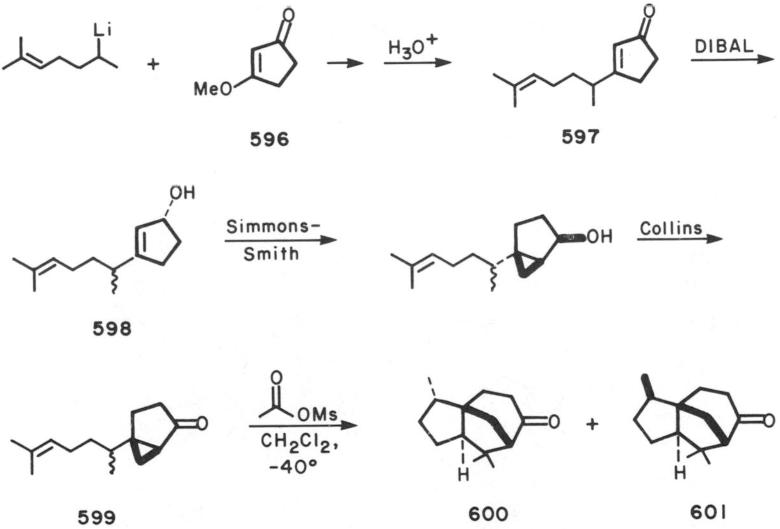

Scheme 66. Corey-Balanson Synthesis
of ($\pm$)-α-Cedrene and ($\pm$)-Cedrol

Lansbury's synthesis of (±)-α-cedrene is summarized in Scheme 67.[525] Addition of lithium reagent **603** to ketone **602** gives alcohol **604** as a diastereomeric mixture. Formolysis gives a mixture of cis and trans ketones **605** and **606**, with the latter predominating. After equilibration with sodium methoxide, the trans isomer **606** constitutes 85% of the mixture. Free-radical chlorination at the less hindered cyclohexyl tertiary hydrogen gives a mixture consisting principally of the diastereomeric chlorides **607**. Regioselective dehydrochlorination is brought about by heating the alkoxide obtained by addition of methyllithium to **607**.[526] The unsaturated alcohol so produced (**608**) is formolyzed to give a complex mixture of hydrocarbons in which (±)-α-cedrene (**587**) is present in 80% yield.

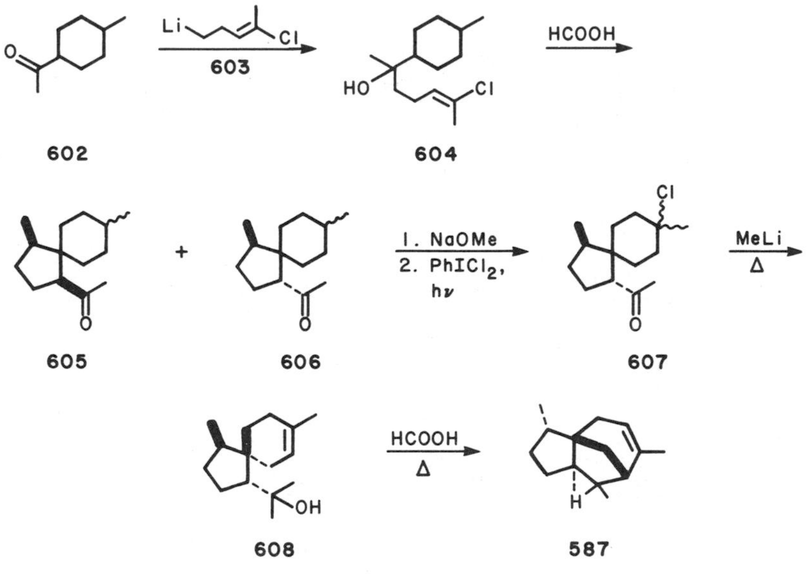

Scheme 67.   Lansbury's Synthesis of (±)-α-Cedrene

Büchi's khusimone synthesis also leads to racemic α-cedrene (Scheme 68).[520] Unsaturated ketone **574**, the major isomer produced in the intramolecular Diels-Alder cyclization of **572** (see Scheme 63), undergoes acid-catalyzed rearrangement to give a mixture consisting of 15% isokhusimone (**575**) and 40% of norcedrenone (**609**). Wittig methylenation of the latter compound gives cedradiene **610**, which is reduced by diimide to a mixture of ($\pm$)-α-cedrene (**587**) and ($\pm$)-2-epi-α-cedrene (**611**).

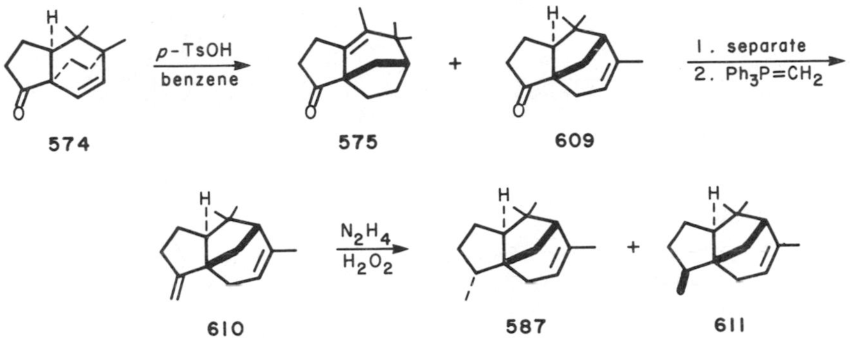

Scheme 68.   Büchi's Synthesis of ($\pm$)-α-Cedrene

Breitholle and Fallis have prepared ($\pm$)-cedrol and ($\pm$)-α-cedrene by a route involving intramolecular Diels-Alder reaction as the key maneuver (Scheme 69).[527] Methylheptenone is reduced and converted into tosylate **612**. Alkylation of sodium cyclopentadienide with this material yields triene **613**, which is heated in xylene (sealed tube) at 155°C for 50 hours. Under these conditions a 1:1 mixture of diastereomeric alkenes **614** is produced in 74% yield. Hydroboration-oxidation affords ketones **615**, which are ring expanded by Demjanov rearrangement. Ketones **616** and **617** are produced in a ratio of 6:1. After addition of methyllithium the isomers are separated chromatographically to obtain ($\pm$)-cedrol (**588**). Dehydration of **588** affords ($\pm$)-α-cedrene (**587**).

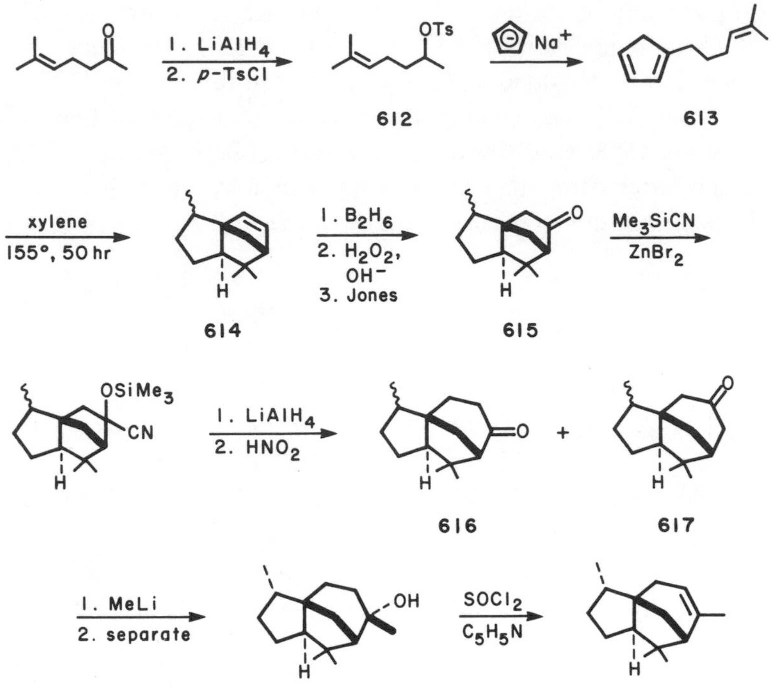

Scheme 69.   Breitholle-Fallis Synthesis
of (±)-Cedrol and (±)-α-Cedrene

(9) Quadrone

The antibiotic quadrone (**618**) is a member of a growing group of sesquiterpenes containing the perhydropentalene nucleus (coriolin, hirsutanes,
isocomene, pentalenolactone, gymnomitrol, etc.). As we have seen earlier in this review, these compounds have captured the imaginations of
synthetic chemists, and a number of successful syntheses have resulted.

618

The only published quadrone synthesis to date is that of Danishefsky and co-workers (Scheme 70).[528] Conjugate addition of vinylmagnesium bromide to enone **620**, followed by alkylation of the resulting enolate with bromide **619**, gives keto ester **621**. Ketalization gives diketal ester **622**. The vinyl group is fashioned into a β-bromoethyl group by the four-step sequence indicated. After liberation of the two carbonyl groups, cyclization is brought about by sodium methoxide in methanol. Mukaiyama addition of 1-*t*-butoxy-1-*t*-butyldimethylsilyloxyethylene to **624** provides keto diester **625** which is hydrolyzed, decarboxylated, reesterified, and ketalized to afford **626**. This material is properly constituted for formation of the third carbocycle. Treatment of the corresponding iodide with lithium hexamethyldisilazide affords, after ketal hydrolysis, keto ester **627**. It is interesting that only the axial methoxycarbonyl isomer is produced. This probably results from a preferred conformation of the enolate derived from ester **626**. The observed product would result from conformation **626a**; **626b** would lead to the diastereomer **628**.* There is no obvious reason for this stereospecificity. With the ester group thus placed in its proper configuration, it is next hydrolyzed to protect against epimerization. It turns out that the keto acid derived from **627** undergoes enolization or enolate formation exclusively toward C-4, probably due to relief of a non-bonded interaction between a C-4 hydrogen and the proximate methyl group. Thus, it is necessary to block this position so that the hydroxymethyl group can be introduced at C-2. This is accomplished by the Sharpless-Reich procedure, which gives enone **629**. Treatment of this

---

*It is known that enolization of such esters in the presence of hexamethylphosphoric triamide affords the enolate in which the alkoxide group is trans to the hydrogen of the double bond.

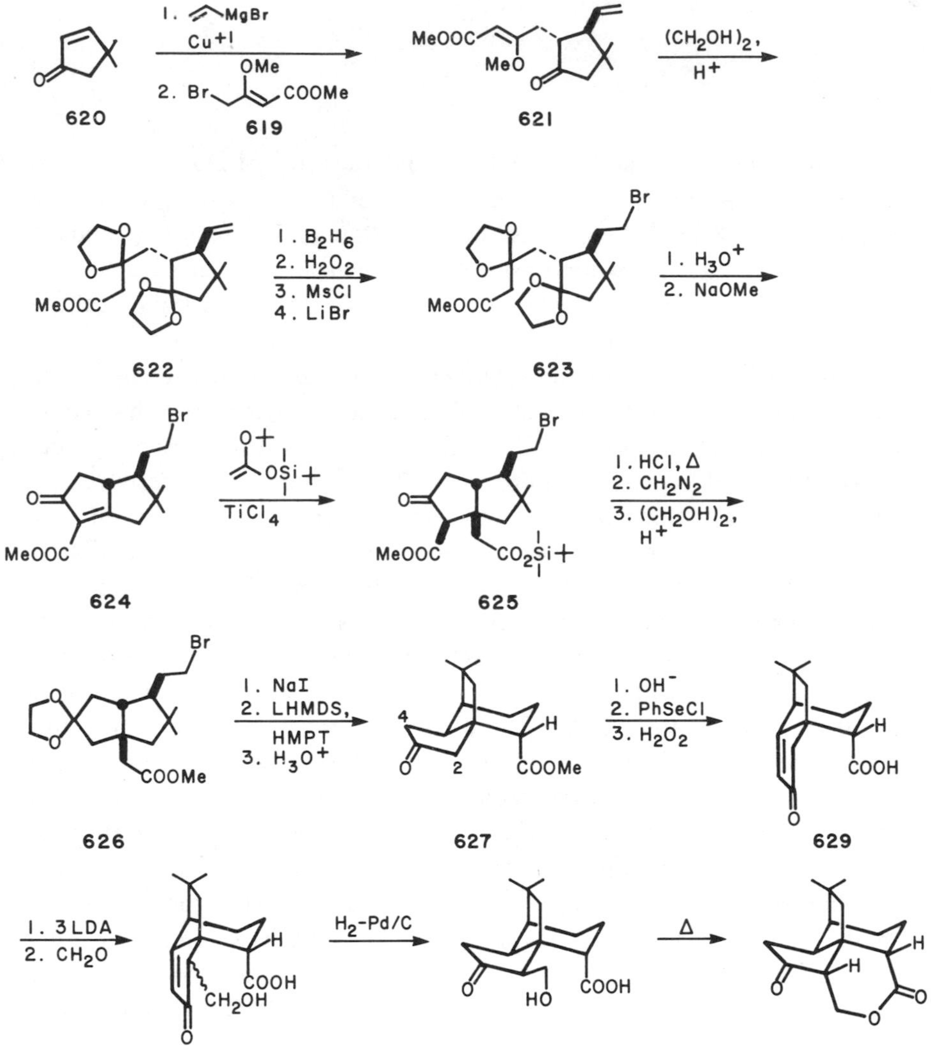

Scheme 70.   Danishefsky's Synthesis of (±)-Quadrone

626a → 627

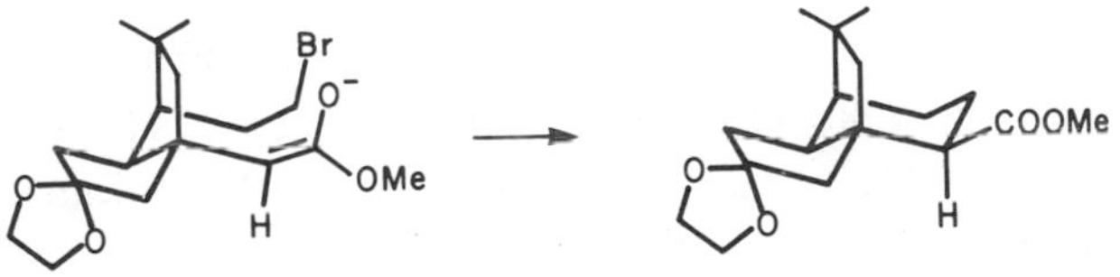

626b → 628

material with three equivalents of lithium diisopropylamide and gaseous formaldehyde affords a hydroxy acid (**630**) which is hydrogenated to obtain **631**. High stereospecificity is observed in the hydrogenation step since reaction from the other face of the double bond would produce a trans-fused perhydropentalene system. Acid-catalyzed dehydration of **631** affords enone **632**, the probable biologically active agent. However,

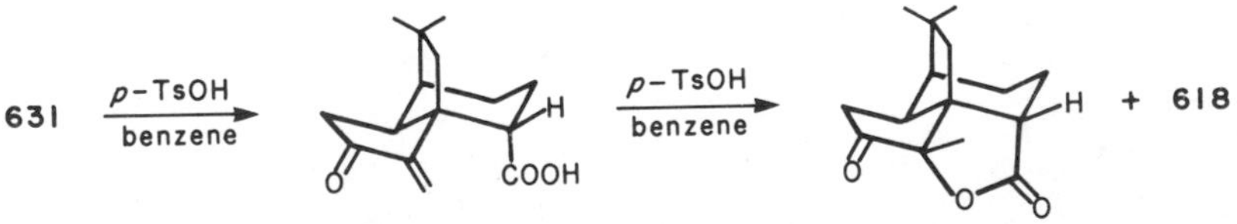

631 $\xrightarrow[\text{benzene}]{p-\text{TsOH}}$ 632 $\xrightarrow[\text{benzene}]{p-\text{TsOH}}$ 633 + 618

further treatment of **632** with acid yields primarily **633**, an isomer of quadrone. It is found that when either enone **632** or β-hydroxy ketone **631** is heated at 190-195°C for 5 minutes, the sole product is (±)-quadrone (**618**).

(10) Isolongifolene

Isolongifolene (634) is actually not a natural product. Instead, it is the product of acid-catalyzed rearrangement of longifolene. In Volume 2,

634

we reported an eight-step synthesis of 634 from camphene-1-carboxylic acid.[529] Piers and Zbozny have recently reported an alternative synthesis of 634 (Scheme 71).[530] The key reaction of the Piers-Zbozny synthesis is

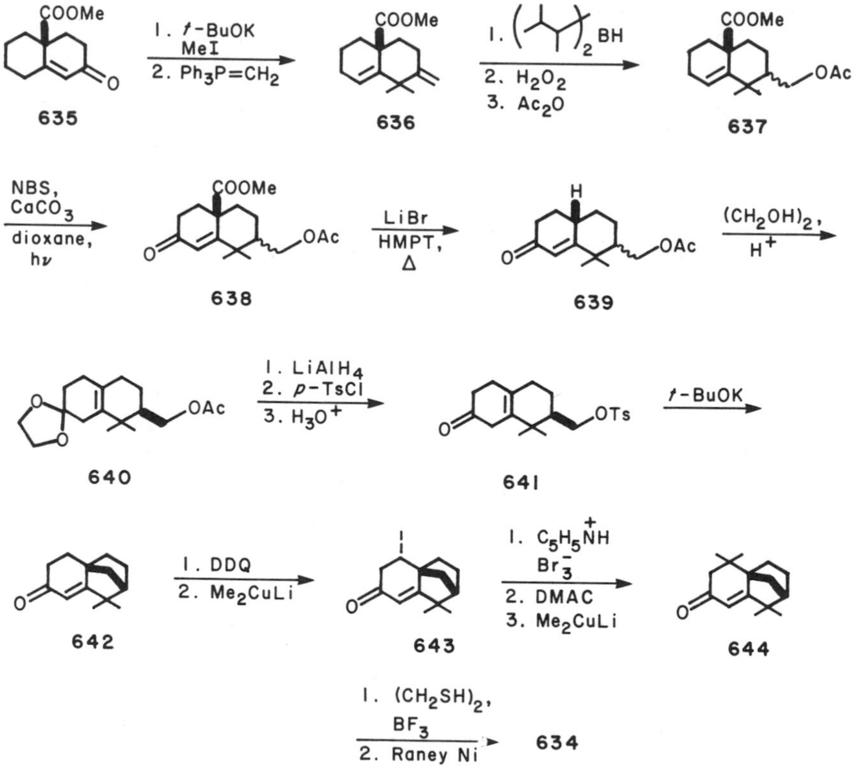

Scheme 71.   Piers-Zbozny Synthesis of (±)-Isolongifolene

self-alkylation of keto tosylate **641**, which occurs solely at the γ-position of the extended enolate to give **642**. Cyclization at the α-position would give a strained bridgehead alkene. Intermediate **641** is prepared by the straightforward sequence of steps shown in the scheme. The most noteworthy transformation is the method used for the allylic oxidation (**637→638**). When oxidation with chromium failed, it was found that the Finucane-Thompson procedure provides enone **638** in 96% yield. In all, the synthesis requires 17 steps and provides **634** in about 1% overall yield. It is considerably less efficient than Dev's eight-step synthesis, which gives an overall yield of 17%.[529]

(11) Ishwarone and Ishwarane

Ishwarone (**645**) and the corresponding hydrocarbon ishwarane (**646**) are two members of the relatively small class of tetracarbocyclic sesquiterpenes. Several syntheses of these interesting compounds have been reported.

645   646

Kelly and co-workers have synthesized (±)-ishwarane as shown in Scheme 72.[531a-c] Enone **647** undergoes photoaddition of allene to give an unsaturated ketone which is ketalized to obtain **648**. Epoxidation gives a mixture of epoxides which is reduced to obtain alcohols **649**. Acid-catalyzed deprotection of the ketone is followed by rearrangement of the resulting aldol to the more stable bicyclo[2.2.0]octane aldol **650**, which is obtained as a diastereomeric mixture. Dehydration and reduction gives a mixture of unsaturated alcohols (**651**). The alcohol is protected as a benzyl ether, the alkene is hydroborated and the resulting organoborane oxidized to the ketone. Deprotection gives hydroxy ketone **652**. The corresponding tosylates are *both* cyclized by treatment

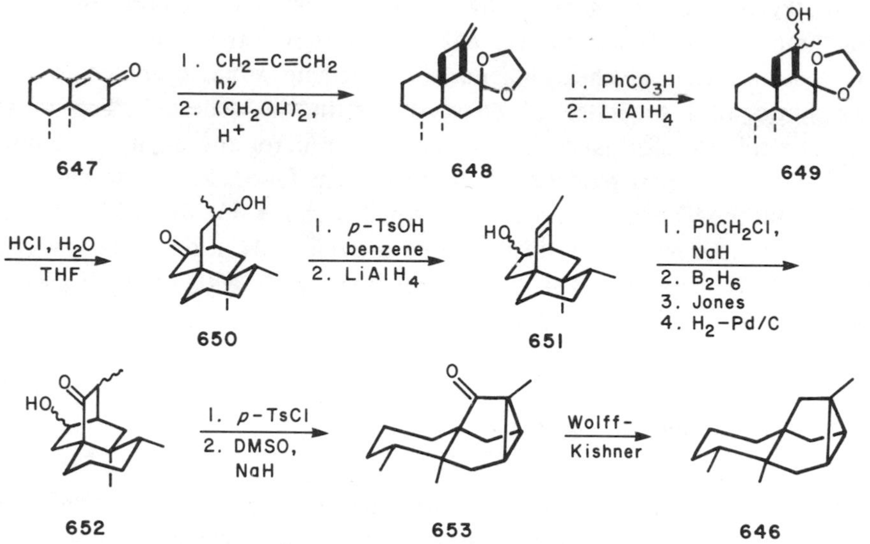

Scheme 72.   Kelly's Synthesis of (±)-Ishwarane

with methylsulfinyl carbanion, yielding a ketone (653) which is reduced to (±)-ishwarane (646). The fact that diastereomeric enolate 652a cyclizes is not surprising, since it is constituted for nucleophilic substitution with inversion of configuration:

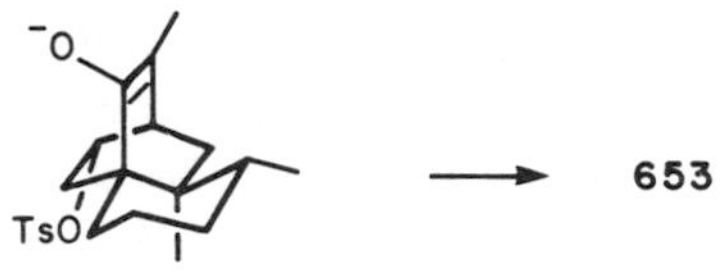

653

652a

However, enolate 652b cyclizes with *retention* of configuration at the tosylate-bearing carbon:

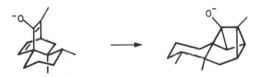

652b → 653

Kelly and co-workers propose that this diastereomer may react by initial displacement of tosylate by a molecule of the solvent, which is in turn displaced by the enolate to give **653**:

→ 653

It is also suggested that elimination may occur first, followed by cyclization to the homoenolate:

Whereas the Kelly synthesis requires 14 steps for the conversion of enone **647** into ishwarane, Cory and McLaren discovered a way that the same conversion can be done in only four steps (Scheme 73).[532] Conjugate addition of lithium dimethylcopper gives ketone **654**, which is converted into alkene **655** by Grignard addition followed by dehydration of the resulting tertiary alcohol. Alkene **655** is accompanied by 20% of its Δ^1-isomer. The key transformation is addition of dibromocarbene to give an intermediate dibromocyclopropane (**656**) which is treated with methyllithium to generate a carbene (**657**). Cyclization occurs regiospecifically, affording (±)-ishwarane (**646**) in 26% yield.

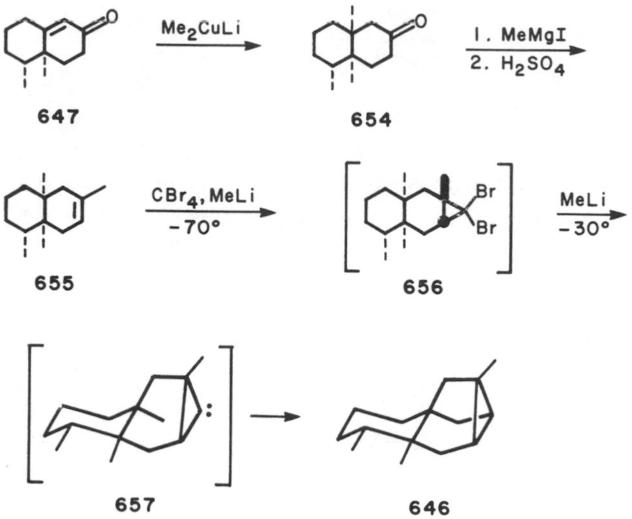

Scheme 73.   Cory's Synthesis of (±)-Ishwarane

An application of this basic approach to the synthesis of (±)-ishwarone is summarized in Scheme 74.[532] Enol ether **658** is converted into enone **659**, which undergoes conjugate addition of the cuprate reagent derived from 4-lithio-2-methyl-1-butene. The product (**660**) is epoxidized and the resulting keto epoxide cyclized by treatment with potassium *t*-butoxide. The product is a mixture of alcohols **661** and **662**. Dehydration with 50% aqueous sulfuric acid gives a 2:1 mixture of unsaturated ketones **663** and **664** in 59% yield. Although this mixture can be separated chromatographically, this is not necessary at this stage, since isomer **664** does not react with dibromocarbene. Thus, treatment of the mixture with bromoform in the presence of 50% aqueous NaOH gives cyclopropane **665**, which is treated with methyllithium in ether at -30°C to obtain racemic ishwarone.

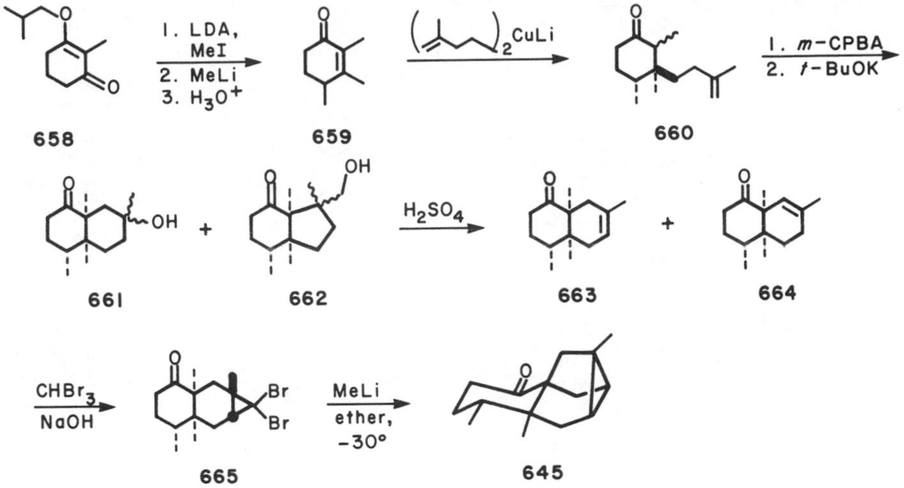

Scheme 74.   Cory's Synthesis of (±)-Ishwarone

Piers and Hall have synthesized (±)-ishwarone as outlined in Scheme 75.[533] Ketal aldehyde **666**, an intermediate in the Ziegler eremophilone synthesis, is converted into acetylenic alcohol **667**, which is deprotected and hydrogenated to the cis-allylic alcohol **668**. The derived mesylate is cyclized to give unsaturated ketone **669**. The ketal is allowed to react with dimethyl diazomalonate to give cyclopropane diester **670**. This compound is converted into dichloro ketone **671** by a standard four-step sequence. Base-catalyzed cyclization gives the chloroishwarone **672**, which is reduced with lithium triethylborohydride. Reoxidation of the resulting alcohol provides racemic ishwarone.

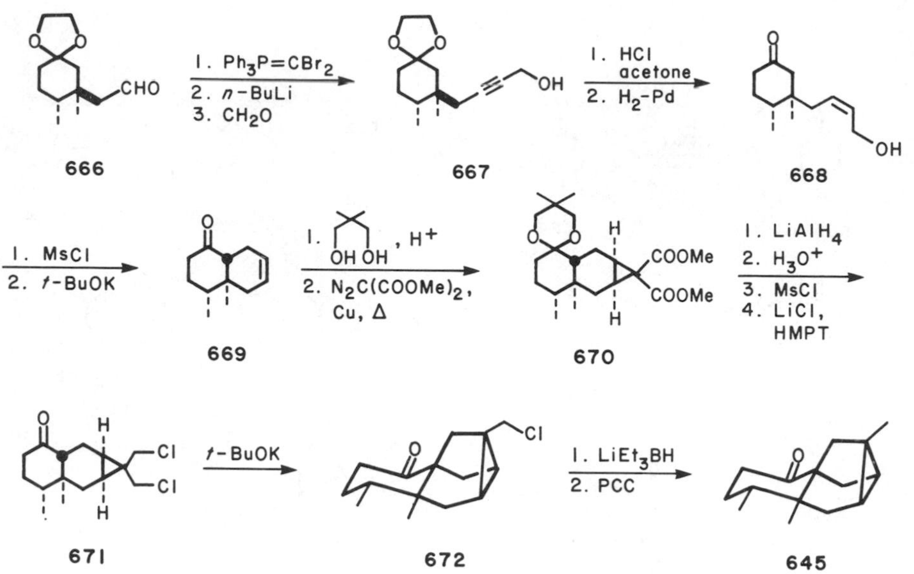

Scheme 75.   Piers-Hall Synthesis of (±)-Ishwarone

7. SESQUITERPENE ALKALOIDS

A. Illudinine

The fungal metabolite illudinine (**1**) has been synthesized by Woodward and Hoye as outlined in Scheme 1.[534] Friedel-Crafts acylation of indane with β-chloropropionyl chloride gives a ketone (**3**) which cyclizes upon treatment with hot sulfuric acid to give a mixture of tricyclic ketones **4** and **5**. After methylation and Clemmensen reduction, the resulting isomeric hydrocarbons are separated by careful distillation. Compound **6**

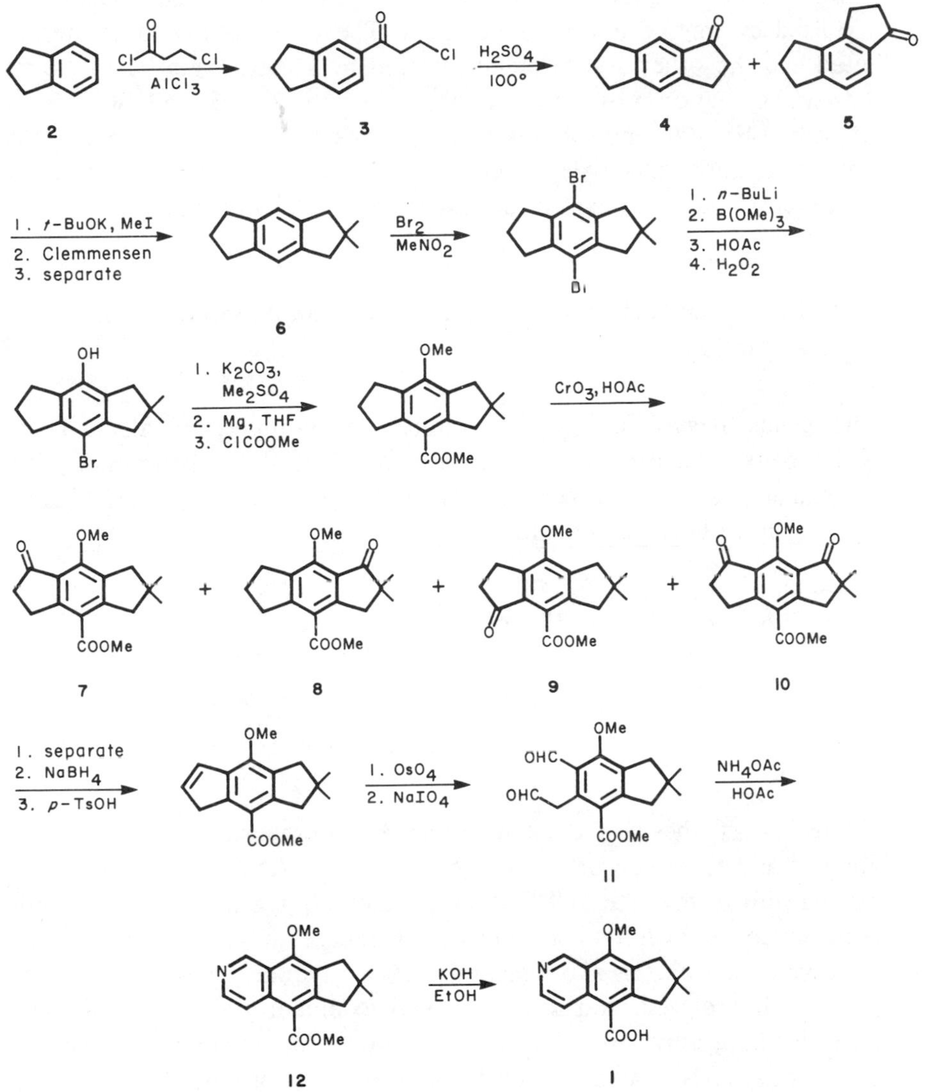

Scheme 1.   Woodward-Hoye Synthesis of Illudinine

is obtained in 44% yield, accompanied by 19% of the isomer derived from ketone 5. After bromination of the ring, one aromatic position is oxidized by way of the boronate ester. The resulting phenol is methylated and the other aryl bromide replaced by a methoxycarbonyl group. Benzylic oxidation provides a mixture of ketones 7, 8, 9, and 10 in yields of 36%, 14%, 8%, and 6%, respectively. The major oxidation product is converted into dialdehyde 11 and thence into pyridine 12 by standard procedures. Saponification of the ester function affords illudinine.

B. Deoxynupharidine, Castoramine, Deoxynupharamine, and Nupharamime

The genus *Nuphar* produces a number of sesquiterpene alkaloids of which deoxynupharidine (13), castoramine (14), and nupharamine (15) are characteristic. These compounds and the deoxy analog of nupharamine (16) have been synthesized.

13   14   15: R = OH
                                           16: R = H

The first synthesis of deoxynupharidine was reported by Kotake and Kaneko and their co-workers (Scheme 2).[535a-d] 2,5-Dimethylpyridine is heated with formalin at 160°C to obtain diol 17, which is acetylated with concomitant dehydration to obtain allylic acetate 18. Alkylation of malonic ester with this substance provides compound 19, which is hydrogenated, hydrolyzed, and decarboxylated to obtain amino acid 20. The corresponding ethyl ester is acetylated with 3-furancarbonyl chloride to 21. When 21 is heated with soda lime, there is produced an enamine (22) which is hydrogenated to a mixture of (±)-deoxynupharidine (13) and its diastereomer 23. Isomers 13 and 23 are readily separated by chromatography on alumina.

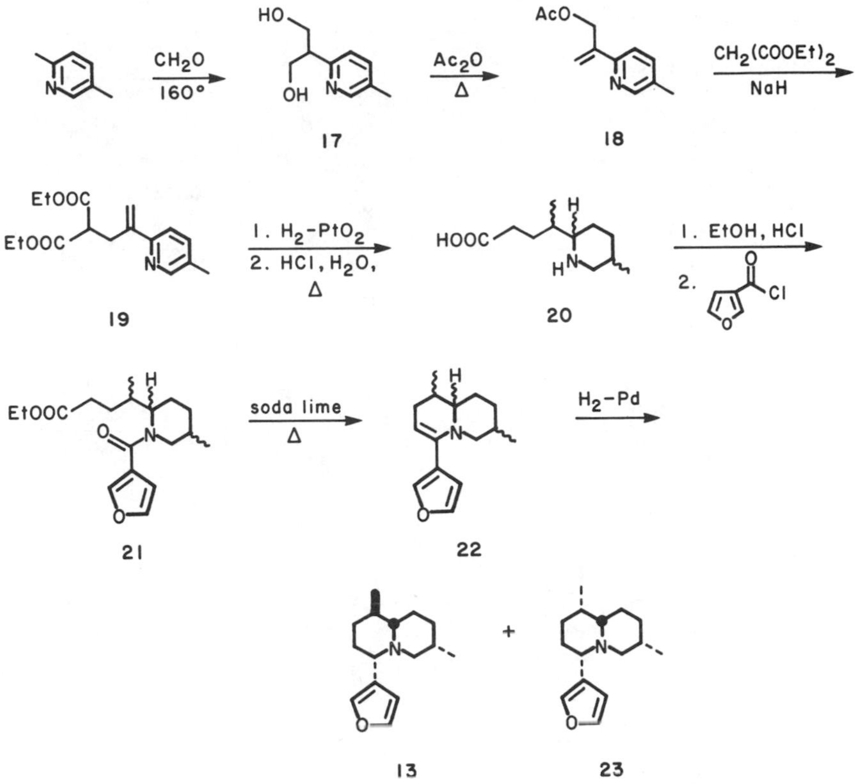

Scheme 2.   Kotake-Kaneko Synthesis of (±)-Deoxynupharidine

Arata's approach to deoxynupharidine is summarized in Scheme 3.[536] Substituted malonate 24 is hydrogenated over Adams' catalyst and the resulting amino diester heated to obtain lactam 25. After hydrolysis, the mixture of diastereomeric amino acids is separated and isomer 26 is converted into lactam 27, of the proper configuration for conversion into racemic deoxynupharidine. Claisen acylation of 31 with ethyl 3-furoate gives β-keto amide 28 which is hydrolyzed and decarboxylated to amino

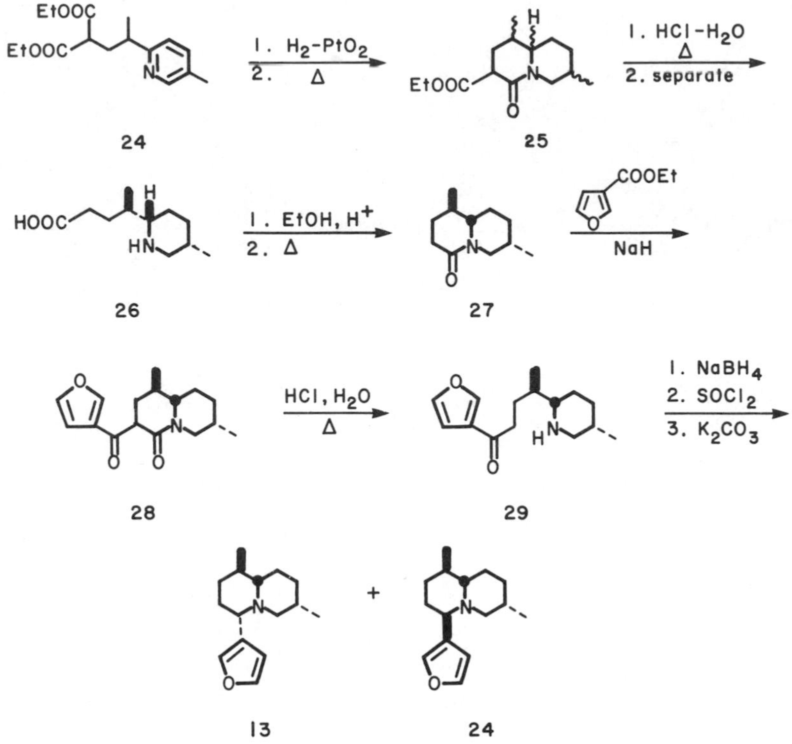

Scheme 3.   Arata's Synthesis of (±)-Deoxynupharidine

ketone **29**. After reduction of the carbonyl group, the resulting alcohol is allowed to react with thionyl chloride. Cyclization is brought about by treatment with potassium carbonate. (±)-Deoxynupharidine (**13**), presumably accompanied by some of its diastereomer **24**, is obtained and purified by recrystallization of the perchlorate salt. A similar approach was reported by Jezo and co-workers.[537]

Bohlmann and co-workers carried out an extensive stereochemical investigation which fully clarified the structure of deoxynupharidine.[538] As shown in Scheme 4, ester **30** is alkylated successively with methyl

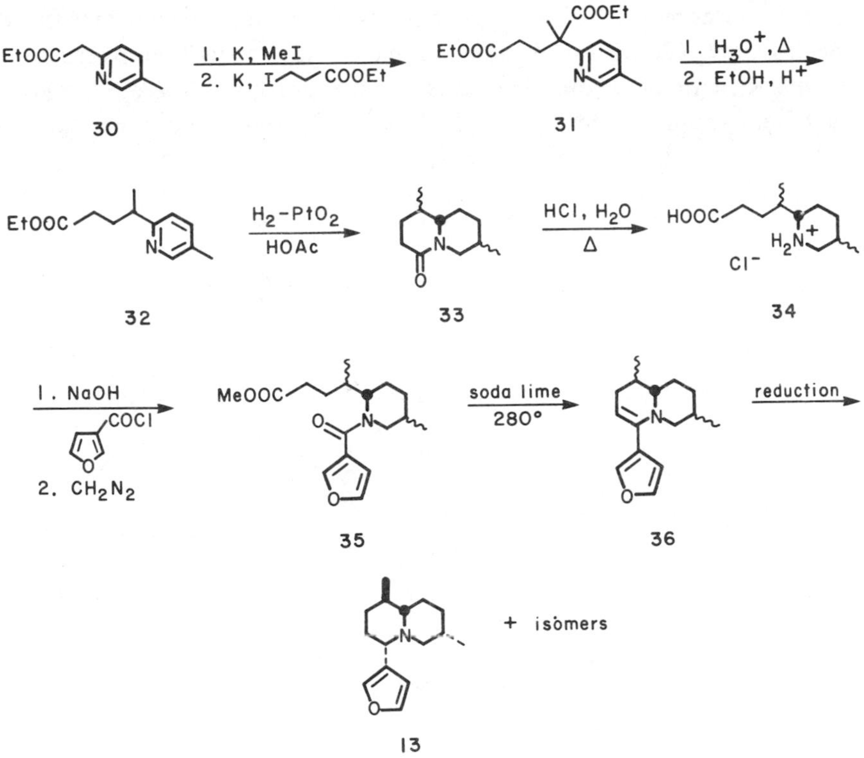

Scheme 4.   Bohlmann's Synthesis of (±)-Deoxynupharidine

iodide and ethyl β-iodopropionate to obtain diester **30**, which is elaborated by straightforward methods into ester **32**. Catalytic hydrogenation gives a mixture of the four diastereomeric lactams **33**, which are
hydrolyzed to obtain the mixture of amino acid salts **34**. Shotten-
Baumann acylation with 3-furancarbonyl chloride gives an amide (**35**),
which is cyclized to enamine **36** by heating with soda lime. Catalytic
hydrogenation of **36** over palladium on barium sulfate gives a mixture of

four diastereomers. All four of these isomers were shown by infrared spectroscopy (Bohlmann bands) to have the "trans" perhydroquinolizidine structure. These structures are presumably **13**, **23**, **37**, and **38**. (±)-Deoxynupharidine (**13**) is obtained in 7% yield by chromatographic

13

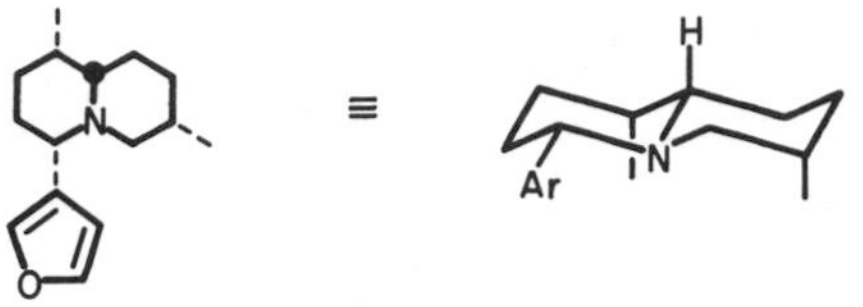

23

37

38

separation of the mixture. When enamine **36** is protonated and the resulting immonium salt reduced with sodium borohydride, six diastereomeric amines are produced. In addition to isomers **13**, **23**, **37**, and **38**, the "cis" perhydroquinolizidine isomers **39** and **40** are also obtained under these conditions.

39

40

Wróbel and Dabrowski synthesized (±)-deoxynupharidine as shown in Scheme 5.[539] Addition of ethyl propiolate to ketone **41** gives an adduct (**42**) which is converted into amino ester **43** and thence into Bohlmann's lactam (**33**). Addition of 3-furyllithium to **33**, followed by dehydration, gives enamine **36**, which is hydrogenated as in the Bohlmann synthesis.

Subsequently, Wróbel and co-workers worked out the first stereoselective synthesis of (±)-deoxynupharidine (Scheme 6).[539] 3-Acetylfuran (**44**) is converted into enol ether **45**, which is converted into quinolizidinium salt **46** under acidic conditions. Reduction of this material with sodium cyanoborohydride provides diastereomeric amino esters **47** and **48** in 56% and 11% yield, respectively. Base-catalyzed equilibration of the mixture gives a single isomer which is converted into (±)-deoxynupharidine (**13**) by standard methods.

The synthesis of (±)-castoramine (Scheme 7)[540] by Bohlmann and co-workers utilizes the Wróbel-Dabrowski method for introduction of the furyl group. The quinolizidine is constructed by a classical sequence starting with the keto diester **50**. Lactam ester **51** is reduced by lithium borohydride to a mixture of diastereomeric hydroxy lactams (**52**). After protection of the hydroxy group, the furyl group is introduced and the resulting enamine hydrogenated. (±)-Castoramine (**14**) is obtained in approximately 10% yield by chromatography. In addition to **14**, three

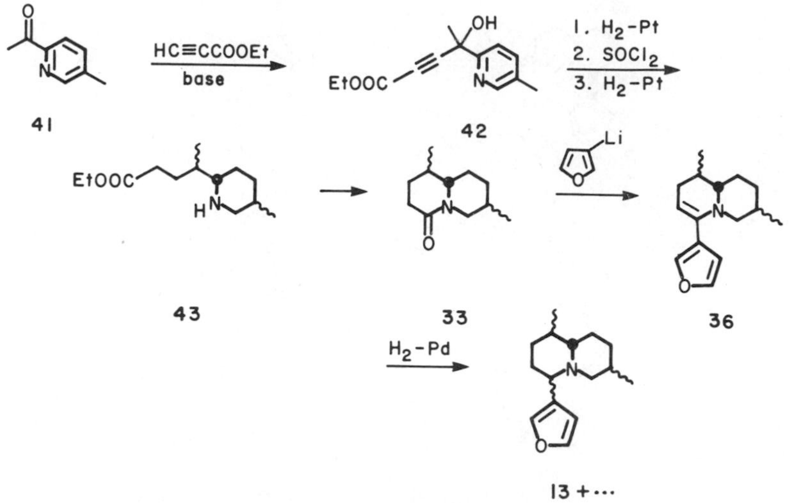

Scheme 5.   Wróbel-Dabrowski Synthesis of (±)-Deoxynupharidine

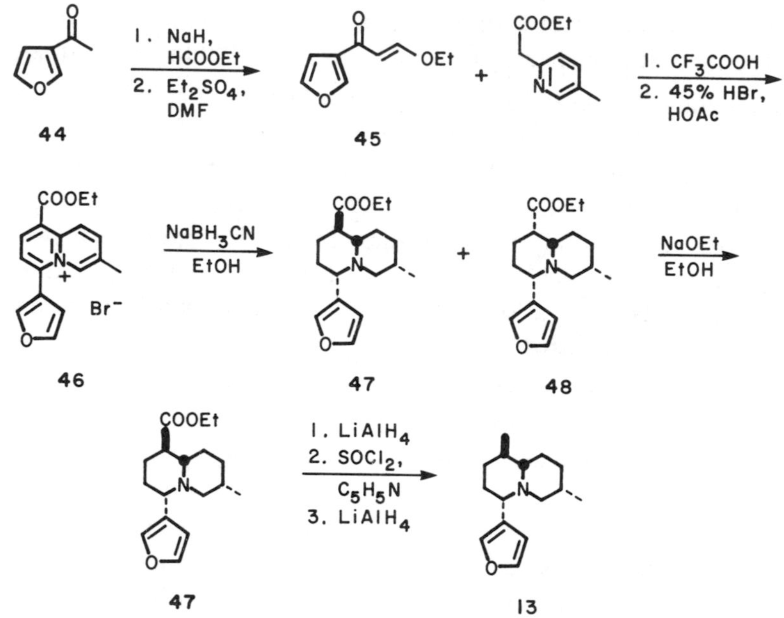

Scheme 6.   Wróbel's Synthesis of (±)-Deoxynupharidine

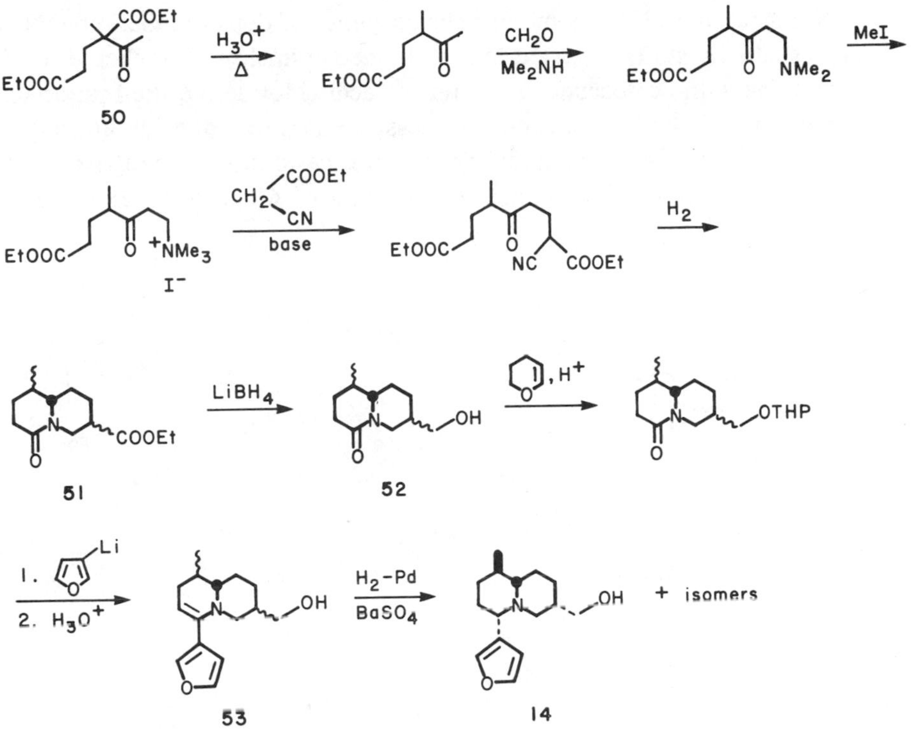

Scheme 7. Bohlmann's Synthesis of (±)-Castoramine

crystalline stereoisomers are obtained in comparable yields. These three isomers are believed to have structures **54-56**.

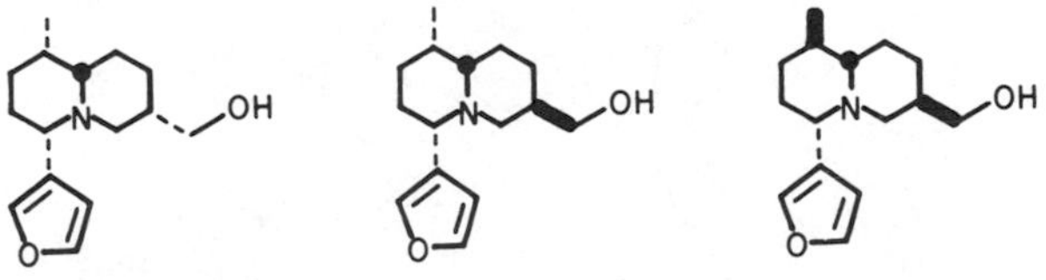

54 55 56

Nupharamine (**15**) is a monocyclic congener of deoxynupharidine. The first synthetic study reported led to the deoxy analog **16** (Scheme 8).[539] Beginning with acetoacetic ester and the acid chloride **57**, the monocyclic lactam (**63**) is built up in more or less the same manner as utilized by Bohlmann in the deoxynupharidine and castoramine syntheses (see Schemes 4 and 7). Hydrolytic ring opening of **63** gives an amino ester

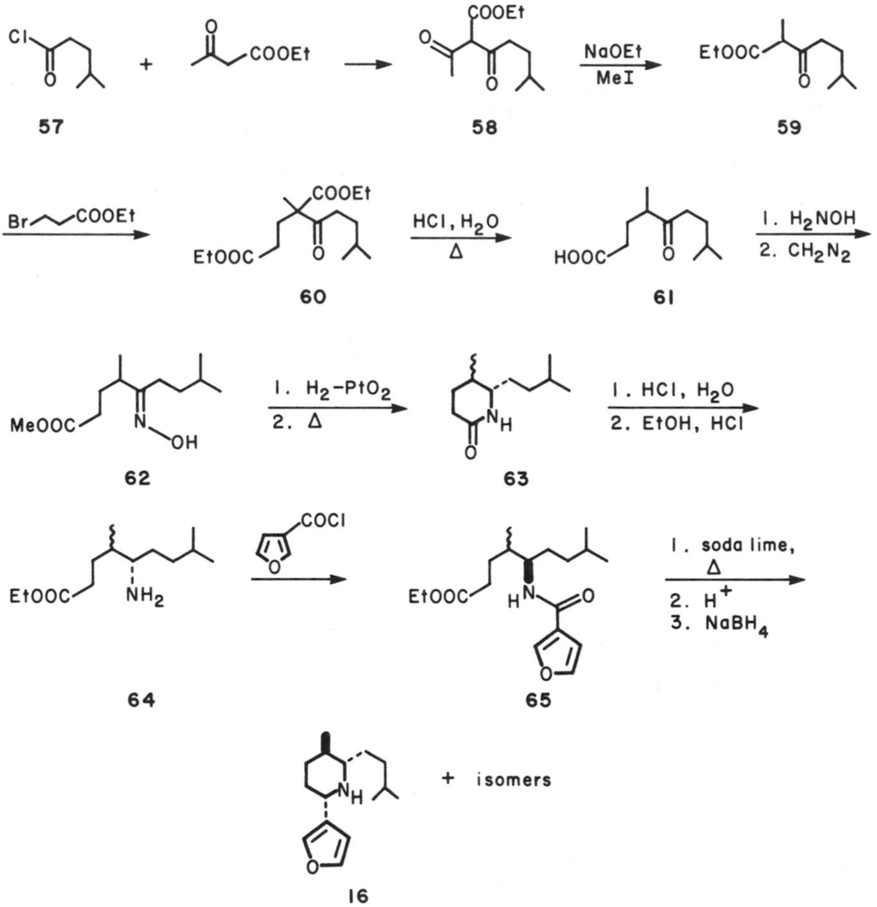

Scheme 8.   Arata's Synthesis of (±)-Deoxynupharamine

(64), which is acylated with 3-furancarbonyl chloride to obtain amide ester 65. Ring closure is accomplished by the Kotake-Kaneko method. After reduction of the resulting enamine, a mixture of (±)-deoxynupharamine and its stereoisomers is obtained.

Wróbel and co-workers have prepared (±)-nupharamine itself (Scheme 9).[539] The synthesis begins with a Hantzsch pyridine construction, leading from 3-acetylfuran to the furylpyridine 67. After conversion of the ethoxycarbonyl group to methyl, the more acidic methyl position is alkylated with methallyl chloride and the pyridine ring reduced with sodium in ethanol. Diastereomers 71 and 72 are obtained as a 3:1 mixture in 12% yield. After hydration of the alkene linkage, the isomers are separated, giving a 3:1 mixture of (±)-3-epinupharamine (73) and (±)-nupharamine (15).

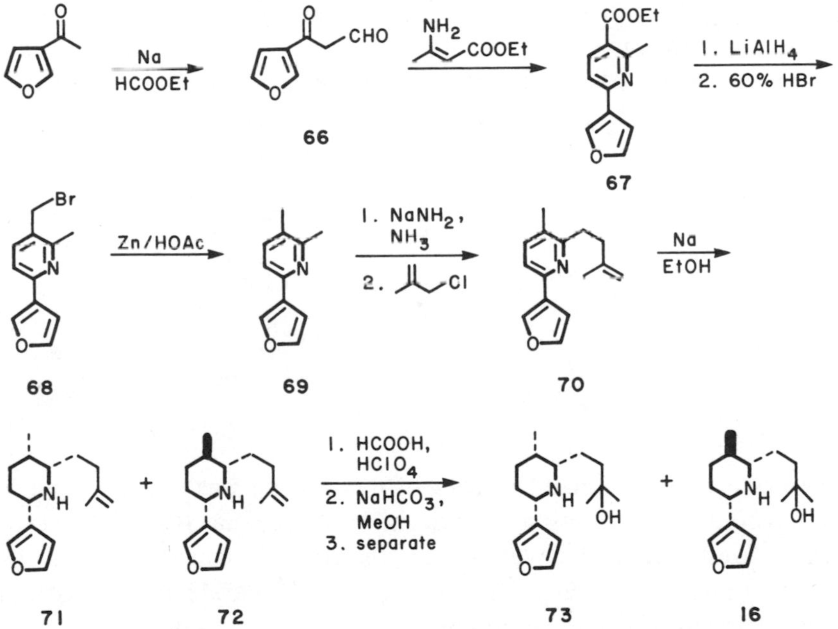

Scheme 9.   Wróbel's Synthesis of (±)-Nupharamine

C. Dendrobine

Dendrobine (74) was first isolated by Suzuki and co-workers from the stem of the ornamental Chinese orchid *Dendrobium nobile* (*Orchidacae*). It is a component of the Chinese herbal drug Chin-Shih-Hu, which is used as a tonic and antipyretic. An interesting feature of the dendrobine

74

molecule is the fact that the cyclohexane ring must exist in a boat conformation, with alkyl groups at the bowsprit positions. At the outset of any dendrobine synthesis, it would appear that this requirement would constitute a formidable hurdle. However, investigations of the alkaloid showed that hydroxy acid 75, obtainable from dendrobine by hydrolysis, undergoes ready acid-catalyzed lactonization to regenerate the alkaloid.

74   **75**

The first successful total synthesis of dendrobine was that of Yamada and co-workers, which was completed in 1972 (Scheme 10).[544] The synthesis utilizes a method developed by Johnson for construction of cis-fused hydrindanones. Thus, treatment of diketo acid 78 with potassium *t*-butoxide results in Michael addition, followed by intramolecular aldolization and provides acids 79 and 80 in a ratio of 8:1. The stereochemistry of this process is interesting. The 8:1 ratio may reflect the ratio of

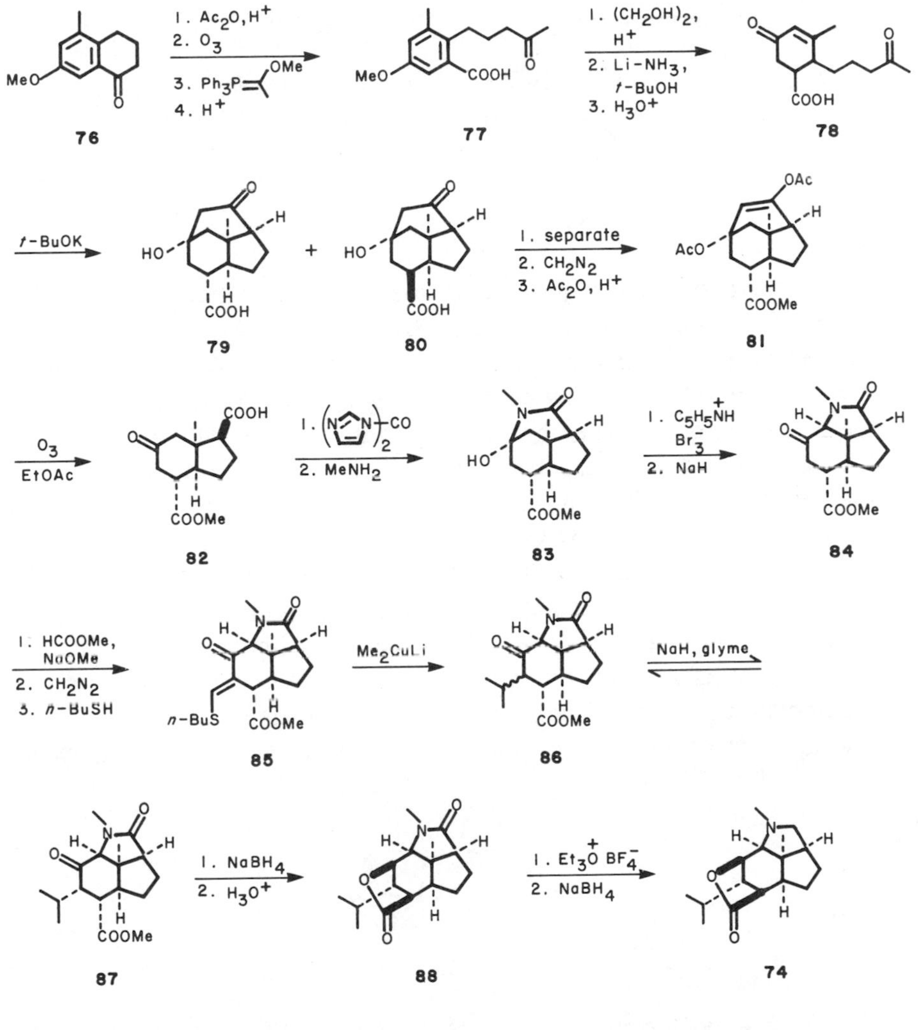

Scheme 10.   Yamada's Synthesis of (±)-Dendrobine

trans and cis isomers of the precursor **78**. Alternatively, the entire process may be under thermodynamic control, and the diastereomer ratio may reflect the relative stability of the products:

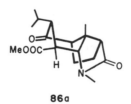

79   80

After separation, the major isomer is esterified and treated with acetic anhydride to obtain an enol acetate (**81**) which is ozonized to obtain **82**. The nitrogen is inserted at this point by reaction of the imidazolide with methylamine. The resulting carbinolamide is brominated and the resulting bromo ketone cyclized by treatment with sodium hydride. Although it is apparently fortuitous, the regiospecific bromination of **83** is perhaps the key step in the Yamada synthesis. The isopropyl group is introduced by a method developed by Coates, that is, by addition of excess lithium dimethylcopper to the α-n-butylthiomethylene derivative **85**. The resulting keto ester, **86**, is equilibrated by treatment with sodium hydride in glyme, leading to a 1:1 mixture of **86** and **87**. Although the full stereochemistry of isomer **86** was not elucidated, it was shown by nmr evidence that the methoxycarbonyl group had retained its configuration. The likely structure of this compound is that in which the cyclohexane

86a

ring is in a boat conformation (**86a**). The synthesis is completed by reduction of keto ester **87** to the hydroxy ester, which is cyclized. The amide carbonyl is removed by Borch's method giving (±)-dendrobine.

Inubushi and co-workers prepared (±)-dendrobine as shown in Scheme 11.[545] The known enedione **89** is stereospecifically reduced and converted into an enone tosylate **(90)** which is treated with cyanide in dimethyl sulfoxide. Although Inubushi and co-workers hoped to obtain the inverted nitrile by this method, the displacement occurs with apparent retention of configuration, undoubtedly as a result of subsequent epimerization of the initial product. The same cyano enone is obtained by Robinson annelation of cyano ketone **92** with methyl vinyl ketone. After saturation of the double bond of **91**, a double bond is reintroduced by bromination-dehydrobromination. Enones **95** and **96** are obtained in a 1:3.5 ratio. Although the undesired major isomer may be recycled to **94**, this is an exceedingly poor step. Nevertheless, the minor isomer **95**, obtainable in 21% yield by silica gel chromatography, is utilized to continue the synthesis. The enone is ketalized and the cyano function hydrolyzed with alkali. Acidic hydrolysis of the ketal provides a 3:1 mixture of enone acid **97** and keto lactone **98**; the latter substance is obtained in 22% yield by chromatography on silica gel. Treatment of the keto lactone with methylamine gives keto lactam **99**, which is allowed to react with isopropylmagnesium bromide. Unfortunately, enolization of the ketone is the sole result at room temperature. At -60°C to -70°C some addition occurs and alcohol **100** may be obtained by fractional crystallization in 17% yield. Resubmission of the mother liquors to the Grignard reaction gives more alcohol. After 4-5 cycles of this procedure, alcohol **100** may be obtained in a total yield of 74%. Dehydration provides a 1:1:1 mixture of three alkenes **(101)**. This mixture is treated under Prevost hydroxylation conditions. After treatment of the crude product with silica gel, allylic acetates **102** and **103** are obtained in yields of 10% and 20%, respectively. The minor isomer is converted into enone **104**. Nagata cyanation of **104** yields a mixture of β-cyano ketones **105**, **106**, and **107** in isolated yields of 18%, 20% and 29%, respectively. The isomers are separated and the major isomer converted into keto ester **87**. The remainder of the Inubushi synthesis is identical to the Yamada synthesis.

Kende's synthesis of dendrobine is outlined in Scheme 12.[546] The known triacetate **108** is saponified and oxidized to hydroxy benzoquinone

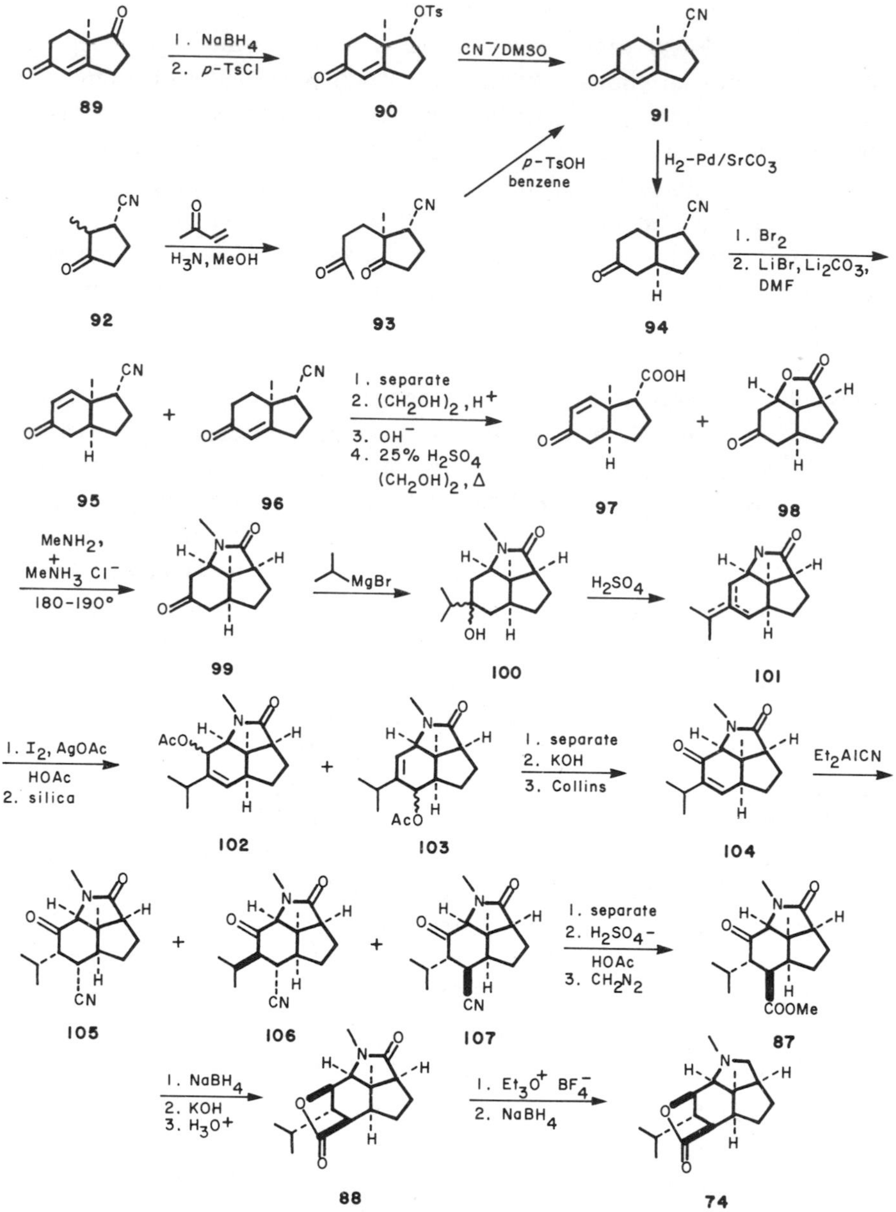

Scheme 11.   Inubushi's Synthesis of (±)-Dendrobine

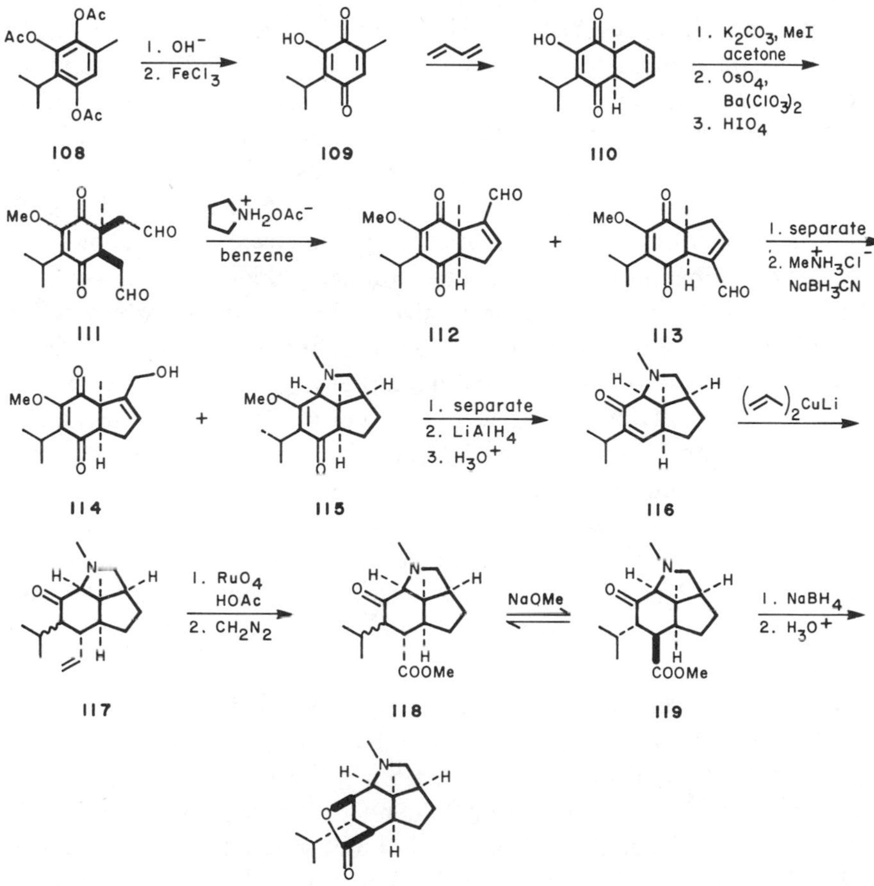

Scheme 12.   Kende's Synthesis of (±)-Dendrobine

109, which undergoes regiospecific Diels-Alder cycloaddition with butadiene to afford adduct **110** in 95% overall yield for the three steps. After methylation of the hydroxy group, the cyclohexene double bond is cleaved to obtain dialdehyde **111**, which is cyclized to a 1:1 mixture of unsaturated aldehydes **112** and **113**, each of which is obtained in 25% overall yield from **110**. The isomers are separated and compound **112** reductively aminated by treatment with methylamine and sodium cyanoborohydride. The resulting product is a mixture of alcohol **114** and tricyclic amine **115**, which are obtained in yields of 26% and 35%, respectively. The unexpected formation of tricyclic **115** in this reaction probably results from initial 1,4-reduction of aldehyde **112**, followed by reductive amination of the resulting saturated aldehyde. The secondary amine so formed must condense to give an immonium ion which suffers reduction to produce **115**. Conversion to enone **116** is routine. Kende and co-workers introduce the methoxycarbonyl group by conjugate addition of lithium divinylcopper, to obtain **117** in 80% yield. Cleavage of the vinyl group is accomplished by oxidation with ruthenium tetroxide. After esterification there is produced a keto ester (stereochemistry of the isopropyl group unknown). Treatment of **118** with sodium methoxide gives a mixture of which the desired keto ester **119** constitutes only 30%. Compound **119** is isolated in 43% yield, based on unrecovered keto ester. Sodium borohydride reduction and acidification provide (±)-dendrobine in 46% yield.

The final successful synthesis of dendrobine, which was reported in 1978,[547] is that of Roush. As shown in Scheme 13, the Roush synthesis begins with a straightforward construction of triene ester **125**, which is silylated and heated in refluxing toluene to accomplish intramolecular Diels-Alder cycloaddition. Isomers **126-129** are obtained in yields of 39%, 36%, 1% and 9%, respectively. The two major isomers, **126** and **127**, both result from endo addition:

126,127

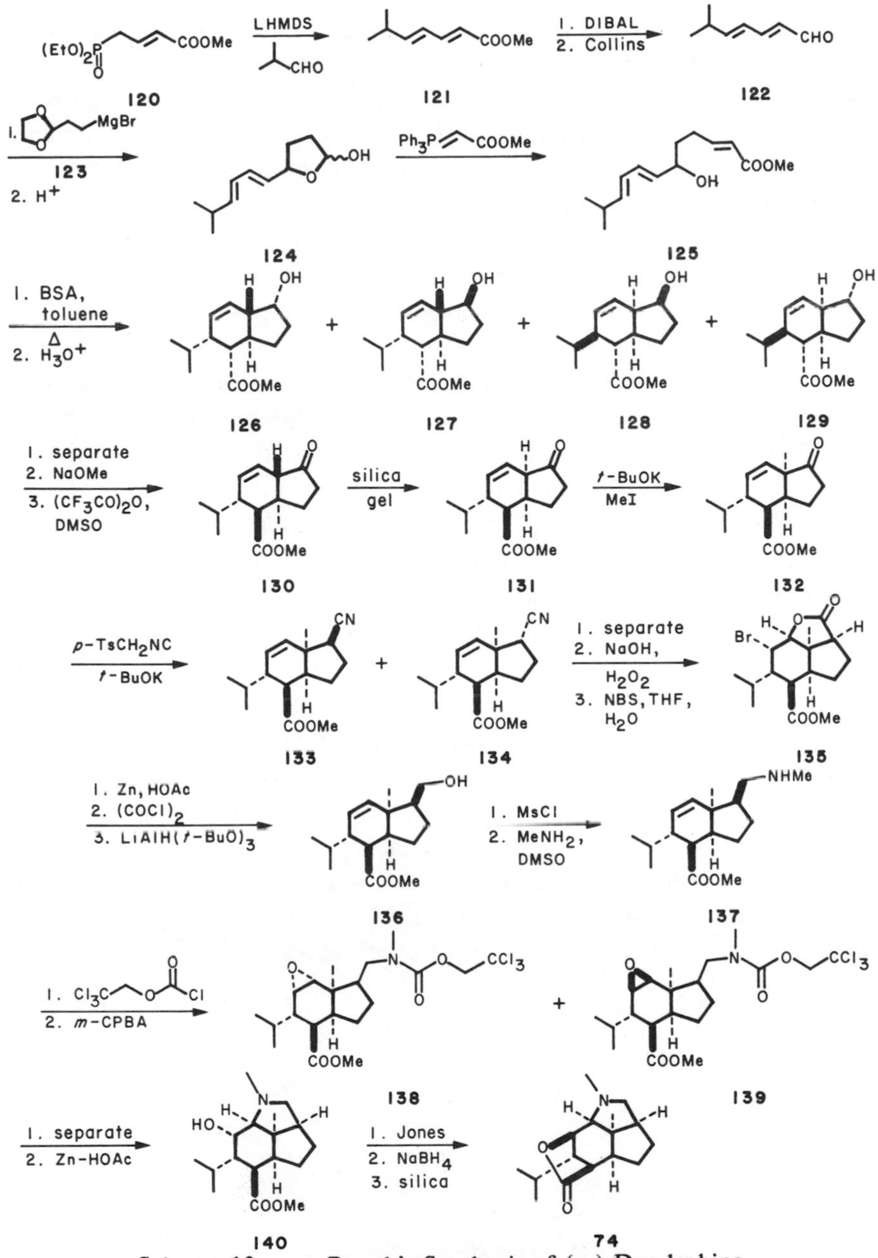

Scheme 13.   Roush's Synthesis of (±)-Dendrobine

After separation, these isomers are oxidized and the hydrindenone isomerized to the more stable cis form by treatment with silica gel. Methylation with potassium *t*-butoxide and methyl iodide provides an enone (131). This is converted into a 4:1 mixture of nitriles 133 and 134 by treatment with *p*-toluenesulfonylisocyanide and potassium *t*-butoxide. After separation, the major isomer is hydrolyzed to an unsaturated amide, which is bromolactonized to obtain 135. Reductive elimination gives an unsaturated acid, which is reduced via its acid chloride to obtain alcohol 136. The methylamino group is introduced by displacement on the derived mesylate and the resulting secondary amine (137) protected as a carbamate. Epoxidation of the double bond is accomplished by *m*-chloroperoxybenzoic acid, but diastereomeric epoxides 138 and 139 are produced in yields of 40 and 47%, respectively. The minor isomer is deprotected and cyclized to obtain 140. Inversion of the secondary alcohol is accomplished by oxidation and reduction of the resulting ketone. Spontaneous lactonization results when the hydroxy ester is chromatographed on silica gel.

At this point, it is instructive to compare the four dendrobine syntheses which have been completed to date (Table 4). The most efficient, from the standpoint of step economy, is the Kende synthesis. However, this synthesis gives a lower overall yield and requires a more unfavorable isomer separation than does the Yamada synthesis. The highest overall yield is obtained in the Roush synthesis, but this synthesis is also quite long. The least efficient synthesis, on all counts, is the Inubushi method. It is clear that there is still room for improvement in the synthesis of this interesting alkaloid. Unsuccessful approaches to dendrobine have emanated from three other laboratories.[548-550]

Table 4
Comparison of Dendrobine Syntheses

Investigator	Starting Point	Overall Steps	Isomer Yield	Separations
Yamada, 1972		19	0.8%	2(8:1,1:1)
Inubushi, 1972		24	0.00008%	4(1:3.5,[a] 1:3,[a] 1:2,[a] 3:2:2)
Kende, 197		15	0.6%	2(1:1.3:7[a])
Roush, 1978		23	1.0%	3(7:1,4:1, 4:5[a])

a. It is necessary to employ the minor isomer from this separation.

REFERENCES

1. L. Ruzicka, *Helv. Chim. Acta*, **6**, 492 (1923).

2. C. H. Heathcock in *The Total Synthesis of Natural Products, Volume 2*, edited by J. ApSimon, John Wiley and Sons, Inc., New York, 1973.

3. E. J. Corey and H. Yamamoto, *J. Am. Chem. Soc.*, **92**, 6637 (1970).

4. (a) E. J. Corey and H. Yamamoto, *J. Am. Chem. Soc.*, **92**, 3523 (1970). (b) E. J. Corey, J. I. Shulman, and H. Yamamoto, *Tetrahedron Lett.*, 447 (1970).

5. B. S. Pitzele, J. S. Baran, and D. H. Steinman, *Tetrahedron*, **32**, 1347 (1976).

6. O. P. Vig, A. K. Vig, and S. D. Kumar, *Indian J. Chem.*, **13**, 1244 (1975).

7. S. Akutagawa, T. Taketomi, and S. Otsuka, *Chem. Lett.*, 485 (1976).

8. L. Ahlquist and S. Ställberg-Stenhagen, *Acta Chem. Scand.*, **25**, 1685 (1971).

9. O. P. Vig, A. K. Vig and S. D. Kumar, *Indian J. Chem.*, **13**, 1003 (1975).

10. F. J. Gottschalk and P. Weyerstahl, *Chem. Ber.*, **108**, 2799, (1975).

11. See ref. 2, Scheme 9, pages 209-210.

12. (a) J. A. Findlay, W. D. MacKay, and W. S. Bowers, *J. Chem. Soc. (C)*, 2631, (1970). (b) See ref. 2, Scheme 15, page 219.

13. See ref. 2, Scheme 16, page 219.

14. J. S. Cochrane and J. R. Hanson, *J. Chem. Soc., Perkin I*, 361 (1972).

15. See ref. 2, Scheme 5, page 205.

16. See ref. 2, Scheme 9, pages 209-210.

17. R. J. Anderson, C. A. Henrick, J. B. Siddall, and R. Zurflüh, *J. Am. Chem. Soc.*, **94**, 5379 (1972).

18. See ref. 2, Scheme 11, pages 213-214.

19. C. A. Henrick, F. Schaub, and J. B. Siddall, *J. Am Chem. Soc.*, **94**, 5374 (1972).

20. G. Büchi and H. Wuest, *J. Am. Chem. Soc.*, **96**, 7573 (1974).

21. M. Baumann, W. Hoffmann, and H. Pommer, *Leibigs Ann. Chem.*, 1626 (1976).

22. See ref. 2, Schemes 2 and 3, pp, 202-203.

23. T. Hiyama, A. Kanakura, H. Yamamoto, and H. Nozaki, *Tetrahedron Lett.*, 3051 (1978).

24. See ref. 2, pages 200-205.

25. O. P. Vig, S. D. Sharma, M. L. Sharma, and S. S. Bari, *Indian J. Chem.*, **14B**, 926 (1976).

26. F. Bohlmann and R. Krammer, *Chem. Ber.*, **109**, 3362 (1976).

27. J. Meinwald, K. Opheim, and T. Eisner, *Tetrahedron Lett.*, 281 (1973).

28. C. H. Miller, J. A. Katzenellenbogen, and S. B. Bowlus, *Tetrahedron Lett.*, 285 (1973).

29. T. Kato, H. Takayanagi, T. Uyehara and Y. Kitahara, *Chem. Lett.*, 1009 (1977).

30. L. J. Altman, L. Ash, and S. Marson, *Synthesis*, 129 (1974).

31. Y. Gopichand, R. S. Prasad, and K. K. Chakravarti, *Tetrahedron Lett.*, 5177 (1973).

32. S. Kumazawa, K. Nishihara, T. Kato, Y. Kitahara, H. Komae, and N. Hayashi, *Bull. Chem. Soc. Japan*, **47**, 1530 (1974).

33. A. Takeda, K. Shinhama, and S. Tsuboi, *Bull. Chem. Soc. Japan*, **50**, 1903 (1977).

34. N. Fukamiya and S. Yasuda, *Chem. Ind. (London)*, 126 (1979).

35. A. F. Thomas and R. Dubini, *Helv. Chem. Acta*, **57**, 2066 (1974).

36. See ref. 2, pages 227-233.

37. S. Takahashi, *Synth. Commun.*, **6**, 331 (1976).

38. K. Kondo and M. Matsumoto, *Tetrahedron Lett.*, 391 (1976).

39. L. T. Burka, B. J. Wilson, and T. M. Harris, *J. Org. Chem.*, **39**, 2212 (1974).

40. K. Kondo and M. Matsumoto, *Tetrahedron Lett.*, 4363 (1976).

41. C. F. Ingham, R. A. Massy-Westropp and G. D. Reynolds, *Aust. J. Chem.*, **27**, 1477 (1974).

42. C. F. Ingham and R. A. Massy-Westropp, *Aust. J. Chem.*, **27**, 1491 (1974).

43. D. W. Knight and G. Pattenden, *J. Chem. Soc., Perkin I*, 641 (1975).

44. A. P. Krapcho and E. G. E. Jahngen, Jr., *J. Org. Chem.*, **39**, 1322 (1974).

45. See ref. 2, Scheme 6, page 243.

46. S. S. Hall, F. J. McEnroe, and H. -J. Shue, *J. Org. Chem.*, **40**, 3306 (1975).

47. See ref. 2, Scheme 6, page 243.

48. O. P. Vig, R. C. Anand, A. Singh, and J. P. Salota, *Indian J. Chem.*, **8**, 953 (1970).

49. O. P. Vig, A. K. Vig, and O. P. Chugh, *Indian J. Chem.*, **10**, 571 (1972).

50. P. A. Grieco and R. S. Finkelhor, *J. Org. Chem.*, **38**, 2909 (1973).

51. O. S. Park, Y. Grillasca, G. A. Garcia, and L. A. Maldonado, *Synth. Commun.*, **7**, 345 (1977).

52. P. Gosselin, S. Masson, and A. Thuillier, *J. Org. Chem.*, **44**, 2807 (1979).

53. See ref. 2, Scheme 13, page 249.

54. G. Gast and Y. -R. Naves, *Helv. Chim. Acta*, **54**, 1369 (1971).

55. See ref. 2, Schemes 15-16, pages 250-251.

56. See ref. 2, Schemes 21-22, pages 224-227.

57. See ref. 2, Scheme 23, page 228.

58. See Section 2 of this article, Scheme 27.

59. O. Ermer and S. Lifson, *Tetrahedron*, 30, 2425 (1974).

60. F. J. McEnroe and W. Fenical, *Tetrahedron*, 34, 1661 (1978).

61. R. B. Mane and G. S. Krishna Rao, *Indian J. Chem.*, 12, 938 (1974).

62. F. Bohlmann and D. Körnig, *Chem. Ber.*, 107, 1777 (1974).

63. N. Dennison, R. N. Mirrington, and A. D. Stuart, *Aust. J. Chem.*, 28, 1339 (1975).

64. O. P. Vig, H. Kumar, J. P. Salota, and S. D. Sharma, *Indian J. Chem.*, 11, 86 (1973).

65. P. R. Vijayasarathy, R. B. Mane, and G. S. Krishna Rao, *J. Chem. Soc., Perkin I*, 34 (1977).

66. M. Ando, S. Ibe, S. Kagabu, T. Nakagawa, T. Asao, and K. Takase, *Chem. Commun.*, 1538 (1970).

67. O. P. Vig, J. Chander, and B. Ram, *Indian J. Chem.*, 12, 1156 (1974).

68. E. Givaudi, M. Plattier and P. Teisseire, *Recherches*, 19, 205 (1974).

69. F. Delay and G. Ohloff, *Helv. Chim. Acta.*, 62, 369 (1979).

70. (a) O. P. Vig, S. D. Sharma, K. L. Matta, and J. M. Sehgal, *J. Indian Chem. Soc.*, 48, 993 (1971). (b) See ref. 2, Scheme 2, page 237.

71. O. P. Vig, B. Ram, C. P. Khera, and J. Chander, *Indian J. Chem.*, 8, 955 (1970).

72. D. J. Faulkner and L. E. Wolinsky, *J. Org. Chem.*, 40, 389 (1975).

73. L. E. Wolinsky and D. J. Faulkner, *J. Org. Chem.*, 41, 697 (1976).

74. R. J. Crawford, W. F. Erman, and C. D. Broaddus, *J. Am. Chem. Soc.*, 94, 4298 (1972).

75. A. Manjarrez, T. Rios, and A. Guzmán, *Tetrahedron*, 20, 333 (1964).

76. B. Cazes and S. Julia, *Tetrahedron Lett.*, 4065 (1978).

77. J. Alexander and G. S. Krishna Rao, *Indian J. Chem.*, 11, 859 (1973).

78. J. H. Babler, D. O. Olsen, and W. H. Arnold, *J. Org. Chem.*, 39, 1656 (1974).

79. See ref. 2, Scheme 1, page 235.

80. A. Manjarrez and A. Guzmán, *J. Org. Chem.*, 31, 348 (1966); see also ref. 2, page 236.

81. G. D. Gutsche, J. R. Maycock, and C. T. Chang, *Tetrahedron*, 24, 859 (1968).

82. W. Rittersdorf and F. Cramer, *Tetrahedron*, 24, 43 (1968).

83. J. M. Forrester and T. Money, *Can. J. Chem.*, 50, 3310 (1972).

84. W. Knöll and C. Tamm, *Helv. Chim. Acta.*, 58, 1162 (1975).

85. M. A. Schwartz and G. C. Swanson, *J. Org. Chem.*, 44, 953 (1979).

86. T. Iwashita, T. Kusumi, and H. Kakisawa, *Chem. Lett.*, 947 (1979).

87. Y. Gopichand and K. K. Chakravarti, *Tetrahedron Lett.*, 3851 (1974).

88. K. Suga, S. Watanabe, and T. Fujita, *Aust. J. Chem.*, **25**, 2393 (1972).

89. O. P. Vig, S. S. Bari, S. D. Sharma, and M. Lal, *Indian J. Chem,* **14B**, 929 (1976).

90. See ref. 2, Schemes 2 and 3, pages 237 and 239.

91. F. L. Malanco and L. A. Maldonado, *Synth. Commun.*, **6**, 515 (1976).

92. A. Kergomard and H. Veschambre, *Tetrahedron Lett.*, 4069 (1976).

93. (a) I. H. Rogers, J. F. Manville, T. Sahota, *Can. J. Chem.*, **52**, 1192 (1974).   (b) J.F. Manville, *ibid,* **53**, 1579 (1979); *ibid,* **54**, 2365 (1976).

94. V. Cerný, L. Dolejš, L. Labler, F. Sórm, and K. Sláma, *Tetrahedron Lett.*, 1053 (1967); *Coll. Czech. Chem. Commun.*, **32**, 3926 (1967).

95. See ref. 2, pages 253-262.

96. J. Ficini, J. d'Angelo, and J. Noiré, *J. Am. Chem. Soc.*, **96**, 1213 (1974).

97. G. Farges and H. Veschambre, *Bull. Soc. Chim. France,* 3172 (1973).

98. E. Negishi, M. Sabanski, J. -J. Katz, and H. C. Brown, *Tetrahedron,* **32**, 925 (1976).

99. D. A. Evans and J. V. Nelson, *J. Am. Chem. Soc.*, **102**, 774 (1980).

100. S. D. Larsen and S. A. Monti, *Synth. Commun.*, **9**, 141 (1979).

101. K. Uneyama and S. Torii, *Tetrahedron Lett.*, 443 (1976).

102. F. W. Sum and L. Weiler, *Tetrahedron Lett.*, 707 (1979).

103. J. W. Cornforth, B. V. Milborrow, and G. Ryback, *Nature,* **206**, 715 (1975)

104. M. Mousseron-Canet, J. -C. Mani, J. P. Dalle, and J. -L. Olivé, *Bull. Soc. Chim. France,* 3874 (1966).

105. D. L. Roberts, R. A. Heckman, B. P. Hege, and S. A. Bellin, *J. Org. Chem.*, **33**, 3566 (1968).

106. J. A. Findlay and W. D. MacKay, *Can. J. Chem.*, **49**, 2369 (1971).

107. K. Ohkuma, *Agr. Biol. Chem.* (Tokyo), **29**, 962 (1965); **30**, 434 (1966).

108. K. Mori, *Tetrahedron,* **30**, 1065 (1974).

109. T. Oritani and K. Yamashita, *Tetrahedron Lett.*, 2521 (1972).

110. F. Kienzle, H. Mayer, R. E. Minder, and W. Thommen, *Helv. Chim. Acta.*, **61**, 2616 (1978).

111. R. C. Cookson and P. Lombardi, *Gazz. Chim. Ital.*, **105**, 621 (1975).

112. P. Lombardi, R. C. Cookson, H. P. Weber, W. Renold, A. Hauser, K. H. Schulte-Elte, B. Willhalm, W. Thomner, and G. Ohloff, *Helv. Chim. Acta,* **59**, 1158 (1976).

113. T. R. Hoye and M. J. Kurth, *J. Org. Chem.*, **44**, 3461 (1979).

114. R. Baker, P. H. Briner, and D. A. Evans, *J. Chem. Soc., Chem. Commun.*, 981 (1978).

115. T. R. Hoye and M. J. Kurth, *J. Am. Chem. Soc.*, **101** 5065 (1979).

116. T. Kato, I. Ichinose, A. Kamoshida, and Y. Kitahara, *J. Chem. Soc. Chem. Commun.*, 518 (1976).

117. A. O. González, J. D. Martín, C. Pérez, and M. A. Ramírez, *Tetrahedron Lett.*, 137 (1976).

118. A. O. González, J. D. Martín, and M. A. Melián, *Tetrahedron Lett.*, 2279 (1976).

119. J. Froborg, G. Magnusson, and S. Thoren, *Acta Chem. Scand.*, **B28**, 265 (1974).

120. M. Kato, H. Kurihara, and A. Yoshikoshi, *J. Chem. Soc., Perkin I*, 2740 (1979).

121. G. Majetich, P. A. Grieco, and M. Nishizawa, *J. Org. Chem.*, **42**, 2327 (1977).

122. (a) K. Kato, Y. Hirata, and S. Yamamura, *Chem. Commun.*, 1324 (1970). (b) K. Kato, Y. Hirata, and S. Yamamura, *Tetrahedron*, **27**, 5987 (1971).

123. G. Fràter, *Helv. Chim. Acta*, **61**, 2709 (1978).

124. J. R. Williams and J. F. Callahan, *J. Chem. Soc., Chem. Commun.*, 404 (1979).

125. (a) C. Alexandre and F. Rouessac, *J. Chem. Soc., Chem. Commun.*, 275 (1975). (b) C. Alexandre and F. Rouessac, *Bull. Soc. Chim. France*, 117 (1977).

126. (a) M. Ando, A. Akahane, and K. Takase, *Bull. Chem. Soc. Japan*, **51**, 283 (1978). (b) M. Ando, K. Tajima, and K. Takase, *Chem. Lett.*, 617 (1978).

127. (a) P. A. Grieco and M. Nishizawa, *J. Chem. Soc., Chem. Commun.*, 582 (1976). (b) M. Nishizawa, P. A. Grieco, S. D. Burke, and W. Metz, *ibid.*, 76 (1978).

128. (a) P. A. Grieco, M. Nishizawa, S. D. Burke, and N. Marinovic, *J. Am. Chem. Soc.*, **98**, 1612 (1976). (b) P. A. Grieco, M. Nishizawa, T. Oguri, S. D. Burke, and N. Marinovic, *ibid.*, **99**, 5773 (1977).

129. (a) S. Danishefsky, T. Kitahara, P. F. Schuda, and S. J. Etheredge, *J. Am. Chem. Soc.*, **98**, 3028 (1976). (b) S. Danishefsky, P. F. Schuda, T. Kitahara, and S. J. Etheredge, *ibid.*, **99**, 6066 (1977).

130. C. G. Chavdarian and C. H. Heathcock, *J. Org. Chem.*, **40**, 2970 (1975).

131. G. R. Kieczykowski and R. H. Schlessinger, *J. Am. Chem. Soc.*, **100**, 1938 (1978).

132. (a) H. Iio, M. Isobe, T. Kawai, and T. Goto, *Tetrahedron*, **35**, 941 (1979). (b) M. Isobe, H. Iio, T. Kawai, and T. Goto, *J. Am. Chem. Soc.*, **100**, 1940 (1978). (c) H. Iio, M. Isobe, T. Kawai, and T. Goto, *ibid*, **101**, 6076 (1979).

133. R. D. Clark and C. H. Heathcock, *Tetrahedron Lett.*, 1713 (1974); *J. Org. Chem.*, **41**, 1396 (1976).

134. P. M. Wege, R. D. Clark, and C. H. Heathcock, *J. Org. Chem.*, **41**, 1396 (1976).

135. C. G. Chavdarian, S. L. Woo, R. D. Clark, and C. H. Heathcock, *Tetrahedron Lett.*, 1769 (1976).

136. S. Torii, T. Okamoto, and S. Kadono, *Chem. Lett.*, 495 (1977).

137. R. Scheffold, L. Révés, J. Aebersold, and A. Schaltegger, *Chimia*, **30**, 57 (1976).

138. F. Zutterman, P. de Clerq, and M. Vandewalle, *Tetrahedron Lett.*, 3191 (1977).

139. J. A. Marshall and G. A. Flynn, *J. Org. Chem.*, **44**, 1391 (1979).

140. J. J. Looker, D. P. Maier, and T. H. Regan, *J. Org. Chem.*, **37**, 3401 (1972).

141. Y. Fukuyama, T. Tokoroyama, and T. Kubota, *Tetrahedron Lett.*, 4869 (1973).

142. P. A. Grieco, T. Oguri, C. -L. J. Wang, and E. Williams, *J. Org. Chem.*, **42**, 4113 (1977).

143. P.A. Grieco, T. Oguri, S. Gilman, and G. T. De Titta, *J. Am. Chem. Soc.*, **100**, 1616 (1978).

144. A. Murai, M. Ono, A. Abiko, and T. Masamune, *J. Am. Chem. Soc.*, **100**, 7751 (1978).

145. P. S. Wharton, C. E. Sundin, D. W. Johnson, and H. C. Kluender, *J. Org. Chem.*, **37**, 34 (1972).

146. (a) M. Kodama, Y. Matsuki, and Shô Itô, *Tetrahedron Lett.*, 1121 (1976). (b) M. Kodama, S. Yokoo, H. Yamada, and Shô Itô, *ibid.*, 3121 (1978).

147. W. C. Still, *J. Am. Chem. Soc.*, **99**, 4186 (1977).

148. P. A. Grieco and M. Nishizawa, *J. Org. Chem.*, **42**, 1717 (1977).

149. Y. Fujimoto, T. Shimizu, and T. Tatsuno, *Tetrahedron Lett.*, 2041 (1976).

150. G. Lange and F. C. McCarthy, *Tetrahedron Lett.*, 4749 (1978).

151. W. C. Still, *J. Am. Chem. Soc.*, **101**, 2493 (1979).

152. O. P. Vig, B. Ram, K. S. Atwal, and S. S. Bari, *Indian J. Chem.*, **14B**, 855 (1976).

153. See ref. 2, Scheme 38, pages 280-281.

154. Y. Kitagawa, A. Itoh, S. Hashimoto, H. Yamamoto, and H. Nozaki, *J. Am. Chem. Soc.*, **99**, 3864 (1977).

155. T. -L. Ho, *Chem. and Ind.*, 487 (1971).

156. T. -L. Ho, *Can. J. Chem.*, **50**, 1098 (1972).

157. T. -L. Ho, *J. Chem. Soc., Perkin I*, 2579 (1973).

158. P. A. Reddy and G. S. Krishna Rao, *J. Chem. Soc., Perkin I*, 237 (1979).

159. (a) N. Katsui, A. Matsunaga, K. Imaizumi, T. Masamune, and K. Tomiyama, *Tetrahedron Lett.*, 83 (1971). (b) N. Katsui, A. Matsunaga, K. Imaizumi, T. Masamune, and K. Tomiyama, *Bull. Chem. Soc. Japan*, **45**, 2871 (1972).

160. F. Bohlmann and E. Eickeler, *Chem. Ber.*, **112**, 2811 (1979).

161. See ref. 2, Scheme 5, page 292.

162. D. Caine and J. T. Gupton, III, *J. Org. Chem.*, **39**, 2654 (1974).

163. R. B. Gammill and T. A. Bryson, *Synth. Commun.*, **6**, 209 (1976).

164. (a) J. W. Huffman and M. L. Mole, *Tetrahedron Lett.*, 501 (1971). (b) J. W. Huffman and M. L. Mole, *J. Org. Chem.*, **37**, 13 (1972).

165. While it is true that the Marshall synthesis provides acid **57** in less than 1% overall yield from 2-methylcyclohexanone (ref. 2, Scheme 4, page 290), the Heathcock-Kelly synthesis affords β-eudesmol in more than 5% overall yield from 2-methylcyclohexanone (ref. 2, Scheme 6, page 293).

166. R. G. Carlson and E. G. Zey, *J. Org. Chem.*, **37**, 2468 (1972).

167. C. H. Heathcock and T. R. Kelly, *Tetrahedron*, **24**, 1801 (1968).

168. B. D. MacKenzie, M. M. Angelo, and J. Wolinsky, *J. Org. Chem.*, **44**, 4042 (1979).

169. M. A. Schwartz, J. D. Crowell, and J. H. Musser, *J. Am. Chem. Soc.*, **94**, 4361 (1972).

170. P. A. Wender and J. C. Lechleiter, *J. Am. Chem. Soc.*, **100**, 4321 (1978).

171. R. B. Kelly, S. J. Alward, K. Suryanarayana Murty, and J. B. Stothers, *Can. J. Chem.*, **56**, 2508 (1978).

172. D.J. Goldsmith and I. Sakano, *J. Org. Chem.*, **41**, 2095 (1976).

173. See also ref. 2, Scheme 57, pages 372-373.

174. G. Büchi and H. Wuest, *J. Org. Chem.*, **44**, 546 (1979).

175. See ref. 2, Scheme 17, page 303.

176. T.R. Kelly, *J. Org. Chem.*, **37**, 3393 (1972).

177. R. Baker, D. A. Evans, and P. G. McDowell, *J. Chem. Soc., Chem. Commun.*, 111 (1977).

178. See also ref. 2, Scheme 22, page 309.

179. A. Murai, K. Nishizakura, N. Katsui, and T. Masamune, *Tetrahedron Lett.*, 4399 (1975).

180. A. Murai, H. Taketsuru, K. Fujisawa, Y. Nakahara, M. Takasugi, and T. Masamune, *Chem. Lett.*, 665 (1977).

181. (a) A. G. Hortmann, 4th Midwest Regional Meeting of the American Chemical Society, Manhattan, Kansas, November 1, 1968. (b) A. G. Hortmann, D. S. Daniel, and J. E. Martinelli, *J. Org. Chem.*, **38**, 728 (1973).

182. See ref. 2, page 309.

183. A. G. Hortmann, J. E. Martinelli, and Y. Wang, *J. Org. Chem.*, **34**, 732 (1969).

184. M. Ando, K. Nanaumi, T. Nakagawa, T. Asao, and K. Takase, *Tetrahedron Lett.*, 3891 (1970).

185. See ref. 2, Scheme 19, pages 305-306.

186. M. Sergent, M. Mongrain, and P. Deslongchamps, *Can. J. Chem.*, **50**, 336 (1972).

187. Y. Amano and C. H. Heathcock, *Can J. Chem.*, **50**, 340 (1972).

188. See ref. 2, Scheme 22, pages 309-310.

189. D. S. Watt and E. J. Corey, *Tetrahedron Lett.*, 4651 (1972).

190. J. A. Marshall and P. G. M. Wuts, *J. Org. Chem.*, **42**, 1794 (1977).

191. See ref. 2, pages 315-326.

192. J. A.. Marshall and P. G. M. Wuts, *J. Org. Chem.*, **43**, 1086 (1978).

193. D. Caine and G. Hasenhuettl, *Tetrahedron Lett.*, 743 (1975).

194. P. A. Grieco and M. Nishizawa, *J. Chem. Soc., Chem. Commun.*, 582 (1976).

195. D. Becker, N. C. Brodsky, and J. Kalo, *J. Org. Chem.*, **43**, 2557 (1978).

196. See ref. 2, Scheme 4, page 290.

197. G. H. Posner and G. L. Loomis, *J. Org. Chem.*, **38**, 4459 (1973).

198. See ref. 2, Scheme 30, pages 326-327.

199. See ref. 2, Scheme 23, pages 311-312.

200. R. B. Miller and E. S. Behare, *J. Am. Chem. Soc.*, **96**, 8102 (1974).

201. (a) J. D. Godfrey and A. G. Schultz, *Tetrahedron Lett.*, 3241 (1979). (b) A. G. Schultz and J. D. Godfrey, *J. Org. Chem.*, **41**, 3494 (1976).

202. See ref. 2, Schemes 4-7, pages 290-294 and Scheme 2, page 405.

203. W. C. Still and M. J. Schneider, *J. Am. Chem. Soc.*, **99**, 948 (1977).

204. (a) F. Kido, R. Maruta, K. Tsutsumi, and A. Yoshikoshi, *Chem. Lett.*, 311 (1979). (b) F. Kido, K. Tsutsumi, R. Maruta, and A. Yoshikoshi, *J. Am. Chem. Soc.*, **101**, 6420 (1979).

205. J. Alexander and G. S. Krishna Rao, *Tetrahedron*, **27**, 645 (1971).

206. V. Viswanatha and G. S. Krishna Rao, *Tetrahedron Lett.*, 243 (1974).

207. V. Viswanatha and G. S. Krishna Rao, *Tetrahedron Lett.*, 247 (1974).

208. V. Viswanatha and G. S. Krishna Rao, *Indian J. Chem.*, **11**, 974 (1973).

209. V. Viswanatha and G. S. Krishna Rao, *J. Chem. Soc., Perkin I*, 450 (1974).

210. J. P. McCormick, J. P. Pachlatko, and T. R. Schafer, *Tetrahedron Lett.*, 3993 (1978).

211. P. B. Talukdar, *J. Org. Chem.*, **21**, 506 (1956).

212. O. P. Vig, S. D. Sharma, G. L. Kad, and M. L. Sharma, *Ind. J. Chem.*, **13**, 764 (1975).

213. See ref. 2, Scheme 38, page 336. For full experimental details see L. A. Burk and M. D. Soffer, *Tetrahedron*, **32**, 2083 (1976).

214. M. D. Soffer and L. A. Burk, *Tetrahedron Lett.*, 211 (1970).

215. (a) R. P. Gregson and R. N. Mirrington, *J. Chem. Soc., Chem. Commun.*, 598 (1973). (b) R. P. Gregson and R. N. Mirrington, *Aus. J. Chem.*, **29**, 2037 (1976).

216. (a) M. Iguchi, M. Niwa, and S. Yamamura, *Bull. Chem. Soc. Japan*, **46**, 2920 (1973). (b) K. Kato, Y. Hirata, and S. Yamamura, *Tetrahedron*, **27**, 5987 (1971).

217. V. K. Belavadi and S. N. Kulkarni, *Indian J. Chem.*, **14B**, 901 (1976).

218. D. Caine and A. A. Frobese, *Tetrahedron Lett.*, 3107 (1977).

219. D. F. Taber and B. P. Gunn, *J. Am. Chem. Soc.*, **101**, 3992 (1979). See also the O. P. Vig 1977 (±)-kushitene synthesis, Scheme 85.

220. J. R. Hlubucek, A. J. Aasen, S -O. Almquist, and C. R. Enzell, *Acta Chem. Scand.*, **B28**, 289 (1974).

221. A. J. Aasen, C. H. G. Vogt, and C. R. Enzell, *Acta. Chem. Scand.*, **B29**, 51 (1975).

222. See ref. 2, Schemes 41-42, pages 340-342.

223. See ref. 2, pages 339-340.

224. See ref. 2, Scheme 46, pages 347-348.

225. H. Akita, T. Naito, and T. Oishi, *Chem. Lett.*, 1365 (1979).

226. See ref. 2, Scheme 44, page 344.

227. (a) T. Suzuki, M. Tanemura, T. Kato, and Y. Kitahara, *Bull Chem. Soc. Japan*, **43**, 1268 (1970). (b) H. Yanagawa, T. Kato, and Y. Kitahara, *Synthesis*, 257 (1970).

228. T. Kato, T. Suzuki, M. Tanemura, A. S. Kumanireng, N. Ototani, and Y. Kitahara, *Tetrahedron Lett.*, 1961 (1971).

229. T. Kato, T. Iida, T. Suzuki, Y. Kitahara, and K. H. Overton, *Tetrahedron Lett.*, 4257 (1972).

230. A. Ohsuka and A. Matsukawa, *Chem. Lett.*, 635 (1979).

231. T. Nakata, H. Akita, T. Naito, and T. Oishi, *J. Am. Chem. Soc.*, **101**, 4400 (1979).

232. S.P. Tanis and K. Nakanishi, *J. Am. Chem. Soc.*, **101**, 4398 (1979).

233. See ref. 2, Scheme 48, page 352.

234. D. Nasipuri and G. Das, *J. Chem. Soc., Perkin I*, 2776 (1979).

235. T. Matsumoto and S. Usui, *Chem. Lett.*, 105 (1978).

236. J. A. Marshall and R. A. Ruden, *J. Org. Chem.*, **36**, 594 (1971).

237. J. A. Marshall and T. M. Warne, Jr. *J. Org. Chem.*, **36**, 178 (1971).

238. See ref. 2, Schemes 53 and 54, pages 364-366.

239. H. C. Odom, Jr. and A. R. Pinder, *J. Chem. Soc., Perkin I*, 2193 (1972).

240. See ref. 2, pages 371-372.

241. Y. Takagi, Y. Nakahara, and M. Matsui, *Tetrahedron*, **34**, 517 (1978).

242. H. M. McGuire, H. C. Odom, Jr., and A.R. Pinder, *J. Chem. Soc., Perkin I*, 1879 (1974).

243. K. P. Dastur, *J. Am. Chem. Soc.*, **95**, 6509 (1973); **96**, 2605 (1974).

244. C. H. Heathcock and R. D. Clark, unpublished results.

245. T. Hiyama, M. Shinoda, and H. Nozaki, *Tetrahedron Lett.*, 3529 (1979).

246. T. Yanami, M. Miyashita, and A. Yoshikoshi, *J. Chem. Soc., Chem. Commun.*, 525 (1979).

247. D. Caine and S. L. Graham, *Tetrahedron Lett.*, 2521 (1976), and private communication.

248. See ref. 2, Schemes 57 and 60, pages 372 and 378.

249. E. Piers and R. D. Smillie, *J. Org. Chem.*, 35, 3997 (1970).

250. A. K. Torrence and A. R. Pinder, *Tetrahedron Lett.*, 745 (1971).

251. J. A. Marshall and G. M. Cohen, *J. Org. Chem.*, 36, 877 (1971).

252. S. Torii, T. Inokuchi, and T. Yamafuji, *Bull. Chem. Soc. Japan*, 52, 2640 (1979).

253. (a) K. Yamakawa, I. Izuta, H. Oka, and R. Sakaguchi, *Tetrahedron Lett.*, 2187 (1974). (b) K. Yamakawa, I. Izuta, H. Oka, R. Sakaquchi, M. Kobayashi, and T. Satoh, *Chem. Pharm. Bull.*, 27, 331 (1979).

254. S. Torii, T. Inokuchi, and K. Kawai, *Bull. Chem. Soc. Japan*, 52, 861 (1979).

255. F. E. Ziegler and P. A. Wender, *Tetrahedron Lett.*, 449 (1974).

256. F. E. Ziegler, G. R. Reid, W. L. Studt, and P. A. Wender, *J. Org. Chem.*, 42, 1991 (1977).

257. J. E. McMurry, J. H. Musser, M. S. Ahmad, and L. C. Blaszczak, *J. Org. Chem.*, 40, 1829 (1975).

258. J. Ficini and A. M. Touzin, *Tetrahedron Lett.*, 1081 (1977).

259. F. Näf, R. Decorzant, and W. Thommen, *Helv. Chim. Acta*, 62, 114 (1979).

260. E. Piers, M. B. Geraghty, and R. D. Smillie, *J. Chem. Soc., Chem. Commun.*, 614 (1971).

261. E. Piers and M. B. Geraghty, *Can. J. Chem.*, 51, 2166 (1973).

262. See ref. 2, Scheme 57, pages 372-373 and Scheme 61 of this section.

263. T. Tatee and T. Takahashi, *Chem. Lett.*, 929 (1973); *Bull. Chem. Soc. Japan*, 48, 281 (1975).

264. I. Nagakura, S. Maeda, M. Ueno, M. Funamizu, and Y. Kitahara, *Chem. Lett.*, 1143 (1975).

265. F. Bohlmann, H. -J. Förster, and C. H. Fischer, *Ann.*, 1487 (1976).

266. K. Yamakawa and T. Satoh, *Chem. Pharm. Bull. (Tokyo)* 25, 2535 (1977); *Heterocycles*, 8, 221 (1977); *Chem. Pharm. Bull. (Tokyo)*, 26, 3704 (1978).

267. M. Miyashita, T. Kumazawa, and A. Yoshikoshi, *Chem. Lett.*, 163 (1979).

268. Y. Inouye, Y. Uchida, and H. Kakisawa, *Chem. Lett.*, 1317 (1975).

269. F. Yuste and F. Walls, *Aust. J. Chem.*, 29, 2333 (1976).

270. J. W. Huffman and R. Pandian, *J. Org. Chem.*, 44, 1851 (1979).

271. D. K. Banerjee and V. B. Angadi, *Ind. J. Chem.*, 11, 511 (1973).

272. G. H. Posner, C. E. Whitten, J. J. Sterling, and D. J. Brunelle, *Tetrahedron Lett.*, 2591 (1974).

273. S. W. Baldwin and R. E. Gawley, *Tetrahedron Lett.*, 3969 (1975).

274. O. P. Vig, S. D. Sharma, O. P. Chugh, and A. K. Vig, *Ind. J. Chem.*, **12**, 1050 (1974).

275. O. P. Vig, I. R. Trehan, and R. Kumar, *Ind. J. Chem.*, **15B**, 319 (1977).

276. R. K. Boeckman, Jr. and S. M. Silver, *Tetrahedron Lett.*, 3497 (1973); *J. Org. Chem.*, **40**, 1755 (1975).

277. R. A. Moss, E. Y. Chen, J. Banger, and M. Matsuo, *Tetrahedron Lett.*, 4365 (1978).

278. See ref. 2, Scheme 4, page 290.

279. See ref. 2, page 286.

280. M. Ando, S. Sayama, and K. Takase, *Chem. Lett.*, 191 (1979).

281. See ref. 2, Scheme 5, page 292.

282. D. Caine, P. Chen, A. S. Frobese, and J. T. Gupton, *J. Org. Chem.*, **44**, 4981 (1979).

283. J. E. McMurry and L. C. Blaszczak, *J. Org. Chem.*, **39**, 2217 (1974).

284. See ref. 2, Scheme 70, page 394.

285. S. J. Branca, R. L. Lock, and A. B. Smith, III., *J. Org. Chem.*, **42**, 3165 (1977).

286. M. Nakayama, S. Ohira, S. Shinke, Y. Matsushita, A. Matsuo, and S. Hayashi, *Chem. Lett.*, 1249 (1975).

287. Y. Hayashi, M. Nishizawa, and T. Sakan, *Chem. Lett.*, 387 (1975).

288. O. P. Vig, R. K. Parti, K. C. Gupta, and M. S. Bhatia, *Ind. J. Chem.*, **11**, 981 (1973).

289. C. W. Bird, Y. C. Yeong, and J. Hudec, *Synthesis*, **No.1**, 27 (1974).

290. P. de Mayo and R. Suau, *J. Chem. Soc., Perkin I*, 2559 (1974).

291. T. Kametani, M. Tsubuki, and H. Nemoto, *Heterocycles*, **12**, 791 (1979).

292. E. Wenkert, B. L. Buckwalter, A. A. Craveiro, E. L. Sanchez, and S. S. Sathe, *J. Am. Chem. Soc.*, **100**, 1267 (1978).

293. R. B. Mane and G. S. Krishna Rao, *J. Chem. Soc., Perkin I*, 1806 (1973).

294. A. Casares and L. A. Maldonado, *Synth. Commun.*, **6**, 11 (1976).

295. T. Irie, T. Suzuki, Y. Yasunari, E. Kurosawa, and T. Masamune, *Tetrahedron*, **25**, 459 (1969).

296. J. E. McMurry and L. A. von Beroldingen, *Tetrahedron*, **30**, 2027 (1974).

297. R. C. Ronald, *Tetrahedron Lett.*, 4413 (1976).

298. (a) E. W. Colvin, R. A. Raphael, and J. S. Roberts, *Chem. Commun.*, 858 (1971).
(b) E. W. Colvin, S. Malchenko, R. A. Raphael, and J. S. Roberts, *J. Chem. Soc.
Perkin I*, 1989 (1973).

299. Y. Fujimoto, S. Yokura, T. Nakamura, T. Morikawa, and T. Tatsuno, *Tetrahedron
Lett.*, 2523 (1974).

300. See ref. 2, Schemes 53 and 54, pages 364-365.

301. N. Masuoka and T. Kamikawa, *Tetrahedron Lett.*, 1691 (1976).

302. Note that the structure of **870**, which is apparently a single diastereomer, is drawn
incorrectly in ref. 301.

303. W. C. Still and M. -Y. Tsai, *J. Am. Chem. Soc.*, **102**, 3654 (1980).

304. (a) S. C. Welch, A. S. C. P. Rao, and R. Y. Wong, *Synth. Commun.*, **6**, 443
(1976). (b) S. C. Welch, A. S. C. P. Rao, and C. G. Gibbs, *ibid.*, 485 (1976).

305. (a). S. C. Welch and R. Y. Wong, *Tetrahedron Lett.*, 1853 (1972). (b). S. C.
Welch and R. Y. Wong, *Synth. Commun.*, **2**, 291 (1972).

306. K. Yamakawa, R. Sakaguchi, T. Nakamura, and K. Watanabe, *Chem. Lett.*, 991,
(1976).

307. G. I. Feutrill, R. N. Mirrington and R. J. Nichols, *Aust. J. Chem.*, **26**, 345 (1973).

308. K. Yamada, H. Yazawa, D. Uemura, M. Toda, and Y. Hirata, *Tetrahedron*, **25**,
3509 (1969).

309. G. L. Hodgson, D. F. MacSweeney and T. Money, *Chem. Commun.*, 766 (1971).

310. G. L. Hodgson, D. F. MacSweeney, and T. Money, *Tetrahedron Lett.*, 3683
(1972).

311. G. L. Hodgson, D. F. MacSweeney, R. W. Mills, and T. Money, *J. Chem. Soc.,
Chem. Commun.*, 235 (1973).

312. E. J. Corey, H. A. Kirst, and J. A. Katzenellenbogen, *J. Am. Chem. Soc.*, **92**, 6314
(1970).

313. See ref. 2, Scheme 6, page 205.

314. M. Julia and P. Ward, *Bull. Soc., Chem. France*, 3065 (1973).

315. K. Sato, S. Inoue, Y. Takagi, and S. Morii, *Bull. Chem. Soc. Japan*, **49**, 3351
(1976).

316. S. A. Monti and S. D. Larsen, *J. Org. Chem.*, **43**, 2282 (1978).

317. S. D. Larsen and S. A. Monti, *J. Am. Chem. Soc.*, **99**, 8015 (1977).

318. M. Bertrand, H. Monti, and K. C. Huong, *Tetrahedron Lett.*, 15 (1979).

319. See ref. 2, Scheme 44, page 483.

320. P. A. Christenson and B. J. Willis, *J. Org. Chem.*, **44**, 2012 (1979).

321. See ref. 2, Scheme 50, page 489.

322. M. Baumann and W. Hoffmann, *Liebigs Ann. Chem.*, 743 (1979).

323. See ref. 2, Schemes 15 and 16, pages 250-251.

324. E. J. Corey, D. E. Cane, and L. Libit, *J. Am. Chem. Soc.*, **93**, 7016 (1971).

325. Y. Bessiere-Chretien and C. Grison, *C. R. Acad. Sci., Ser C,* **275**, 503 (1972).

326. See ref. 2, Scheme 26, page 447.

327. (a) P. A. Grieco and Y. Masaki, *J. Org. Chem.*, **40**, 150 (1975). (b) P. A. Grieco and J. J. Reap, *Synth. Commun.*, **5**, 347 (1975).

328. J. A. Marshall, N. H. Andersen and P. C. Johnson, *J. Am. Chem. Soc.*, **89**, 2748 (1967).

329. J. A. Marshall and S. F. Brady, *J. Org. Chem.*, **35**, 4068 (1970).

330. See ref. 2, Scheme 37, pages 469-470.

331. G. Stork, R. L. Danheiser, and B. Ganem, *J. Am. Chem. Soc.*, **95**, 3414 (1973).

332. (a) P. M. McCurry, Jr., and R. K. Singh, *Tetrahedron Lett.*, 3325 (1973). (b) P. M. McCurry, Jr., R. K. Singh, and S. Link, *ibid.*, 1155 (1973).

333. K. Yamada, H. Nagase, Y. Hayakawa, K. Aoki, and Y. Hirata, *Tetrahedron Lett.*, 4963 (1973).

334. K. Yamada, K. Aoki, H. Nagase, Y. Hayakawa, and Y. Hirata, *Tetrahedron Lett.*, 4967 (1973).

335. K. Yamada, S. Goto, H. Nagase, and A. T. Christensen, *J. Chem. Soc., Chem. Commun.*, 554 (1977).

336. D. Caine and C -Y. Chu, *Tetrahedron Lett.*, 703 (1974).

337. G. Bozzato, J -P. Bachmann, and M. Pesaro, *J. Chem. Soc., Chem. Commun.*, 1005 (1974).

338. B. M. Trost, M. Preckel, and L. M. Leichter, *J. Am. Chem. Soc.*, **97**, 2224 (1975).

339. (a) D. Buddhasukh and P. Magnus, *J. Chem. Soc., Chem. Commun.*, 952 (1975). (b) D. A. Chass, D. Buddhasukh, and P. D. Magnus, *J. Org. Chem.*, **43**, 1750 (1978).

340. W. G. Dauben and D. J. Hart, *J. Am. Chem. Soc.*, **97**, 1622 (1975); **99**, 7307 (1977).

341. M. Deighton, C. R. Hughes, and R. Ramage, *J. Chem. Soc., Chem. Commun.*, 662 (1975).

342. D. Caine, A. A. Boucugnani, S. T. Chao, J. B. Dawson, and P. F. Ingwalson, *J. Org. Chem.*, **41**, 1539 (1976).

343. D. Caine, A. A. Boucugnani and W. R. Pennington, *J. Org. Chem.*, **41**, 3632 (1976).

344. G. Büchi, D. Berthet, R. Decorzant, A. Grieder, and A. Hauser, *J. Org. Chem.*, **41**, 3208 (1976).

345. (a) K. Uneyama, K. Okamoto, and S. Torii, *Chem. Lett.*, 493 (1977). (b) S. Torii, K. Uneyama, and K. Okamoto, *Bull. Chem. Soc. Japan*, **51**, 3590 (1978).

346. See ref. 2, Scheme 36, page 467.

347. (a) T. Ibuka, K. Hayashi, H. Minakata, and Y. Inubushi, *Tetrahedron Lett.*, 159 (1979). (b) T. Ibuka, K. Hayashi, H. Minakata, Y. Ito, and Y. Inubushi, *Can. J. Chem.*, **57**, 1579 (1979).

348. J. M. Conia, J. P. Drouet, and J. Gore, *Tetrahedron*, **27**, 2481 (1971).

349. A. R. Pinder, S. J. Price, and R. M. Rice, *J. Org. Chem.*, **37**, 2202 (1972).

350. T. R. Kasturi and M. Thomas, *Indian J. Chem.*, **10**, 777 (1972).

351. P. Naegeli and R. Kaiser, *Tetrahedron Lett.*, 2013 (1972).

352. I. G. Guest, C. R. Hughes, R. Ramage, and A. Sattar, *J. Chem. Soc., Chem. Commun.*, 526 (1973).

353. (a) J. N. Marx and L. R. Norman, *Tetrahedron Lett.*, 4375 (1973). (b) J. N. Marx and L. R. Norman, *J. Org. Chem.*, **40**, 1602 (1975).

354. (a) W. Oppolzer, *Helv. Chim. Acta*, **56**, 1812 (1973). (b) W. Oppolzer and K. K. Mahalanabis, *Tetrahedron Lett.*, 3411 (1975). (c) W. Oppolzer, K. K. Mahalanabis and K. Bättig, *Helv. Chim. Acta*, **60**, 2388 (1977).

355. H. Wolf and M. Kolleck, *Tetrahedron Lett.*, 451 (1975).

356. H. Wolf, M. Kotteck, K. Claussen and W. Rascher, *Chem. Ber.*, **109**, 41 (1976).

357. B. M. Trost, K. Hiroi, and N. Holy, *J. Am. Chem. Soc.*, **97**, 5873 (1975).

358. J. F. Ruppert, M. A. Avery, and J. D. White, *J. Chem. Soc., Chem. Commun.*, 978 (1976).

359. D. A. McCrae and L. Dolby, *J. Org. Chem.*, **42**, 1607 (1977).

360. (a) G. L. Lange, W. J. Orrom, and D. J. Wallace, *Tetrahedron Lett.*, 4479 (1977). (b) G. L. Lange, E. E. Neidert, W. J. Orrom, and D. J. Wallace, *Can. J. Chem.*, **56**, 1628 (1978).

361. M. Pesaro and J -P. Bachmann, *J. Chem. Soc., Chem. Commun.*, 203 (1978).

362. S. F. Martin and T. Chou, *J. Org. Chem.*, **43**, 1027 (1978).

363. D. Caine and H. Deutsch, *J. Am. Chem. Soc.*, **100**, 8030 (1978).

364. See ref. 2, page 286.

365. J. D. White, S. Torii, and J. Nogami, *Tetrahedron Lett.*, 2879 (1974).

366. G. Fráter, *Helv. Chim. Acta*, **60**, 515 (1977).

367. C. Iwata, M. Yamada, and Y. Shinoo, *Chem. Pharm. Bull.*, **27**, 274 (1979).

368. L. E. Wolinsky and D. J. Faulkner, *J. Org. Chem.*, **41**, 597 (1976).

369. W. Fenical and B. M. Howard, *Tetrahedron Lett.*, 2519 (1976).

370. I. Ichinose and T. Kato, *Chem. Lett.*, 61 (1979).

371. See ref. 2, Scheme 29, pages 452-453.

372. O. P. Vig, B. Ram, S. S. Bari, and S. D. Sharma, *Indian J. Chem.*, **14B**, 852 (1976).

534   References

373. See ref. 2, Scheme 32, page 272.

374. See ref. 2, Scheme 33, page 273.

375. See ref. 2, Scheme 20, pages 435-437.

376. J. J. Plattner and H. Rapoport, *J. Am. Chem. Soc.*, **93**, 1758 (1971).

377. C. F. Garbers, J. A. Steenkamp, and H. E. Visagie, *Tetrahedron Lett.*, 3753 (1975).

378. K. Kitatani, T. Hiyama, and H. Nozaki, *J. Am. Chem. Soc.*, **98**, 2362 (1976).

379. See ref. 2, Scheme 16 pages 429-431.

380. S. Danishefsky, M. Hirama, K. Gombatz, T. Harayama, E. Berman, and P. Schuda, *J. Am. Chem. Soc.*, **100**, 6536 (1978); *ibid.*, **101**, 7020 (1979).

381. W. H. Parsons, R. H. Schlessinger, and M. L. Quesada, *J. Am. Chem. Soc.*, **102**, 889 (1980).

382. Y. Hayashi, M. Nishizawa, S. Harita, and T. Sakan, *Chem. Lett.*, 375 (1972).

383. M. E. N. Nambudiry and G. S. Krishna Rao, *J. Chem Soc., Perkin I*, 317 (1974).

384. Y. Kitahara, N. Abe, T. Kato, and K. Shirahata, *Nippon Kagaku Zasshi*, **90**, 221 (1969).

385. K. Hayashi, H. Nakamura, and H. Mitsuhashi, *Chem. Pharm. Bull.*, **21**, 2806 (1973).

386. K. Naya, M. Hayashi, I. Takagi, S. Nakamura, and M. Kobayashi, *Bull. Chem. Soc. Japan*, **45**, 3673 (1972).

387. (a) D. A. Evans and C. L. Sims *Tetrahedron Lett.*, 4691 (1973). (b) D. A. Evans, C. L. Sims, and G. C. Andrews, *J. Am. Chem. Soc.*, **99**, 5453 (1977).

388. See ref. 2, Scheme 65, page 386.

389. S. Sarma and B. Sarkar, *Indian J. Chem.*, **10**, 950 (1972).

390. D. Caine and F. N. Tuller, *J. Org. Chem.*, **38**, 3663 (1973).

391. D. F. Taber and R. W. Korsmeyer, *J. Org. Chem.*, **43**, 4925 (1978).

392. K. Sisido, S. Kurozumi, K. Utimoto, and T. Isida, *J. Org. Chem*, **31**, 2795 (1966).

393. E. J. Corey and H. L. Pearce, *J. Am. Chem. Soc.*, **101**, 5841 (1979).

394. N. H. Andersen and H. Uh, *Synth. Commun.*, **3**, 115 (1973).

395. (a) See ref. 2, pages 404-407. (b) C. H. Heathcock and R. Ratcliffe, *J. Am. Chem. Soc.*, **93**, 1746 (1971).

396. G. Mehta and B. P. Singh, *Tetrahedron Lett.*, 4495 (1975).

397. See ref. 2, pages 402-410.

398. (a) J. A. Marshall and A. E. Greene, *J. Org. Chem.*, **37**, 982 (1972). (b) J. A. Marshall, A. E. Greene, and R. Ruden, *Tetrahedron Lett.*, 855 (1971). (c) J. A. Marshall and A. E. Greene, *ibid.*, 859 (1971).

399. G. L. Buchanan and G. A. R. Young, *J. Chem. Soc., Perkin I,* 2404 (1973); *J. Chem. Soc., Chem. Commun.,* 643 (1971).

400. N. H. Andersen and H. S. Uh, *Tetrahedron Lett.,* 2079 (1973).

401. H -J. Liu and S. P. Lee, *Tetrahedron Lett.,* 3699 (1977).

402. N. H. Andersen and F. A. Golec, Jr., *Tetrahedron Lett.,* 3783 (1977).

403. J. N. Marx and S. M. McGaughey, *Tetrahedron,* **28,** 3583 (1972).

404. See ref. 2, pages 413-416.

405. M. T. Edgar, A. E. Greene, and P. Crabbé, *J. Org. Chem.,* **44,** 159 (1979).

406. See ref. 2, pages 417-423.

407. D. Caine and P. F. Ingwalson, *J. Org. Chem.,* **37,** 3751 (1972).

408. J. A. Marshall and J. A. Ruth, *J. Org. Chem.,* **39,** 1971 (1974).

409. See ref. 2, Scheme 4, page 408.

410. D. Caine and J. T. Gupton, *J. Org. Chem.,* **40,** 809 (1975).

411. J. A. Marshall and W. R. Snyder, *J. Org. Chem.,* **40,** 1656 (1975).

412. R. A. Kretchmer and W. J. Thompson, *J. Am. Chem. Soc.,* **98,** 3379 (1976).

413. P. DeClercq and M. Vandewalle, *J. Org. Chem.,* **42,** 3447 (1977).

414. P. Kok, P. DeClercq, and M. Vandewalle, *Bull. Soc., Chem. Belg.,* **87,** 615 (1978).

415. P. A. Grieco, Y. Ohfune, and G. Majetich, *J. Am. Chem. Soc.,* **99,** 7393 (1977).

416. (a) R. Peel and J. K. Sutherland, *J. Chem. Soc., Chem. Commun.,* 151 (1974). (b) J. S. Bindra, A. Grodski, T. K. Schaaf, and E. J. Corey, *J. Am. Chem. Soc.,* **95,** 7522 (1973).

417. P. A. Grieco, T. Oguri, S. Burke, E. Rodriguez, G. T. DeTitta, and S. Fortier, *J. Org. Chem.,* **43,** 4552 (1978).

418. (a) M. R. Roberts and R. H. Schlessinger *J. Am. Chem. Soc.,* **101,** 7626 (1979). (b) G. J. Quallich and R. H. Schlessinger, *ibid.,* **101,** 7627 (1979).

419. P. T. Lansbury and A. K. Serelis, *Tetrahedron Lett.,* 1909 (1978).

420. P. A. Wender, M. A. Eissenstat, and M. P. Filosa, *J. Am. Chem. Soc.,* **101,** 2196 (1979).

421. J. A. Marshall and R. H. Ellison, *J. Am. Chem. Soc.,* **98,** 4312 (1976).

422. M. F. Semmelhack, A. Yamashita, J. C. Tomesch, and K. Hirotsu *J. Am. Chem. Soc.,* **100,** 5565 (1978).

423. C. H. Heathcock, E. G. Delmar, and S. L. Graham, *J. Am. Chem. Soc.,* **104,** 1907 (1982).

424. M. Demuynck, P. DeClercq, and M. Vandewalle, *J. Org. Chem.,* **44,** 4863 (1979).

425. C. H. Heathcock, T. C. Germroth and C. M. Tice, *J. Am. Chem. Soc.,* **104,** 0000 (1982).

426. (a) Y. Ohfune, P. A. Grieco, C. -L. Wang, G. Majetich, *J. Am. Chem. Soc.*, **100**, 5946 (1978). (b) P. A. Grieco, Y. Ohfune, and G. Majetich, *J. Org. Chem.*, **44**, 3092 (1979).

427. P. A. Grieco, Y. Ohfune, and G. Majetich, *Tetrahedron Lett.*, 3265 (1979).

428. P. Kok, P. DeClercq, and M. Vandewalle, *J. Org. Chem.*, **44**, 4553 (1979).

429. (a) P. T. Lansbury and D. Hangauer, *Tetrahedron Lett.*, 3623 (1979). (b) P. T. Lansbury, D. G. Hangauer, and J. P. Vacca, *J. Am. Chem. Soc.*, **102**, 3964 (1980).

430. M. Soucek, *Coll. Czech. Chem. Commun.*, **27**, 2929 (1962).

431. M. Yamasaki, *J. Chem. Soc., Chem. Commun.*, 606 (1972).

432. (a) H. deBroissia, J. Levisalles, and H. Rudler, *J. Chem. Soc., Chem. Commun.*, 855 (1972). (b) *idem.*, *Bull. Soc. Chim. Fr.*, 4314 (1972).

433. (a) J. Froborg, G. Magnusson, and S. Thorén, *J. Org. Chem.*, **40**, 1595 (1975). (b) T. Fex, J. Froborg, G. Magnusson, and S. Thorén, *ibid*, **41**, 3518 (1976). (c) J. Froborg and G. Magnusson, *J. Am. Chem. Soc.*, **100**, 6728 (1978).

434. See ref. 2, pages 396-402.

435. E. Wenkert and K. Naemura, *Synth. Commun.*, **3**, 45 (1973).

436. G. Mehta and S. K. Kapoor, *J. Org. Chem.*, **39**, 2618 (1974).

437. E. Piers and E. H. Ruediger, *J. Chem. Soc., Chem. Commun.*, 166 (1979).

438. A. G. González, J. Darias, J. D. Martin, and M. A. Melián, *Tetrahedron Lett.*, 481 (1978).

439. S. Danishefsky and K. Tsuzuki, *J. Am. Chem. Soc.*, **102**, 6891 (1980).

440. See ref. 2, Scheme 42, page 480.

441. M. Bertrand and J. -L. Gras, *Tetrahedron*, **30**, 793 (1974).

442. A. Kumar, A. Singh, and D. Devaprabhakara, *Tetrahedron Lett.*, 2177 (1976).

443. A. Kumar and D. Devaprabhakara, *Synthesis*, 461 (1976).

444. E. Piers, R. W. Britton, and W. de Waal, *Can. J. Chem.*, **49**, 12, 1971.

445. A. Tanaka, R. Tanaka, H. Uda, and A. Yoshikoshi, *J. Chem. Soc., Perkin I*, 1721 (1972).

446. S. Torii and T. Okamoto, *Bull. Chem. Soc. Japan*, **49**, 771 (1976).

447. See ref. 2, Scheme 17, page 538.

448. T. Matsumoto, K. Miyano, S. Kagawa, S. Yü, J. Ogawa, and A. Ichihara, *Tetrahedron Lett.*, 3521 (1971).

449. Y. Ohfune, H. Shirahama, and T. Matsumoto, *Tetrahedron Lett.*, 4377 (1975).

450. H. Takeshita, H. Iwabuchi, I. Kouno, M. Iino, and D. Nomura, *Chem. Lett.*, 649 (1979).

451. D. Helmlinger, P. de Mayo, M. Nye, L. Westfelt, and R. B. Yeats, *Tetrahedron Lett.*, 349 (1970).

452. S. R. Wilson and R. B. Turner, *J. Org. Chem.*, **38**, 2870 (1973).

453. W. J. Greenlee and R. B. Woodward, *J. Am. Chem. Soc.*, **98**, 6075 (1976).

454. R. K. Boeckman, Jr. and S. S. Ko, *J. Am. Chem. Soc.*, **102**, 7146 (1980).

455. (a) P. T. Lansbury, N. Y. Wang, and J. E. Rhodes, *Tetrahedron Lett.*, 1829 (1971). (b) P. T. Lansbury and N. Nazarenko, *ibid.*, 1833 (1971).

456. P. T. Lansbury, N. Y. Wang, and J. E. Rhodes, *Tetrahedron Lett.*, 2053 (1972).

457. H. Hashimoto, K. Tsuzuki, F. Sakan, H. Shirahama, and T. Matsumoto, *Tetrahedron Lett.*, 3745 (1974).

458. F. Sakan, H. Hashimoto, A. Ichihara, H. Shirahama, and T. Matsumoto, *Tetrahedron Lett.*, 3703 (1971).

459. B. M. Trost, C. D. Shuey, F. DiNinno, Jr. and S. S. McElvain, *J. Am. Chem. Soc.*, **101**, 1284 (1979).

460. K. Tatsuta, K. Akimoto, and M. Kinoshita, *J. Am. Chem. Soc.*, **101**, 6116 (1979).

461. K. Tatsuta, K. Akimoto, and M. Kinoshita, *J. Antibiot.*, **33**, 100 (1980).

462. (a) S. Danishefsky, R. Zamboni, M. Kahn, and S. J. Etheredge, *J. Am. Chem. Soc.*, **102**, 2097 (1980). (b) S. Danishefsky and R. Zamboni, *Tetrahedron Lett.*, 3439 (1980).

463. M. Shibasaki, K. Iseki, and S. Ikegami, *Tetrahedron Lett.*, 3587 (1980).

464. A. E. Greene, *Tetrahedron Lett.*, 3059 (1980).

465. R. D. Little and G. W. Miller, *J. Am. Chem. Soc.*, **101**, 7129 (1979).

466. W. Oppolzer, K. Bättig, and T. Hudlicky, *Helv. Chim. Acta*, **62**, 1493 (1979).

467. M. C. Pirrung, *J. Am. Chem. Soc.*, **101**, 7130 (1979); **103**, 82 (1981).

468. L. A. Paquette and Y. K. Han, *J. Org. Chem.*, **44**, 4014 (1979).

469. W. G. Dauben and D. M. Walker, *J. Org. Chem.*, **46**, 1103 (1981).

470. S. Chatterjee, *J. Chem. Soc., Chem. Commun.*, 620 (1979).

471. Ref. 468, footnote 16. See also J. Cornforth, *Tetrahedron Lett.*, 709 (1980).

472. R. M. Coates, S. K. Shah, and R. W. Mason, *J. Am. Chem. Soc.*, **101**, 6765 (1979).

473. G. Büchi and P. -S. Chu, *J. Am. Chem. Soc.*, **101**, 6767 (1979).

474. S. C. Welch and S. Chayabunjonglerd, *J. Am. Chem. Soc.*, **101**, 6768 (1979).

475. Y. -K. Han and L. A. Paquette, *J. Org. Chem.*, **44**, 3731 (1979).

476. See ref. 2, Scheme 11, pages 526-527.

477. J. E. McMurry, *J. Org. Chem.*, **36**, 2826 (1971).

478. See ref. 2, page 523.

479. (a) E. Piers, R. W. Britton, R. J. Keziere, and R. D. Smillie, *Can. J. Chem.*, **49**, 2620 (1971). (b) E. Peirs, R. W. Britton, R. J. Keziere, and R. D. Smillie, *ibid.*,

49, 2623 (1971). (c) E. Piers, M. B. Geraghty, F. Kido, and M. Soucy, *Synth. Commun.*, **3**, 39 (1973). (d) E. Piers, M. B. Geraghty, and M. Soucy, *ibid.*, **3**, 401 (1973). (e) E. Piers, R. W. Britton, M. B. Geraghty, R. J. Keziere, and R. D. Smillie, *Can. J. Chem.*, **53**, 2827 (1975). (f) E. Peirs, R. W. Britton, M. B. Geraghty, R. J. Keziere, and F. Kido, *ibid.*, **53**, 2838 (1975). (g) E. Piers, M. B. Geraghty, R. D. Smillie, and M. Soucy, *ibid.*, **53**, 2849 (1975).

480. (a) G. L. Hodgson, D. F. MacSweeney, and T. Money, *Tetrahedron Lett.*, 3683 (1972). (b) G. L. Hodgson, D. F. MacSweeney, and T. Money, *J. Chem. Soc., Perkin I*, 2113 (1973). (c) C. R. Eck, G. L. Hodgson, D. F. MacSweeney, R. W. Mills, and T. Money, *ibid.*, 1938 (1974).

481. S. N. Baldwin and J. C. Tomesch, *Tetrahedron Lett.*, 1055 (1975).

482. P. Bakuzis, O. O. S. Campos, and M. L. F. Bakuzis, *J. Org. Chem.*, **41**, 3261 (1976).

483. M. Yanagiya, K. Kaneko, T. Kaji, and T. Matsumoto, *Tetrahedron Lett.*, 1761 (1979).

484. (a) E. Piers and H. -P. Isenring, *Synth. Commun.*, **6**, 221 (1976). (b) E. Piers and H. -P. Isenring, *Can. J. Chem.*, **55**, 1039 (1977).

485. J. E. McMurry and M. G. Silvestri, *J. Org. Chem.*, **41**, 3953 (1976).

486. P. A. Collins and D. Wege, *Aust. J. Chem.*, **32**, 1819 (1979).

487. See ref. 2, Scheme 9, pages 518-519.

488. J. E. McMurry and S. J. Isser, *J. Am. Chem. Soc.*, **94**, 7132 (1972).

489. R. A. Volkmann, G. C. Andrews, and W. A. Johnson, *J. Am. Chem. Soc.*, **97**, 4777 (1975).

490. W. Oppolzer and T. Godel, *J. Am. Chem. Soc.*, **100**, 2583 (1978).

491. (a) S. C. Welch and R. L. Walters, *Synth. Commun.*, **3**, 15 (1973). (b) S. C. Welch and R. L. Walters, *ibid.*, **3**, 419 (1973). (c) S. C. Welch and R. L. Walters, *J. Org. Chem.*, **39**, 2665 (1974).

492. See ref. 2, pages 521-525.

493. E. J. Corey and D. S. Watt, *J. Am. Chem. Soc.*, **95**, 2302 (1973).

494. (a) M. Miyashita and A. Yoshikoshi, *Chem. Commun.*, 1091 (1971). (b) M. Miyashita and A. Yoshikoshi, *J. Chem. Soc., Chem. Commun.*, 1173 (1972). (c) M. Miyashita and A. Yoshikoshi, *J. Am. Chem. Soc.*, **96**, 1917 (1974).

495. See also C. H. Heathcock and R. A. Badger, *Chem. Commun.*, 1510 (1968) and C. H. Heathcock, R. A. Badger, and R. A. Starkey, *J. Org. Chem.*, **37**, 231 (1972).

496 E. J. Corey, M. Behforouz, and M. Ishiguro, *J. Am. Chem. Soc.*, **101**, 1608 (1979).

497. H. Yamamoto and H. L. Sham, *J. Am. Chem. Soc.*, **101**, 1609 (1979).

498. E. J. Corey and M. Ishiguro, *Tetrahedron Lett.*, 2745 (1979).

499. See ref. 2, pages 492-503.

500. E. Piers, W. de Waal and R. W. Britton, *J. Am. Chem. Soc.*, **93**, 5113 (1971).

501. (a) K. J. Schmalzl and R. N. Mirrington, *Tetrahedron Lett.*, 3219 (1970). (b) R. N. Mirrington and K. J. Schmalzl, *J. Org. Chem.*, **37**, 2871 (1972). (c) R. N. Mirrington and K. J. Schmalzl, *ibid.*, **37**, 2877 (1972).

502. See ref. 2, Scheme 2, pages 497-498.

503. See ref. 2, Scheme 3, pages 501-502.

504. (a) N. Fukamiya, M. Kato, and A. Yoshikoshi, *Chem. Commun.*, 1120 (1971). (b) N. Fukamiya, M. Kato, and A. Yoshikoshi, *J. Chem. Soc., Perkin I*, 1843 (1973).

505. F. Näf and G. Ohloff, *Helv. Chim. Acta*, **57**, 1868 (1974).

506. G. Fráter *Helv. Chim. Acta*, **57**, 172 (1974).

507. M. E. Jung and C. A. McCombs, *J. Am. Chem. Soc.*, **100**, 5207 (1978).

508. D. Spitzner, *Tetrahedron Lett.*, 3349 (1978).

509. K. Yamada, Y. Kyotani, S. Manabe, and M. Suzuki, *Tetrahedron*, **35**, 293 (1979).

510. P. Teisseire, P. Pesnelle, B. Corbies, M. Plattier, and P. Manpetit, *Recherches*, **19**, 69 (1974).

511. W. Oppolzer and R. L. Snowden, *Tetrahedron Lett.*, 3505 (1978).

512. J. -L. Gras, *Tetrahedron Lett.*, 4117 (1977).

513. F. Kido, H. Uda and A. Yoshikoshi, *J. Chem. Soc., Perkin I*, 1755 (1972).

514. D. F. MacSweeney and R. Ramage, *Tetrahedron*, **27**, 1481 (1971).

515. See ref. 2, Scheme 8, pages 515-516.

516. See ref. 2, Scheme 10, pages 417-418.

517. A. Deljac, W. D. MacKay, C. S. J. Pan, K. J. Wiesner, and K. Wiesner, *Can. J. Chem.*, **50**, 726 (1972).

518. R. M. Coates and R. L. Sowerby, *J. Am. Chem. Soc.*, **94**, 5386 (1972).

519. E. Piers and J. Banville, *J. Chem Soc., Chem. Commun.*, 1138 (1979).

520. G. Büchi, A. Hauser, and J. Limacher, *J. Org. Chem.*, **42**, 3323 (1977).

521. H. -J. Liu and W. H. Chan, *Can. J. Chem.*, **57**, 708 (1979).

522. P. Naegeli and R. Kaiser, *Tetrahedron Lett.*, 2013 (1972).

523. E. J. Corey and R. D. Balanson, *Tetrahedron Lett.*, 3153 (1973).

524. See ref. 2, Scheme 6, page 510.

525. P. T. Lansbury, V. R. Haddon, and R. C. Stewart, *J. Am. Chem. Soc.*, **96**, 896 (1974).

526. See also ref. 2, Scheme 7 and accompanying discussion, pages 512-514.

540

527. (a) E. G. Breitholle and A. G. Fallis, *Can. J. Chem.*, **54**, 1991 (1976). (b) E. G. Breitholle and A. G. Fallis, *J. Org. Chem.*, **43**, 1964 (1978).

528. S. Danishefsky, K. Vaughn, R. Gadwood, and K. Tsuzukim *J. Am. Chem. Soc.*, **102**, 4262 (1980).

529. See ref. 2, Scheme 19, pages 542-543.

530. E. Piers and M. Zbozny, *Can. J. Chem.*, **57**, 2249 (1979).

531. (a) R. B. Kelly and J. Zamecnik, *Chem. Commun.*, 1102 (1970). (b) R. B. Kelly, J. Zamecnik, and B. A. Beckett, *ibid.* 479 (1971). (c) R. B. Kelly, J. Zamecnik, and B. A. Beckett, *Can. J. Chem.*, **50**, 3455 (1972).

532. (a) R. M. Cory and F. R. McLaren, *J. Chem. Soc., Chem. Commun.*, 587 (1977). (b) R. M. Cory, D. M. T. Chan, F. R. McLaren, M. H. Rasmussen, and R. M. Renneboog, *Tetrahedron Lett.*, 4133 (1979).

533. E. Piers and T. -W. Hall, *J. Chem Soc., Chem. Commun.*, 880 (1977).

534. R. B. Woodward and T. R. Hoye, *J. Am. Chem. Soc.*, **99**, 8007 (1977).

535. (a) T. Kaneko, I. Kawasaki and T. Okamoto, *Chem. and Ind. (London)*, 1191 (1959). (b) M. Kotake, I. Kawasaki, T. Okamoto, S. Kusumoto and T. Kaneko, *Bull. Chem. Soc. Japan.*, 32, 892 (1959). (c) M. Kotake, I. Kawasaki, T. Okamoto, S. Kusumoto and T. Kaneko, *Ann. Chem.*, **636**, 158 (1960). (d) M. Kotake, I. Kawasaki, S. Matsutani, S. Kusumoto and T. Kaneko, *Bull. Chem. Soc. Japan*, **35**, 1494 (1962).

536. (a) Y.. Arata and T. Nakanishi, *J. Pharm. Soc. Japan*, **80**, 855 (1960). (b) Y. Arata, T. Nakanishi, and Y. Asaoka, *Chem. Pharm. Bull. (Tokyo)*, **10**, 675 (1962).

537. I. Jezo, M. Karvas and K. Tihlarik, *Chem. Zvesti*, **15**, 283 (1961).

538. F. Bohlmann, E. Winterfeldt, P. Studt, H. Laurent, G. Boroschewski, and K. -M. Kleine, *Chem. Ber.*, **94**, 3151 (1961).

539. (a) J. T. Wróbel and Z. Dabrowski, *Roczniki Chem.*, **39**, 1239 (1965); (b) J. Szychowski, A. Leniewski and J. T. Wróbel, *Chem. and Ind. (London)*, 273 (1978).

540. F. Bohlmann, E. Winterfeldt, H. Laurent, and W. Ude, *Tetrahedron*, **19**, 195 (1963).

541. Y. Arata, T. Ohashi, Z. Okomura, Y. Wada, and M. Ishikawa, *J. Pharm. Soc. Japan*, **82**, 782 (1962); **83**, 79 (1963).

542. J. Szychowski, J. T. Wróbel and A. Leniewski, *Can. J. Chem.*, **55**, 3105 (1977).

543. H. Suzuki, I. Keimatsu and M. Ito, *J. Pharm. Soc. Japan*, **52**, 162 (1932).

544. (a) T. Hayakawa, H. Nakamura, K. Aoki, M. Suzuki, K. Yamada, and Y. Hirata, *Tetrahedron*, **27**, 5157 (1971). (b) K. Yamada, M. Suzuki, Y. Hayakawa, K. Aoki, H. Nakamura, H. Nagase, and Y. Hirata, *J. Am. Chem. Soc.*, **94**, 8278 (1972). (c) M. Suzuki, Y. Hayakawa, K. Aoki, H. Nagase, H. Nakamura, K. Yamada, and Y. Hirata, *Tetrahedron Lett.*, 331 (1973).

545. (a) Y. Inubushi, T. Kikuchi, T. Ibuka, K. Tanaka, I. Saji, and K. Tokane, *J. Chem. Soc., Chem. Commun.*, 1251 (1972). (b) Y. Inubushi, T. Kikuchi, T. Ibuka, K. Tanaka, I. Saji, and K. Tokane, *Chem. Pharm. Bull.*, **22**, 349 (1974).

546. A. S. Kende, T. J. Bentley, R. A. Mader, and D. Ridge, *J. Am. Chem. Soc.*, **96**, 4332 (1974).

547. W. R. Roush, *J. Am. Chem. Soc.*, **100**, 3599 (1978).

548. K. Yamamoto, I. Kawasaki, and T. Kaneko, *Tetrahedron Lett.*, 4859 (1970).

549. D. N. Brattesani and C. H. Heathcock, *J. Org. Chem.*, **40**, 2165 (1975).

550. R. F. Borch, A. J. Evans, and J. J. Wade, *J. Am. Chem. Soc.*, **99**, 1612 (1977).

Index